WORKED PROBLEMS
IN APPLIED MATHEMATICS

WORKED PROBLEMS
IN APPLIED MATHEMATICS

N. N. LEBEDEV
I. P. SKALSKAYA
Y. S. UFLYAND

A. F. Ioffe Physico-Technical Institute
Academy of Sciences, U. S. S. R.

Revised, Enlarged and Corrected English Edition
Translated and Edited by

Richard A. Silverman

With a Supplement by

Edward L. Reiss

Courant Institute of Mathematical Sciences
New York University

DOVER PUBLICATIONS, INC.
NEW YORK

Published in Canada by General Publishing Com-
pany, Ltd., 30 Lesmill Road, Don Mills, Toronto,
Ontario.
Published in the United Kingdom by Constable
and Company, Ltd., 10 Orange Street, London
WC2H 7EG.

This Dover edition, first published in 1979, is
an unabridged republication of the work originally
published by Prentice-Hall, Inc., in 1965 under the
title *Problems of Mathematical Physics*.

International Standard Book Number: 0-486-63730-1
Library of Congress Catalog Card Number: 78-67857

Manufactured in the United States of America
Dover Publications, Inc.
180 Varick Street
New York, N.Y. 10014

AUTHORS' PREFACE

The aim of the present book is to help the reader acquire the proficiency needed to successfully apply the methods of mathematical physics to a variety of problems drawn from mechanics, the theory of heat conduction, and the theory of electric and magnetic phenomena. A wide range of topics is covered, including not only problems of the simpler sort, but also problems of a more complicated nature involving such things as curvilinear coordinates, integral transforms, certain kinds of integral equations, etc. The book is intended both for students concomitantly studying the corresponding topics in courses of mathematical physics, and for research scientists who in their work find it necessary to carry out calculations using the methods described here. We also think that quite apart from its value as a tool for acquiring technique, the book can also serve as a handbook, especially in view of the fact that answers to the problems are included.

A rather solid background in applied mathematics is needed to profit from the book in its entirety. However, most of the problems appearing in Chapters 2 to 5 will be accessible to those who have taken only the usual first course in methods of mathematical physics. Chapters 6 to 8 are more specialized, and presuppose some familiarity with special functions, integral transforms, integral equations, and so on.

To make the book easier to use, each section begins with a brief introduction describing its contents and presenting a certain amount of relevant background information. However, it is not claimed that this information is complete in any sense, and the reader desiring further details must consult the literature, e.g., the books and monographs cited at the end of each chapter.

The majority of problems in this collection are accompanied by hints, facilitating the choice of meaningful methods of solution. In addition, certain problems, whose numbers are equipped with asterisks (e.g., *52, *148, etc.), are solved in detail in a special section at the end of the book. The problems singled out in this way have been selected either because they illustrate the application of certain specific methods, or because of their special difficulty or particular importance in the applications. Because of the applied character of the book, we restrict ourselves to formal solutions, whose rigorous justification can be supplied by the interested reader.

In compiling the collection, we have consulted not only the classic works on mathematical physics, but also a number of journal articles. Material accumulated during years of teaching and research in the Department of Mathematical Physics at the Leningrad Polytechnic Institute, as well as work done in connection with industrial projects, plays a role in the material presented here.

It would be impractical, and in many cases impossible, to cite the original source where a given problem was solved for the first time. Thus references to the literature have been confined to cases we find particularly relevant.

We would like to take this opportunity to thank Prof. G. A. Grinberg for many valuable suggestions made in the course of writing the book.

N. N. L.

I. P. S.

Y. S. U.

TRANSLATOR'S PREFACE

The present edition differs from the Russian original in various respects, of which three merit particular mention:

1. The Bibliography has been expanded and updated. For example, the original sources of works translated into Russian have been tracked down, all references have been equipped with titles, further references (especially, later editions and English translations) have been added, and so on. As in other volumes of this series, the system of references is in "letter-number form." Thus L10 refers to the tenth paper (or book) whose (first) author's surname begins with the letter L, where the entire Bibliography is arranged in lexicographic order, and chronological order as well, whenever there are several papers by the same author.

2. Working from an extensive list of errata sent me by the authors, I have corrected numerous misprints and mistakes present in the Russian edition. I am particularly grateful for their help, since the task of eliminating errors from a book of this type (consisting primarily of problems and answers) is both imperative and one which only the authors themselves can perform in finite time! The authors have also been kind enough to answer a number of specific questions that arose in the course of the translation.

3. It was felt that the English-language edition would benefit greatly by the addition of material on the approximate solution of problems of mathematical physics, since the emphasis of the Russian

edition is on exact solutions. This led to the writing of a Supplement on variational and related methods by Professor Edward L. Reiss of the Courant Institute of Mathematical Sciences of New York University. The Supplement is independent of the rest of the book, even to the extent of having its own references.

R. A. S.

CONTENTS

WORKED PROBLEMS IN APPLIED MATHEMATICS

Part 1

PROBLEMS

1

DERIVATION OF EQUATIONS AND
FORMULATION OF PROBLEMS

Chapter 1 is devoted to problem material on the derivation of the equations of mathematical physics and the formulation of appropriate initial and boundary conditions. It also serves as a convenient place to list the basic equations appearing later in the book. Throughout, we assume that the reader is familiar with the physical laws underlying the mathematical formulation of the problems which arise in various branches of physics.

The chapter consists of three sections devoted in turn to problems of mechanics, heat conduction and the theory of electric and magnetic phenomena. Each section starts with the basic equations governing the corresponding set of problems, with appropriate references to sources where the derivations can be found. Special attention is devoted to the formulation of problems of electrodynamics, since this subject is inadequately covered in the available literature.[1]

I. Mechanics

This section contains problems on the derivation of equations of motion and formulation of initial and boundary conditions for vibrating strings, membranes, rods and plates, as well as some examples pertaining to the statics of deformable media. It will be assumed that the reader has already

---

[1] Those particularly interested in mathematical aspects of the formulation of physical problems can find relevant material in C5, G1, L1, P2, S1 and S13. (The reference scheme is explained in the Translator's Preface.)

encountered the basic equations in a first course on mathematical physics.[2] Thus we shall merely list the equations concisely, at the same time explaining the notation to be used in the book.

1. The equation of a vibrating string is

$$\frac{\partial^2 u}{\partial x^2} - \frac{1}{v^2}\frac{\partial^2 u}{\partial t^2} = -\frac{q(x, t)}{T}, \qquad v = \sqrt{\frac{T}{\rho}},$$

where $u(x, t)$ is the displacement of the point of the string with abscissa x at the time t, $q(x, t)$ is the external load per unit length, T is the tension, and ρ is the linear density.

2. The equation for longitudinal oscillations of a rod of constant cross section is

$$\frac{\partial^2 u}{\partial x^2} - \frac{1}{v^2}\frac{\partial^2 u}{\partial t^2} = 0, \qquad v = \sqrt{\frac{E}{\rho}},$$

where $u(x, t)$ is the displacement of the cross section of the rod with abscissa x at the time t, E is Young's modulus, and ρ is the density.

3. The equation for transverse oscillations of a rod (beam) is

$$\frac{\partial^4 u}{\partial x^4} + \frac{1}{a^4}\frac{\partial^2 u}{\partial t^2} = \frac{q(x, t)}{EJ}, \qquad a^2 = \sqrt{\frac{EJ}{\rho S}},$$

where $u(x, t)$ is the displacement of the points along the midline of the rod, $q(x, t)$ is the external load per unit length, E is Young's modulus, J is the moment of inertia of a transverse cross section, ρ is the density, and S is the cross-sectional area.

4. The equation of a vibrating membrane is

$$\frac{\partial^2 u}{\partial x^2} + \frac{\partial^2 u}{\partial y^2} - \frac{1}{v^2}\frac{\partial^2 u}{\partial t^2} = -\frac{q(x, y, t)}{T}, \qquad v = \sqrt{\frac{T}{\rho}},$$

where $u(x, y, t)$ is the displacement of the point (x, y) of the membrane at the time t, $q(x, y, t)$ is the external load per unit area, T is the tension per unit length of the boundary of the membrane, and ρ is the surface density.

5. The equation for transverse oscillations of a thin elastic plate is

$$\Delta^2 u + \frac{1}{b^4}\frac{\partial^2 u}{\partial t^2} = \frac{q(x, y, t)}{D}, \qquad b^2 = \sqrt{\frac{D}{\rho h}},$$

---

[2] See S6 (Vol. II), S14, T1 and T2. Concerning the derivation of the equations of vibrating plates, see T4.

where $u(x, t)$ is the displacement of the point (x, y) of the midplane of the plate at the time t, $q(x, y, t)$ is the density of the external load, D is the flexural rigidity, h is the thickness, ρ is the density, and

$$\Delta = \frac{\partial^2}{\partial x^2} + \frac{\partial^2}{\partial y^2}$$

is the two-dimensional Laplacian operator.

The above equations lead to corresponding equations for static deflections, if we regard the external load q and the unknown displacement u as independent of the time t. For example, the equilibrium equation for the membrane is

6.

$$\frac{\partial^2 u}{\partial x^2} + \frac{\partial^2 u}{\partial y^2} = -\frac{q(x, y)}{T},$$

the static deflection of the plate satisfies the equation

7.

$$\Delta^2 u = \frac{q(x, y)}{D},$$

and so on.

Among the other equations governing the statics of elastic bodies which will figure in this book, we cite the familiar equation

8.

$$\frac{\partial^2 u}{\partial x^2} + \frac{\partial^2 u}{\partial y^2} = -2,$$

for twisting of a prismatic rod, where $u(x, y)$ is the torsion function.

We now give some problems on the formulation of initial and boundary conditions for these equations, and also some problems on the derivation of other differential equations.

1. Describe the initial and boundary conditions for a vibrating string with fixed ends $(0 \leqslant x \leqslant l)$, which is stretched at the point $x = c$ and time $t = 0$ to a height h, and then released without initial velocity.

Ans.

$$u\big|_{t=0} = f(x) = \begin{cases} \dfrac{hx}{c}, & 0 \leqslant x \leqslant c, \\[2mm] \dfrac{h(l - x)}{l - c}, & c \leqslant x \leqslant l, \end{cases} \qquad \frac{\partial u}{\partial t}\bigg|_{t=0} = 0;$$

$$u\big|_{x=0} = u\big|_{x=l} = 0.$$

2. A concentrated load of mass m_0 is fastened at the point $x = c$ of a string $0 \leqslant x \leqslant l$ of length l. Find the equations describing vibrations of the string with arbitrary initial conditions, assuming that the ends of the string are fastened.

Ans.

$$u = \begin{cases} u_1, & 0 \leqslant x \leqslant c, \\ u_2, & c \leqslant x \leqslant l, \end{cases} \quad \frac{\partial^2 u_i}{\partial x^2} - \frac{1}{v^2} \frac{\partial^2 u_i}{\partial t^2} = 0 \quad (i = 1, 2),$$

with initial conditions

$$u|_{t=0} = f(x), \qquad \frac{\partial u}{\partial t}\bigg|_{t=0} = g(x),$$

and boundary conditions

$$u_1|_{x=0} = u_2|_{x=l} = 0, \qquad u_1|_{x=c} = u_2|_{x=c}, \qquad \left(\frac{\partial u_2}{\partial x} - \frac{\partial u_1}{\partial x}\right)_{x=c} = \frac{m_0}{T} \frac{\partial^2 u}{\partial t^2}\bigg|_{x=c}.$$

3. Formulate initial and boundary conditions for the problem of longitudinal oscillations of a rod in the following special cases:

a) A rod of length l is clamped at the end $x = 0$ and stretched by a force F applied to the other end; at the time $t = 0$ the force is suddenly discontinued;

b) A tensile force $F(t)$ is applied at the time $t = 0$ to the end $x = l$ of a cantilever in equilibrium;

c) A cantilever clamped at the point $x = 0$, with a load of mass M_0 at the free end $x = l$, undergoes longitudinal oscillations subject to arbitrary initial conditions.

Ans.

a) $u|_{t=0} = \dfrac{Fx}{ES}$, $\quad \dfrac{\partial u}{\partial t}\bigg|_{t=0} = 0$, $\quad u|_{x=0} = 0$, $\quad \dfrac{\partial u}{\partial x}\bigg|_{x=l} = 0$;

b) $u|_{t=0} = 0$, $\quad \dfrac{\partial u}{\partial t}\bigg|_{t=0} = 0$, $\quad u|_{x=0} = 0$, $\quad \dfrac{\partial u}{\partial x}\bigg|_{x=l} = \dfrac{F(t)}{ES}$;

c) $u|_{t=0} = f(x)$, $\quad \dfrac{\partial u}{\partial t}\bigg|_{t=0} = g(x)$, $\quad u|_{x=0} = 0$, $\quad \dfrac{\partial u}{\partial x}\bigg|_{x=l} = -\dfrac{M_0}{ES} \dfrac{\partial^2 u}{\partial t^2}\bigg|_{x=l}.$

4. Derive the differential equation for longitudinal oscillations of a thin rod of variable cross section $S = S(x)$. As an example, derive the equation for oscillations of a conical rod.

Ans.

$$\frac{1}{S(x)} \frac{\partial}{\partial x}\left[S(x) \frac{\partial u}{\partial x}\right] - \frac{1}{v^2} \frac{\partial^2 u}{\partial t^2} = 0, \qquad v = \sqrt{\frac{E}{\rho}}.$$

5. Derive the equation for torsional oscillations of a shaft of circular cross section.

Ans.

$$\frac{\partial^2 \theta}{\partial x^2} - \frac{1}{v^2} \frac{\partial^2 \theta}{\partial t^2} = 0,$$

where $\theta(x, t)$ is the angular displacement of the cross section x relative to the equilibrium position, $v = \sqrt{G/\rho}$, ρ is the density, and G is the shear modulus.

Hint. The torque at the cross section x is given by the expression

$$M = GJ \frac{\partial \theta}{\partial x},$$

where J is the polar moment of inertia of a cross section of the shaft.

6. Formulate initial and boundary conditions for the problem of torsional oscillations of a shaft of circular cross section and length l, where the end $x = 0$ is clamped and a disk-shaped mass with moment of inertia J_0 is attached to the other end. At the time $t = 0$, the disk is rotated through a given angle α and then released without initial velocity.

Ans.

$$\theta\big|_{t=0} = \alpha \frac{x}{l}, \qquad \frac{\partial \theta}{\partial t}\bigg|_{t=0} = 0,$$

$$\theta\big|_{x=0} = 0, \qquad \frac{\partial \theta}{\partial x}\bigg|_{x=l} = -\frac{J_0}{GJ} \frac{\partial^2 \theta}{\partial t^2}\bigg|_{x=l}.$$

7. A cantilever of length l is clamped at one end $x = 0$ and loaded by a force F at the other end. At the time $t = 0$, the action of the force is suddenly discontinued. Formulate initial and boundary conditions for the corresponding oscillations.

Ans. Initial conditions

$$u\big|_{t=0} = \frac{F}{6EJ}(3lx^2 - x^3), \qquad \frac{\partial u}{\partial t}\bigg|_{t=0} = 0,$$

and boundary conditions

$$u\big|_{x=0} = \frac{\partial u}{\partial x}\bigg|_{x=0} = 0, \qquad \frac{\partial^2 u}{\partial x^2}\bigg|_{x=l} = \frac{\partial^3 u}{\partial x^3}\bigg|_{x=l} = 0.$$

8. Describe initial and boundary conditions for the problem of free oscillations of a disk-shaped plate with clamped edge, whose initial deformation is due to a concentrated force F applied at the center of the disk.

Ans.

$$u\big|_{t=0} = \frac{Fr^2}{8\pi D} \ln \frac{r}{a} + \frac{F}{16\pi D}(a^2 - r^2), \qquad \frac{\partial u}{\partial t}\bigg|_{t=0} = 0;$$

$$u\big|_{r=a} = 0, \qquad \frac{\partial u}{\partial r}\bigg|_{r=a} = 0.$$

Hint. To determine the static deflection due to the concentrated force, consider the force as the limiting case of a load of density $F/\pi\varepsilon^2$ uniformly distributed over a small disk of radius ε.

9. Show that the problem of the deflection of a plate with a simply supported polygonal boundary reduces to the solution of Poisson's equation

$$\Delta w = f(x, y),$$

with boundary condition $w|_\Gamma = 0$ (f is a known function).

Hint. Note that in the present case, the boundary conditions on the supported edge can be written in the form $u|_\Gamma = 0$, $\Delta u|_\Gamma = 0$.

10. Show that the velocity potential for the three-dimensional flow of an ideal incompressible fluid containing no sources is described by Laplace's equation

$$\Delta u = 0.$$

Hint. Use the condition

$$\int_S \mathbf{v} \cdot \mathbf{n} \, dS = 0$$

($\mathbf{v}$ is the vector describing the velocity of fluid particles at a given point, S is an arbitrary closed surface inside the flow, and $\mathbf{n}$ is the exterior normal to the surface S) and the condition

$$\mathbf{v} = -\text{grad } u$$

for potential flow.

11. Formulate mathematically the problem of the flow of an ideal fluid past an object bounded by a surface S, where fluid emanates from a point source of strength m located at a point M_0 in the region exterior to S.

Ans. The problem reduces to finding a solution of the equation

$$\Delta u = 0$$

which is regular (i.e., has no singularities) in the region exterior to S, except at the point M_0. In a neighborhood of M_0,

$$u = \frac{m}{4\pi\rho \, |MM_0|} + \text{a regular function}$$

where M is a point near M_0 and ρ is the density of the fluid ($|MM_0|$ denotes the distance between M and M_0). The desired function u must satisfy the boundary condition

$$\frac{\partial u}{\partial n}\bigg|_S = 0$$

and the condition

$$u = O(R^{-1}), \qquad R \to \infty$$

at infinity.

2. Heat Conduction

As proved in courses on mathematical physics (see S1, T1), the flow of heat in a body of thermal conductivity k, specific heat c and density ρ is governed by *Fourier's equation*

$$\Delta T = \frac{c\rho}{k} \frac{\partial T}{\partial t} - \frac{Q}{k},$$

where $T(M, t)$ is the temperature at the point M, and Q is the density of heat sources within the body.[3] The boundary conditions to be satisfied on the surface of the body (or its parts) depend on the particular problem under consideration. Most often it is assumed that the surface of the body has a given temperature $T|_S = f(P, t)$, where P is a point of the surface S, or that the body radiates heat into the surrounding medium according to *Newton's law*, which states that the amount of heat radiated by a unit area of the surface per unit time is proportional to the difference between the temperature of the surface and that of the surrounding medium. In the latter case, the boundary condition takes the form

$$\left(\frac{\partial T}{\partial n} + hT \right) \Bigg|_S = hT_{\text{med}},$$

where $\partial/\partial n$ indicates differentiation with respect to the exterior normal to S, T_{med} is the temperature of the surrounding medium, and h is the heat exchange coefficient or emissivity. Without loss of generality, we can assume that $T_{\text{med}} = 0$; this assumption is made in all the problems involving heat conduction except Prob. 155.[4]

We now give a few problems on the formulation of initial and boundary conditions for the equation of heat conduction (and for the related diffusion equation).

12. Let the temperature of a conductor in the form of an infinite cylinder of radius a be initially the same as that of the surrounding medium. Suppose that starting from the time $t = 0$, the conductor is heated by a constant

---

[3] The density of heat current (i.e., the heat flux) is described by the vector

$$\mathbf{q} = -k \operatorname{grad} T.$$

[4] Examples of other boundary conditions encountered in the applications are given in Probs. 365, 367 and 370.

electric current releasing an amount of heat Q per unit volume of the conductor. Give a mathematical formulation of the corresponding problem of heat conduction, assuming that the heat exchange at the surface of the conductor obeys Newton's law.[5]

Ans. The temperature $T(r, t)$ satisfies the equation

$$\frac{1}{r}\frac{\partial}{\partial r}\left(r\frac{\partial T}{\partial r}\right) = \frac{\partial T}{\partial \tau} - \frac{Q}{k}, \qquad \tau = \frac{kt}{c\rho},$$

with initial condition

$$T\big|_{\tau=0} = 0$$

and boundary condition

$$\left(\frac{\partial T}{\partial r} + hT\right)\bigg|_{r=a} = 0.$$

13. A homogeneous sphere of radius a is heated for a long time by heat sources uniformly distributed throughout its volume with density Q. Write the equations which describe the cooling of the sphere after the sources are turned off, assuming that the heat exchange between the surface of the sphere and the surrounding medium, during both the heating and cooling, obeys Newton's law.

Ans.

$$\frac{1}{r^2}\frac{\partial}{\partial r}\left(r^2\frac{\partial T}{\partial r}\right) = \frac{\partial T}{\partial \tau}, \qquad \left(\frac{\partial T}{\partial r} + hT\right)\bigg|_{r=a} = 0,$$

$$T\big|_{\tau=0} = \frac{Q}{6k}(a^2 - r^2) + \frac{Qa}{3kh}.$$

14. Two slabs of thicknesses a_1 and a_2, made from different materials and heated to temperatures T_1^0 and T_2^0, are put into contact with each other at the time $t = 0$. Write the equations governing the resulting process of temperature equalization, assuming that the free surfaces are thermally insulated from the surrounding medium.

Ans.

$$\frac{\partial^2 T_1}{\partial x^2} = \frac{c_1\rho_1}{k_1}\frac{\partial T_1}{\partial t} \quad (0 < x < a_1), \quad \frac{\partial^2 T_2}{\partial x^2} = \frac{c_2\rho_2}{k_2}\frac{\partial T_2}{\partial t} \quad (a_1 < x < a_1 + a_2),$$

with initial conditions

$$T_1\big|_{t=0} = T_1^0, \qquad T_2\big|_{t=0} = T_2^0,$$

---

[5] It is recommended that the problem be solved directly from underlying physical principles, without regarding Fourier's equation as known.

and boundary conditions

$$\frac{\partial T_1}{\partial x}\bigg|_{x=0} = 0, \quad T_1\big|_{x=a_1} = T_2\big|_{x=a_1}, \quad k_1 \frac{\partial T_1}{\partial x}\bigg|_{x=a_1} = k_2 \frac{\partial T_2}{\partial x}\bigg|_{x=a_1} \quad \frac{\partial T_2}{\partial x}\bigg|_{x=a_1+a_2} = 0.$$

15. A nonuniformly heated body in the form of a circular ring of radius a with a small cross section cools by giving off heat from its lateral surface. Write the equations describing the corresponding process of temperature equalization, assuming that the temperature drop inside the ring can be neglected and that the surface cooling obeys Newton's law.

Ans.

$$\frac{1}{a^2} \frac{\partial^2 T}{\partial \varphi^2} = \frac{\partial T}{\partial \tau} + \frac{hp}{S} T, \quad \tau = \frac{kt}{c\rho},$$

where p is the perimeter, S the cross-sectional area and h the heat exchange coefficient. The temperature, which must be a periodic function of the angular coordinate φ, satisfies the initial condition

$$T\big|_{\tau=0} = f(\varphi),$$

where f is a given function.

16. Show that the concentration $C(x, y, z, t)$ of a substance diffusing in a gas or liquid obeys the differential equation

$$\Delta C = \frac{1}{D} \frac{\partial C}{\partial t} - \frac{Q}{D},$$

where Q is the source density of the diffusing substance and D is the diffusion coefficient.

Hint. Starting from Nernst's law $\mathbf{q} = -\operatorname{grad} C$ (where the vector $\mathbf{q}$ is the density of flow of the diffusing substance), write a conservation equation for an arbitrary volume element.

3. Electricity and Magnetism

An important class of problems of mathematical physics involves integration of the differential equations arising in various branches of electromagnetic theory. Assuming that the reader has previously encountered this subject (see G5, J6, P1), we shall regard the following basic equations as known:

1. The equations of electrostatics

$$\Delta u = -\frac{4\pi\rho}{\varepsilon}, \quad \mathbf{E} = -\operatorname{grad} u,$$

where u is the potential of the electrostatic field $\mathbf{E}$, $\rho = \rho(M)$ is the volume density of charge at the point M, ε is the dielectric constant of the medium, and Δ is the Laplacian operator.

2. The equations

$$\Delta u = -\frac{Q}{\sigma}, \qquad \mathbf{j} = -\sigma \operatorname{grad} u, \tag{1}$$

for the distribution of d-c current density inside a homogeneous conductor, where u is the potential of the current field, $\mathbf{j}$ is the current density vector, $Q = Q(M)$ is the volume density of current sources (in particular, Q may vanish), and σ is the conductivity.

3. The equations

$$\Delta \mathbf{A} = -\frac{4\pi\mu}{c}\mathbf{j}^{(e)}, \qquad \mathbf{H} = \frac{1}{\mu}\operatorname{curl}\mathbf{A}$$

for the magnetic field due to d-c currents, where $\mathbf{A}$ is the vector potential of the magnetic field $\mathbf{H}$, the vector $\mathbf{j}^{(e)}$ is the density of the (external) currents producing the magnetic field, μ is the magnetic permeability of the medium, c is the velocity of light in vacuum, and Δ is the Laplacian operator.[6]

4. *Maxwell's equations*

$$\operatorname{curl}\mathbf{H} = \frac{\varepsilon}{c}\frac{\partial \mathbf{E}}{\partial t} + \frac{4\pi\sigma}{c}\mathbf{E} + \frac{4\pi}{c}\mathbf{j}^{(e)},$$

$$\operatorname{curl}\mathbf{E} = -\frac{\mu}{c}\frac{\partial \mathbf{H}}{\partial t},$$

$$\operatorname{div}\mathbf{E} = \frac{4\pi\rho}{\varepsilon},$$

$$\operatorname{div}\mathbf{H} = 0$$

for the electromagnetic field in a homogeneous isotropic medium, where $\mathbf{E}$ and $\mathbf{H}$ are the electric and magnetic field vectors, ε, μ and σ are the dielectric constant, the magnetic permeability and the conductivity of the medium, c is the velocity of light, and ρ and $\mathbf{j}^{(e)}$ are the charge and current densities producing the field.[7]

---

[6] The components of the vector $\Delta\mathbf{A}$ in a Cartesian coordinate system are ΔA_x, ΔA_y and ΔA_z. To calculate the components of the vector $\Delta\mathbf{A}$ in other coordinate systems, one should use the relation

$$\Delta\mathbf{A} = \operatorname{grad}\operatorname{div}\mathbf{A} - \operatorname{curl}\operatorname{curl}\mathbf{A}.$$

Expressions for the components of $\Delta\mathbf{A}$ in cylindrical and spherical coordinates are given on p. 389–390.

[7] It should be noted that if $\mathbf{j}^{(e)}$ is given, then ρ cannot be chosen arbitrarily, but must satisfy the differential equation

$$\frac{\partial\rho}{\partial t} + \frac{4\pi\sigma}{\varepsilon}\rho = -\operatorname{div}\mathbf{j}^{(e)}$$

implied by the first and third Maxwell equations.

If we use the relations

$$\mathbf{H} = \frac{1}{\mu}\,\text{curl }\mathbf{A}, \qquad \mathbf{E} = -\text{grad } u - \frac{1}{c}\frac{\partial \mathbf{A}}{\partial t}$$

to introduce the vector and scalar potentials A and u,[8] the problem of determining the electromagnetic field reduces to integrating the system of equations

$$\Delta\mathbf{A} - \frac{\varepsilon\mu}{c^2}\frac{\partial^2\mathbf{A}}{\partial t^2} - \frac{4\pi\sigma\mu}{c^2}\frac{\partial\mathbf{A}}{\partial t} = -\frac{4\pi\mu}{c}\,\mathbf{j}^{(e)},$$

$$\Delta u - \frac{\varepsilon\mu}{c^2}\frac{\partial^2 u}{\partial t^2} - \frac{4\pi\mu\sigma}{c^2}\frac{\partial u}{\partial t} = -\frac{4\pi\rho}{\varepsilon}.$$

We now consider the mathematical formulation of various problems involving electric and magnetic fields (both static and variable), as well as some problems on transformations of the differential equations of electrodynamics which are useful in special cases.

17. Formulate mathematically the problem of finding the three-dimensional electrostatic field between N conductors of arbitrary shape at given potentials V_i $(i = 1, \ldots, N)$.

Ans. In the region D bounded by the surfaces S_i $(i = 1, \ldots, N)$ of the conductors, the potential u satisfies Laplace's equation

$$\Delta u = 0.$$

The boundary conditions have the form

$$u\big|_{S_i} = V_i, \qquad i = 1, \ldots, N,$$

where, in the case where the point at infinity belongs to D, these conditions must be supplemented by the requirement that at infinity the potential u approach zero uniformly in all directions.

Comment. If none of the surfaces S_i extends to infinity, then the products Ru and R^2 grad u (where $R^2 = x^2 + y^2 + z^2$) remain uniformly bounded as $R \to \infty$. However, these conditions need not be included in the formulation of the problem, since the uniqueness of the solution is guaranteed by the above requirement that the potential u approach zero uniformly as $R \to \infty$.

18. A charge Q is placed at the point $M_0 = (x_0, y_0, z_0)$ near a conductor at potential V, bounded by a surface S. Formulate the corresponding problem of electrostatics.

---

[8] The quantities A and u are not independent, but are connected by the relation

$$\text{div }\mathbf{A} + \frac{\varepsilon\mu}{c}\frac{\partial u}{\partial t} + \frac{4\pi\mu}{c}\,u = 0.$$

Ans. The potential u satisfies Laplace's equation at every point of the region surrounding the conductor, except at the point M_0, near which

$$u = \frac{q}{R} + \text{a regular function,}$$

($R = |M_0 M|$ is the distance between the points M and M_0). Moreover, the potential satisfies the boundary condition $u|_S = V$ and the condition that $u|_\infty \to 0$ uniformly in all directions.

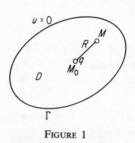

19. A thin charged wire of charge q per unit length is placed inside a grounded cylindrical shell whose generators are parallel to the wire (see Figure 1). Formulate the corresponding two-dimensional electrostatic problem.

Ans. The potential u satisfies the two-dimensional Laplace equation

FIGURE 1

$$\Delta u = \frac{\partial^2 u}{\partial x^2} + \frac{\partial^2 u}{\partial y^2} = 0$$

in the whole region D except at the point M_0, where the potential has a logarithmic singularity

$$u = -2q \ln R + \text{a regular function.}$$

The boundary condition is $u|_\Gamma = 0$.

20. Reformulate the preceding problem for the case where the charged wire is placed outside the conductor, and the total charge per unit length of the conductor is specified instead of its potential.

Ans. The boundary condition is now

$$\left.\frac{\partial u}{\partial s}\right|_\Gamma = 0$$

($\partial/\partial s$ denotes differentiation along the tangent to the contour Γ), while the condition at infinity becomes

$$u|_\infty = -2(Q + q) \ln R + \text{a bounded function,}$$

where Q is the charge per unit length of the conductor.

21. Show that the problem of the current distribution in a thin conducting shell (see Figure 2) reduces to integration of the equation

$$\Delta u = \frac{j_n^{(e)}}{\sigma h},$$

where u is the potential of the current distribution in the shell ($\mathbf{j} = -\sigma \,\mathrm{grad}\, u$ is the current density vector), σ is the conductivity and h the thickness of the shell, $j_n^{(e)}$ is the normal component of the density of current applied to the shell by using suitable electrodes, and Δ is the appropriate two-dimensional Laplacian, i.e.,

$$ds^2 = H_\alpha^2 \, d\alpha^2 + H_\beta^2 \, d\beta^2, \qquad \Delta u = \frac{1}{H_\alpha H_\beta}\left[\frac{\partial}{\partial\alpha}\left(\frac{H_\beta}{H_\alpha}\frac{\partial u}{\partial\alpha}\right) + \frac{\partial}{\partial\beta}\left(\frac{H_\alpha}{H_\beta}\frac{\partial u}{\partial\beta}\right)\right]$$

(ds is the element of arc length on the surface of the shell).

Hint. Average equation (1), p. 12 (giving the volume distribution of current) over the thickness of the shell.

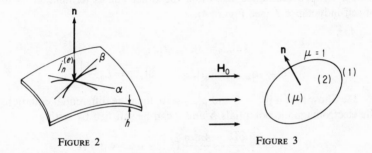

FIGURE 2   FIGURE 3

22. Suppose an object of arbitrary shape, made of magnetic material of permeability μ, is magnetized by being introduced into a homogeneous magnetic field $\mathbf{H}_0$ (see Figure 3). Formulate mathematically the corresponding problem of magnetostatics.

Ans. If u is the potential of the magnetic field $\mathbf{H}$ (i.e., $\mathbf{H} = -\mathrm{grad}\, u$), the problem reduces to integration of the equation

$$\Delta u_i = 0, \qquad i = 1, 2,$$

with the boundary conditions

$$u_1\big|_S = u_2\big|_S, \qquad \frac{\partial u_1}{\partial n}\bigg|_S = \mu \frac{\partial u_2}{\partial n}\bigg|_S$$

and the following conditions at infinity

$$-\mathrm{grad}\, u_1\big|_\infty = \mathbf{H}_0, \qquad u_1\big|_\infty \text{ is bounded.}$$

Hint. It helps to keep in mind that in the source-free part of space, the magnetic field $\mathbf{H}$ satisfies the equations

$$\mathrm{curl}\, \mathbf{H} = 0, \qquad \mathrm{div}\, \mathbf{H} = 0,$$

which imply

$$\mathbf{H} = -\mathrm{grad}\, u, \qquad \Delta u = 0.$$

23. The differential equations for wave propagation down a long transmission line, with self-inductance L, capacitance C, resistance R and leakage conductance G per unit length, take the form

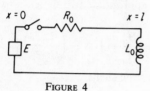

FIGURE 4

$$-\frac{\partial u}{\partial x} = L\frac{\partial I}{\partial t} + RI, \qquad -\frac{\partial I}{\partial x} = C\frac{\partial u}{\partial t} + Gu,$$

where $u(x, t)$ and $I(x, t)$ are the values of the voltage and current at the point x at the time t (see T1, p. 18). Formulate initial and boundary conditions for wave propagation along such a line, if at one end a constant voltage E is switched on in series with a lumped resistance R_0, while the other end is terminated by a coil of self-inductance L (see Figure 4).

Ans.

$$u|_{t=0} = 0, \qquad I|_{t=0} = 0,$$

$$E = u|_{x=0} + R_0 I|_{x=0}, \qquad u|_{x=l} = L_0 \frac{\partial I}{\partial t}\bigg|_{x=l}.$$

24. Show that if $\mathbf{j}^{(e)} = 0$ and $\rho|_{t=0} = 0$, then the differential equations for the electromagnetic potentials $\mathbf{A}$ and u can be satisfied by setting

$$\mathbf{A} = \frac{\varepsilon\mu}{c}\frac{\partial \mathbf{\Pi}}{\partial t} + \frac{4\pi\mu\sigma}{c}\mathbf{\Pi}, \qquad u = -\operatorname{div}\mathbf{\Pi},$$

where $\mathbf{\Pi}$ is the Hertz vector satisfying the equation

$$\Delta\mathbf{\Pi} - \frac{\varepsilon\mu}{c^2}\frac{\partial^2\mathbf{\Pi}}{\partial t^2} - \frac{4\pi\sigma\mu}{c^2}\frac{\partial\mathbf{\Pi}}{\partial t} = 0.$$

Derive expressions for the vectors $\mathbf{E}$ and $\mathbf{H}$ in terms of the vector $\mathbf{\Pi}$.

Ans.

$$\mathbf{H} = \frac{\varepsilon}{c}\operatorname{curl}\left(\frac{\partial\mathbf{\Pi}}{\partial t} + \frac{4\pi\sigma}{\varepsilon}\mathbf{\Pi}\right), \qquad \mathbf{E} = \operatorname{curl}\operatorname{curl}\mathbf{\Pi}.$$

Hint. According to footnote 7, p. 12, it follows from $\mathbf{j}^{(e)} = 0$ and $\rho|_{t=0} = 0$ that $\rho = 0$ for arbitrary t.

25. Verify that if $\mathbf{j}^{(e)} = 0$ and $\rho|_{t=0} = 0$, then the vectors $\mathbf{E}$ and $\mathbf{H}$ satisfy the same differential equation as the Hertz vector, i.e.,[9]

$$\Delta\mathbf{E} - \frac{\varepsilon\mu}{c^2}\frac{\partial^2\mathbf{E}}{\partial t^2} - \frac{4\pi\mu\sigma}{c^2}\frac{\partial\mathbf{E}}{\partial t} = 0,$$

$$\Delta\mathbf{H} - \frac{\varepsilon\mu}{c^2}\frac{\partial^2\mathbf{H}}{\partial t^2} - \frac{4\pi\mu\sigma}{c^2}\frac{\partial\mathbf{H}}{\partial t} = 0.$$

---

[9] In some problems it is more convenient to start from these equations than from the equations for the electromagnetic potentials or for the Hertz vector.

26. Show that for steady-state harmonic oscillations of frequency ω, in which the time dependence of the quantities defining the electromagnetic field (i.e., the vectors $\mathbf{E}$ and $\mathbf{H}$, the charge and current densities which are the sources of the field, etc.) is characterized by a factor $e^{i\omega t}$, Maxwell's equations (see p. 12) take the form

$$\Delta \mathbf{A}^* + k^2 \mathbf{A}^* = -\frac{4\pi\mu}{c}\mathbf{j}^{(e)*},$$

$$\mathbf{H}^* = \frac{1}{\mu}\operatorname{curl}\mathbf{A}^*, \qquad \mathbf{E}^* = -\frac{i\omega}{ck^2}\left(\operatorname{curl}\operatorname{curl}\mathbf{A}^* - \frac{4\pi\mu}{c}\mathbf{j}^{(e)*}\right),$$

where f^* denotes the complex amplitude of a scalar or vector function f, and[10]

$$k = \sqrt{\frac{\varepsilon\mu\omega^2 - 4\pi\mu\sigma i\omega}{c^2}}, \qquad \operatorname{Im} k < 0.$$

Comment. The importance of this problem consists in showing that only one unknown function (rather than two) is needed to calculate the electromagnetic field in the case of harmonic time dependence.

27. Starting from Maxwell's equations for the case of steady-state harmonic oscillations, deduce the corresponding differential equations for the two-dimensional (planar) electromagnetic problem, where

$$j_x^{(e)*} = j_y^{(e)*} = 0, \qquad j_z^{(e)*} = j(x, y),$$
$$E_x^* = E_y^* = 0, \qquad E_z^* = E(x, y),$$
$$H_x^* = H_x(x, y), \qquad H_y^* = H_y(x, y), \qquad H_z^* = 0.$$

Ans.

$$\Delta E + k^2 E = \frac{4\pi\mu i\omega}{c^2}j,$$

$$H_x = -\frac{c}{\mu i\omega}\frac{\partial E}{\partial y}, \qquad H_y = \frac{c}{\mu i\omega}\frac{\partial E}{\partial x}.$$

Hint. For harmonic time dependence, the connection between $\mathbf{j}^{(e)*}$ and ρ^* is given by

$$\rho^* = \frac{\varepsilon\mu i\omega}{c^2 k^2}\operatorname{div}\mathbf{j}^{(e)*},$$

which in the present case implies $\rho^* = 0$.

---

[10] For example, if $\mathbf{j}^{(e)} = \mathbf{j}_0^{(e)}\sin\omega t$, where $\mathbf{j}_0^{(e)}$ is real, then the actual values of $\mathbf{E}$ and $\mathbf{H}$ are given by the imaginary parts of the expressions $\mathbf{E}^* e^{i\omega t}$ and $\mathbf{H}^* e^{i\omega t}$.

28. Derive the equations for steady-state harmonic electromagnetic oscillations for the case of axial symmetry, where

$$j_r^{(e)*} = j_\varphi^{(e)*} = 0, \qquad j_z^{(e)*} = j(r, z)$$
$$E_r^* = E_r(r, z), \quad E_\varphi^* = 0, \quad E_z^* = E_z(r, z),$$
$$H_r^* = H_z^* = 0, \qquad H_\varphi^* = H(r, z).$$

Ans.

$$\Delta H + \left(k^2 - \frac{1}{r^2}\right)H = \frac{4\pi}{c}\frac{\partial j}{\partial r},$$

$$E_r = \frac{\mu i \omega}{ck^2}\frac{\partial H}{\partial z},$$

$$E_z = -\frac{\mu i \omega}{ck^2}\left[\frac{1}{r}\frac{\partial}{\partial r}(rH) - \frac{4\pi}{c}j\right].$$

Hint. Note that

$$\rho^* = \frac{\varepsilon \mu i \omega}{c^2 k^2}\frac{\partial j}{\partial z}.$$

References

Courant and Hilbert (C5), Frank and von Mises (F6), Garabedian (G1), Grinberg (G5), Morse and Feshbach (M9), Petrovski (P2), Smirnov (S6, Vol. II), Sommerfeld (S14), Tikhonov and Samarski (T1), Timoshenko (T2), Webster (W5).

2

SOME SPECIAL METHODS
FOR SOLVING HYPERBOLIC
AND ELLIPTIC EQUATIONS

This chapter deals with some special methods which, unlike those considered later, can only be used to solve problems pertaining to partial differential equations of a particular type, e.g., of the hyperbolic or elliptic type.[1] Among such methods, we mention Riemann's method for solving the Cauchy problem for hyperbolic equations, the Green's function method for solving boundary value problems involving elliptic equations, complex variable methods for solving the two-dimensional problems of potential theory and so on. There are a great many such special methods, which in some cases belong to the more difficult problems of the theory of partial differential equations. Thus it will be impossible to go into very much detail here. Instead we confine ourselves to a few simple problems illustrating the methods most frequently encountered in practice.

I. Hyperbolic Equations

It will be recalled that problems of mathematical physics involving the propagation of various kinds of waves (elastic, electromagnetic, etc.) in one, two or three dimensions lead to the consideration of partial differential equations of the hyperbolic type, subject to extra conditions. Depending on the character of these conditions, the problem is classified as a Cauchy problem or as a mixed problem. By the *mixed problem* for an equation of the

---

[1] Concerning the classification of partial differential equations, see C5, G1, T1, etc.

hyperbolic type, we mean a problem where the unknown function u must satisfy not only initial conditions, but also boundary conditions specified on the boundary of some spatial region D, while by the *Cauchy problem* we mean a problem where D is an unbounded region and u need only satisfy initial conditions.

In particular, for a second-order differential equation of the form

$$\sum_{i=1}^{4} \sum_{k=1}^{4} a_{ik} \frac{\partial^2 u}{\partial x_i \, \partial x_k} + f\left(x_1, \ldots, x_4, u, \frac{\partial u}{\partial x_1}, \ldots, \frac{\partial u}{\partial x_4}\right) = 0, \tag{1}$$

where x_1, x_2, x_3 are spatial coordinates and $x_4 = t$ is the time, the Cauchy problem reduces to finding a solution satisfying the initial conditions

$$u\big|_{t=0} = \varphi(x_1, x_2, x_3), \qquad \frac{\partial u}{\partial t}\bigg|_{t=0} = \psi(x_1, x_2, x_3),$$

where φ and ψ are given functions of the coordinates. A more general problem, also known in the literature as the Cauchy problem, consists in finding a solution of (1) subject to conditions of the form

$$u\big|_S = \varphi(P), \qquad \frac{\partial u}{\partial n}\bigg|_S = \psi(P),$$

where S is a given hyperplane in four-dimensional space-time, n is the normal to S, and $\varphi(P)$, $\psi(P)$ are given functions of a point P on the surface S.

In many cases, Cauchy's problem can be solved by resorting to special methods such as the wave-propagation method (which can be used to construct general integrals of certain partial differential equations of the hyperbolic type), Riemann's method and its generalizations, etc., as illustrated by the following set of problems:[2]

29. Solve the problem of oscillations of an infinite string under the action of a distributed load $q(x, t)$, subject to the initial conditions

$$u\big|_{t=0} = \varphi(x), \qquad \frac{\partial u}{\partial t}\bigg|_{t=0} = \psi(x).$$

Ans.

$$u(x, t) = \frac{1}{2}\left[\varphi(x - vt) + \varphi(x + vt)\right] + \frac{1}{2v} \int_{x-vt}^{x+vt} \psi(\xi) \, d\xi$$

$$+ \frac{v}{2T} \iint_D q(\xi, \tau) \, d\xi \, d\tau,$$

---

[2] These methods can sometimes be used to solve the mixed problem (see Prob. 41 and T1, Sec. 2.2). However, as a rule, the mixed problem requires the use of the methods considered in Chaps. 4–7, among which the method of the Laplace transform (see Chap. 6, Sec. 3) is particularly effective.

where $v = \sqrt{T/\rho}$, T is the tension and ρ the linear density of the string, and the region of integration D is shown in Figure 5.

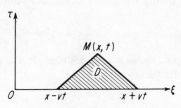

FIGURE 5

Hint. Introduce new variables

$$x - vt = \alpha, \qquad x + vt = \beta.$$

30. Study the oscillations of an infinite string produced by a concentrated load $Q = Q(t)$ moving along the string in the direction of the positive x-axis with velocity $v_0 < v$ (where v is the velocity of propagation of oscillations along the string). Solve the problem, assuming that at the time $t = 0$ the string is at rest and the source of oscillations is at the point $x = 0$.

Ans. The displacement at a fixed point $x > 0$ is given by the formulas

$$u = \begin{cases} 0, & 0 < t < \dfrac{x}{v}, \\[2mm] \dfrac{v}{2T} \displaystyle\int_0^{(vt-x)/(v-v_0)} Q(\tau)\, d\tau, & \dfrac{x}{v} < t < \dfrac{x}{v_0}, \\[2mm] \dfrac{v}{2T} \displaystyle\int_0^{(vt+x)/(v+v_0)} Q(\tau)\, d\tau, & \dfrac{x}{v_0} < t < \infty. \end{cases}$$

For $x < 0$ we have

$$u = \begin{cases} 0, & 0 < t < -\dfrac{x}{v}, \\[2mm] \dfrac{v}{2T} \displaystyle\int_0^{(vt+x)/(v+v_0)} Q(\tau)\, d\tau, & -\dfrac{x}{v} < t < \infty. \end{cases}$$

Hint. Regard the concentrated load as the limit of a distributed load, and use the result of the previous problem.

31. Investigate the nature of the oscillations observed at a fixed point $x > 0$ of the string considered in the preceding problem, assuming that the moving source undergoes harmonic oscillations

$$Q(t) = Q_0 \sin \omega t$$

of frequency ω.

Ans. The displacement at the point $x > 0$ is given by the formula

$$u = \begin{cases} 0, & 0 < t < \dfrac{x}{v}, \\[2mm] \dfrac{vQ_0}{2\omega T}\left(1 - \cos \omega\, \dfrac{vt - x}{v - v_0}\right), & \dfrac{x}{v} < t < \dfrac{x}{v_0}, \\[2mm] \dfrac{vQ_0}{2\omega T}\left(1 - \cos \omega\, \dfrac{vt + x}{v + v_0}\right), & \dfrac{x}{v_0} < t < \infty. \end{cases}$$

Before the time $t = x/v$, the point x is at rest. During the interval

$$\frac{x}{v} < t < \frac{x}{v_0},$$

i.e., as the source approaches, the point x undergoes harmonic oscillations of frequency

$$\omega_1 = \frac{\omega v}{v - v_0} > \omega.$$

For $t > x/v_0$, i.e., as the source recedes, the frequency of the observed osciilations is

$$\omega_2 = \frac{\omega v}{v + v_0} < \omega$$

(the Doppler effect).

32. A semi-infinite rod, clamped at the end $x = \infty$ and free from forces at the end $x = 0$, undergoes longitudinal oscillations. Investigate the nature of these oscillations, assuming that the initial conditions are of the form

$$u\big|_{t=0} = f(x), \qquad \frac{\partial u}{\partial t}\bigg|_{t=0} = 0,$$

and that $f(x) \to 0$ as $x \to \infty$.

Ans.

$$u(x, t) = \begin{cases} \frac{1}{2}[f(vt - x) + f(vt + x)], & 0 < x < vt, \\ \frac{1}{2}[f(x - vt) + f(x + vt)], & vt < x < \infty. \end{cases}$$

Hint. Make the even extension of the function $f(x)$ to the negative x-axis, and use the solution of the Cauchy problem for an infinite string.

33. Find the distribution of voltage and current along an infinite transmission line with parameters L, C, R and G, given the initial conditions

$$u\big|_{t=0} = \varphi(x), \quad I\big|_{t=0} = \psi(x), \qquad -\infty < x < \infty,$$

assuming that the parameters of the line are connected by the relation

$$\frac{R}{L} = \frac{G}{C}$$

(a distortionless line).

Ans.

$$u(x, t) = e^{-\alpha t}\left\{ \frac{1}{2}\left[\varphi(x - vt) + \varphi(x + vt)\right] + \frac{1}{2Cv}\left[\psi(x - vt) - \psi(x + vt)\right] \right\},$$

$$I(x, t) = e^{-\alpha t}\left\{ \frac{1}{2}\left[\psi(x - vt) + \psi(x + vt)\right] + \frac{1}{2Lv}\left[\varphi(x - vt) - \varphi(x + vt)\right] \right\},$$

where $\alpha = R/L$, $v = 1/\sqrt{LC}$.

Hint. Use Prob. 23 to write the differential equation for u, introduce a new unknown function w by writing $u = e^{-\gamma t}w$, and choose γ such that the coefficient of $\partial w/\partial t$ vanishes.

34. Show that the solution of the wave equation

$$\Delta u - \frac{1}{v^2}\frac{\partial^2 u}{\partial t^2} = 0,$$

with radially symmetric initial conditions

$$u\big|_{t=0} = \varphi(r), \qquad \frac{\partial u}{\partial t}\bigg|_{t=0} = \psi(r), \qquad 0 \leqslant r < \infty,$$

is given by the formula

$$u(r, t) = \frac{(r - vt)\varphi(r - vt) + (r + vt)\varphi(r + vt)}{2r} + \frac{1}{2vr}\int_{r-vt}^{r+vt} \rho\psi(\rho)\,d\rho,$$

where the values of the functions φ and ψ for negative arguments are given by the relations

$$\varphi(-r) = \varphi(r), \qquad \psi(-r) = \psi(r).$$

Hint. Transform the equation by setting $ru = w$, where w is a new unknown function. Then bear in mind that u remains bounded as $r \to 0$.

35. Study the oscillations occurring in a gas initially at rest when a local condensation s_0 is formed inside a sphere of radius a contained in the gas.

Ans. The condensation of the gas at an arbitrary point r is

$$s = \begin{cases} s_0, & r < a \\ 0 & r > a \end{cases} \qquad 0 < t < \frac{|r-a|}{v},$$

$$s = s_0\frac{r - vt}{2r}, \qquad \frac{|r-a|}{v} < t < \frac{r+a}{v},$$

$$s = 0, \qquad \frac{r+a}{v} < t < \infty,$$

where v is the velocity of wave propagation in the gas.

Hint. By the condensation is meant the quantity

$$s = \frac{\rho - \rho_0}{\rho_0}$$

(i.e., the relative change in density of the oscillating gas), which satisfies the differential equation

$$\Delta s - \frac{1}{v^2}\frac{\partial^2 s}{\partial t^2} = 0,$$

where $v = \sqrt{c_p p_0/c_v \rho_0}$, c_p and c_v are the specific heats at constant pressure and volume, and p_0 and ρ_0 are the initial values of the pressure and density. This problem is a special case of the preceding problem, corresponding to initial conditions

$$s\Big|_{t=0} = \begin{cases} s_0, & 0 \leqslant r < a, \\ 0, & r > a, \end{cases} \qquad \frac{\partial s}{\partial t}\Big|_{t=0} = 0.$$

36. In a gas initially at rest, a condensation $s = s_0$ localized in the volume bounded by a surface σ is created at the time $t = 0$. Show that the condensation at the point $M = (x, y, z)$ at an arbitrary time t is given by the expression

$$s(x, y, z, t) = s_0 \frac{\partial}{\partial t}\left(t \frac{\sigma_t}{S_{vt}}\right),$$

where S_{vt} is the sphere of radius vt with center at the point M, and σ_t is the part cut out of S_{vt} by the surface σ.

Hint. Use the general solution of the homogeneous wave equation

$$\Delta u - \frac{1}{v^2}\frac{\partial^2 u}{\partial t^2} = 0$$

for arbitrary initial conditions (see G1, p. 197).

37. The solution of the Cauchy problem for the three-dimensional inhomogeneous wave equation

$$\Delta u - \frac{1}{v^2}\frac{\partial^2 u}{\partial t^2} = -4\pi\rho(x, y, z, t),$$

with initial conditions

$$u\Big|_{t=0} = \varphi(x, y, z), \qquad \frac{\partial u}{\partial t}\Big|_{t=0} = \psi(x, y, z),$$

is of the form

$$u(x, y, z, t) = \frac{\partial}{\partial t}(t\overline{\varphi}_{vt} + t\overline{\psi}_{vt}) + \int_{G_{vt}} \frac{\rho\left(\xi, \eta, \zeta, t - \dfrac{r}{v}\right)}{r}\, d\xi\, d\eta\, d\zeta,$$

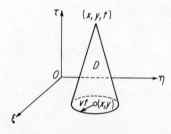

FIGURE 6

where $r = \sqrt{(x - \xi)^2 + (y - \eta)^2 + (z - \zeta)^2}$, $\overline{\varphi}_{vt}$ and $\overline{\psi}_{vt}$ are the average values of the functions φ and ψ over a sphere of radius vt with its center at the point $M = (x, y, z)$, and the integration is over the region G_{vt} bounded by this sphere (see K4, p. 101). Starting from this fact, solve the corresponding problem for the two-dimensional inhomogeneous equation

$$\frac{\partial^2 u}{\partial x^2} + \frac{\partial^2 u}{\partial y^2} - \frac{1}{v^2}\frac{\partial^2 u}{\partial t^2} = -4\pi\rho(x, y, t),$$

with initial conditions

$$u\big|_{t=0} = \varphi(x, y), \qquad \frac{\partial u}{\partial t}\bigg|_{t=0} = \psi(x, y).$$

Ans.

$$u(x, y, t) = \frac{\partial}{\partial t}(t\varphi^* + t\psi^*) + 2v\int_D \frac{\rho(\xi, \eta, \tau)}{\sqrt{v^2(t - \tau)^2 - r^2}}\,d\xi\,d\eta\,d\tau,$$

where

$$f^* = \frac{1}{2\pi vt}\int_S \frac{f(\xi, \eta)}{\sqrt{v^2 t^2 - r^2}}\,d\xi\,d\eta, \qquad r^2 = (x - \xi)^2 + (y - \eta)^2,$$

S is the disk of radius vt with center at the point (x, y), and D is the right circular cone shown in Figure 6.

38. Show that one solution of the two-dimensional wave equation

$$\frac{\partial^2 u}{\partial x^2} + \frac{\partial^2 u}{\partial y^2} - \frac{1}{v^2}\frac{\partial^2 u}{\partial t^2} = 0$$

is the function

$$u = \operatorname{Re} f(\theta),$$

where f is an arbitrary analytic function of the argument θ, related to the variables x, y, and t by the relation

$$t - \theta\frac{x}{v} + \sqrt{1 - \theta^2}\frac{y}{v} = 0.$$

Comment. This class of solutions of the wave equation is widely used in diffraction theory and other applications (see K4, p. 114 and S6, Vol. III, Pt. 2, p. 176).

39. Applying Riemann's method (see T1, p. 116), solve the hyperbolic equation

$$\frac{\partial^2 u}{\partial x^2} - \frac{1}{v^2}\frac{\partial^2 u}{\partial t^2} - cu = 0$$

(where v and c are given constants), with arbitrary initial conditions

$$u\big|_{t=0} = \varphi(x), \qquad \frac{\partial u}{\partial t}\bigg|_{t=0} = \psi(x).$$

Ans.

$$u(x, t) = \frac{1}{2} \left[\varphi(x - vt) + \varphi(x + vt) \right]$$

$$+ \frac{1}{2v} \int_{x-vt}^{x+vt} \left[\psi(\xi) J_0(\sqrt{c} \, R) - \frac{v^2 \sqrt{c} \, t}{R} \varphi(\xi) J_1(\sqrt{c} \, R) \right] d\xi,$$

where

$$R = \sqrt{(vt)^2 - (x - \xi)^2}.$$

40. Find the distribution of current along an infinite transmission line with parameters L, C, R and G, assuming that at the time $t = 0$ the current vanishes while the voltage is nonzero and equal to a constant V on the section of the line $|x| < a$.

Ans.

$$I(x, t)\Big|_{vt < a} = \begin{cases} 0, & 0 < x < a - vt, \\ \dfrac{V}{2Z} e^{-\alpha t} I_0\left(|\beta|\sqrt{t^2 - \dfrac{(x - a)^2}{v^2}}\right), & a - vt < x < a + vt, \\ 0, & a + vt < x < \infty, \end{cases}$$

$$I(x, t)\Big|_{vt > a} = \begin{cases} \dfrac{V}{2Z} e^{-\alpha t}\left[I_0\left(|\beta|\sqrt{t^2 - \dfrac{(x - a)^2}{v^2}}\right) - I_0\left(|\beta|\sqrt{t^2 - \dfrac{(x + a)^2}{v^2}}\right) \right], \\ \qquad\qquad 0 < x < vt - a, \\ \dfrac{V}{2Z} e^{-\alpha t} I_0\left(|\beta|\sqrt{t^2 - \dfrac{(x - a)^2}{v^2}}\right), & vt - a < x < vt + a, \\ 0, & vt + a < x < \infty, \end{cases}$$

where

$$v = \frac{1}{\sqrt{LC}}, \quad Z = Lv, \quad \alpha = \frac{RC + LG}{2LC}, \quad \beta = \frac{RC - LG}{2LC},$$

and $I_0(x)$ is the Bessel function of imaginary argument.

Hint. Use the result of the preceding problem.

41. A semi-infinite rod of variable cross section $S(x) = S(0)e^{-\alpha x}$, where the end $x = 0$ is clamped, undergoes longitudinal oscillations with initial conditions

$$u\Big|_{t=0} = f(x), \qquad \frac{\partial u}{\partial t}\Big|_{t=0} = 0.$$

Find the displacement of an arbitrary cross section of the rod at an arbitrary time t.

Ans.

$$u(x, t) = \frac{1}{2} e^{\alpha x/2} [e^{-\alpha(x+vt)/2} f(vt + x) - e^{\alpha(x-vt)/2} f(vt - x)]$$

$$+ \frac{\alpha vt}{4} e^{\alpha x/2} \left[\int_0^{vt-x} J_1 \left(\frac{\alpha R_2}{2} \right) f(\xi) e^{-\alpha\xi/2} \frac{d\xi}{R_2} - \int_0^{vt+x} J_1 \left(\frac{\alpha R_1}{2} \right) f(\xi) e^{-\alpha\xi/2} \frac{d\xi}{R_1} \right]$$

$$\text{if } 0 < x < vt,$$

$$u(x, t) = \frac{1}{2} [e^{-\alpha vt/2} f(x + vt) + e^{\alpha vt/2} f(x - vt)]$$

$$- \frac{\alpha vt}{4} e^{\alpha x/2} \int_{x-vt}^{x+vt} J_1 \left(\frac{\alpha R_1}{2} \right) f(\xi) e^{-\alpha\xi/2} \frac{d\xi}{R_1} \quad \text{if } vt < x < \infty,$$

where $R_{1,2} = \sqrt{(vt)^2 - (x \mp \xi)^2}$.

Hint. By introducing a new unknown function $w = \sqrt{S(x)}u$, reduce this problem to the integration of the equation in Prob. 39.

2. Elliptic Equations: The Green's Function Method

A typical problem of the kind to be considered in this section is to find a solution of a partial differential equation of the elliptic type which is well-behaved in a given spatial region D and satisfies certain conditions on the boundary S of D. The simplest such problem is to find a function u which is harmonic in a region D with boundary S,[3] and satisfies one of the following boundary conditions

$$u\big|_S = f(P), \tag{2a}$$

$$\frac{\partial u}{\partial n_i}\bigg|_S = f(P), \tag{2b}$$

$$\left(\frac{\partial u}{\partial n_i} - hu \right)\bigg|_S = f(P), \tag{2c}$$

where $f(P)$ is a given function of a variable point P of S, n_i is the interior normal to S at P, and h is a positive constant. The problem is called the *first boundary value problem (of potential theory)* or the *Dirichlet problem* if the boundary condition is of the form (2a), the *second boundary value problem* or the *Neumann problem* if the boundary condition is of the form (2b), and the *third boundary value problem* or the *Robin problem* if the boundary condition

---

[3] A function u is said to be *harmonic* in a (two- or three-dimensional) region D if u and its first and second partial derivatives are continuous and satisfy Laplace's equation $\Delta u = 0$ in D. If D is unbounded, certain extra requirements must be imposed on the behavior of u at infinity (see T1, p. 265).

is of the form (2c). Similar problems can also be formulated for *Helmholtz's equation*

$$\Delta u + k^2 u = 0$$

and other equations of elliptic type encountered in mathematical physics.

One of the special methods for solving boundary value problems of this kind is based on the use of the *Green's function* (see S6, Vol. IV and T1, Chap. 4). The key result of this theory is that the solution of the boundary value problem for *Poisson's equation*

$$\Delta u = -F(M), \tag{3}$$

subject to any of the boundary conditions (2a)–(2c), can be written in quadratures, once we know the Green's function. The Green's function does not depend on the form of the functions $f(P)$ and $F(M)$, and can be found by considering a special boundary value problem (see below).

Thus, for example, the solution of the first boundary value problem for the equation (3) can be written in the form

$$u(M_0) = \int_S f(P) \frac{\partial G}{\partial n_i} \, d\sigma + \int_D f(M) G \, d\tau, \tag{4}$$

where M is a variable point and M_0 a fixed point of the region D ($d\sigma$ is the element of surface area and $d\tau$ the element of volume). Here the Green's function $G(M, M_0)$ is the function such that

1. G is harmonic in D except at the point M_0, near which G is of the form

$$G(M, M_0) = \frac{1}{4\pi \, |MM_0|} + v,$$

where the function v is regular (i.e., has no singularities) in D;

2. G satisfies the boundary condition

$$G|_S = 0.$$

It follows that v is harmonic and satisfies the boundary condition

$$v|_S = -\frac{1}{4\pi \, |PM_0|},$$

i.e., v is the solution of a special case of the Dirichlet problem.

The same formula (4) gives the solution of the first boundary value problem in two dimensions, if by the Green's function we now mean a function such that

1. G is harmonic in a planar region D except at the point M_0, near which G is of the form

$$G(M, M_0) = \frac{1}{2\pi} \ln \frac{1}{|MM_0|} + \text{a regular function};$$

2. G satisfies the boundary condition

$$G|_S = 0$$

on the contour bounding D.

Formulas of a similar kind can be found giving solutions of other boundary value problems, involving Laplace's equation, Poisson's equation, Helmholtz's equation, etc.

Green's functions for regions of various shapes can be found by using the methods considered in Chaps. 4–7, and also by using certain special techniques, like the method of images and the method of inversion.[4] The method of images allows us to construct the Green's function for a half-space and for a sphere (or, in two dimensions, for a half-plane and a circle) and for certain regions of a more complicated shape, e.g., the layer bounded by two parallel planes or the interior of an angle of π/n radians ($n = 1, 2, \ldots$). Starting from the Green's function for a region D and using the method of inversion, we can find the Green's function for the region D^* obtained by inverting D in a sphere lying outside D. Thus, for example, we can find the Green's function for a sphere from a knowledge of the Green's function for a half-space, the Green's function for the region bounded by two intersecting spheres from the Green's function for the region bounded by two intersecting planes, and so on.

We now give some problems illustrating these methods of constructing Green's functions, and also a few problems of a more theoretical nature.

42. Construct the Green's function for the two-dimensional Dirichlet problem in the case where the region D is the first quadrant $x \geqslant 0, y \geqslant 0$.

Ans.

$$G(M, M_0) = \frac{1}{4\pi}\left[\frac{1}{|MM_0|} - \frac{1}{|MM_2|} + \frac{1}{|MM_3|} - \frac{1}{|MM_4|}\right],$$

where $M = (x, y, z), M_0 = (x_0, y_0, z_0), M_2 = (-x_0, y_0, z_0), M_3 = (-x_0, -y_0, z_0)$ and $M_4 = (x_0, -y_0, z_0)$.

Hint. Use the method of images.

43. Using the method of images, construct the Green's function for the Dirichlet problem in the case where the region D is the part of space lying between two parallel planes $z = \pm l/2$.

---

[4] See T1, and in particular G5, which contains a number of interesting applications of the method of inversion to problems of electrostatics.

Ans.

$$G(M, M_0) = \frac{1}{4\pi} \sum_{n=-\infty}^{\infty} \left\{ \frac{1}{\sqrt{R^2 + [z - z_0 - 2nl]^2}} \right.$$
$$\left. - \frac{1}{\sqrt{R^2 + [z + z_0 - (2n + 1)l]^2}} \right\},$$

where

$$R = \sqrt{(x - x_0)^2 + (y - y_0)^2}.$$

44. Use the method of inversion to deduce the Green's function for the Dirichlet problem in the case where D is a sphere of radius a with its center at the origin O, assuming that the expression for the Green's function of a half-space is known.

Ans.

$$G(M, M_0) = \frac{1}{4\pi} \left[\frac{1}{|MM_0|} - \frac{a}{|OM_0|} \frac{1}{|MM_1|} \right],$$

where M_1 is the image of the point M_0 in the sphere.

45. Find the Green's function for a hemisphere of radius a.

Ans.

$$G(M, M_0) = \frac{1}{4\pi} \left[\frac{1}{|MM_0|} - \frac{a}{|OM_0|} \frac{1}{|MM_1|} - \frac{1}{|MM_2|} + \frac{a}{|OM_0|} \frac{1}{|MM_3|} \right],$$

where M_1 is the image of M_0 in the corresponding full sphere, M_2 is the image of M_0 in the diametral plane of the hemisphere, and M_3 is the image of M_2 in the full sphere.

46. The Green's function $G = G(M, M_0)$ for the Neumann problem[5]

$$\Delta u = -F(M), \qquad \frac{\partial u}{\partial n_i}\bigg|_S = f(P)$$

is defined by the conditions

1. G is harmonic in D except at the fixed point M_0, near which G is of the form

$$G = \frac{1}{4\pi |MM_0|} + \text{a regular function};$$

---

[5] Here M is a point of the three-dimensional region D, P is a point of the surface S bounding D, and the functions f and F satisfy the condition

$$\int_S f \, d\sigma = \int_D F \, d\tau$$

for the solvability of the Neumann problem. If the Green's function is known, the solution is given by the formula

$$u(M_0) = -\int_S f(P)G \, d\sigma + \int_D F(M)G \, d\tau + \text{const.}$$

2.

$$\frac{\partial G}{\partial n_i}\bigg|_S = \frac{1}{S_0},$$

where S_0 is the area of the surface S;

3.

$$\int_S G \, d\sigma = 0.$$

Verify that in the special case where D is a sphere of radius a with center at the origin O, the Green's function is

$$G(M, M_0) = \frac{1}{4\pi \, |MM_0|} + \frac{a}{4\pi \, |OM_0| \, |MM_1|}$$

$$+ \frac{1}{4\pi a} \ln \frac{2}{1 - \dfrac{|OQ|}{|OM_1|} + \dfrac{|MM_1|}{|OM_1|}} - \frac{1}{2\pi a},$$

where M_1 is the image of the point M_0 in the sphere, and Q is the foot of the perpendicular dropped from the point M onto the line OM_1.

47. A conductor bounded by a closed surface S and held at a given potential V is introduced into an arbitrary external field with potential u_0. Suppose we know the charge density $\rho(P, M_0)$ at the point P of the surface S in the case where the surface is grounded and the external field is due to a unit charge at an arbitrary point M_0 outside the conductor. Show that the potential distribution in the general case is given by the formula

$$u(M_0) = V + u_0(M_0) + \int_S u_0(P)\rho(P, M_0) \, d\sigma.$$

Hint. The formula

$$w(M_0) = \int_S f(P) \frac{\partial G}{\partial n} \, d\sigma$$

represents the solution of the boundary value problem

$$\Delta w = 0 \text{ outside } S,$$

$$W|_S = f(P), \qquad W|_\infty = 0$$

in terms of the Green's function. Apply this formula to the function $w = u - u_0$, bearing in mind the electrostatic interpretation of the Green's function.

48. Find the Green's function for the two-dimensional Dirichlet problem, assuming that we know the function $\zeta = \zeta(z)$ mapping a given region in the z-plane conformally onto the upper half of the ζ-plane (Im $\zeta > 0$). Use the result to construct the Green's function for the half-strip $x \geqslant 0$, $0 \leqslant y \leqslant \pi$.

Ans.

$$G(M, M_0) = -\frac{1}{2\pi} \operatorname{Re} \left\{ \ln \frac{\zeta - \zeta_0}{\zeta - \bar{\zeta}_0} \right\},$$

where ζ and ζ_0 are the points of the half-plane corresponding to $M = (x, y)$ and $M_0 = (x_0, y_0)$. In the case of the half-strip,

$G(M, M_0)$

$$= -\frac{1}{4\pi} \ln \frac{[\cosh(x + x_0) - \cos(y + y_0)][\cosh(x - x_0) - \cos(y - y_0)]}{[\cosh(x + x_0) - \cos(y - y_0)][\cosh(x - x_0) - \cos(y + y_0)]}.$$

Hint. The conformal mapping of the half-strip onto the half-plane is accomplished by the function $\zeta = \cosh z$.

49. The boundary value problem

$$\Delta^2 u = F(M),$$

$$u|_S = 0, \qquad \frac{\partial u}{\partial n}\bigg|_S = 0$$

(Δ is the two-dimensional Laplacian, and M is a point of a planar region D bounded by a contour S) is encountered in the theory of bending of thin elastic plates. The solution of this problem can be written in the form

$$u(M_0) = \int_D F(M) G \, d\tau$$

($d\tau$ is an element of area), where the Green's function $G = G(M, M_0)$ is defined by the conditions

1. G is the solution of the biharmonic equation $\Delta^2 u = 0$ which is regular (i.e., free of singularities) in D, except at the fixed point M_0, near which G is of the form

$$G = \frac{1}{8\pi} |MM_0|^2 \ln |MM_0| + \text{a regular function};$$

2.

$$G|_S = 0, \qquad \frac{\partial G}{\partial n}\bigg|_S = 0.$$

Verify that in the special case where D is a disk of radius a with its center at the origin O, the Green's function is given by

$$G(M, M_0) = \frac{a^2}{16\pi} \left[\frac{|MM_0|^2}{a^2} \ln \frac{a^2 |MM_0|^2}{|OM_0|^2 |MM_1|^2} + \left(1 - \frac{|OM_0|^2}{a^2}\right)\left(1 - \frac{|OM|^2}{a^2}\right) \right],$$

where M_1 is the image of the point M_0 in the circle bounding D.

3. Elliptic Equations: The Method of Conformal Mapping

In mathematical physics one often encounters the problem of finding a function which is harmonic in a two-dimensional region D and satisfies the boundary condition

$$u\big|_S = f \tag{5a}$$

or

$$\frac{\partial u}{\partial n}\bigg|_S = f, \tag{5b}$$

where f is a given function and n is the normal to the contour S bounding D. For example, such problems arise in studying electrostatics, magnetostatics, heat conduction, flow of ideal fluids, filtration phenomena, and so on. An effective method of solving problems of this kind is to construct a function of a complex variable $\zeta = F(z)$ such that $F(z)$ is analytic in D and maps D conformally onto a region D^* (with boundary S^*) of a special form for which the solution of the given problem is either known or can be found more simply than for the original region D. Here it is assumed that $F'(z)$ is non-zero in the region D, a condition which guarantees that the mapping is one-to-one. In asserting that this method leads to a solution of the boundary value problem, we rely on the fact that the Laplacian and the boundary conditions (5a) and (5b) preserve their form[6] under the transformation from the variables x and y to the new variables ξ and η defined by the relation

$$\xi + i\eta = F(x + iy).$$

The method of conformal mapping can also be used to deal with more complicated boundary value problems, e.g., problems where the value of the unknown function u is specified on parts of the contour S while the value of $\partial u/\partial n$ is specified on the rest of S, problems of jet flow of an ideal flow where the form of S is not known in advance but is determined in the course of solving the problem, and so on.

In many cases, the construction of the function $\zeta = F(z)$ mapping the region D onto the region D^* can be accomplished by consecutive application of several mappings which involve elementary functions. Of particular importance in applied work is the case where D is a polygon and D^* is the upper half-plane. Then the function effecting the mapping can be found by using the familiar Schwarz-Christoffel transformation (see W1). The use of conformal mapping to solve problems of mathematical physics, involving

---

[6] In the case of the boundary condition (5b), the value taken by the normal derivative on the contour S^* is

$$\frac{1}{|F'(z)|}\bigg|_S f.$$

the biharmonic equation as well as Laplace's equation, is amply discussed in books on complex variable theory and in special monographs (see B3, F10, M10, S7, etc.). Hence we confine ourselves here to a few typical problems which illustrate the technique of the method, assuming that the reader is already familiar with the elementary theory of conformal mapping.

In most of the problems, the required conformal mapping can be found by using the Schwarz-Christoffel transformation. Problems 51, 57 and 59 require knowledge of the properties of elliptic integrals and Jacobian elliptic functions. In connection with Probs. 50–54, the following remarks will be found helpful: If φ is the potential of a stationary plane-parallel flow of an ideal fluid, described by the velocity field $\mathbf{v} = -\text{grad } \varphi$, then by the *complex potential* $w = w(z)$ is meant a function of the complex variable $z = x + iy$ whose real part equals φ. In other words, $w = \varphi + i\psi$, where ψ is related to φ by the Cauchy-Riemann equations

$$\frac{\partial \varphi}{\partial x} = \frac{\partial \psi}{\partial y}, \qquad \frac{\partial \varphi}{\partial y} = -\frac{\partial \psi}{\partial x}.$$

The lines of flow or *streamlines* are described by the family of curves $\psi = $ const, and hence the function χ is called the *stream function*. The amount of fluid Q flowing per unit time between two streamlines $\psi = \psi_1$ and $\psi = \psi_2$ (in a slab of unit thickness parallel to the xy-plane) is given by

$$Q = |\psi_1 - \psi_2|.$$

The components of the velocity vector $\mathbf{v} = v_x + iv_y$ are related to the derivative of the complex potential by the formula

$$v_x - iv_y = -\frac{dw}{dz}.$$

The complex potential is a valuable tool for studying plane-parallel flows.[7]

50. An ideal fluid, whose velocity at infinity equals $v_x = v_\infty$, $v_y = 0$, flows past an obstacle in the shape of an elliptical cylinder

$$\frac{x^2}{a^2} + \frac{y^2}{b^2} = 1.$$

Use the method of conformal mapping to find the complex potential of the flow.

---

[7] Similarly, in the theory of stationary heat flow and in electrostatics, one can introduce complex potentials, with the role of $\mathbf{v}$ and φ being played by $\mathbf{q}/k$ (the ratio of the heat flow density to the thermal conductivity) and the temperature T in the first case, and by the electric vector $\mathbf{E}$ and the electrostatic potential φ in the second case.

Ans. The relation between the complex potential w and the variable z is given in parametric form by the equations

$$w = -\frac{v_\infty(a+b)}{2a}\left(t + \frac{a^2}{t}\right), \qquad z = \frac{a+b}{2}\frac{t}{a} + \frac{a-b}{2}\frac{a}{t},$$

where t belongs to the region $|t| \geqslant a$, $0 \leqslant \arg t \leqslant \pi$.

Hint. First make a conformal mapping of the part of the region occupied by the flow and lying above the axis of symmetry onto the half-plane with a semi-circular cut of radius a, and then map this region conformally onto the upper half of the ζ-plane in such a way that the semi-circular arc of radius a goes into the interval $(-a, a)$ of the real axis.

51. Solve the preceding problem for the case where the obstacle is a cylinder $-a \leqslant x \leqslant a$, $-b \leqslant y \leqslant b$ of rectangular cross section.

Ans. The complex potential has the parametric representation

$$w = -\frac{v_\infty A}{k}\zeta, \qquad z = A\int_0^\zeta \sqrt{\frac{1-\zeta^2}{1-k^2\zeta^2}}\,d\zeta + ib \qquad (0 \leqslant \arg \zeta \leqslant \pi),$$

where

$$A = \frac{ak^2}{E(k) - k'^2 K(k)},$$

and the modulus of the elliptic integrals is determined from the condition

$$\frac{b}{a} = \frac{E(k') - k^2 K(k')}{E(k) - k'^2 K(k)}.$$

Hint. Use the Schwarz-Christoffel transformation to map the region occupied by the flow and lying above the axis of symmetry $y = 0$ onto the half-plane in such a way that the vertices $\pm a$, $\pm a \pm ib$ go into the points $\pm 1/k$, ± 1.

*52. Study the two-dimensional motion of an ideal fluid in the channel of variable cross section shown in Figure 7, assuming that at infinity the direction of the flow coincides with the x-axis and has the values

$$v_x\big|_{x\to-\infty} = v_a, \qquad v_x\big|_{x\to+\infty} = v_b$$

$$(av_a = bv_b).$$

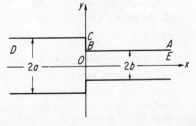

FIGURE 7

Find the distribution of velocity along the axis of symmetry of the flow.[8]

Ans. The velocity distribution in parametric form is given by the equations

$$\frac{v_x|_{y=0}}{v_a} = t, \qquad \frac{x}{a} = \frac{1}{\pi}\left[\ln\frac{t-1}{t+1} + \frac{b}{a}\ln\frac{1+(b/a)t}{1-(b/a)t}\right] \qquad \left(1 \leqslant t \leqslant \frac{a}{b}\right).$$

Hint. Using the Schwarz-Christoffel transformation, map the domain

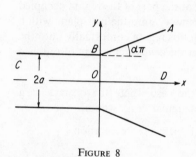

$ABCDE$ onto the upper half-plane of the complex variable $\zeta = \xi + i\eta$, requiring the points B, C and D to go into the points $\zeta = -1$, $\zeta = -\lambda$ and $\zeta = 0$, where λ is a number between 0 and 1 which subsequent calculations show to be equal to the ratio b^2/a^2.

FIGURE 8

53. Solve the preceding problem for the case where the channel has the form shown in Figure 8, assuming that

$$v_x|_{x\to-\infty} = v_a.$$

Ans.

$$\frac{v_x|_{y=0}}{v_a} = \frac{1}{(1+\xi)^\alpha}, \qquad \frac{x}{a} = \frac{1}{\pi}\left[\ln\xi + \int_0^{1+\xi}\frac{t^\alpha-1}{t-1}dt\right] \qquad (\xi \geqslant 0).$$

Hint. Transform the region bounded by the wall of the channel and the axis of symmetry of the flow into the upper half-plane of the complex variable ζ, making the vertices B and C go into the points -1 and 0.

*54. Investigate the jet flow of a liquid through a slot of width $2a$ in a plane wall (see Figure 9), assuming that the amount of fluid flowing through the slot per unit time (in a slab of unit thickness parallel to the xy-plane) equals Q. Find the form of the jet.

Ans. In parametric form, the equation of the curve bounding the jet is

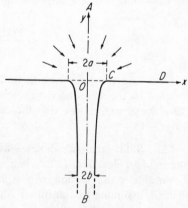

FIGURE 9

---

[8] In Probs. 52–54, where the flow is symmetric with respect to an axis, it is convenient to assume that $\psi = 0$ along this axis. The value of ψ along any other streamline can be found by using the formula $Q = |\psi_1 - \psi_2|$.

given by

$$\frac{x}{a} = \frac{\sqrt{t} + (\pi/2)}{1 + (\pi/2)}, \qquad \frac{y}{a} = \frac{1}{1 + (\pi/2)}\left[\sqrt{1 - t} - \frac{1}{2}\ln\frac{1 + \sqrt{1 - t}}{1 - \sqrt{1 - t}}\right]$$

$$(0 \leqslant t \leqslant 1).$$

The width of the jet at a great distance from the slot is

$$2b = \frac{2a}{1 + (2/\pi)}.$$

Hint. Use Kirkhhoff's method (see K2, p. 332 ff.).

55. A pipe of radius a lies below the ground at depth h (see Figure 10). Find the stationary temperature distribution in the region surrounding the pipe, assuming that the temperature of the earth's surface is zero, while the temperature of the pipe is T_0.

Ans.

$$T = \frac{T_0}{\ln\left[(h + c)/a\right]}$$
$$\times \ln\frac{\sqrt{(x^2 - c^2 + y^2)^2 + 4c^2 y^2}}{(x - c)^2 + y^2},$$

where $c = \sqrt{h^2 - a^2}$.

Hint. Use a fractional linear transformation

$$\zeta = \frac{z - c}{z + c}$$

to map the given region into a circular ring.

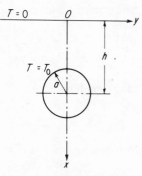

FIGURE 10

56. Find the stationary temperature distribution in a wall of thickness a near the corner of a building (see Figure 11), assuming that the temperature of the inside surface of the wall is T_0, while the temperature of the outside surface is zero.

Ans.

$$T = T_0 \operatorname{Re}\left\{\frac{1}{\pi i}\ln\zeta\right\},$$

$$\frac{z}{a} = \frac{2}{\pi}\left[\frac{1}{2}\ln\frac{1 + t}{1 - t} - \arctan t\right] + 1 + i,$$

where

$$t = \left(\frac{\zeta + 1}{\zeta - 1}\right)^{1/2},$$

and ln and arc tan denote the branches which go to zero as $t \to 0$.

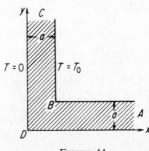

FIGURE 11

Hint. Use the Schwarz-Christoffel transformation to map the figure $ABCDA$ onto the upper half-plane of the complex variable ζ, making the points B, C and D go into the points -1, 0, λ, where λ is to be determined (a calculation shows that $\lambda = 1$).

57. Solve the problem of the stationary temperature distribution in a homogeneous slab $-\infty < x < \infty$, $-b \leqslant y \leqslant b$ of thickness $2b$, inside which there is another thin slab of thickness $2a$ $(a < b)$ sharing the same midplane and held at temperature T_0. It is assumed that the temperature of the outside surfaces of the slab equals zero. Calculate the flow of heat Q given off by the source per unit time.

Ans.

$$w = T + i\psi = T_0\left(1 - \frac{\zeta}{iK'}\right),$$

where the relation between the complex variables ζ and z is given by the equation

$$\operatorname{sn} \zeta = \frac{1}{k} \tanh \frac{\pi z}{2b},$$

and the modulus of the elliptic function is

$$k = \tanh \frac{\pi a}{2b}.$$

Moreover,

$$Q = \frac{4T_0}{\varkappa} \frac{K}{K'},$$

where $\varkappa$ is the thermal conductivity of the slab, while K and K' are the complete elliptic integrals with moduli k and $k' = \sqrt{1 - k^2}$.

Hint. Using the transformation

$$t = \frac{1}{k} \tanh \frac{\pi z}{2b},$$

where k has the value indicated above, map the strip $-\infty < x < \infty$, $0 \leqslant y \leqslant b$ onto the upper half-plane of the variable t. Then use the Schwarz-Christoffel transformation to transform this strip into a rectangle with vertices at the points $\pm K$, $\pm iK'$ in the ζ-plane.

58. A wire with charge q per unit length is located near the rectangular edge of a grounded conductor (see Figure 12). Find the distribution of the electric

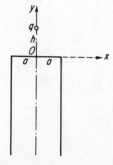

FIGURE 12

field in the symmetry plane of the region between the wire and the conductor.

Ans. In parametric form, the field is given by the formulas

$$E_y|_{x=0} = -\frac{\pi q \eta_q}{a(\eta_q^2 - \eta^2)\sqrt{1 + \eta^2}},$$

$$\frac{\pi y}{2a} = \eta\sqrt{1 + \eta^2} + \ln(\eta + \sqrt{1 + \eta^2}) \qquad (0 \leqslant \eta \leqslant \eta_q),$$

where η_q is the value of the parameter η corresponding to $y = h$.

Hint. Map the part of the z-plane lying outside the conductor onto the upper half-plane of the variable $\zeta = \xi + i\eta$, making the corners go into the points $\zeta = \pm 1$.

*59. On the axis of a box $-a \leqslant x \leqslant a, 0 \leqslant y \leqslant b$ of rectangular cross section with grounded walls, there is a thin wire with charge q per unit length. Write an expression for an appropriate complex potential, and calculate the distribution of charge density on the walls of the box.

Ans.

$$w = -2q \operatorname{sn} \frac{\operatorname{sn}\dfrac{Kz}{a} - \dfrac{i}{\sqrt{k}}}{\operatorname{sn}\dfrac{Kz}{a} + \dfrac{i}{\sqrt{k}}},$$

where $\operatorname{sn} z$ is a Jacobian elliptic function with modulus k. The modulus k is determined from the equation

$$\frac{b}{a} = \frac{K'}{K},$$

where $K = K(k)$ is the complete elliptic integral of the first kind and $K' = K(\sqrt{1 - k^2})$. The distribution of charge density on the wall $-a \leqslant x \leqslant a$, $y = 0$ is given by the formulas

$$\frac{\sigma}{\sigma_0} = \frac{\operatorname{cn}\dfrac{Kx}{a} \operatorname{dn}\dfrac{Kx}{a}}{1 + k \operatorname{sn}^2 \dfrac{Kx}{a}}, \qquad \sigma_0 = -\frac{q}{a}\frac{K\sqrt{k}}{\pi},$$

where $\operatorname{cn} z$ and $\operatorname{dn} z$ are Jacobian elliptic functions.

Hint. Use the Schwarz-Christoffel transformation to map the interior of the rectangle onto the upper half-plane, making the vertices $\pm a, \pm a + ib$ of the rectangle go into the points $\pm 1, \pm 1/k$. During the calculations, bear in mind that

$$\operatorname{sn}\frac{iK'}{2} = \frac{i}{\sqrt{k}}.$$

60. Find the electrostatic field on the axis of an electronic lens made of two pairs of plates at potentials $+V$ and $-V$, separated by a space $2a$ (see Figure 13).

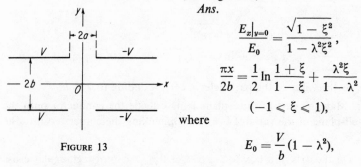

Ans.

$$\frac{E_x|_{y=0}}{E_0} = \frac{\sqrt{1 - \xi^2}}{1 - \lambda^2 \xi^2},$$

$$\frac{\pi x}{2b} = \frac{1}{2} \ln \frac{1 + \xi}{1 - \xi} + \frac{\lambda^2 \xi}{1 - \lambda^2}$$

$$(-1 \leqslant \xi \leqslant 1),$$

where

$$E_0 = \frac{V}{b}(1 - \lambda^2),$$

FIGURE 13

and λ is a number between 0 and 1 determined from the equation

$$\frac{\pi a}{2b} = \frac{1}{2} \ln \frac{1 - \lambda}{1 + \lambda} - \frac{\lambda}{1 - \lambda^2}.$$

Hint. Map the upper half-plane of the variable $z = x + iy$ cut along the line segments $(-\infty + ib, -a + ib)$ and $(a + ib, \infty + ib)$ onto the upper half-plane of the variable ζ, in such a way that the corners go into the points $\pm 1, \pm 1/\lambda$. Then transform the half-plane onto the half-strip

$$-\frac{\pi}{2} \leqslant \text{Re } t \leqslant \frac{\pi}{2}, \qquad \text{Im } t \geqslant 0.$$

61. Find the field on the axis of the electronic lens shown in Figure 14.

Ans.

$$\frac{E_x|_{y=0}}{E_0} = \frac{\sqrt{1 - \xi^2}}{1 - (\xi^2/\lambda^2)},$$

$$\frac{\pi x}{a + b} = \frac{1 + \gamma \lambda}{\lambda^2 - 1} (\lambda + \xi) + \frac{b}{a + b} \ln \frac{1 + \lambda}{1 - \xi}$$

$$+ \frac{a}{a + b} \ln \frac{1 + \xi}{\lambda - 1}$$

$$(-1 < \xi < 1),$$

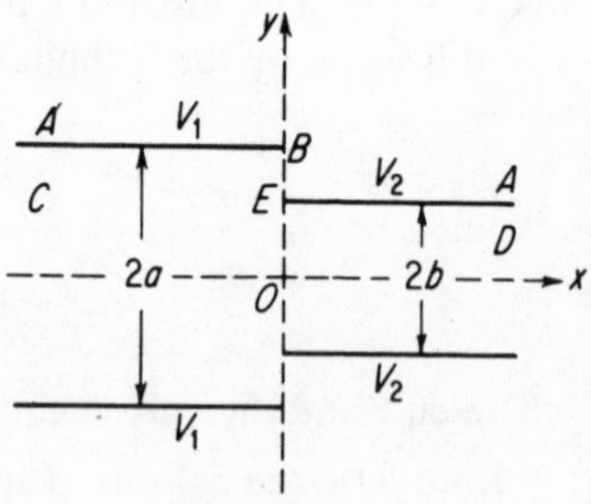

where λ is determined from the equation

FIGURE 14

$$\frac{\gamma \lambda^2 + 2\lambda + \gamma}{\lambda^2 - 1} + \ln \frac{\lambda + 1}{\lambda - 1} - \gamma \ln \frac{1 + \gamma \lambda}{2} = \gamma \ln \left(1 + \frac{b}{a}\right),$$

and we introduce the abbreviations

$$\frac{a-b}{a+b} = \gamma, \qquad E_0 = \frac{V_1 - V_2}{a+b} \frac{1 - (1/\lambda^2)}{1 + \gamma\lambda}.$$

Hint. Map the domain $ABCDEA$ onto the upper half-plane of the variable $\zeta = \xi + i\eta$, making the points B, C, D and E go into the points $-\lambda$, -1, 1 and μ. After determining the function $z = z(\zeta)$, carry out the transformation $\zeta = \sin t$.

62. Find the magnetic field in the midplane of the magnet whose poles have the rectangular shape shown in Figure 15, assuming that the magnet is made of iron with infinite magnetic permeability ($\mu = \infty$).

Ans.

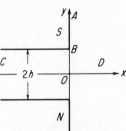

$$\frac{H_y|_{y=0}}{H_\infty} = t, \qquad \frac{\pi x}{2h} = \frac{1}{t} + \frac{1}{2}\ln\frac{1-t}{1+t}$$

$$(0 < t < 1),$$

where H_∞ is the homogeneous field in the midplane of the magnet at a great distance from the edge.

FIGURE 15

Hint. Map the region $ABCD$ onto the upper half-plane, making the points B and C go into the points -1 and 0.

63. The region $x \geqslant 0$, $y \leqslant 0$ is filled with iron of magnetic permeability $\mu = \infty$. Find the magnetic field due to a linear current source J passing through the point $(-a, 0)$.

Ans. The components of the field are determined by the relation

$$H_x - iH_y = \frac{4iJ}{3cz^{1/3}} \frac{a^{2/3} - 2z^{2/3}}{z^{4/3} + z^{2/3}a^{2/3} + a^{4/3}},$$

where c is the velocity of light.

Hint. Bear in mind that near the current source, the complex potential of the magnetic field has a logarithmic singularity:

$$w = \frac{2iJ}{c}\ln(z - z_0) + \text{a regular function}.$$

References

Betz (B2), Courant and Hilbert (C5), Kupradze (K5), Morse and Feshbach (M9), Petrovski (P2), Smythe (S7), Sneddon (S11), Sternberg and Smith (S16), Tikhonov and Samarski (T1), Walker (W1).

3

STEADY-STATE HARMONIC

OSCILLATIONS

A solution of a partial differential equation is said to be a *steady-state harmonic oscillation* if its time dependence is described by the factor $e^{i\omega t}$, where ω is the frequency.[1] Problems involving steady-state harmonic oscillations are among the simplest and the most important encountered in mathematical physics. Because of the particularly simple form of the time dependence, we can eliminate the variable t from the original equation, thereby reducing the problem to the determination of complex amplitudes depending only on the spatial coordinates. In the special case where the solution depends only on a single spatial coordinate, the equation for the complex amplitude reduces to an ordinary differential equation. This category, to which most of the problems in the present chapter belong, is of considerable interest because of its numerous applications to concrete problems of mechanics, electromagnetic theory, etc. Moreover, such problems are very important from a methodological standpoint, since they serve as the best introduction to the technique of particular solutions to be considered in Chapter 4. Thus, for example, the problem of determining natural frequencies anticipates the problem of determining eigenvalues, and the problem of forced oscillations gives insight into ways of overcoming difficulties associated with the application of the Fourier method to inhomogeneous equations.

---

[1] In using complex quantities in intermediate steps of our calculations, we rely on the fact that the equations of mathematical physics (at least, those considered here) are linear. Thus, to obtain the final answer, we need only take the real or imaginary part of some expression (depending on the conditions of the particular problem).

This chapter contains three sections. The first is devoted to problems on the determination of natural frequencies of vibrating systems (strings, rods, membranes and plates), while the second deals with forced oscillations of such systems.[2] The third section is concerned with problems on steady-state electromagnetic oscillations in transmission lines and cavity resonators, certain related problems on the propagation of electromagnetic waves in waveguides of given cross section, etc.

I. Elastic Bodies: Free Oscillations

64. Find the natural frequencies for longitudinal oscillations of a cantilever beam of length l.

Ans.

$$\omega_n = \frac{2n+1}{2l}\pi v, \qquad n = 0, 1, 2, \ldots,$$

where $v = \sqrt{E/\rho}$, E is Young's modulus, and ρ is the density.

65. Solve the preceding problem, assuming that the free end of the beam is loaded by a mass M_0.

Ans.

$$\omega_n = \frac{v}{l}\gamma_n, \qquad n = 1, 2, \ldots,$$

where the γ_n are consecutive positive roots of the equation

$$\gamma \tan \gamma = \frac{M}{M_0}$$

(i.e., $0 < \gamma_1 < \ldots < \gamma_n < \ldots$), and M is the mass of the beam.

66. Determine the natural frequencies for torsional oscillations of a rod of length l, one end of which is clamped, while the other end is attached to a disk whose moment of inertia with respect to the axis of rotation is J_0.

Ans.

$$\omega_n = \frac{v}{l}\gamma_n, \qquad n = 1, 2, \ldots,$$

where $v = \sqrt{G/\rho}$, G is the shear modulus, ρ is the density, the γ_n are consecutive positive roots of the equation

$$\gamma \tan \gamma = \frac{J}{J_0},$$

and J is the moment of inertia of the rod.

---

[2] The forced oscillations studied in this chapter will always have the same frequency as the perturbing force itself.

67. Find the natural frequencies for transverse oscillations of a beam of length l with simply supported ends.

Ans.

$$\omega_n = \frac{a^2}{l^2} n^2 \pi^2, \qquad n = 1, 2, \ldots,$$

where $a^2 = \sqrt{EJ/\rho S}$, E is Young's modulus, J is the moment of inertia and S the area of a cross section, and ρ is the density.

68. Find the natural frequencies for transverse oscillations of a beam of length l with clamped ends.

Ans.

$$\omega_n = \frac{a^2}{l^2} \gamma_n^2, \qquad n = 1, 2, \ldots,$$

where the constant a is the same as in the preceding problem and the γ_n are consecutive positive roots of the equation $\cosh \gamma \cos \gamma = 1$.

*69. Solve the preceding problem, assuming that one end of the beam (of mass M) is clamped, while the other is loaded by a mass M_0. Using the method of successive approximations, calculate the values of the first three frequencies, given that

$$\frac{M}{M_0} = 5.$$

Ans.

$$\omega_n = \frac{a^2}{l^2} \gamma_n^2, \qquad n = 1, 2, \ldots,$$

where the γ_n are consecutive positive roots of the equation

$$1 + \cosh \gamma \cos \gamma = \frac{M_0}{M} \gamma(\sin \gamma \cosh \gamma - \cos \gamma \sinh \gamma).$$

70. Find the natural frequencies for radial oscillations of a circular membrane of radius a.

Ans.

$$\omega_n = \frac{v}{a} \gamma_n, \qquad n = 1, 2, \ldots,$$

where $v = \sqrt{T/\rho}$, T is the tension per unit length of the boundary, ρ is the surface density, and the γ_n are consecutive positive roots of the equation $J_0(\gamma) = 0$ involving the Bessel function of order zero.

71. Find the natural frequencies for oscillations of a rectangular membrane with sides a and b.

Ans.

$$\omega_{mn} = \pi v \sqrt{\frac{m^2}{a^2} + \frac{n^2}{b^2}}, \qquad m, n = 1, 2, \ldots,$$

where v is the same as in the preceding problem.

72. Find the natural frequencies for transverse radial oscillations of a circular plate of radius a whose edge is clamped. Calculate the first three roots of the transcendental equation determining the frequencies.

Ans.

$$\omega_n = \frac{b^2}{a^2} \gamma_n^2, \qquad n = 1, 2, \ldots,$$

where $b^2 = \sqrt{D/\rho h}$, D is the flexural rigidity, ρ the density and h the thickness of the plate, and the γ_n are consecutive positive roots of the equation

$$J_0(\gamma) I_1(\gamma) + J_1(\gamma) I_0(\gamma) = 0$$

(the notation is the same as in the theory of cylinder functions). Numerical calculations show that the first three roots are $\gamma_1 = 3.20$, $\gamma_2 = 6.30$, $\gamma_3 = 9.44$.

73. Find the maximum wavelength $\lambda_{\max}$ of a nonplanar sound wave[3] which can propagate inside a hollow cylinder tube with perfectly reflecting walls, whose cross section is a) a rectangle with sides a and b; b) a circle of radius a.

Ans.

a) $\qquad \lambda_{\max} = 2a, \qquad a > b;$

b) $\qquad \lambda_{\max} = \dfrac{2\pi a}{\gamma_1},$

where $\gamma_1 = 3.832$ is the smallest positive root of the equation $J_1(\gamma) = 0$ (for waves which are symmetric with respect to the diametral plane).

74. Find the natural frequencies for acoustic oscillations in an enclosure shaped like a rectangular parallelepiped with sides a, b and c.

Ans.

$$\omega_{mnp} = \pi v \sqrt{\left(\frac{m}{a}\right)^2 + \left(\frac{n}{b}\right)^2 + \left(\frac{p}{c}\right)^2}, \qquad m, n, p = 0, 1, 2, \ldots,$$

where v is the velocity of wave propagation (m, n, p cannot all vanish simultaneously).

---

[3] A *plane* sound wave $f\left(t - \dfrac{z}{v}\right)$ can propagate unimpeded in a tube of arbitrary cross section.

75. Find the natural frequencies of an acoustic resonator,[4] where the oscillations have axial symmetry and the resonator is a) a sphere of radius a; b) a circular cylinder of radius a and height l.

Ans.

a) $$\omega_{mn} = \frac{v}{a}\gamma_{mn}, \qquad m = 0, 1, 2, \ldots, \qquad n = 1, 2, \ldots,$$

where the γ_{mn} are consecutive positive roots of the equation

$$2\gamma_m J'_{m+\frac{1}{2}}(\gamma_m) = J_{m+\frac{1}{2}}(\gamma_m),$$

$J_{m+\frac{1}{2}}(x)$ is the Bessel function of order $m + \frac{1}{2}$, and v is the velocity of wave propagation;

b) $$\omega_{mn} = v\sqrt{\left(\frac{m\pi}{l}\right)^2 + \left(\frac{\gamma n}{a}\right)^2}, \qquad m = 0, 1, 2, \ldots,$$

where the γ_n are consecutive positive roots of the equation $J_1(\gamma) = 0$.

2. Elastic Bodies: Forced Oscillations

76. A string of length l with ends fastened at the points $x = 0$ and $x = l$ undergoes oscillations under the action of a concentrated force $A \sin(\omega t + \varphi)$ applied at some point $x = c$ of the string. Find the form of the forced oscillations.

Ans.

$$u(x, t) = \frac{Av}{\omega T}\frac{\sin(\omega t + \varphi)}{\sin(\omega l/v)} \times \begin{cases} \sin\dfrac{\omega x}{v}\sin\dfrac{\omega(l - c)}{v}, & 0 \leqslant x \leqslant c, \\[2ex] \sin\dfrac{\omega c}{v}\sin\dfrac{\omega(l - x)}{v}, & c \leqslant x \leqslant l, \end{cases}$$

where $v = \sqrt{T/\rho}$, T is the tension and ρ the linear density of the string.

77. Solve the preceding problem for the case where the external force is uniformly distributed over the whole length of the string.

Ans.

$$u(x, t) = \frac{2qv^2}{\omega^2 T}\frac{\sin\dfrac{\omega x}{2v}\sin\dfrac{\omega(l - x)}{2v}}{\cos\dfrac{\omega l}{2v}}\sin(\omega l + \varphi),$$

where q is the amplitude of the load per unit length of the string.

---

[4] An *acoustic resonator* is a device used to amplify acoustic oscillations, and consists of an enclosure whose walls reflect sound.

78. Find the forced longitudinal oscillations of a rod of length l, if the end $x = 0$ is clamped while the end $x = l$ is acted upon by a force $A \sin(\omega t + \varphi)$.

Ans.

$$u(x, t) = \frac{Av}{ES\omega} \frac{\sin \dfrac{\omega x}{v}}{\cos \dfrac{\omega l}{v}} \sin(\omega t + \varphi),$$

where $v = \sqrt{E/\rho}$, E is Young's modulus, ρ is the density and S the cross-sectional area of the rod.

79. Find the forced oscillations of a beam simply supported at the ends $x = 0$ and $x = l$, under the action of a uniformly distributed pulsating load $q \sin \omega t$.

Ans.

$$u(x, t) = \frac{qa^4}{EJ\omega^2} \left[\frac{\cosh \dfrac{\sqrt{\omega}}{a}\left(x - \dfrac{l}{2}\right)}{2 \cosh \dfrac{\sqrt{\omega}}{a}\dfrac{l}{2}} + \frac{\cos \dfrac{\sqrt{\omega}}{a}\left(x - \dfrac{l}{2}\right)}{2 \cos \dfrac{\sqrt{\omega}}{a}\dfrac{l}{2}} - 1 \right] \sin \omega t,$$

where $a^2 = \sqrt{EJ/\rho S}$, E is Young's modulus, J is the moment of inertia, ρ the density and S the cross-sectional area of the beam.

80. Solve the preceding problem under the assumption that the oscillations are due to a concentrated force $A \sin \omega t$ applied to the point $x = c$.

Ans.

$$u(x, t) = \frac{Aa^3 \sin \omega t}{2EJ\omega\sqrt{\omega} \sin \dfrac{\sqrt{\omega}l}{a} \sinh \dfrac{\sqrt{\omega}l}{a}}$$

$$\times \begin{cases} \sinh \dfrac{\sqrt{\omega}l}{a} \sin \dfrac{\sqrt{\omega}(l-c)}{a} \sin \dfrac{\sqrt{\omega}x}{a} - \sin \dfrac{\sqrt{\omega}l}{a} \sinh \dfrac{\sqrt{\omega}(l-c)}{a} \sinh \dfrac{\sqrt{\omega}x}{a}, \\ \qquad\qquad\qquad\qquad\qquad\qquad\qquad 0 \leqslant x \leqslant c, \\ \sinh \dfrac{\sqrt{\omega}l}{a} \sin \dfrac{\sqrt{\omega}(l-x)}{a} \sin \dfrac{\sqrt{\omega}c}{2} - \sin \dfrac{\sqrt{\omega}l}{a} \sinh \dfrac{\sqrt{\omega}(l-x)}{a} \sinh \dfrac{\sqrt{\omega}c}{a}, \\ \qquad\qquad\qquad\qquad\qquad\qquad\qquad c \leqslant x \leqslant l \end{cases}$$

81. A cantilever is clamped at one end $x = 0$ and loaded at the other end $x = l$ by a force $A \sin \omega t$. Find the resulting forced oscillations.

Ans. With the notation of the preceding problem,

$$u(x, t) = \frac{Aa^3 \sin \omega t}{2EJ\omega\sqrt{\omega}}\left[\frac{\left(\cosh\dfrac{\sqrt{\omega}x}{a} - \cos\dfrac{\sqrt{\omega}x}{a}\right)\left(\sinh\dfrac{\sqrt{\omega}l}{a} + \sin\dfrac{\sqrt{\omega}l}{a}\right)}{1 + \cos\dfrac{\sqrt{\omega}l}{a}\cosh\dfrac{\sqrt{\omega}l}{a}}\right.$$

$$\left. - \frac{\left(\sinh\dfrac{\sqrt{\omega}x}{a} - \sin\dfrac{\sqrt{\omega}x}{a}\right)\left(\cosh\dfrac{\sqrt{\omega}l}{a} + \cos\dfrac{\sqrt{\omega}l}{a}\right)}{1 + \cos\dfrac{\sqrt{\omega}l}{a}\cosh\dfrac{\sqrt{\omega}l}{a}}\right].$$

82. Find the forced oscillations of a circular membrane of radius a due to a pulsating load $q \sin (\omega t + \varphi)$ uniformly distributed over the membrane.

Ans.

$$u(r, t) = -\frac{q}{\rho\omega^2}\left[1 - \frac{J_0(\omega r/v)}{J_0(\omega a/v)}\right] \sin (\omega t + \varphi),$$

where $v = \sqrt{T/\rho}$, T is the tension per unit length of the contour, ρ is the surface density of the membrane, and $J_0(x)$ is the Bessel function of order zero.

***83.** Solve the preceding problem, assuming that the load is uniformly distributed over a disk of radius $b < a$.

Ans.

$$u(r, t) = -\frac{\pi bqv}{2\omega T} \sin (\omega t + \varphi)$$

$$\times \begin{cases} \dfrac{2v}{\pi\omega b} - \dfrac{J_0(\omega r)/v}{J_0(\omega a)/v}\left[Y_0\left(\dfrac{\omega a}{v}\right)J_1\left(\dfrac{\omega b}{v}\right) - J_0\left(\dfrac{\omega a}{v}\right)Y_1\left(\dfrac{\omega b}{v}\right)\right], & 0 \leqslant r \leqslant b \\[4mm] \dfrac{J_1(\omega b/v)}{J_0(\omega a/v)}\left[Y_0\left(\dfrac{\omega r}{v}\right)J_0\left(\dfrac{\omega a}{v}\right) - J_0\left(\dfrac{\omega r}{v}\right)Y_0\left(\dfrac{\omega a}{v}\right)\right], & b \leqslant r \leqslant a, \end{cases}$$

where $J_n(x)$ and $Y_n(x)$ are cylinder functions.

84. Study the forced oscillations of a circular plate of radius a with a clamped edge under the action of a uniformly distributed pulsating load $q \sin (\omega t + \varphi)$.

Ans. With the usual notation from the theory of cylinder functions,

$$u(r, t) = \frac{q}{\omega^2 \rho h} \left[\frac{I_1\left(\frac{\sqrt{\omega}a}{b}\right) J_0\left(\frac{\sqrt{\omega}r}{b}\right) + J_1\left(\frac{\sqrt{\omega}a}{b}\right) I_0\left(\frac{\sqrt{\omega}r}{b}\right)}{I_1\left(\frac{\sqrt{\omega}a}{b}\right) J_0\left(\frac{\sqrt{\omega}a}{b}\right) + J_1\left(\frac{\sqrt{\omega}a}{b}\right) I_0\left(\frac{\sqrt{\omega}a}{b}\right)} - 1 \right] \sin(\omega t + \varphi),$$

where $b^2 = \sqrt{D/\rho h}$, D is the flexural rigidity, h the thickness and ρ the density of the plate.

85. Find the steady-state harmonic oscillations of frequency ω inside a spherical resonator due to a point source of sound located at the center of the sphere, bearing in mind that the potential of a point source of frequency ω in free space is given by

$$u_0 = A \frac{\sin(\omega t - kR)}{R},$$

where $k = \omega/v$ is the wave number and R the distance from the source.

Ans. The velocity potential is

$$u(r, t) = A \frac{\sin(\omega t - kr)}{r} + A \frac{ka \cos(\omega t - ka) + \sin(\omega t - ka)}{ka \cos ka - \sin ka} \frac{\sin kr}{r},$$

where a is the radius of the sphere.

3. Electromagnetic Oscillations

86. Find the steady-state harmonic oscillations of voltage in a long transmission line with parameters L, C and R, if the end $x = 0$ is attached to a source of variable voltage $E \sin(\omega t + \varphi)$, while the end $x = l$ is terminated by a resistance R_0.

Ans.

$$u(x, t) = \text{Im} \left\{ E e^{i(\omega t + \varphi)} \frac{\sin \frac{\omega(l - x)}{v^*} + \frac{R_0}{iZ^*} \cos \frac{\omega(l - x)}{v^*}}{\sin \frac{\omega l}{v^*} + \frac{R_0}{iZ^*} \cos \frac{\omega l}{v^*}} \right\},$$

where

$$v^* = \frac{1}{\sqrt{LC}} \frac{1}{\sqrt{1 - \frac{iR}{\omega L}}}, \qquad Z^* = \sqrt{\frac{L}{C}} \sqrt{1 - \frac{iR}{\omega L}}$$

are the complex propagation velocity and wave resistance of the line.

87. Solve the preceding problem, assuming that the load terminating the line is a concentrated inductance L_0, instead of a resistance R_0.

Ans.

$$u(x, t) = \text{Im} \left\{ E e^{i(\omega t + \varphi)} \frac{\sin \dfrac{\omega(l - x)}{v^*} + \dfrac{\omega L_0}{Z^*} \cos \dfrac{\omega(l - x)}{v^*}}{\sin \dfrac{\omega l}{v^*} + \dfrac{\omega L_0}{Z^*} \cos \dfrac{\omega l}{v^*}} \right\}.$$

88. Find the components of the electromagnetic field in a transverse magnetic wave propagating in a waveguide whose cross section is a rectangle with sides a and b.[5] Calculate the corresponding cutoff wavelength $\lambda_{\max}$ (i.e., the maximum wavelength passed by the waveguide).

Ans.

$$E_x = -i\nu \frac{m\pi}{a} \cos \frac{m\pi x}{a} \sin \frac{n\pi y}{b} e^{-i(\nu z - \omega t)}.$$

$$E_y = -i\nu \frac{n\pi}{b} \sin \frac{m\pi x}{a} \cos \frac{n\pi y}{b} e^{-i(\nu z - \omega t)},$$

$$E_z = \pi^2 \left(\frac{m^2}{a^2} + \frac{n^2}{b^2} \right) \sin \frac{m\pi x}{a} \sin \frac{n\pi y}{b} e^{-i(\nu z - \omega t)},$$

$$H_x = ik \frac{n\pi}{b} \sin \frac{m\pi x}{a} \cos \frac{n\pi y}{b} e^{-i(\nu z - \omega t)},$$

$$H_y = -ik \frac{m\pi}{a} \cos \frac{m\pi x}{a} \sin \frac{n\pi y}{b} e^{-i(\nu z - \omega t)},$$

$$H_z = 0,$$

$$\lambda_{\max} = \frac{2ab}{\sqrt{a^2 + b^2}}, \qquad \nu = \sqrt{k^2 - \frac{m^2 \pi^2}{a^2} - \frac{n^2 \pi^2}{b^2}}, \qquad m, n = 1, 2, \ldots,$$

where $k = \omega/c$ is the wave number. An arbitrary constant factor has been omitted in all the expressions for the components of the electromagnetic field.

89. Solve the preceding problem for a transverse electric wave.

---

[5] By a *transverse magnetic wave* (TM-wave) is meant a wave in which the magnetic field vector **H** is perpendicular to the direction of wave propagation. Similarly, a *transverse electric wave* (TE-wave) is a wave in which the electric field vector **E** is perpendicular to the direction of propagation, and a *transverse electromagnetic wave* (TEM-wave) is a wave in which both vectors **E** and **H** are perpendicular to the direction of propagation (see S3, p. 154).

Ans.

$$E_x = ik\frac{n\pi}{b}\cos\frac{m\pi x}{a}\sin\frac{n\pi y}{b}\,e^{-i(vz-\omega t)},$$

$$E_y = -ik\frac{m\pi}{a}\sin\frac{m\pi x}{a}\cos\frac{n\pi y}{b}\,e^{-i(vz-\omega t)},$$

$$E_z = 0,$$

$$H_x = iv\frac{m\pi}{a}\sin\frac{m\pi x}{a}\cos\frac{n\pi y}{b}\,e^{-i(vz-\omega t)},$$

$$H_y = iv\frac{n\pi}{b}\cos\frac{m\pi x}{a}\sin\frac{n\pi y}{b}\,e^{-i(vz-\omega t)},$$

$$H_z = \pi^2\left(\frac{m^2}{a^2}+\frac{n^2}{b^2}\right)\cos\frac{m\pi x}{a}\cos\frac{n\pi y}{b}\,e^{-i(vz-\omega t)},$$

$$v = \sqrt{k^2 - \frac{m^2\pi^2}{a^2} - \frac{n^2\pi^2}{b^2}}\,, \qquad m, n = 0, 1, 2, \dots,$$

$$\lambda_{\max} = 2a \quad \text{if} \quad a > b$$

(m and n cannot vanish simultaneously).

90. Find the components of the electromagnetic field in a transverse magnetic wave propagating in a waveguide whose cross section is a circle of radius a, and determine the corresponding cut-off frequency $\lambda_{\max}$.

Ans.

$$H_r = -ik\frac{m}{r}\sin m\varphi\, J_m\!\left(\gamma_{mn}\frac{r}{a}\right)e^{-i(vz-\omega t)},$$

$$H_\varphi = -ik\frac{\gamma_{mn}}{a}\cos m\varphi\, J'_m\!\left(\gamma_{mn}\frac{r}{a}\right)e^{-i(vz-\omega t)},$$

$$H_z = 0,$$

$$E_r = -iv\frac{\gamma_{mn}}{a}\cos m\varphi\, J'_m\!\left(\gamma_{mn}\frac{r}{a}\right)e^{-i(vz-\omega t)},$$

$$E_\varphi = iv\frac{m}{r}\sin m\varphi\, J_m\!\left(\gamma_{mn}\frac{r}{a}\right)e^{-i(vz-\omega t)},$$

$$E_z = \left(\frac{\gamma_{mn}}{a}\right)^2\cos m\varphi\, J_m\!\left(\gamma_{mn}\frac{r}{a}\right)e^{-i(vz-\omega t)},$$

$$v = \sqrt{k^2 - \frac{\gamma_{mn}^2}{a^2}}\,,$$

$$\lambda_{\max} = \frac{2\pi a}{\gamma_{01}} \qquad (\gamma_{01} = 2.405),$$

where $k = \omega/c$ is the wave number, and the γ_{mn} are consecutive positive roots of the equation $J_m(\gamma) = 0$ ($m = 0, 1, 2, \dots$) involving the Bessel function of order m.

*91. Calculate the cutoff wavelength $\lambda_{\max}$ for a *TM*-wave propagating in a waveguide whose cross section is a circular sector of radius a and central angle α.

Ans.

$$\lambda_{\max} = \frac{2\pi a}{\gamma_0},$$

where γ_0 is the smallest positive root of the equations

$$J_{m\pi/\alpha}(\gamma) = 0, \qquad m = 1, 2, \ldots$$

involving the Bessel function $J_\nu(x)$.

92. Describe the free harmonic oscillations in an electromagnetic resonator in the form of a rectangular parallelepiped with sides a, b, c and perfectly conducting walls.

Ans.

$$E_x = A \cos \frac{m\pi x}{a} \sin \frac{n\pi y}{b} \sin \frac{p\pi z}{c}, \qquad H_x = M \sin \frac{m\pi x}{a} \cos \frac{n\pi y}{b} \cos \frac{p\pi z}{c},$$

$$E_y = B \sin \frac{m\pi x}{a} \cos \frac{n\pi y}{b} \sin \frac{p\pi z}{c}, \qquad H_y = N \cos \frac{m\pi x}{a} \sin \frac{n\pi y}{b} \cos \frac{p\pi z}{c},$$

$$E_z = C \sin \frac{m\pi x}{a} \sin \frac{n\pi y}{b} \cos \frac{p\pi z}{c}, \qquad H_z = P \cos \frac{m\pi x}{a} \cos \frac{n\pi y}{b} \sin \frac{p\pi z}{c},$$

where m, n and p are integers, and the constants A, B, C, M, N and P are connected by the relations

$$\frac{m\pi}{a} A + \frac{n\pi}{b} B + \frac{p\pi}{c} C = 0$$

$$M = -\frac{1}{ik}\left(\frac{n\pi}{b} C - \frac{p\pi}{c} B\right),$$

$$N = -\frac{1}{ik}\left(\frac{p\pi}{c} A - \frac{m\pi}{a} C\right),$$

$$P = -\frac{1}{ik}\left(\frac{m\pi}{a} B - \frac{n\pi}{b} A\right),$$

(k is the wave number).

93. Solve the preceding problem for a resonator in the form of a circular cylinder of radius a and length l.

Ans.

$$E_r = A J'_m\left(\gamma_{mn}\frac{r}{a}\right)\frac{\cos m\varphi}{\sin m\varphi}\sin\frac{n\pi z}{l},$$

$$E_\varphi = B\frac{1}{r}J_m\left(\gamma_{mn}\frac{r}{a}\right)\frac{\sin m\varphi}{\cos m\varphi}\sin\frac{n\pi z}{l},$$

$$E_z = C J_m\left(\gamma_{mn}\frac{r}{a}\right)\frac{\cos m\varphi}{\sin m\varphi}\cos\frac{n\pi z}{l},$$

$$H_r = M\frac{1}{r}J_m\left(\gamma_{mn}\frac{r}{a}\right)\frac{\sin m\varphi}{\cos m\varphi}\cos\frac{n\pi z}{l},$$

$$H_\varphi = N J'_m\left(\gamma_{mn}\frac{r}{a}\right)\frac{\cos m\varphi}{\sin m\varphi}\cos\frac{n\pi z}{l},$$

$$H_z = P\frac{1}{r}J'_m\left(\gamma_{mn}\frac{r}{a}\right)\frac{\sin m\varphi}{\cos m\varphi}\sin\frac{n\pi z}{l},$$

where m and n are integers, the constants A, B, C, M, N and P are connected by the relations

$$A\frac{\gamma_{mn}}{a} + C\frac{n\pi}{l} = 0,$$

$$A\frac{\gamma_{mn}}{a} \pm B\frac{1}{m} = 0,$$

$$M = -\frac{1}{ik}\left(\mp Cm - B\frac{n\pi}{l}\right),$$

$$N = -\frac{1}{ik}\left(A\frac{n\pi}{l} - C\frac{\gamma_{mn}}{a}\right),$$

$$P = -\frac{1}{ik}\left(B\frac{\gamma_{mn}}{a} \pm Am\right),$$

and the γ_{mn} are consecutive roots of the equation $J_m(\gamma) = 0$.

94. A high-frequency current $I\sin\omega t$ flows along a cylindrical conductor of radius a, made of material of conductivity σ and magnetic permeability μ. Find the distribution of current density along the cross section of the wire, and calculate the active resistance of the conductor at the frequency ω (the *skin effect* problem).

Ans. The complex amplitude of the current density is given by the formula

$$\frac{j(r)}{j(a)} = \frac{J_0(kr)}{J_0(ka)},$$

where

$$j(a) = \frac{Ik}{2\pi a}\frac{J_0(ka)}{J_1(ka)}.$$

The resistance per unit length of the conductor is

$$R_\omega = \frac{|k|^2}{2\pi a\sigma(k^2 - \bar{k}^2)}\left[k\frac{J_0(\bar{k}a)}{J_1(\bar{k}a)} - \bar{k}\frac{J_0(ka)}{J_1(ka)}\right],$$

(the overbar denotes the complex conjugate), where

$$k = \sqrt{-\frac{4\pi\sigma\omega\mu i}{c^2}} = \frac{1}{c}\sqrt{2\pi\sigma\omega\mu}\,(1 - i).$$

Taking account of the asymptotic behavior of the Bessel functions for large values of the argument, we find that

$$\left|\frac{j(r)}{j(a)}\right| \approx \sqrt{\frac{a}{r}}\,e^{-(a-r)/\delta}, \qquad \frac{R_\omega}{R_0} \approx \frac{a}{2\delta},$$

where

$$\delta = \frac{c}{\sqrt{2\pi\sigma\omega\mu}}$$

and

$$R_0 = \frac{1}{\pi a^2\sigma}$$

is the d-c resistance.

95. Solve the skin effect problem for a conductor whose cross section is a strip of width $2a$. Find the corresponding current distribution and resistance.

Ans.

$$\frac{j(x)}{j(a)} = \frac{\cos kx}{\cos ka}, \qquad j(a) = \frac{Ik}{2}\cot ka,$$

$$R_\omega = \frac{|k|^2}{2\sigma(k^2 - \bar{k}^2)}\,(k\cot \bar{k}a - \bar{k}\cot ka),$$

where I is the amplitude of the total current. For high frequencies,

$$\left|\frac{j(x)}{j(a)}\right| \approx e^{-(a-x)/\delta}, \qquad \frac{R_\omega}{R_0} \approx \frac{a}{\delta}, \qquad R_0 = \frac{1}{2a\sigma},$$

where δ is the same as in the preceding problem.

References

Marcuvitz (M4), Morse (M8), Rayleigh (R1), Schelkunoff (S3), Stratton (S17), Timoshenko (T2).

4

THE FOURIER METHOD

The Fourier method is one of the most general techniques of mathematical physics, and is effective in solving a very wide class of problems. Its use, which is not restricted to equations of any particular type (e.g., hyperbolic or elliptic), relies on the fact that linear problems obey the *superposition principle*, i.e., any linear combination of solutions of a homogeneous linear partial differential equation is itself a solution of the equation. Thus, if a linear equation $Lu = 0$ has a certain set of particular solutions

$$u = u_n, \qquad n = 1, 2, \ldots,$$

the sum of the series

$$u = \sum_{n=1}^{\infty} u_n$$

is also a solution, provided the convergence of the series permits interchanging the operations L and Σ. Similarly, if $Lu = 0$ has a set of particular solutions

$$u = u_\lambda, \qquad \mu \leqslant \lambda \leqslant \nu,$$

which depend continuously on the parameter λ in the interval (μ, ν), then the integral

$$u = \int_\mu^\nu u_\lambda \, d\lambda$$

is also a solution, provided the operations L and $\int$ can be interchanged. Given a problem of mathematical physics involving the integration of a differential equation $Lu = 0$ subject to certain initial and boundary conditions, the basic idea of the Fourier method is to construct a solution by superposition of particular solutions. If the operator $Lu = 0$ has an appropriate structure, we can "separate variables," i.e., the particular solutions can be written as products of factors, each involving only one independent variable and satisfying an ordinary differential equation. By suitably choosing some of the parameters figuring in this relatively simple problem, it is usually possible to satisfy all the homogeneous boundary conditions, thereby singling out a countable or uncountable set of particular solutions of the required type. Then, after making a superposition of these solutions, we choose the remaining parameters in such a way as to satisfy the inhomogeneous boundary conditions.

Having made these general remarks, we now confine ourselves in this chapter to problems of mathematical physics which lead to integration of the differential equation

$$\frac{1}{r(x)}\left\{\frac{\partial}{\partial x}\left[p(x)\frac{\partial u}{\partial x}\right] - q(x)u\right\} + M_y u = 0 \qquad (a < x < b, c < y < d), \quad (1)$$

where M_y is a differential operator of the form

$$M_y = A\frac{\partial^2}{\partial y^2} + B\frac{\partial}{\partial y} + C,$$

A, B and C are given constants, and $p(x)$, $q(x)$ and $r(x)$ are given continuous functions such that $p(x)$ and $q(x)$ are positive, and $p(x)$ is continuously differentiable.[1] [In the next chapter, we shall consider the inhomogeneous case, where the right-hand side of (1) is a given function $F(x, y)$.] For the time being, we assume that the interval (a, b) is finite and that the behavior of the functions p, q and r at the end points a and b is such that the ratios all

$$\frac{p'(x)}{p(x)}, \quad \frac{q'(x)}{p(x)}, \quad \frac{r'(x)}{p(x)} \tag{2}$$

approach finite limits as $x \to a$ and $x \to b$. Moreover, we require the solution of (1) to satisfy homogeneous boundary conditions at the end points of (a, b), of the form

$$\alpha_a\frac{\partial u}{\partial x}\Big|_{x=a} + \beta_a u\big|_{x=a} = 0, \qquad \alpha_b\frac{\partial u}{\partial x}\Big|_{x=b} + \beta_b u\big|_{x=b} = 0, \tag{3}$$

---

[1] Equation (1) is not the most general second-order equation with two independent variables which permits separation of variables, but it includes as special cases most of the commonly encountered equations of mathematical physics.

where α_a, α_b, β_a and β_b are given constants, some of which may equal zero,[2] and inhomogeneous boundary conditions at the end points of (c, d), whose form depends on whether the differential equation is of hyperbolic, parabolic or elliptic type (cf. footnote 1, p. 20). If the equation is of elliptic type, it will be sufficiently general for our purposes to assume that these conditions are of the form

$$\gamma_c \frac{\partial u}{\partial y}\bigg|_{y=c} + \delta_c u\big|_{y=c} = g_c(x), \qquad \gamma_d \frac{\partial u}{\partial y}\bigg|_{y=d} + \delta_d u\big|_{y=d} = g_d(x), \qquad (4)$$

where γ_c, γ_d, δ_c and δ_d are given constants, while $g_c(x)$ and $g_d(x)$ are given functions defined in the interval (c, d). On the other hand, if the equation is of the hyperbolic or parabolic type, which corresponds to problems of mathematical physics where the variable y plays the role of a time varying over an *infinite* interval (c, ∞), then the inhomogeneous boundary conditions take the form of initial conditions, i.e.,

$$u\big|_{y=c} = f(x), \qquad \frac{\partial u}{\partial y}\bigg|_{y=c} = g(x) \qquad (4')$$

in the hyperbolic case, and

$$u\big|_{y=c} = f(x) \qquad (4'')$$

in the parabolic case.

We now look for a function $u = u(x, y)$ satisfying the differential equation (1) and the boundary conditions (3) and (4) [or (4'), (4'')]. Following the basic procedure already mentioned, we consider particular solutions of equation (1) of the form

$$u = X(x)Y(y). \qquad (5)$$

After substituting (5) into (1), the variables separate, and the result is a pair of ordinary differential equations

$$(pX')' + (\lambda r - q)X = 0, \qquad (6)$$

$$M_y Y - \lambda Y = 0 \qquad (7)$$

---

[2] In particular, we obtain boundary conditions of the first kind

$$u\big|_{x=a} = u\big|_{x=b} = 0$$

if $\alpha_a = \alpha_b = 0$, $\beta_a = \beta_b = 1$, boundary conditions of the second kind

$$\frac{\partial u}{\partial x}\bigg|_{x=a} = \frac{\partial u}{\partial x}\bigg|_{x=b} = 0$$

if $\alpha_a = \alpha_b = 1$, $\beta_a = \beta_b = 0$, and so on. In the applications, one also encounters boundary conditions of the form

$$u\big|_{x=a} = u\big|_{x=b}, \qquad \frac{\partial u}{\partial x}\bigg|_{x=c} = \frac{\partial u}{\partial x}\bigg|_{x=b}, \qquad (3')$$

which are not comprised in the formulas (3).

determining the factors X and Y, where λ is an arbitrary parameter. The requirement that the particular solutions (5) satisfy the homogeneous boundary conditions (3) leads to corresponding homogeneous boundary conditions for the function X:

$$\alpha_a X'(a) + \beta_a X(a) = 0, \qquad \alpha_b X'(b) + \beta_b X(b) = 0. \tag{8}$$

The problem of solving equation (6) subject to the boundary conditions (8) is called the *Sturm-Liouville problem.* For arbitrary λ, this problem will in general have no solution other than the trivial solution $X \equiv 0$. However, for certain values of λ, called *eigenvalues,* there are nontrivial solutions, called *eigenfunctions.* In the theory of the Sturm-Liouville problem, it is shown that with our assumptions concerning the interval (a, b) and the functions p, q and r, the *spectrum* (i.e., the set of all eigenvalues) is discrete, consisting of countably many real eigenvalues $\lambda = \lambda_n$ $(n = 1, 2, \ldots)$, each associated with a single eigenfunction $X = X_n(x)$ which is uniquely defined (except for a constant factor). The eigenfunctions $X_n(x)$ are found to be *orthogonal* on the interval (a, b) with *weight* $r(x)$, i.e.,

$$\int_a^b r X_m(x) X_n(x)\, dx = 0 \qquad \text{if} \quad m \neq n.$$

Moreover, under certain conditions,[3] a function $f(x)$ defined in (a, b) can be expanded as a series of the form

$$f(x) = \sum_{n=1}^{\infty} f_n X_n(x), \qquad a < x < b,$$

with coefficients

$$f_n = \frac{\int_a^b r f(x) X_n(x)\, dx}{\int_a^b r X_n^2(x)\, dx}.$$

The calculation of the eigenvalues and the corresponding eigenfunctions is easily carried out in the case where the linearly independent solutions and hence the general solution of (6) are known for arbitary λ. In fact, substitution of the general solution into the boundary conditions (8) then gives a homogeneous linear system for the arbitrary constants, and the condition that the determinant of this system vanish leads at once to a transcendental equation for the permissible values of λ. After the eigenvalues and eigenfunctions have been determined, we find the second factor $Y(y)$ in (5) by solving (7), with $\lambda = \lambda_n$. If the original equation is of hyperbolic or elliptic type, the general solution of (7) can be written in the form

$$Y(y) = c_n^{(1)} Y_n^{(1)}(y) + c_n^{(2)} Y_n^{(2)}(y),$$

---

[3] For example, if $f(x)$ is piecewise smooth in (a, b).

where $Y_n^{(1)}$ and $Y_n^{(2)}$ are linearly independent solutions of (7) and $c_n^{(1)}$, $c_n^{(2)}$ are arbitrary constants.[4] In this way, we arrive at a set of particular solutions

$$u_n = [c_n^{(1)}Y_n^{(1)}(y) + c_n^{(2)}Y_n^{(2)}(y)]X_n(x), \qquad n = 1, 2, \ldots,$$

and the solution of the problem is then constructed in the form of a series

$$u = \sum_{n=1}^{\infty} u_n,$$

where the coefficients $c_n^{(i)}$ are found by substituting this series into the boundary conditions (4).[5]

If the function u satisfies boundary conditions of the type (3′) instead of (3), the above method carries over virtually without change, except that now two linearly independent eigenfunctions may correspond to the same eigenvalue. Things become more complicated if the interval (a, b) is infinite, or if one (or both) of the end points of (a, b) is *singular*, i.e., if one of the ratios (2) becomes infinite as we approach the given end point. In such cases, which are among the most interesting encountered in practice, the boundary condition involving the singular end point or the point $x = b = \infty$ cannot be prescribed arbitrarily, but rather is replaced by a condition whose formulation in concrete situations usually presents no special difficulties (most often, the condition consists in the requirement that the solution remain bounded as the singular point is approached). In the case where the interval (a, b) is finite and only one end point is singular, the eigenfunctions are found as the nontrivial solutions of equation (6) satisfying some condition of the type just mentioned at the singular point and a condition like (3) at the other end point. The same approach can be used to find the eigenfunctions for a finite interval with two singular end points, for an infinite interval, and so on. The essential difference between these cases and the case analyzed above is that the spectrum may now be either discrete or not, depending on the structure of the differential equation and the nature of the boundary conditions. If the spectrum is still discrete, despite the presence of an infinite interval or of a singular end point, the Fourier method can be applied with no essential changes. On the other hand, if the spectrum is no longer discrete, the character of the solution changes. In the case of a continuous spectrum,[6]

---

[4] If the equation is of parabolic type, the general solution is of the form

$$Y(y) = c_n Y_n(y).$$

[5] For rigorous justification of the application of the Fourier method to problems of mathematical physics of this or more complicated types, see L1, T1, T7, etc. In many cases, however, it is an easy matter to verify directly the validity of the solution found formally by the procedure just described.

[6] Chapters 4–5 are devoted exclusively to problems with discrete spectra. Problems with continuous spectra will be considered in Chapter 6.

the solution is constructed from particular solutions by integrating instead of summing with respect to the parameter λ, and the unknown functions appearing in the integrand are determined by using the theory of integral transforms instead of the theory of expansions in series of eigenfunctions.[7]

The problems in the present chapter can all be solved by the Fourier method (in most cases by the method just described), and are grouped into five sections, two devoted to mechanics, two to heat conduction (including a few problems on diffusion), and one to electricity and magnetism. We also include a few problems involving inhomogeneous equations and inhomogeneous boundary conditions, which can be solved by the Fourier method after being reduced to homogeneous problems by the use of appropriate tricks. However, inhomogeneous problems will for the most part be considered in Chap. 5, where they are studied systematically. Since an entire chapter (Chap. 7) will be devoted to the less familiar special coordinate systems, we confine ourselves here to rectangular and polar coordinates (both cylindrical and spherical). To illustrate further extensions of the Fourier method, we include a few problems of a more complicated type, e.g., problems involving three variables, elasticity theory, fourth-order differential equations, etc.

I. Mechanics: Vibrating Systems, Acoustics

*96. At the time $t = 0$, a string with ends fastened at the points $x = 0$ and $x = l$ is plucked at the point $x = c$, and then released without initial velocity. Find the displacement $u(x, t)$ of an arbitrary point of the string if $u(c, 0) = h$.

Ans.

$$u(x, t) = \frac{2h}{\pi^2} \frac{l^2}{c(l - c)} \sum_{n=1}^{\infty} \frac{\sin (n\pi c/l)}{n^2} \sin \frac{n\pi x}{l} \cos \frac{n\pi vt}{l} ,$$

where $v = \sqrt{T/\rho}$, T is the tension and ρ is the linear density.

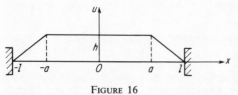

FIGURE 16

97. Find the vibrations of a string if the initial displacement has the form shown in Figure 16, while the initial velocity is zero at every point of the string.

---

[7] For information concerning such integral expansions and Sturm-Liouville theory in general (especially the singular case), see A1, L13, S6, Vol. V and T6.

Ans.

$$u(x, t) = \frac{8hl}{(l - a)\pi^2} \sum_{n=0}^{\infty} \frac{\cos\left[(2n + 1)\pi a/2l\right]}{(2n + 1)^2} \cos \frac{(2n + 1)\pi x}{2l} \cos \frac{(2n + 1)\pi vt}{2l}.$$

98. Solve the preceding problem, assuming that the initial form of the string is a parabola symmetric with respect to the center of the string and that the maximum initial displacement from equilibrium is h.

Ans.

$$u(x, t) = \frac{32h}{\pi^3} \sum_{n=0}^{\infty} \frac{(-1)^n}{(2n + 1)^3} \cos \frac{(2n + 1)\pi x}{2l} \cos \frac{(2n + 1)\pi vt}{2l}.$$

99. At the time $t = 0$, the center of a string of length $2l$ fastened at the points $x = -l$ and $x = l$ receives an impulse P. Find the subsequent vibrations of the string

$$u(x, t) = \frac{2P}{\pi v \rho} \sum_{n=0}^{\infty} \frac{\cos\left[(2n + 1)\pi x/2l\right]}{2n + 1} \sin \frac{(2n + 1)\pi vt}{2l}.$$

Hint. Consider the vibrations of the string subject to the initial conditions

$$u\big|_{t=0} = 0, \quad \frac{\partial u}{\partial t}\bigg|_{t=0} = \begin{cases} v_0 = \dfrac{P}{2\rho\varepsilon}, & |x| < \varepsilon, \\[2mm] 0, & \varepsilon < |x| \leqslant l, \end{cases}$$

and then take the limit as $\varepsilon \to 0$.

100. Study the vibrations of a string fastened at the points $x = 0$ and $x = l$ due to a suddenly applied load distributed along the string with constant density q which subsequently remains constant. The string is assumed to be at rest at the time $t = 0$.

Ans.

$$u(x, t) = \frac{ql^2}{T}\left\{\frac{x}{2l}\left(1 - \frac{x}{l}\right) - \frac{4}{\pi^3} \sum_{n=0}^{\infty} \frac{\sin[(2n + 1)\pi x/l]}{(2n + 1)^3} \cos \frac{(2n + 1)\pi vt}{l}\right\}.$$

Hint. Before applying the Fourier method, make the problem homogeneous by subtracting out the static deflection of the string under the uniform load.

101. Find the vibrations of a string $-l \leqslant x \leqslant l$ of mass m loaded at the point $x = 0$ by a concentrated mass m_0. In solving the problem, assume that the load is initially displaced by a small amount h, and that the initial velocity of the string is zero.

Ans.

$$u(x, t) = 2h\alpha \sum_{n=1}^{\infty} \frac{\cos \gamma_n}{\gamma_n^3} \frac{\sin \gamma_n \left(1 - \dfrac{|x|}{l}\right)}{1 + \dfrac{\sin 2\gamma_n}{2\gamma_n}} \cos \frac{\gamma_n vt}{l},$$

where $\alpha = m/m_0$ and the γ_n are consecutive positive roots of the equation $\tan \gamma = \alpha/\gamma$.

102. A rod of length l, density ρ and cross-sectional area S is clamped at the end $x = 0$ and stretched by a force F applied at the other end $x = l$. Study the longitudinal oscillations of the rod if the force is suddenly discontinued at the time $t = 0$.

Ans.

$$u(x, t) = \frac{8Fl}{\pi^2 ES} \sum_{n=0}^{\infty} \frac{(-1)^n}{(2n+1)^2} \sin \frac{(2n+1)\pi x}{2l} \cos \frac{(2n+1)\pi vt}{2l},$$

where $v = \sqrt{E/\rho}$ and E is Young's modulus.

103. Find the general solution of the problem of longitudinal oscillations of a rod of length l with arbitrary initial conditions

$$u\big|_{t=0} = f(x), \qquad \frac{\partial u}{\partial t}\bigg|_{t=0} = g(x)$$

if the end $x = 0$ is clamped and the end $x = l$ is free.

Ans.

$$u(x, t) = \frac{2}{l} \sum_{n=0}^{\infty} \left[\cos \frac{(2n+1)\pi vt}{2l} \int_0^l f(\xi) \sin \frac{(2n+1)\pi \xi}{2l} \, d\xi + \frac{2l}{(2n+1)\pi v} \right.$$

$$\left. \times \sin \frac{(2n+1)\pi vt}{2l} \int_0^l g(\xi) \sin \frac{(2n+1)\pi \xi}{2l} \, d\xi \right] \sin \frac{(2n+1)\pi x}{2l}.$$

104. Investigate the longitudinal oscillations of a cantilever of length l and mass M if the end $x = 0$ is clamped while the end $x = l$ is loaded by a concentrated mass M_0, which at the time $t = 0$ experiences a displacement δ without acquiring any initial velocity.

Ans.

$$u(x, t) = 2\delta \frac{M}{M_0} \sum_{n=1}^{\infty} \frac{\cos \gamma_n}{\gamma_n^3} \frac{\sin \dfrac{\gamma_n x}{l}}{1 + \dfrac{\sin 2\gamma_n}{2\gamma_n}} \cos \frac{\gamma_n vt}{l},$$

where the γ_n are consecutive positive roots of the equation

$$\gamma \tan \gamma = \frac{M}{M_0}.$$

105. Find the longitudinal oscillations of a rod of length l if the end $x = 0$ is clamped while the end $x = l$ receives an impulse P at the time $t = 0$. The rod is assumed to be at rest before the impulse acts.

Ans. In the notation of the preceding problem,

$$u(x, t) = \frac{4Pl}{\pi v M} \sum_{n=0}^{\infty} \frac{(-1)^n}{2n + 1} \sin \frac{(2n + 1)\pi x}{2l} \sin \frac{(2n + 1)\pi v t}{2l}.$$

Hint. Solve the problem of oscillations with the initial conditions

$$u\big|_{t=0} = 0, \quad \frac{\partial u}{\partial t}\bigg|_{t=0} = \begin{cases} 0, & 0 < x < l - \varepsilon, \\ v_0 = \dfrac{P}{\rho \varepsilon S}, & l - \varepsilon < x < l, \end{cases}$$

where S is the cross-sectional area of the rod, and then take the limit as $\varepsilon \to 0$.

106. Find the displacement of the points of a rod of length l clamped at the end $x = 0$, which undergoes longitudinal oscillations under the action of a pulsating force $A \sin \omega t$ applied to the free end $x = l$. The rod is assumed to be at rest before the force begins to act.

Ans.

$$u(x, t) = \frac{Av}{ES\omega} \left\{ \frac{\sin \dfrac{\omega x}{v}}{\cos \dfrac{\omega l}{v}} \sin \omega t \right.$$
$$\left. - \frac{4}{\pi} \left(\frac{\omega l}{v}\right)^2 \sum_{n=0}^{\infty} \frac{(-1)^n}{2n + 1} \frac{\sin \dfrac{(2n + 1)\pi x}{2l} \sin \dfrac{(2n + 1)\pi v t}{2l}}{\left[\dfrac{(2n + 1)\pi}{2}\right]^2 - \left(\dfrac{\omega l}{v}\right)^2} \right\}.$$

Hint. To make the problem homogeneous, represent the displacement as a sum of free and forced oscillations (see Prob. 78). Another method of solution is given in Chap. 5 (see Prob. 211).

107. A conical cantilever with the dimensions shown in Figure 17 is stretched by a force F applied at the end $x = l$. Study the longitudinal oscillations which result when the force is suddenly discontinued.

Ans.

$$u(x, t) = \frac{2F \cot \alpha}{\pi E(a - x \tan \alpha)} \sum_{n=1}^{\infty} \frac{\cos \gamma_n}{\gamma_n [(\sin 2\gamma_n / 2\gamma_n) - 1]} \sin \frac{\gamma_n x}{l} \cos \frac{\gamma_n v t}{l},$$

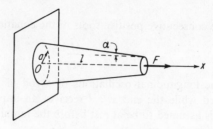

FIGURE 17

where the γ_n are consecutive positive roots of the equation

$$\tan \gamma = \left(1 - \frac{a}{l} \cot \alpha\right) \gamma.$$

*108. Solve the problem of the longitudinal oscillations of the pyramid-shaped cantilever of rectangular cross section and constant thickness shown in Figure 18, subject to a given initial deformation $u|_{t=0} = f(x)$.

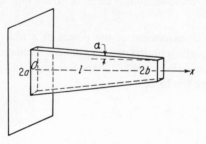

FIGURE 18

Ans.

$$u(x, t) = \frac{2 \tan \alpha}{b^2} \sum_{n=1}^{\infty} \frac{X_{\gamma_n}(a - x \tan \alpha)}{(4a^2/\pi^2 \gamma_n^2 b^2) - X_{\gamma_n}^2(b)} \cos \frac{vt\gamma_n \tan \alpha}{a}$$

$$\times \int_0^l f(\xi)(a - \xi \tan \alpha) X_{\gamma_n}(a - \xi \tan \alpha) \, d\xi,$$

where

$$X_\gamma(y) = Y_0(\gamma) J_0\left(\frac{\gamma y}{a}\right) - J_0(\gamma) Y_0\left(\frac{\gamma y}{a}\right),$$

$J_0(x)$ and $Y_0(x)$ are cylinder functions of order zero), and the γ_n are consecutive positive roots of the equation $X_\gamma'(b) = 0$.

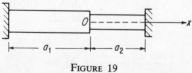

FIGURE 19

*109. Find the general solution of the problem of longitudinal oscillations of a rod consisting of two rigidly fastened sections with different dimensions and elastic properties (see Figure 19). It is assumed that the ends of the rods are clamped and that the initial state of the rod is characterized by the conditions

$$u|_{t=0} = f(x), \qquad \frac{\partial u}{\partial t}\bigg|_{t=0} = 0.$$

Ans.

$$u(x, t) = 2 \sum_{n=1}^{\infty} \frac{S_1 \rho_1 \int_{-a_1}^{0} f(\xi) u_n^{(1)}(\xi)\, d\xi + S_2 \rho_2 \int_{0}^{a_2} f(\xi) u_n^{(2)}(\xi)\, d\xi}{a_1 S_1 \rho_1 \sin^2 \frac{\gamma_n a_2 v_1}{a_1 v_2} + a_2 S_2 \rho_2 \sin^2 \gamma_n} u_n(x) \cos \frac{\gamma_n v_1 t}{a_1},$$

where

$$u_n(x) = \begin{cases} u_n^{(1)}(x) = \sin \dfrac{\gamma_n(x + a_1)}{a_1} \sin \dfrac{\gamma_n a_2 v_1}{a_1 v_2}, & -a_1 \leqslant x \leqslant 0, \\[3mm] u_n^{(2)}(x) = \sin \dfrac{\gamma_n(a_2 - x)v_1}{a_1 v_2} \sin \gamma_n, & 0 \leqslant x \leqslant a_2, \end{cases}$$

the γ_n are consecutive positive roots of the equation

$$S_2 \sqrt{E_2 \rho_2} \tan \gamma + S_1 \sqrt{E_1 \rho_1} \tan \frac{\gamma a_2 v_1}{a_1 v_2} = 0,$$

the two sections have Young's moduli E_i, cross-sectional areas S_i and densities ρ_i ($i = 1, 2$), and $v_i = \sqrt{E/\rho_i}$.

110. A pointer is fastened to the free end of a rod of length l clamped at the end $x = 0$. Study the torsional oscillations which result if at the time $t = 0$ the pointer is twisted through an angle α and then released without initial velocity, given that the moment of inertia of the pointer with respect to the axis of rotation is J_0.

Ans.

$$\theta(x, t) = 2\alpha \frac{J}{J_0} \sum_{n=1}^{\infty} \frac{\cos \gamma_n}{\gamma_n^3} \frac{\sin \dfrac{\gamma_n x}{l}}{1 + \dfrac{\sin 2\gamma_n}{2\gamma_n}} \cos \frac{\gamma_n v t}{l},$$

where the γ_n are consecutive positive roots of the equation

$$\gamma \tan \gamma = \frac{J}{J_0},$$

J is the moment of inertia, G the shear modulus and ρ the density of the rod, and $v = \sqrt{G/\rho}$.

111. Solve the preceding problem with arbitrary initial conditions

$$\theta|_{t=0} = f(x), \qquad \frac{\partial \theta}{\partial t}\bigg|_{t=0} = g(x).$$

Ans.

$$\theta(x, t) = 2 \sum_{n=1}^{\infty} \frac{\sin \dfrac{\gamma_n x}{l}}{\gamma_n \left(1 + \dfrac{\sin 2\gamma_n}{2\gamma_n}\right)}$$

$$\times \left[\cos \frac{\gamma_n vt}{l} \int_0^l f'(\xi) \cos \frac{\gamma_n \xi}{l}\, d\xi + \frac{l}{\gamma_n v} \sin \frac{\gamma_n vt}{l} \int_0^l g'(\xi) \cos \frac{\gamma_n \xi}{l}\, d\xi\right],$$

with the previous notation.

*112. A disk with moment of inertia J_0 is fastened to the point $x = c$ of a cylindrical shaft with clamped ends $x = 0$ and $x = l$. Find the torsional oscillations of the shaft if the disk is twisted through the angle α at the time $t = 0$ and then released without initial velocity.

Ans.

$$\theta(x, t) = -\frac{2\alpha l}{a\left(1 - \dfrac{a}{l}\right)} \sum_{n=1}^{\infty} \frac{\sin \gamma_n\left(1 - \dfrac{a}{l}\right) \sin \dfrac{\gamma_n a}{l}\, \theta_n(x) \cos \dfrac{\gamma_n vt}{l}}{D_n}$$

where

$$D_n = \gamma_n^2 \left\{\left[\frac{a}{l} + \frac{\sin(2\gamma_n a/l)}{2\gamma_n}\right] \sin^2 \gamma_n\left(1 - \frac{a}{l}\right)\right.$$

$$+ \left.\left[1 - \frac{a}{l} + \frac{\sin^2 2\gamma_n(1 - (a/l))}{2\gamma_n}\right] \sin^2 \frac{\gamma_n a}{l}\right\},$$

$$\theta_n(x) = \begin{cases} \sin \dfrac{\gamma_n x}{l} \sin \gamma_n\left(1 - \dfrac{a}{l}\right), & 0 \leqslant x \leqslant a, \\[2mm] \sin \dfrac{\gamma_n a}{l} \sin \gamma_n\left(1 - \dfrac{x}{l}\right), & a \leqslant x \leqslant l, \end{cases}$$

the γ_n are consecutive positive roots of the equation

$$\sin \gamma = \frac{J_0 \gamma}{J} \sin \frac{\gamma a}{l} \sin \gamma\left(1 - \frac{a}{l}\right),$$

and J is the moment of inertia of the shaft.

113. A disk with moment of inertia J_0 is fastened to one end $x = 0$ of a circular shaft, and another disk with moment of inertia J_l is fastened to the other end $x = l$. Find the torsional oscillations of the shaft if the relative angle of rotation of the disks equals α.

Ans.

$$\theta(x, t) = 2\alpha \frac{J}{J_0} \sum_{n=1}^{\infty} \frac{\frac{J}{J_0}(1 - \cos \gamma_n) + \gamma_n \sin \gamma_n}{\gamma_n^2 \left[\left(\frac{J^2}{J_0^2} + \gamma_n^2 \right) - \left(\frac{J^2}{J_0^2} - \gamma_n^2 \right) \frac{\sin 2\gamma_n}{2\gamma_n} + \frac{J}{J_0}(1 - \cos 2\gamma_n) \right]}$$
$$\times \left[1 - \left(\cos \frac{\gamma_n x}{l} - \gamma_n \frac{J_0}{J} \sin \frac{\gamma_n x}{l} \right) \cos \frac{\gamma_n v t}{l} \right],$$

where the γ_n are consecutive positive roots of the equation

$$\tan \gamma = \frac{\left(\frac{J_0}{J} + \frac{J_l}{J} \right) \gamma}{\frac{J_0}{J} \frac{J_l}{J} \gamma^2 - 1},$$

and J is the moment of inertia of the shaft.

*114. Find the general solution of the problem of transverse oscillations of a beam of length l, simply supported at its ends $x = 0$ and $x = l$, with arbitrary initial conditions

$$u|_{t=0} = f(x), \qquad \frac{\partial u}{\partial t}\bigg|_{t=0} = g(x).$$

Ans.

$$u(x, t) = \frac{2}{l} \sum_{n=1}^{\infty} \left[\cos \frac{n^2 \pi^2 a^2 t}{l^2} \int_0^l f(\xi) \sin \frac{n\pi\xi}{l} \, d\xi \right.$$
$$\left. + \frac{l^2}{n^2 \pi^2 a^2} \sin \frac{n^2 \pi^2 a^2 t}{l^2} \int_0^l g(\xi) \sin \frac{n\pi\xi}{l} \, d\xi \right] \sin \frac{n\pi x}{l},$$

where $a^2 = \sqrt{EJ/\rho S}$, E is Young's modulus, J the moment of inertia of a cross section, ρ the density and S the cross-sectional area of the beam.

115. Investigate the transverse oscillations of a beam of length l, simply supported at its ends $x = 0$ and $x = l$, under the action of an impulse P applied to the point $x = c$ at the time $t = 0$.

Ans.

$$u(x, t) = \frac{2lP}{\pi^2 \sqrt{EJ\rho S}} \sum_{n=1}^{\infty} \frac{\sin (n\pi c/l)}{n^2} \sin \frac{n\pi x}{l} \sin \frac{n^2 \pi^2 a^2 t}{l^2}.$$

Hint. Solve the problem of the oscillations of the beam with the initial conditions

$$u|_{t=0} = 0, \qquad \frac{\partial u}{\partial t}\bigg|_{t=0} = \begin{cases} v_0 = \dfrac{P}{2\rho S \varepsilon}, & c - \varepsilon < x < c + \varepsilon, \\ 0, & \text{otherwise,} \end{cases}$$

and then take the limit as $\varepsilon \to 0$.

116. A beam of length $2l$, clamped at its ends $x = \pm l$, undergoes transverse oscillations with initial conditions

$$u\big|_{t=0} = f(x), \qquad \frac{\partial u}{\partial t}\bigg|_{t=0} = 0.$$

Find the oscillations of the beam, assuming that the initial deflection of the beam is symmetric with respect to the center of the beam and that there is no initial velocity.

Ans.

$$u(x, t) = \frac{1}{l}\sum_{n=1}^{\infty} \frac{\cos(\gamma_n^2 a^2 t/l^2)}{\cos^2 \gamma_n \cosh^2 \gamma_n}\left(\cosh \gamma_n \cos \frac{\gamma_n x}{l} - \cos \gamma_n \cosh \frac{\gamma_n x}{l}\right)$$

$$\times \int_0^l f(\xi)\left(\cosh \gamma_n \cos \frac{\gamma_n \xi}{l} - \cos \gamma_n \cosh \frac{\gamma_n \xi}{l}\right) d\xi,$$

where the γ_n are consecutive positive roots of the equation $\tan \gamma + \tanh \gamma = 0$.

117. A beam simply supported at the points $x = 0$ and $x = l$ is in equilibrium under the action of a concentrated force F applied at the point $x = c$. Find the transverse oscillations which result if the force is suddenly removed.

Ans.

$$u(x, t) = \frac{2Fl^3}{\pi^4 EJ}\sum_{n=1}^{\infty} \frac{\sin(n\pi c/l)}{n^4}\sin\frac{n\pi x}{l}\cos\frac{n^2\pi^2 a^2 t}{l^2}.$$

*118. Find the transverse oscillations of a cantilever of length l if the initial deflection is due to a concentrated force F applied to the free end $x = l$ and is suddenly removed at the time $t = 0$.

Ans.

$$u(x, t) = \frac{2Fl^3}{EJ}\sum_{n=1}^{\infty} \frac{X_n(x)\cos(\gamma_n^2 a^2 t/l^2)}{(\cos \gamma_n \sinh \gamma_n - \sin \gamma_n \cosh \gamma_n)\gamma_n^4},$$

where

$$X_n(x) = (\sin \gamma_n + \sinh \gamma_n)\left(\cos\frac{\gamma_n x}{l} - \cosh\frac{\gamma_n x}{l}\right)$$

$$- (\cos \gamma_n + \cosh \gamma_n)\left(\sin\frac{\gamma_n x}{l} - \sinh\frac{\gamma_n x}{l}\right),$$

and the γ_n are consecutive positive roots of the equation $\cos \gamma \cosh \gamma + 1 = 0$.

119. A beam of length l is simply supported at the end $x = 0$ and clamped at the end $x = l$. Find the transverse oscillations of the beam under the action of a suddenly applied uniformly distributed load q.

Ans.

$$u(x, t) = \frac{qx(l - x)}{48EJ}(l^2 + lx - 2x^2)$$

$$+ \frac{ql^4}{EJ} \sum_{n=1}^{\infty} \frac{\sinh \gamma_n - 2 \cosh \gamma_n \sin \gamma_n + \sin \gamma_n}{\gamma_n^5 \sinh^2 \gamma_n \sin^2 \gamma_n} X_n(x) \cos \frac{\gamma_n^2 a^2 t}{l^2},$$

$$X_n(x) = \sinh \gamma_n \sin \frac{\gamma_n x}{l} - \sin \gamma_n \sinh \frac{\gamma_n x}{l},$$

where the γ_n are consecutive positive roots of the equation $\tan \gamma = \tanh \gamma$.

Hint. Make the problem homogeneous by subtracting out the static deflection of the beam.

*120. Study the axially symmetric vibrations of a circular membrane of radius a due to an impulse P applied at the time $t = 0$ and distributed over a disk of radius ε.

Ans.

$$u(r, t) = \frac{2Pv}{\pi \varepsilon T} \sum_{n=1}^{\infty} \frac{J_1(\gamma_n \varepsilon / a)}{\gamma_n^2 J_1^2(\gamma_n)} J_0\left(\frac{\gamma_n r}{a}\right) \sin \frac{\gamma_n v t}{a},$$

where $J_0(x)$ and $J_1(x)$ are Bessel functions, the γ_n are consecutive positive roots of the equation $J_0(\gamma) = 0$, T is the tension per unit length of the boundary, ρ is the surface density of the membrane, and $v = \sqrt{T/\rho}$,

Hint. The initial conditions have the form

$$u\big|_{t=0} = 0, \qquad \frac{\partial u}{\partial t}\bigg|_{t=0} = \begin{cases} v_0 = \dfrac{P}{\pi \varepsilon^2 \rho}, & 0 < r < \varepsilon, \\ 0 & \varepsilon < r < a. \end{cases}$$

121. Find the general solution of the problem of vibrations of a ring-shaped membrane fastened to the circles $r = a$ and $r = b$, and subject to arbitrary initial conditions

$$u\big|_{t=0} = f(r), \qquad \frac{\partial u}{\partial t}\bigg|_{t=0} = g(r).$$

Ans.

$$u(r, t) = \frac{2}{b^2} \sum_{n=1}^{\infty} \frac{\gamma_n^2 R_{\gamma_n}(r)}{(4/\pi^2) - a^2 R_{\gamma_n}'^2(a)} \left[\cos \frac{\gamma_n v t}{b} \int_a^b \rho f(\rho) R_{\gamma_n}(\rho) \, d\rho \right.$$

$$\left. + \frac{b}{\gamma_n v} \sin \frac{\gamma_n v t}{b} \int_a^b \rho g(\rho) R_{\gamma_n}(\rho) \, d\rho \right],$$

where

$$R_\gamma(r) = Y_0(\gamma) J_0\left(\frac{\gamma r}{b}\right) - J_0(\gamma) Y_0\left(\frac{\gamma r}{b}\right)$$

is a linear combination of Bessel functions of the first and second kinds, and the γ_n are consecutive positive roots of the equation $R_\gamma(a) = 0$.

122. Determine the axially symmetric vibrations of a circular membrane of radius a due to a pulsating load $q \sin \omega t$ which is uniformly distributed over the whole membrane and begins to act at the time $t = 0$.

Ans.

$$u(r, t) = -\frac{q}{\rho\omega^2}\left[1 - \frac{J_0(\omega r/v)}{J_0(\omega a/v)}\right]\sin \omega t + \frac{2vaq}{\omega T}\sum_{n=1}^{\infty}\frac{J_0(\gamma_n r/a)}{\gamma_n^2 J_1(\gamma_n)}\frac{\sin (\gamma_n vt/a)}{1 - (v\gamma_n/\omega a)^2},$$

where the γ_n are consecutive positive roots of the equation $J_0(\gamma) = 0$.

Hint. Make the problem homogeneous by subtracting out the forced oscillations (see Prob. 82).

123. Find the vibrations of a rectangular membrane $-a \leqslant x \leqslant a$, $-b \leqslant y \leqslant b$ with initial conditions

$$u\big|_{t=0} = f(x, y), \qquad \frac{\partial u}{\partial t}\bigg|_{t=0} = 0,$$

where f is a given function which is even in each of the variables.

Ans.

$$u(x,y,t) = \sum_{m=0}^{\infty}\sum_{n=0}^{\infty}A_{mn}\cos\frac{(2m + 1)\pi x}{2a}\cos\frac{(2n + 1)\pi y}{2b}$$
$$\times \cos\sqrt{\left(\frac{2m + 1}{2a}\right)^2 + \left(\frac{2n + 1}{2b}\right)^2}\,\pi vt,$$

where

$$A_{mn} = \frac{4}{ab}\int_0^a\int_0^b f(x, y)\cos\frac{(2m + 1)\pi x}{2a}\cos\frac{(2n + 1)\pi y}{2b}\,dx\,dy.$$

*124. Study the transverse oscillations of a circular plate of radius a with a clamped edge, for arbitrary initial conditions

$$u\big|_{t=0} = f(r), \qquad \frac{\partial u}{\partial t}\bigg|_{t=0} = g(r).$$

Ans.

$$u(r, t) = \frac{1}{a^2}\sum_{n=1}^{\infty}\frac{R_{\gamma_n}(r)}{I_0^2(\gamma_n)J_0^2(\gamma_n)}\left[\cos\frac{\gamma_n^2 b^2 t}{a^2}\int_0^a \rho f(\rho)R_{\gamma_n}(\rho)\,d\rho\right.$$
$$\left. + \frac{a^2}{b^2\gamma_n^2}\sin\frac{\gamma_n^2 b^2 t}{a^2}\int_0^a \rho g(\rho)R_{\gamma_n}(\rho)\,d\rho\right],$$

where

$$R_\gamma(r) = I_0(\gamma)J_0\left(\frac{\gamma r}{a}\right) - J_0(\gamma)I_0\left(\frac{\gamma r}{a}\right)$$

is a linear combination of cylinder functions, the γ_n are consecutive roots of the equation $R'_\gamma(a) = 0$, $b^2 = \sqrt{D/\rho h}$, and D is the flexural rigidity, h the thickness and ρ the density of the plate.

125. Solve the preceding problem for the case where the oscillations are due to an impulse P applied at the center of the plate at the time $t = 0$.

Ans.

$$u(r, t) = \frac{Pb^2}{2\pi D} \sum_{n=1}^{\infty} \frac{[I_0(\gamma_n) - J_0(\gamma_n)]R_{\gamma_n}(r)}{\gamma_n^2 I_0^2(\gamma_n)J_0^2(\gamma_n)} \sin \frac{\gamma_n^2 b^2 t}{a^2}.$$

Hint. Solve the problem with the initial conditions

$$u\big|_{t=0} = 0, \qquad \frac{\partial u}{\partial t}\bigg|_{t=0} = \begin{cases} v_0 = \dfrac{P}{\pi \varepsilon^2 \rho h}, & 0 \leqslant r < \varepsilon, \\ 0, & \varepsilon < r \leqslant a, \end{cases}$$

and then take the limit as $\varepsilon \to 0$.

126. Investigate the transverse oscillations of a circular plate of radius a with a clamped edge under the action of a concentrated force F applied to the center of the plate. The plate is assumed to be at rest at the time $t = 0$.

Ans.

$$u(r,t) = \frac{F}{16\pi D}\left(r^2 \ln \frac{r^2}{a^2} + a^2 - r^2\right)$$

$$- \frac{Fa^2}{2\pi D} \sum_{n=1}^{\infty} \frac{I_0(\gamma_n) - J_0(\gamma_n)}{\gamma_n^4 I_0^2(\gamma_n)J_0^2(\gamma_n)} R_{\gamma_n}(r)\cos \frac{\gamma_n^2 b^2 t}{a^2},$$

with the notation of Prob. 124.

Hint. Make the problem homogeneous by subtracting out the static deflection of the plate.

127. Study the radial oscillations of a gas confined in a spherical resonator,[8] assuming that the initial values of the velocity potential and its time derivative are

$$u\big|_{t=0} = f(r), \qquad \frac{\partial u}{\partial t}\bigg|_{t=0} = 0.$$

---

[8] The velocity potential of an oscillating gas satisfies the wave equation

$$\Delta u - \frac{1}{v^2}\frac{\partial^2 u}{\partial t^2}$$

(see T1, p. 25). In Probs. 127–130 it is assumed that the walls are perfectly reflecting, i.e. that

$$\frac{\partial u}{\partial n}\bigg|_{s} = 0.$$

Ans.

$$u(r, t) = \frac{2}{a} \sum_{n=1}^{\infty} \frac{\sin(\gamma_n r/a)}{r} \frac{\cos(\gamma_n vt/a)}{\sin^2 \gamma_n} \int_0^a \rho f(\rho) \sin \frac{\gamma_n \rho}{a} \, d\rho + \frac{3}{a^2} \int_0^a \rho^2 f(\rho) \, d\rho,$$

where the γ_n are consecutive positive roots of the equation $\tan \gamma = \gamma$, a is the radius of the sphere and v is the velocity of wave propagation in the gas.

128. Investigate the steady-state acoustic oscillations in a semi-infinite cylindrical pipe of radius a, assuming that the distribution of the normal component of the velocity of the air particles in the plane $z = 0$ is a given function

$$(v_\Gamma)_z|_{z=0} = f(r) \sin \omega t.$$

Consider the special cases

$$\text{a)} \quad f(r) = v_0, \qquad \text{b)} \quad f(r) = v_0 J_0\left(\frac{\gamma r}{a}\right),$$

where γ is the smallest positive root of the equation $J_1(\gamma) = 0$.

Ans. The velocity potential is given by the formula

$$u(r, z, t) = \text{Im}\left\{ \frac{2e^{i\omega t}}{ia^2} \sum_{n=0}^{\infty} \frac{e^{-i\sqrt{k^2-(\gamma_n^2/a^2)}z}}{\sqrt{k^2 - (\gamma_n^2/a^2)}} \frac{J_0(\gamma_n r/a)}{J_0^2(\gamma_n)} \int_0^a \rho f(\rho) J_0\left(\frac{\gamma_n \rho}{a}\right) d\rho \right\},$$

where $k = \omega/v$, the γ_n are consecutive nonnegative roots of the equation $J_1(\gamma) = 0$ ($\gamma_0 = 0$), and $J_0(x), J_1(x)$ are Bessel functions. In the special cases,

$$u(z, t) = -\frac{v_0}{k} \cos(\omega t - kz),$$

$$u(r, z, t) = \text{Im}\left\{ v_0 J_0\left(\frac{\gamma r}{a}\right) \frac{e^{-i\sqrt{k^2-(\gamma^2/a^2)}z}}{i\sqrt{k^2 - (\gamma^2/a^2)}} e^{i\omega t} \right\}.$$

129. Find the steady-state harmonic oscillations of sound inside a conical horn $a \leqslant r < \infty$, $0 \leqslant \theta \leqslant \alpha$, assuming that the velocity distribution along the base of the horn is given by

$$(v_\Gamma)_r|_{r=a} = f(\theta) \sin \omega t.$$

Consider the special case $f(\theta) = v_0$.

Ans.

$$u(r, \theta, t) = \text{Im}\left\{ -\frac{e^{i\omega t}}{k \sin^2 \alpha} \sum_{n=1}^{\infty} \frac{2\nu_n + 1}{P_{\nu_n}(\cos \alpha)} \frac{H^{(2)}_{\nu_n+\frac{1}{2}}(kr)}{H^{(2)'}_{\nu_n+\frac{1}{2}}(ka)} \right.$$

$$\times \left. \frac{P_{\nu_n}(\cos \theta)}{\left.\dfrac{\partial P'_{\nu_m}(\cos \alpha)}{\partial \nu_m}\right|_{\nu_m=\nu_n}} \int_0^a f(\theta) P_{\nu_n}(\cos \theta) \sin \theta \, d\theta \right.$$

$$\left. - \frac{e^{i\omega t}}{k(1 - \cos \alpha)} \frac{H^{(2)}_{1/2}(kr)}{H^{(2)}_{1/2}(ka)} \int_0^\alpha f(\theta) \sin \theta \, d\theta \right\},$$

where the ν_n are consecutive roots of the equation $P'_{\nu_n}(\cos \alpha) = 0$, $P_\nu(x)$ is the Legendre function and $H^{(2)}_{\nu+\frac{1}{2}}(x)$ the second Hankel function. In the special case,

$$u(r, t) = 2v_0 a \sqrt{\frac{a}{r}} \frac{\sin [\omega t - k(r - a)] - 2ka \cos [\omega t - k(r - a)]}{1 + 4k^2 a^2}.$$

130. Solve the problem of diffraction of a plane sound wave $u_0 e^{i(\omega t - kz)}$ by a spherical obstacle of radius a.

Ans.

$$u = u_0 e^{i(\omega t - kz)} - u_0 \sqrt{\frac{\pi}{2kr}} e^{i\omega t}$$

$$\times \sum_{n=0}^{\infty} (2n + 1) e^{-in\pi/2} \frac{2ka J'_{n+\frac{1}{2}}(ka) - J_{n+\frac{1}{2}}(ka)}{2ka H^{(2)'}_{n+\frac{1}{2}}(ka) - H^{(2)}_{n+\frac{1}{2}}(ka)} H^{(2)}_{n+\frac{1}{2}}(kr) P_n(\cos \theta),$$

where $J_{n+\frac{1}{2}}(x)$ is the Bessel function of the first kind, $H(x)^{(2)}_{n+\frac{1}{2}}$ the second Hankel function, and $P_n(x)$ the Legendre polynomial.

Hint. If the velocity potential is written as a sum

$$u = (u_0 e^{-ikz} + u_1) e^{i\omega t},$$

then solving the problem reduces to integrating Helmholtz's equation

$$\Delta u_1 + k^2 u_1 = 0$$

with the boundary condition

$$-\frac{\partial u_1}{\partial r}\bigg|_{r=a} = \frac{\partial}{\partial r} (e^{-ikz} u_0)\big|_{r=a},$$

where u_1 must satisfy the radiation condition at infinity.

2. Mechanics: Statics of Deformable Media, Fluid Dynamics

131. Find the equilibrium shape of a rectangular membrane with sides $2a$ and $2b$ under the action of a uniformly distributed load q, choosing the origin at the center of the membrane. Calculate the deflection of the center of the membrane, assuming that the ratio b/a takes the values 1, 2 and 3.

Ans.

$$u(x, y)$$

$$= \frac{qa^2}{T} \left\{ \frac{1}{2}\left(1 - \frac{x^2}{a^2}\right) + \frac{16}{\pi^3} \sum_{n=0}^{\infty} \frac{(-1)^{n+1}}{(2n+1)^3} \frac{\cosh [(2n+1)\pi y/2a]}{\cosh [(2n+1)\pi b/2a]} \cos \frac{(2n+1)\pi x}{2a} \right\},$$

where T is the tension per unit length of the boundary. Numerical calculations show that $u(0, 0) = kQ/T$, where $Q = qab$ is the total load, and

$$k|_{b/a=1} = 0.295, \qquad k|_{b/a=2} = 0.228, \qquad k|_{b/a=3} = 0.164.$$

Hint. Make the problem homogeneous by subtracting out the particular solution of the equation for equilibrium of the membrane which depends only on the coordinate x and satisfies the boundary conditions on the sides $x = \pm a$.

*132. Find the equilibrium shape of a semicircular membrane of radius a (see Figure 20) under a uniformly distributed load q.

Ans.

$$u(r, \varphi) = \frac{qa^2}{2T} \left\{ \frac{1}{\pi} \left[\left(\frac{r}{a} - \frac{a}{r} \right) \sin \varphi \right. \right.$$

$$- \frac{1}{2} \left(\left(\frac{r^2}{a^2} + \frac{a^2}{r^2} \right) \cos 2\varphi - 2 \right) \arctan \frac{2ar \sin \varphi}{a^2 - r^2}$$

$$\left. \left. + \frac{1}{4} \left(\frac{r^2}{a^2} - \frac{a^2}{r^2} \right) \sin 2\varphi \ln \frac{a^2 + r^2 - 2ar \cos \varphi}{a^2 + r^2 + 2ar \cos \varphi} \right] - \frac{r^2 \sin^2 \varphi}{a^2} \right\},$$

where T is the tension per unit length of the boundary.

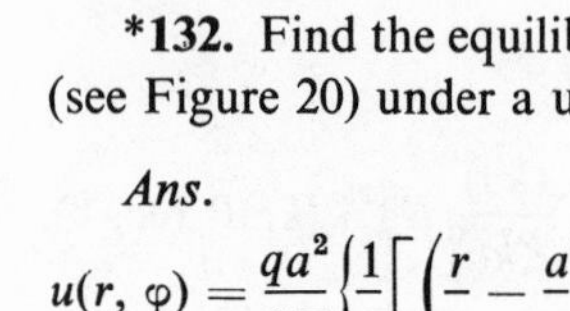

FIGURE 20

Hint. To apply the Fourier method, subtract out the particular solution

$$u_1 = -\frac{qr^2}{2T} \sin^2 \varphi$$

of the equilibrium equation. To write the solution in closed form, it is necessary to sum a series (this has been done in the answer).

133. Study the twisting of a rod whose cross section is a rectangle with sides a and b. Find the torsion function and the torsional rigidity.

Ans. The torsion function is

$$u(x, y) = x(a - x) - \frac{8a^2}{\pi^3} \sum_{n=0}^{\infty} \frac{\sin [(2n + 1)\pi x/a]}{(2n + 1)^3} \frac{\cosh [(2n + 1)(\tfrac{1}{2}b - y)\pi/a]}{\cosh [(2n + 1)\pi b/2a]},$$

$$(0 \leqslant x \leqslant a, 0 \leqslant y \leqslant b).$$

and the torsional rigidity is

$$C = Ga^4 \left\{ \frac{b}{3a} - \frac{64}{\pi^5} \sum_{n=0}^{\infty} \frac{\tanh [(2n + 1)\pi b/2a]}{(2n + 1)^5} \right\},$$

where G is the shear modulus.

Hint. Make the problem homogeneous by subtracting out the particular solution of the differential equation for the torsion function which depends only on the coordinate x and satisfies the boundary conditions for $x = 0$ and $x = a$.

134. A rectangular plate with sides $2a$ and $2b$, simply supported on its edges, is acted upon by a uniformly distributed load q. Find the deformation of the plate, choosing the origin at the center of the plate. Derive an expression for the deflection of the plate.

Ans.

$$u(0, 0) = \frac{qa^4}{D}\left[\frac{5}{24} - \frac{64}{\pi^5}\sum_{n=0}^{\infty}\frac{(-1)^n}{(2n+1)^5}\frac{2 + \frac{(2n+1)\pi b}{2a}\tanh\frac{(2n+1)\pi b}{2a}}{2\cosh\frac{(2n+1)\pi b}{2a}}\right],$$

where D is the flexural rigidity of the plate.

Hint. Subtract out the particular solution of the deflection equation which depends only on the coordinate x and satisfies the boundary conditions for $x = \pm a$.

135. Solve the preceding problem, assuming that the boundaries $x = \pm a$ are simply supported, while the boundaries $y = \pm b$ are free. Calculate the deflection at the center of the plate.

Ans.

$$u(0, 0) = \frac{qa^4}{D}\left[\frac{5}{24} + \frac{64\nu}{\pi^5}\sum_{n=0}^{\infty}\frac{(-1)^n}{(2n+1)^5}\right.$$

$$\left.\times\frac{\frac{1+\nu}{1-\nu}\sinh\frac{(2n+1)\pi b}{2a} - \frac{(2n+1)\pi b}{2a}\cosh\frac{(2n+1)\pi b}{2a}}{(3+\nu)\sinh\frac{(2n+1)\pi b}{2a}\cosh\frac{(2n+1)\pi b}{2a} - (1-\nu)\frac{(2n+1)\pi b}{2a}}\right]$$

where ν is Poisson's ratio.

*136. A semicircular plate of radius a is clamped along the semicircular arc and simply supported along its rectilinear edge. Find the deflection of the plate under a uniform load. Write a formula for the deflection of the axis of symmetry of the plate, and represent the result in the form of a graph.

Ans.

$$u\left(r, \frac{\pi}{2}\right) = \frac{qa^4}{24D}\left[\frac{r^4}{a^4} - \frac{1}{8\pi}\left(6 + 12\frac{r^2}{a^2} + 5\frac{r^4}{a^4} - 4\frac{a^2}{r^2} - 3\frac{a^4}{r^4}\right)\arctan\frac{2ar}{a^2 - r^2}\right.$$

$$\left. - \frac{1}{4\pi}\left(5\frac{r^3}{a^3} - 11\frac{r}{a} + 3\frac{a}{r} + 3\frac{a^3}{r^3}\right)\right]$$

(see Figure 21).

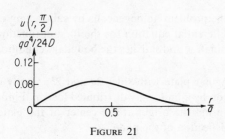

FIGURE 21

Hint. Make the problem homogeneous by subtracting out the particular solution

$$\frac{q}{24D} r^4 \sin^4 \varphi, \qquad 0 \leqslant \varphi \leqslant \pi$$

of the deflection equation satisfying the boundary conditions on the rectilinear edge.

137. An infinite cylinder of radius a is placed in a plane-parallel flow of an ideal fluid. Find the velocity potential, choosing the origin at the center of the cylinder and the direction of the x-axis opposite to the direction of flow (see Figure 22).

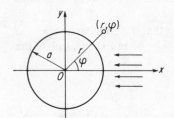

FIGURE 22

Ans.

$$u(r, \varphi) = v_\infty \left(r + \frac{a^2}{r} \right) \cos \varphi + \text{const},$$

where v_∞ is the value of the flow velocity far from the cylinder.

138. Find the velocity potential for flow of an ideal fluid emanating from a source of strength m and flowing past an infinite cylinder of radius a, where the configuration of the cylinder and the source is shown in Figure 23.

Ans.

$$u(r, \varphi) = \frac{m}{2\pi} \ln \frac{rb}{\rho \bar{\rho}} + \text{const},$$

where $b\bar{b} = a^2$ and the meaning of the various symbols is indicated in the figure.

Hint. Subtract the source potential

$$u_1 = -\frac{m}{2\pi} \ln \rho + \text{const}$$

from the solution.

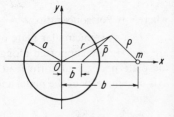

FIGURE 23

139. Solve the problem of plane-parallel flow of an ideal fluid past a sphere of radius a, choosing the origin of a system of spherical coordinates

r, θ, φ at the center of the sphere, with the direction of the z-axis opposite to that of the flow.

Ans.

$$u(r, \theta) = v_\infty \left(r + \frac{a^3}{2r^2} \right) \cos \theta + \text{const},$$

where v_∞ is the value of the flow velocity far from the sphere.

*140. Solve the problem of flow past a sphere of radius a due to a source of strength m at a distance b from the center.

Ans.

$$u(r, \theta) = \frac{m}{4\pi} \left[\frac{1}{\rho} + \frac{a}{b\bar{\rho}} - \frac{1}{a} \ln \frac{r(1 + \cos \theta)}{\bar{\rho} + r \cos \theta - b} \right],$$

with the same notation as in Figure 23, except that the x-axis now becomes the z-axis.

3. Heat Conduction: Nonstationary Problems

141. A slab of thickness $2a$, thermal conductivity k, specific heat c and density ρ is heated to temperature T_0, and its faces are then held at temperature T_0, starting from the time $t = 0$ (see Figure 24). Find the temperature distribution $T(x, t)$ in the slab.

Ans.

$$T(x, t) = \frac{4T_0}{\pi} \sum_{n=0}^{\infty} \frac{(-1)^n}{2n + 1}$$
$$\times \, e^{-(2n+1)^2 \pi^2 \tau / 4a^2} \cos \frac{(2n + 1)\pi x}{2a},$$

where $\tau = kt/c\rho$.

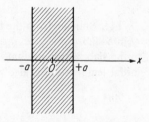

FIGURE 24

142. Describe the equalization of a given initial temperature distribution $T(x, 0) = f(x)$ in a slab whose faces $x = 0$ and $x = a$ do not transmit heat.

Ans.

$$T(x, t) = \frac{1}{a} \int_0^a f(\xi) \, d\xi + \frac{2}{a} \sum_{n=1}^{\infty} e^{-n^2 \pi^2 \tau / a^2} \cos \frac{\pi n x}{a} \int_0^a f(\xi) \cos \frac{n \pi \xi}{a} \, d\xi.$$

143. Starting from the time $t = 0$, a slab $-a \leqslant x \leqslant a$ of thickness $2a$ with a given initial temperature distribution $T(x, 0) = f(x)$ radiates heat into the surrounding medium, whose temperature is taken to be zero. Assuming that the radiation obeys Newton's law, find the temperature distribution in the slab for arbitrary time t.

Ans.

$$T(x, t) = \frac{1}{a} \sum_{n=1}^{\infty} \frac{\cos \frac{\gamma_n^{(1)} x}{a}}{1 + \frac{\sin 2\gamma_n^{(1)}}{2\gamma_n^{(1)}}} e^{-(\gamma_n^{(2)}/a)^2 \tau} \int_{-a}^{a} f(\xi) \cos \frac{\gamma_n^{(1)} \xi}{a} d\xi$$

$$+ \frac{1}{a} \sum_{n=1}^{\infty} \frac{\sin \frac{\gamma_n^{(2)} x}{a}}{1 - \frac{\sin 2\gamma_n^{(2)}}{2\gamma_n^{(2)}}} e^{-(\gamma_n^{(2)}/a)^2 \tau} \int_{-a}^{a} f(\xi) \sin \frac{\gamma_n^{(2)} \xi}{a} d\xi,$$

where $\tau = kt/c\rho$, the $\gamma_n^{(1)}$ are consecutive positive roots of the equation

$$\tan \gamma^{(1)} = \frac{ah}{\gamma^{(1)}},$$

h is the heat exchange coefficient, and the $\gamma_n^{(2)}$ are the corresponding roots of the equation

$$\tan \gamma^{(2)} = -\frac{\gamma^{(2)}}{ah}.$$

144. Starting from the time $t = 0$, heat is produced with constant density Q in a slab $-a \leqslant x \leqslant a$ of thickness $2a$. Find the temperature distribution in the slab, assuming that its faces are held at temperature zero and that the initial temperature is also zero.

Ans.

$$T(x, t) = \frac{Qa^2}{2k}\left(1 - \frac{x^2}{a^2}\right) - \frac{16Qa^2}{\pi^3 k} \sum_{n=0}^{\infty} \frac{(-1)^n}{(2n+1)^3} e^{-(2n+1)^2 \pi^2 \tau/4a^2} \cos \frac{(2n+1)\pi x}{2a}.$$

Hint. Make the problem homogeneous by subtracting out the solution of the corresponding stationary problem.

*145. An inhomogeneous slab consisting of two layers with different thermal properties is heated to a certain temperature T_0, and then cooled by having its faces held at temperature zero starting from the time $t = 0$. Assuming that the faces of the slab are at $x = 0$ and $x = a_1 + a_2$ (where a_1 and a_2 are the thicknesses of the two layers), find the temperature distribution in the slab.

Ans.

$$T(x, t) = 2T_0 \sum_{n=1}^{\infty} \frac{\tan \dfrac{a_2\sqrt{b_2}\gamma_n}{a_1\sqrt{b_1}} \left[\cos \dfrac{a_2\sqrt{b_2}\gamma_n}{a_1\sqrt{b_1}} - \cos \gamma_n \right]}{\gamma_n \left[\sin^2 \dfrac{a_2\sqrt{b_2}\gamma_n}{a_1\sqrt{b_1}} + \dfrac{a_2 b_2 k_2}{a_1 b_1 k_1} \sin^2 \gamma_n \right]} e^{-\gamma_n^2 t/b_1 a_1^2} \, X_n(x),$$

$$X_n(x) = \begin{cases} \sin \dfrac{a_2\sqrt{b_2}\gamma_n}{a_1\sqrt{b_1}} \sin \dfrac{\gamma_n x}{a_1}, & 0 \leqslant x \leqslant a_1, \\[3mm] \sin \gamma_n \sin \dfrac{\sqrt{b_2}\gamma_n(a_1 + a_2 - x)}{\sqrt{b_1}a_1}, & a_1 \leqslant x \leqslant a_1 + a_2, \end{cases}$$

where the γ_n are consecutive positive roots of the equation

$$\sqrt{b_2}\, k_2 \tan \gamma + \sqrt{b_1}\, k_1 \tan \frac{a_2\sqrt{b_2}\gamma}{a_1\sqrt{b_1}} = 0,$$

the two layers have specific heats c_i, densities ρ_i and thermal conductivities k_i ($i = 1, 2$), and $b_i = c_i \rho_i / k_i$.

146. The ends of a thin rod of length l are held at different temperatures, while the lateral surface of the rod gives off heat into the surrounding medium according to Newton's law. Find the temperature distribution along the rod, assuming that the ends of the rod $x = 0$ and $x = l$ have temperatures zero and T_0, respectively, and that the initial temperature equals zero.

Ans.

$$T(x, t) = T_0 \left[\frac{\sinh \sqrt{\mu}\, x}{\sinh \sqrt{\mu}\, l} + \frac{2}{\pi} e^{-\mu\tau} \sum_{n=1}^{\infty} \frac{(-1)^n n}{n^2 + (\mu l^2/\pi^2)} \sin \frac{n\pi x}{l} e^{-n^2\pi^2\tau/l^2} \right],$$

where p and S are the perimeter and cross-sectional area of the rod, h is the heat exchange coefficient figuring in Newton's law, and $\mu = ph/S$.

Hint. The problem reduces to integration of the differential equation

$$\frac{\partial^2 T}{\partial x^2} = \frac{\partial T}{\partial \tau} + \mu T$$

(see C3, p. 134).

147. A cylinder of radius a is heated to temperature T_0 and then cooled by having its surface held at temperature zero starting from the time $t = 0$. Find the subsequent temperature distribution in the cylinder, assuming that all cross sections have the same temperature distribution.[9]

Ans.

$$T(r, t) = 2T_0 \sum_{n=1}^{\infty} \frac{J_0(\gamma_n r/a)}{\gamma_n J_1(\gamma_n)} e^{-\gamma_n^2 \tau/a^2},$$

---

[9] This corresponds to a long cylinder (theoretically, infinitely long).

where r is the distance from the axis of the cylinder, $J_0(x)$ and $J_1(x)$ are Bessel functions, the γ_n are consecutive roots of the equation $J_0(\gamma) = 0$, $\tau = kt/c\rho$ where k is the thermal conductivity, c the specific heat and ρ the density of the cylinder.

*148. Describe the equalization of a given axially symmetric initial temperature distribution $T(r, 0) = f(r)$ in an infinite cylinder of radius a, whose lateral surface does not transmit heat.

Ans.

$$T(r, t) = \frac{2}{a^2}\left[\int_0^a f(\rho)\rho \, d\rho + \sum_{n=1}^{\infty} \frac{J_0(\gamma_n r/a)}{J_0^2(\gamma_n)} e^{-\gamma_n^2 \tau/a^2} \int_0^a f(\rho)\rho J_0\left(\frac{\gamma_n \rho}{a}\right) d\rho\right],$$

where the γ_n are consecutive positive roots of the equation $J_1(\gamma) = 0$.

149. An infinite cylinder of radius a, initially heated to the temperature T_0, subsequently cools off by radiating heat into the surrounding medium according to Newton's law. Describe the cooling process.

Ans.

$$T(r, t) = 2T_0 \sum_{n=1}^{\infty} \frac{J_1(\gamma_n)J_0(\gamma_n r/a)}{J_0^2(\gamma_n) + J_1^2(\gamma_n)} \frac{e^{-\gamma_n^2 \tau/a^2}}{\gamma_n},$$

where the γ_n are consecutive positive roots of the equation

$$\gamma J_1(\gamma) = ah J_0(\gamma).$$

150. Starting from the time $t = 0$, Joule heat is produced with density Q in a cylindrical conductor of radius a. Find the temperature distribution over a cross section, assuming that both the initial temperature and the surface temperature equal zero.

Ans.

$$T(r, t) = \frac{Qa^2}{4k}\left[1 - \left(\frac{r}{a}\right)^2 - 8\sum_{n=1}^{\infty} \frac{J_0(\gamma_n r/a)}{\gamma_n^3 J_1(\gamma_n)} e^{-\gamma_n^2 \tau/a^2}\right],$$

where the γ_n are consecutive positive roots of the equation $J_0(\gamma) = 0$.

Hint. Make the problem homogeneous by subtracting out the particular solution corresponding to the stationary distribution of temperature in the cylinder.

151. A cylindrical conductor of radius a is heated for a long time by an electric current producing heat in the conductor with density Q. Study the process of cooling that ensues after the current is turned off, assuming that the cooling from the surface always obeys Newton's law and that the temperature of the surrounding medium equals zero.

Ans.

$$T(r, t) = \frac{2Qa^3 h}{k} \sum_{n=1}^{\infty} \frac{J_0(\gamma_n r/a)e^{-\gamma_n^2 \tau/a^2}}{\gamma_n^4 J_0(\gamma_n)[1 + (ah/\gamma_n)^2]},$$

where the γ_n are consecutive positive roots of the equation

$$\gamma J_1(\gamma) = ah J_0(\gamma),$$

and h is the heat exchange coefficient.

Hint. To determine the initial condition for the cooling problem, find the stationary distribution of temperature during the period of heating.

152. Find the temperature distribution in a cylindrical pipe $a \leqslant r \leqslant b$ if there is a constant heat current of density q through the inner surface $r = a$, while the outer surface $r = b$ is held at temperature zero. The initial temperature of the pipe is assumed to be zero.

Ans.

$$T(r, t) = \frac{qa}{k}\left\{\ln\frac{b}{r} - \frac{\pi b}{a}\sum_{n=1}^{\infty}\frac{J_0(\gamma_n)J_1(\gamma_n a/b)}{\gamma_n[J_0^2(\gamma_n) - J_1^2(\gamma_n a/b)]}R_{\gamma_n}(r)e^{-\gamma_n^2\tau/b^2}\right\},$$

where

$$R_\gamma(r) = Y_0(\gamma)J_0\left(\frac{\gamma r}{b}\right) - J_0(\gamma)Y_0\left(\frac{\gamma r}{b}\right),$$

where $J_0(x)$ and $Y_0(x)$ are Bessel functions, and the γ_n are consecutive positive roots of the equation $R'_\gamma(a) = 0$.

Hint. Subtract out the particular solution corresponding to the stationary distribution of temperature in the pipe.

*153. Find the general solution of the problem of the cooling of a sphere of radius a, given that the initial temperature distribution of the sphere is $T(r, 0) = f(r)$, while the surface temperature equals zero.

Ans.

$$T(r, t) = \frac{2}{ar}\sum_{n=1}^{\infty}e^{-n^2\pi^2\tau/a^2}\sin\frac{n\pi r}{a}\int_0^a f(\rho)\sin\frac{n\pi\rho}{a}\,\rho\,d\rho,$$

where k is the thermal conductivity, c the specific heat and ρ the density of the sphere, and $\tau = kt/c\rho$.

154. Find the temperature distribution in a sphere of radius a whose surface radiates heat starting from the time $t = 0$ according to Newton's law, if the initial temperature is T_0.

Ans.

$$T(r, t) = \frac{2T_0a}{r}\frac{ah}{1 - ah}\sum_{n=1}^{\infty}\frac{\cos\gamma_n}{\gamma_n}\frac{\sin(\gamma_n r/a)}{1 - (\sin 2\gamma_n/2\gamma_n)}e^{-\gamma_n^2\tau/a^2},$$

where h is the heat exchange coefficient, and the γ_n are consecutive positive roots of the equation

$$\tan\gamma = \frac{\gamma}{1 - ah}.$$

155. A spherical object of radius a is heated for a long time by a source producing heat with volume density Q. Study the process of cooling that ensues after the heating is stopped, assuming that the cooling is due to radiation from the surface and that the temperature of the air in the chamber where the heating occurred is T_0.

Ans.

$$T(r, t) = T_0 + \frac{2Qa^4h}{(1 - ah)kr} \sum_{n=1}^{\infty} \frac{\cos \gamma_n}{\gamma_n^3[1 - (\sin 2\gamma_n/2\gamma_n)]} e^{-\gamma_n^2 \tau/a^2} \sin \frac{\gamma_n r}{a},$$

where the γ_n are consecutive positive roots of the equation

$$\tan \gamma = \frac{\gamma}{1 - ah}.$$

Hint. To determine the initial temperature distribution in the sphere, solve the corresponding stationary problem.

156. The region between two parallel planes $x = 0$ and $x = a$ is occupied by a solution with a given initial concentration $C(x, 0) = f(x)$. Describe the subsequent equalization of concentration, assuming that the walls are impermeable. Examine the special case

$$f(x) = \begin{cases} 0, & 0 \leqslant x < c, \\ C_0, & c < x \leqslant a. \end{cases}$$

Ans. In the special case,

$$C(x, t) = C_0 \left[1 - \frac{c}{a} - \frac{2}{\pi} \sum_{n=1}^{\infty} \frac{\sin (n\pi c/a)}{n} e^{-n^2\pi^2 Dt/a^2} \cos \frac{n\pi x}{a} \right],$$

where D is the diffusion coefficient.

157. Find the concentration in a solution inside a cylindrical pipe $a \leqslant r \leqslant b$ with impermeable walls, if the initial concentration distribution is

$$C|_{t=0} = f(r) = \begin{cases} C_0, & a \leqslant r < c, \\ 0, & c < r \leqslant b. \end{cases}$$

Ans.

$$C = C_0 \left\{ \frac{c^2 - a^2}{b^2 - a^2} + \frac{\pi^2 c}{2} \sum_{n=1}^{\infty} \frac{R'_{\gamma_n}(c)R_{\gamma_n}(r)}{1 - [J_1^2(\gamma_n)/J_1^2(\gamma_n b/a)]} e^{-\gamma_n^2 Dt/a^2} \right\},$$

where

$$R_\gamma(r) = J_1(\gamma)Y_0\left(\frac{\gamma r}{a}\right) - Y_1(\gamma)J_0\left(\frac{\gamma r}{a}\right),$$

where $J_n(x)$ and $Y_n(x)$ are Bessel functions, and the γ_n are consecutive positive roots of the equation $R'_\gamma(b) = 0$.

158. Find the concentration of a gas inside a cylindrical metal object of radius a, assuming that the initial concentration of the gas is $C(t, 0) = f(r)$ and that the object is surrounded by a medium in which the gas is maintained at constant concentration C_1. Consider the special case $f(r) = C_0$.

Ans.

$$C(r, t) = \frac{\alpha_1}{\alpha} C_1 + \frac{2\alpha^2}{D^2} \sum_{n=1}^{\infty} A_n \frac{e^{-\gamma_n^2 Dt/a^2} J_0(\gamma_n r/a)}{J_1^2(\gamma_n)[\gamma_n^2 + (\alpha^2 a^2/D^2)]},$$

$$A_n = \int_0^a \rho f(\rho) J_0\left(\frac{\gamma_n \rho}{a}\right) d\rho - \frac{\alpha_1}{\alpha} \frac{a^2 C_1 J_1(\gamma_n)}{\gamma_n},$$

where the γ_n are consecutive positive roots of the equation

$$\gamma J_1(\gamma) = \frac{\alpha a}{D} J_0(\gamma).$$

In the special case,

$$C(r, t) = \frac{\alpha_1}{\alpha} C_1 + \frac{2\alpha^2 a^2 [C_0 - (\alpha_1/\alpha) C_1]}{D^2} \sum_{n=1}^{\infty} \frac{e^{-\gamma_n^2 Dt/a^2} J_0(\gamma_n r/a)}{\gamma_n J_1(\gamma_n)[\gamma_n^2 + (\alpha^2 a^2/D^2)]}.$$

Hint. The problem reduces to solving the differential equation

$$\Delta C = \frac{1}{D} \frac{\partial C}{\partial t},$$

with initial condition

$$C\big|_{t=0} = f(r)$$

and boundary condition

$$-D \frac{\partial C}{\partial r}\bigg|_{r=a} = \alpha C\big|_{r=a} - \alpha_1 C_1,$$

where α and α_1 are the coefficients characterizing the emission and re-absorption of the gas by the surface of the metal (see G3).

4. Heat Conduction: Stationary Problems

159. Find the stationary temperature distribution $T(x, y)$ in an infinite bar of rectangular cross section (see Figure 25) if three faces are held at temperature zero, while a given temperature distribution $T(x, b) = f(x)$ is maintained on the fourth side. Apply the resulting general formulas to the special case $f(x) = T_0$.

Ans.

$$T(x, y) = \frac{2}{a} \sum_{n=1}^{\infty} \frac{\sinh (n\pi y/a)}{\sinh (n\pi b/a)} \sin \frac{n\pi x}{a}$$

$$\times \int_0^a f(\xi) \sin \frac{n\pi \xi}{a} d\xi.$$

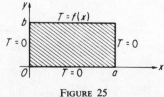

FIGURE 25

In the special case,

$$T(x, y) = \frac{4T_0}{\pi} \sum_{n=0}^{\infty} \frac{\sinh \left[(2n + 1)\pi y/a\right]}{\sinh \left[(2n + 1)\pi b/a\right]} \frac{\sin \left[(2n + 1)\pi x/a\right]}{2n + 1}.$$

160. Find the distribution of temperature in a bar of rectangular cross section if the two opposite faces $y = 0$ and $y = b$ are held at temperatures zero and T_0, respectively, while the other two faces $x = \pm a$ radiate heat into the surrounding medium (assumed to have temperature zero) according to Newton's law.

Ans.

$$T(x, y) = 2T_0 \sum_{n=1}^{\infty} \frac{\sin \gamma_n}{\gamma_n + \sin \gamma_n \cos \gamma_n} \frac{\sinh (\gamma_n y/a)}{\sinh (\gamma_n b/a)} \cos \frac{\gamma_n x}{a},$$

where the γ_n are consecutive positive roots of the equation

$$\tan \gamma = \frac{ah}{\gamma}$$

and h is the heat exchange coefficient.

161. Find the stationary temperature distribution in a conductor of rectangular cross section $-a \leqslant x \leqslant a$, $-b \leqslant y \leqslant b$, heated by an electric current producing Joule heat Q per unit volume, if the faces of the conductor are held at temperature zero.

Ans.

$$T(x, y) = \frac{Qa^2}{2k} \left[1 - \left(\frac{x}{a_1}\right)^2 + \frac{32}{\pi^3} \sum_{n=1}^{\infty} \frac{(-1)^{n+1}}{(2n + 1)^3} \right.$$

$$\left. \times \frac{\cosh \left[(2n + 1)\pi y/2a\right]}{\cosh \left[(2n + 1)\pi b/2a\right]} \cos \frac{(2n + 1)\pi x}{2a} \right],$$

where k is the thermal conductivity of the conductor.

Hint. Subtract out the particular solution of the inhomogeneous heat conduction equation which depends only on the coordinate x and satisfies the boundary conditions for $x = \pm a$.

162. Solve the preceding problem, assuming that all faces of the conductor radiate heat into the surrounding medium (assumed to have temperature zero) according to Newton's law.

Ans.

$$T(x, y) = \frac{Qa^2}{k}\left\{\frac{1}{2}\left(1 - \frac{x^2}{a^2}\right) + \frac{1}{ah}\right.$$

$$\left. - 2ah \sum_{n=1}^{\infty} \frac{\sin \gamma_n \cos (\gamma_n x/a) \; \cosh (\gamma_n y/a)}{\gamma_n^3 [1 + (\sin 2\gamma_n/2\gamma_n)][\gamma_n \sinh (\gamma_n b/a) + ah \cosh (\gamma_n b/a)]}\right\},$$

where the γ_n are consecutive positive roots of the equation

$$\tan \gamma = \frac{ah}{\gamma}.$$

163. A bar of rectangular cross section $0 \leqslant x \leqslant a, 0 \leqslant y \leqslant b$ is heated by a constant thermal current of density q incident on one face $y = b$ of the bar. Find the stationary temperature distribution over a cross section of the bar, assuming that heat is lost by radiation into the surrounding medium according to Newton's law.

Ans.

$$T(x, y) = \frac{q}{kh} \sum_{n=1}^{\infty} \frac{\gamma_n \sin \gamma_n + ah(1 - \cos \gamma_n)}{2ah + \gamma_n^2 + (ah)^2}$$

$$\times \frac{\cosh \dfrac{\gamma_n y}{a} + \dfrac{ah}{\gamma_n} \sinh \dfrac{\gamma_n y}{a}}{\cosh \dfrac{\gamma_n b}{a} + \dfrac{\gamma_n^2 + (ah)^2}{2ah\gamma_n} \sinh \dfrac{\gamma_n b}{a}} \left(\cos \dfrac{\gamma_n x}{a} + \dfrac{ah}{\gamma_n} \sin \dfrac{\gamma_n x}{a}\right),$$

where the γ_n are consecutive positive roots of the equation

$$\tan \gamma = \frac{2ah\gamma}{\gamma^2 - (ah)^2}.$$

164. A rectangular bar consists of two sections with different thermal conductivities k_1 and k_2, respectively (see Figure 26). Find the temperature distribution in the bar, assuming that two opposite faces $y = \pm b$ are at temperature T_0, while the other two sides are at temperature zero.

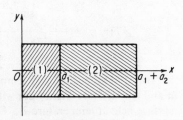

FIGURE 26

Ans.

$$T(x, y) = 2T_0 \sum_{n=1}^{\infty} \frac{\tan \gamma_n \left(\cos \gamma_n - \cos \dfrac{\gamma_n a_2}{a_1}\right)}{\gamma_n \left[\dfrac{k_1}{k_2} \sin^2 \dfrac{\gamma_n a_2}{a_1} + \dfrac{a_2}{a_1} \sin^2 \gamma_n\right]} \frac{\cosh \dfrac{\gamma_n y}{a_1}}{\cosh \dfrac{\gamma_n b}{a_1}} T_n(x),$$

$$T_n(x) = \begin{cases} \sin \dfrac{\gamma_n x}{a_1} \sin \dfrac{\gamma_n a_2}{a_1}, & 0 \leqslant x \leqslant a_1, \\[2ex] \sin \dfrac{\gamma_n(a_1 + a_2 - x)}{a_1} \sin \gamma_n, & a_1 \leqslant x \leqslant a_1 + a_2, \end{cases}$$

where the γ_n are consecutive positive roots of the equation

$$\tan \gamma + \frac{k_1}{k_2} \tan \frac{\gamma a_2}{a_1} = 0.$$

165. Determine the stationary temperature distribution in a bar whose cross section is a "curvilinear rectangle," with two faces consisting of arcs of concentric circles and the other two faces of segments of radii of the larger circle (see Figure 27). It is assumed that one of the curved faces $r = b$ has temperature T_0, while the other faces are held at temperature zero.

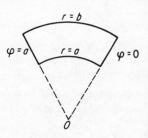

FIGURE 27

Ans.

$$T(r, \varphi) = \frac{4T_0}{\pi} \sum_{n=0}^{\infty} \frac{\left(\dfrac{r}{a}\right)^{(2n+1)\pi/\alpha} - \left(\dfrac{a}{r}\right)^{(2n+1)\pi/\alpha}}{\left(\dfrac{b}{a}\right)^{(2n+1)\pi/\alpha} - \left(\dfrac{a}{b}\right)^{(2n+1)\pi/\alpha}} \frac{\sin \dfrac{(2n+1)\pi\varphi}{\alpha}}{2n+1}.$$

166. Solve the preceding problem, assuming that one of the plane faces $\varphi = \alpha$ is held at temperature T_0, while the other faces are held at temperature zero.

Ans.

$$T(r, \varphi) = \frac{4T_0}{\pi} \sum_{n=0}^{\infty} \frac{\sin \dfrac{(2n+1)\pi \ln(r/a)}{\ln(b/a)}}{2n+1} \frac{\sinh \dfrac{(2n+1)\pi\varphi}{\ln(b/a)}}{\sinh \dfrac{(2n+1)\pi\alpha}{\ln(b/a)}}.$$

167. Find the stationary temperature distribution in a cylinder of radius a and length l (see Figure 28) with ends held at temperature zero and lateral surface held at temperature T_0. Calculate the temperature distribution along the axis of the cylinder, assuming that the ratio a/l equals 0.5, 1, 2.

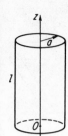

Ans.

$$T(r, z) = \frac{4T_0}{\pi} \sum_{n=0}^{\infty} \frac{I_0[(2n + 1)\pi r/l]}{I_0[(2n + 1)\pi a/l]} \frac{\sin [(2n + 1)\pi z/l]}{2n + 1},$$

where $I_0(x)$ is the Bessel function of imaginary argument. The results of numerical calculations of the quantity

$$\frac{T|_{r=0}}{T_0}$$

FIGURE 28   are given in the following table:

$\dfrac{a}{l}$ \ $\dfrac{z}{l}$	0.1	0.2	0.3	0.4	0.5
0.5	0.246	0.458	0.611	0.698	0.726
1	0.072	0.134	0.188	0.221	0.232
2	0.005	0.009	0.012	0.014	0.015

168. Solve the preceding problem, assuming that the ends of the cylinder do not transmit heat, while a given temperature distribution

$$T|_{r=a} = f(z)$$

is maintained along the lateral surface of the cylinder.

Ans.

$$T(r, z) = \frac{1}{l} \int_0^l f(\zeta) \, d\zeta + \frac{2}{l} \sum_{n=1}^{\infty} \frac{I_0(n\pi r/l)}{I_0(n\pi a/l)} \cos \frac{n\pi z}{l} \int_0^l f(\zeta) \cos \frac{n\pi\zeta}{l} \, d\zeta.$$

*169.** Solve Prob. 167, assuming that the ends of the cylinder cool off according to Newton's law and choosing the origin at the center of the cylinder.

Ans.

$$T(r, z) = 2T_0 \sum_{n=1}^{\infty} \frac{\sin \gamma_n}{\gamma_n + \sin \gamma_n \cos \gamma_n} \frac{I_0(2\gamma_n r/l)}{I_0(2\gamma_n a/l)} \cos \frac{2\gamma_n z}{l},$$

where the γ_n are consecutive positive roots of the equation

$$\tan \gamma = \frac{hl}{2\gamma}$$

and h is the heat exchange coefficient.

170. The walls of a cylindrical hole drilled in an infinite slab of thickness h (see Figure 29) are held at a given temperature T_0. Find the stationary temperature distribution in the slab, if its plane faces have temperature zero.

Ans.

$$T(r, z) = \frac{4T_0}{\pi} \sum_{n=0}^{\infty} \frac{\sin \left[(2n + 1)\pi z/h \right]}{2n + 1}$$
$$\times \frac{K_0[(2n + 1)\pi r/h]}{K_0[(2n + 1)\pi a/h]},$$

where $K_0(x)$ is Macdonald's function.

171. Find the stationary temperature distribution in a cylinder $0 \leqslant r \leqslant a$, $0 \leqslant z \leqslant l$ if the upper end is at temperature T_0 while the rest of the surface is at temperature zero (cf. Prob. 167).

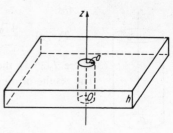

FIGURE 29

Ans.

$$T(r, z) = 2T_0 \sum_{n=1}^{\infty} \frac{J_0(\gamma_n r/a)}{\gamma_n J_1(\gamma_n)} \frac{\sinh (\gamma_n z/a)}{\sinh (\gamma_n l/a)},$$

where the γ_n are consecutive positive roots of the equation $J_0(\gamma) = 0$.

172. Heat is produced with constant density Q in a cylinder of radius a, length l and thermal conductivity k. Find the stationary temperature distribution if heat leaves the cylinder through the part of the upper end bounded by the circle $r = b < a$, but not through the rest of the surface of the cylinder. It is assumed that the flow of heat out of the cylinder is uniformly distributed over the disk $r \leqslant b$.

Ans.

$$T(r, z) = -\frac{Qa^2}{k} \left[\frac{z^2}{2a^2} + \frac{2l}{b} \sum_{n=1}^{\infty} \frac{J_1(\gamma_n b/a)J_0(\gamma_n r/a)}{\gamma_n^2 J_0^2(\gamma_n)} \frac{\cosh (\gamma_n z/a)}{\sinh (\gamma_n l/a)} \right] + \text{const},$$

where the γ_n are consecutive positive roots of the equation $J_1(\gamma) = 0$.

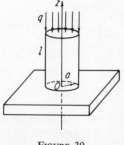

FIGURE 30

Hint. Subtract out a particular solution of the inhomogeneous heat conduction problem which depends only on the coordinate z.

173. A cylinder standing on a thermally insulating slab is heated from above by a uniformly distributed thermal current (see Figure 30), and radiates heat from its lateral surface into the surrounding medium (assumed to be at temperature zero) according to Newton's law. Find the stationary distribution of heat in the cylinder.

Ans.

$$T(r, z) = \frac{2qa^2h}{k} \sum_{n=1}^{\infty} \frac{J_0(\gamma_n r/a)}{\gamma_n^3 J_0(\gamma_n)[1 + (ah/\gamma_n)^2]} \frac{\cosh(\gamma_n z/a)}{\sinh(\gamma_n l/a)},$$

where the γ_n are consecutive positive roots of the equation

$$\gamma J_1(\gamma) = ah J_0(\gamma),$$

h is the heat exchange coefficient and q is the density of the incident heat current.

174. A semi-infinite cylindrical pipe $a \leqslant r \leqslant b$, $0 \leqslant z < \infty$ is heated at the end $z = 0$ held at temperature T_0, and cooled at its lateral surfaces $r = a$ and $r = b$ held at temperature zero. Find the stationary temperature distribution in the pipe.

Ans.

$$T(r, z) = \pi T_0 \sum_{n=1}^{\infty} \frac{Z_0\left(\dfrac{\gamma_n r}{a}\right) e^{-\gamma_n z/a}}{1 + \dfrac{J_0(\gamma_n)}{J_0(\gamma_n b/a)}} \qquad (a \leqslant r \leqslant b, 0 \leqslant z < \infty),$$

where

$$Z_0\left(\frac{\gamma r}{a}\right) = Y_0(\gamma) J_0\left(\frac{\gamma r}{a}\right) - J_0(\gamma) Y_0\left(\frac{\gamma r}{a}\right)$$

is a linear combination of Bessel functions, and the γ_n are consecutive positive roots of the equation

$$Z_0\left(\frac{\gamma b}{a}\right) = 0.$$

*175. An inhomogeneous cylinder formed of two sections with different thermal conductivities k_1 and k_2 (see Figure 31) is heated at its lateral surface held at temperature T_0 and cooled at its ends held at temperature zero. Find the stationary temperature distribution $T(r, z)$ in the cylinder.

Ans.

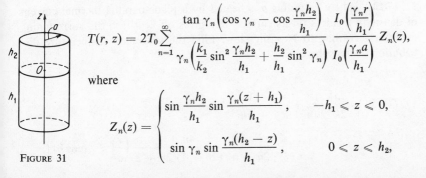

$$T(r, z) = 2T_0 \sum_{n=1}^{\infty} \frac{\tan \gamma_n \left(\cos \gamma_n - \cos \dfrac{\gamma_n h_2}{h_1}\right)}{\gamma_n \left(\dfrac{k_1}{k_2} \sin^2 \dfrac{\gamma_n h_2}{h_1} + \dfrac{h_2}{h_1} \sin^2 \gamma_n\right)} \frac{I_0\left(\dfrac{\gamma_n r}{h_1}\right)}{I_0\left(\dfrac{\gamma_n a}{h_1}\right)} Z_n(z),$$

where

$$Z_n(z) = \begin{cases} \sin \dfrac{\gamma_n h_2}{h_1} \sin \dfrac{\gamma_n(z + h_1)}{h_1}, & -h_1 \leqslant z \leqslant 0, \\[2ex] \sin \gamma_n \sin \dfrac{\gamma_n(h_2 - z)}{h_1}, & 0 \leqslant z \leqslant h_2, \end{cases}$$

FIGURE 31

and the γ_n are consecutive positive roots of the equation

$$\tan \gamma + \frac{k_1}{k_2} \tan \frac{\gamma h_2}{h_1} = 0.$$

*176. Find the stationary temperature distribution in a sphere of radius a, if one part of its surface S_1 is held at constant temperature T_0, while the remaining part S_2 is held at temperature zero (see Figure 32).

Ans.

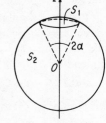

$$T(r, \theta) = \frac{T_0}{2} \Big\{ 1 - \cos \alpha - \sum_{n=1}^{\infty} [P_{n+1}(\cos \alpha)$$
$$- P_{n-1}(\cos \alpha)] \Big(\frac{r}{a} \Big)^n P_n(\cos \theta) \Big\},$$

in terms of the Legendre polynomials $P_n(x)$.

FIGURE 32

177. Solve the preceding problem, assuming that heat is produced in the sphere with volume density Q, and that heat leaves the sphere through the surface S_1 flowing in the normal direction with constant density (the surface S_2 does not transmit heat).

Ans.

$$T(r, \theta) = \frac{Qa^2}{3k} \Big[-\frac{r^2}{2a^2} + \frac{1}{1 - \cos \alpha}$$
$$\times \sum_{n=1}^{\infty} \frac{P_{n+1}(\cos \alpha) - P_{n-1}(\cos \alpha)}{n} \Big(\frac{r}{a} \Big)^n P_n(\cos \theta) \Big] + \text{const},$$

where k is the thermal conductivity of the sphere.

Hint. Subtract out a particular solution of the inhomogeneous heat conduction equation which depends only on the variable r.

178. A sphere of radius a is heated by a plane-parallel thermal current of density q incident on its surface, and gives off heat into the surrounding medium according to Newton's law. Find the stationary temperature distribution in the sphere.

Ans.

$$T(r, \theta) = \frac{qa}{2k} \Big[\frac{1}{2ah} + \frac{r}{a} \frac{\cos \theta}{1 + ah}$$
$$- \sum_{n=1}^{\infty} \frac{P_{2n}(0)}{2n + ah} \frac{4n + 1}{(2n - 1)(2n + 2)} \Big(\frac{r}{a} \Big)^{2n} P_{2n}(\cos \theta) \Big],$$

in terms of the Legendre polynomials $P_n(x)$. Note that

$$P_0(0) = 1, \quad P_{2n}(0) = (-1)^n \frac{1 \cdot 3 \cdot 5 \cdots (2n-3)(2n-1)}{2 \cdot 4 \cdot 6 \cdots (2n-2)2n}, \quad n = 1, 2, \ldots$$

Hint. Here the boundary condition takes the form

$$\left(\frac{\partial T}{\partial r} + hT\right)\bigg|_{r=a} = \begin{cases} \dfrac{q}{k} \cos\theta, & 0 \leqslant \theta \leqslant \dfrac{\pi}{2}, \\[4mm] 0, & \dfrac{\pi}{2} \leqslant \theta \leqslant \pi. \end{cases}$$

5. Electricity and Magnetism

179. Find the electrostatic potential $u(x, y)$ inside an elongated box of rectangular cross section (see Figure 33), if two opposite sides are at potential V and the other two sides are grounded.

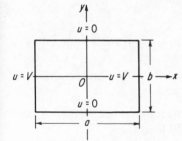

FIGURE 33

Ans.

$$u(x, y) = \frac{4V}{\pi} \sum_{n=0}^{\infty} (-1)^n \frac{\cosh\left[(2n+1)\pi x/b\right]}{\cosh\left[(2n+1)\pi a/b\right]}$$
$$\times \frac{\cos\left[(2n+1)\pi y/b\right]}{2n+1}.$$

180. Find the electrostatic potential $u(x, y)$ inside a semi-infinite rectangular box (see Figure 34), if the vertical wall is held at potential V and the horizontal walls are held at potential zero.

Ans.

$$u(x, y) = \frac{2V}{\pi} \arctan \frac{\sin(\pi y/b)}{\sinh(\pi x/b)}.$$

Hint. To represent the solution in closed form, use the expansion

$$\sum_{n=0}^{\infty} \frac{e^{-(2n+1)x}}{2n+1} \sin(2n+1)y = \frac{1}{2} \arctan \frac{\sin y}{\sinh x},$$
$$x > 0.$$

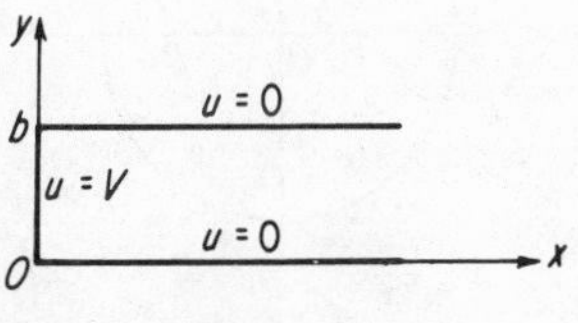

181. Find the electrostatic potential $u(x, y)$ between two infinite parallel sheets if one sheet $y = 0$ is at potential zero, while a given

FIGURE 34

periodic potential

$$u\big|_{y=b} = f(x)$$

is maintained on the other sheet (where f is a function with a given period $2a$).

Ans.

$$u(x, y) = \frac{1}{a} \sum_{n=1}^{\infty} \frac{\sinh(n\pi y/a)}{\sinh(n\pi b/a)} \left[\cos \frac{n\pi x}{a} \int_0^{2a} f(\xi) \cos \frac{n\pi \xi}{a} \, d\xi \right.$$

$$\left. + \sin \frac{n\pi x}{a} \int_0^{2a} f(\xi) \sin \frac{n\pi \xi}{a} \, d\xi \right] + \frac{y}{2ab} \int_0^{2a} f(\xi) \, d\xi.$$

182. A thin charged wire with linear charge density q is placed inside and parallel to a conducting cylinder of radius a held at potential zero. Use the familiar method of images to solve the corresponding electrostatic problem, assuming that the wire is a distance b from the axis of the cylinder.

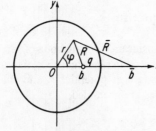

Ans.

$$u(r, \varphi) = -2q \ln \frac{R}{\bar{R}} + 2q \ln \frac{b}{a},$$

where R and $\bar{R}$ are the distances shown in Figure 35, and $a^2 = b\bar{b}$.

FIGURE 35

183. Solve the preceding problem if the wire is placed outside the cylinder, and if the cylinder has total charge Q per unit length.

Ans.

$$u(r, \varphi) = -2q \ln \frac{R}{\bar{R}} - 2(q + Q) \ln r + \text{const},$$

where R and $\bar{R}$ are the distances shown in Figure 36, and $a^2 = b\bar{b}$.

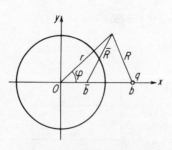

FIGURE 36

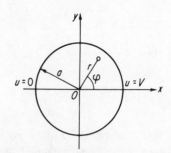

FIGURE 37

184. Find the electrostatic potential $u(r, \varphi)$ in the space between two conducting infinite half-cylinders, one of which is held at potential V and the other at potential zero (see Figure 37). It is assumed that the half-cylinders are separated by thin layers of insulating material along the lines where they meet.

Ans.

$$u(r, \varphi) = \frac{V}{2}\left[1 + \frac{2}{\pi}\arctan\frac{2ar\cos\varphi}{a^2 - r^2}\right].$$

Hint. To solve the problem in closed form, use the expansion

$$\sum_{n=0}^{\infty}\frac{(-1)^n}{2n+1}\left(\frac{r}{a}\right)^{2n+1}\cos(2n+1)\varphi = \frac{1}{2}\arctan\frac{2ar\cos\varphi}{a^2 - r^2}, \qquad r < a.$$

185. A cylinder of radius a made from material with dielectric constant ε is introduced into a plane-parallel electric field with components $E_x = -E_0$, $E_y = E_z = 0$. Find the resulting potential distribution, and show that the field inside the cylinder is homogeneous.

Ans. The potential distribution is

$$u = E_2 x\left[1 - \frac{\varepsilon - 1}{\varepsilon + 1}\left(\frac{a}{r}\right)^2\right] + \text{const} \quad \text{outside the cylinder,}$$

$$u = \frac{2}{\varepsilon + 1}E_0 x + \text{const} \qquad\qquad \text{inside the cylinder.}$$

The field inside the cylinder is

$$E = -\frac{2E_0}{\varepsilon + 1}.$$

186. Find the electrostatic potential $u(r, z)$ inside a closed cylindrical surface of length l and radius a, if the base and lateral surface are held at potential V, while the upper surface is held at potential zero.

Ans.

$$u(r, z) = V\left[1 - 2\sum_{n=1}^{\infty}\frac{J_0(\gamma_n r/a)}{\gamma_n J_1(\gamma_n)}\frac{\sinh(\gamma_n z/a)}{\sinh(\gamma_n l/a)}\right],$$

where the γ_n are consecutive positive roots of the equation $J_0(\gamma) = 0$.

Hint.

$$\int_0^a rJ_0\left(\frac{\gamma_n r}{a}\right)dr = \frac{a^2 J_1(\gamma_n)}{\gamma_n}.$$

187. Two metallic hemispheres of radius a, separated by a thin insulating washer, are held at potential V and zero, respectively, corresponding to the boundary condition

$$u\big|_{r=a} = \begin{cases} V, & 0 \leqslant \theta < \dfrac{\pi}{2} \\ 0, & \dfrac{\pi}{2} < \theta \leqslant \pi. \end{cases}$$

Find the electrostatic potential $u(r, \theta)$ in the space between the hemispheres.

Ans.

$$u(r, \theta) = \frac{V}{2}\left[1 + \sum_{n=0}^{\infty} \frac{4n+3}{2n+2} P_{2n}(0)\left(\frac{r}{a}\right)^{2n+1} P_{2n+1}(\cos \theta)\right],$$

in terms of the Legendre polynomials $P_n(x)$, where

$$P_0(0) = 1, \quad P_{2n}(0) = (-1)^n \frac{1 \cdot 3 \cdot 5 \cdots (2n-1)}{2 \cdot 4 \cdot 6 \cdots 2n}, \quad n = 1, 2, \ldots$$

188. Find the electrostatic field of a point charge q placed at distance b from the center of a conducting sphere of radius a $(a < b)$ held at potential zero.[10]

Ans. The electrostatic potential is

$$u(r, \theta) = \frac{q}{R} + \frac{\bar{q}}{\bar{R}},$$

where

$$R = \sqrt{b^2 + r^2 - 2br \cos \theta}, \qquad \bar{R} = \sqrt{\bar{b}^2 + r^2 - 2\bar{b}r \cos \theta},$$

$$b\bar{b} = a^2, \qquad \bar{q} = -\frac{qa}{b}.$$

189. Solve the preceding problem, assuming that the sphere is made from material of dielectric constant ε.

Ans. The potential is

$$u(r, \theta) = \frac{q}{b} \sum_{n=0}^{\infty} \frac{2n+1}{(\varepsilon+1)n+1}\left(\frac{r}{b}\right)^n P_n(\cos \theta)$$

inside the sphere and

$$u(r, \theta) = \frac{q}{R} - \frac{qa}{br}(\varepsilon-1) \sum_{n=0}^{\infty} \frac{n}{(\varepsilon+1)n+1}\left(\frac{a^2}{br}\right)^n P_n(\cos \theta)$$

outside the sphere, in terms of the Legendre polynomials $P_n(x)$.

---

[10] This problem can either be solved by the method of images or by the method of inversion (starting from the familiar solution of the problem of a point charge placed over a conducting plane).

*190. Find the distribution of d-c current in a thin rectangular sheet, if the current is applied by electrodes at the points $x = -a$, $y = 0$ and $x = a$, $y = 0$ (see Figure 38).

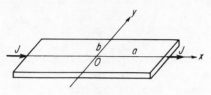

FIGURE 38

Ans. The potential of the current distribution in the sheet is

$$u(x, y) = -\frac{J}{2\sigma h}\left[\frac{x}{b} + \frac{2}{\pi}\sum_{n=1}^{\infty}\frac{\sinh (n\pi x/b)}{n \cosh (n\pi a/b)}\cos\frac{n\pi y}{b}\right] + \text{const},$$

where σ is the conductivity and h the thickness of the sheet, and J is the total current flowing through the sheet.

Hint. The differential equation for the potential of the current distribution in a thin conducting shell is given in Prob. 21.

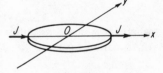

FIGURE 39

191. Find the distribution of d-c current in a thin disk of radius a, if the current is applied by electrodes at the points $r = a$, $\varphi = 0$ and $r = a$, $\varphi = \pi$ (see Figure 39).

Ans.

$$u(r, \varphi) = \frac{J}{2\pi\sigma h}\ln\frac{1 - \dfrac{2r}{a}\cos\varphi + \dfrac{r^2}{a^2}}{1 + \dfrac{2r}{a}\cos\varphi + \dfrac{r^2}{a^2}} + \text{const}.$$

*192. Find the distribution of d-c current in a cylindrical shell of radius a, height $2l$ and thickness h, if the current is applied by electrodes at the points $r = a$, $\varphi = 0$, $z = \pm l$.

Ans.

$$u(\varphi, z) = \frac{J}{2\pi\sigma h}\left[\frac{z}{a} + 2\sum_{n=1}^{\infty}\frac{\sinh (nz/a)}{\cosh (nl/a)}\frac{\cos n\varphi}{n}\right] + \text{const}.$$

193. Find the distribution of d-c current in a hemispherical cap of radius a, if the current is applied by electrodes at the points $r = a$, $\theta = \pi/2$, $\varphi = 0$ and $r = a$, $\theta = \pi/2$, $\varphi = \pi$ (see Figure 40).

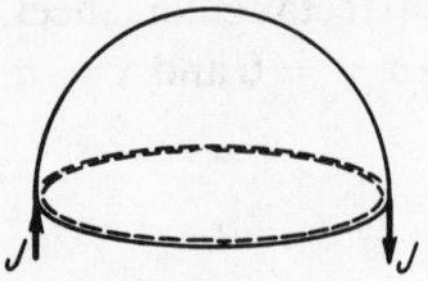

Ans.

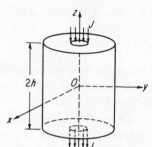

$$u(\theta, \varphi) = \frac{J}{2\pi\sigma h} \ln \frac{1 + 2\tan\frac{\theta}{2}\cos\varphi + \tan^2\frac{\theta}{2}}{1 - 2\tan\frac{\theta}{2}\cos\varphi + \tan^2\frac{\theta}{2}} + \text{const.}$$

FIGURE 40

194. A d-c current J enters one end of a cylindrical conductor of radius a made from material of conductivity σ and leaves the other end, via electrodes in the shape of disks of radius $r < a$ (see Figure 41). Find the current distribution inside the conductor, assuming that the current is uniformly distributed over the electrodes.

Ans.

$$u(r, z) = \frac{Jz}{\pi a^2 \sigma} + \frac{2J}{\pi b\sigma} \sum_{n=1}^{\infty} \frac{\sinh(\gamma_n z/a)}{\cosh(\gamma_n h/a)}$$
$$\times \frac{J_1(\gamma_n b/a)J_0(\gamma_n r/a)}{\gamma_n^2 J_0^2(\gamma_n)} + \text{const,}$$

where the γ_n are consecutive positive roots of the equation $J_1(\gamma) = 0$.

Hint.

$$\int_0^b rJ_0\left(\frac{\gamma_n r}{a}\right) dr = \frac{ab}{\gamma_n} J_1\left(\frac{\gamma_n b}{a}\right).$$

FIGURE 41

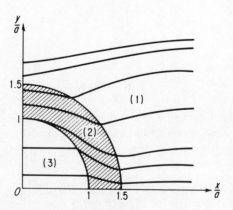

FIGURE 42

195. Find the current distribution in a homogeneous conductor in the form of a rectangular parallelepiped $-a \leqslant x \leqslant a$, $-b \leqslant y \leqslant b$, $-c \leqslant z \leqslant c$, assuming that current enters and leaves via rectangular electrodes of dimensions $2\delta \times 2\varepsilon$ applied at the boundaries $z = \pm c$. The current distribution is assumed to be uniform over the area of the electrodes.

Ans.

$$u(x, y, z) = \frac{Jz}{4\sigma ab} + \frac{Ja}{2\pi^2\sigma\delta\varepsilon} \sum_{m=1}^{\infty} \frac{\sin\dfrac{m\pi\delta}{a}}{m^2} \frac{\sinh\dfrac{m\pi z}{a}}{\cosh\dfrac{m\pi c}{a}} \cos\frac{m\pi x}{a}$$

$$+ \frac{Jb}{2\pi^2\sigma\varepsilon a} \sum_{n=1}^{\infty} \frac{\sin\dfrac{n\pi\varepsilon}{b}}{n^2} \frac{\sinh\dfrac{n\pi z}{b}}{\cosh\dfrac{n\pi c}{b}} \cos\frac{n\pi y}{b}$$

$$+ \frac{J}{\pi^3\sigma\delta\varepsilon} \sum_{m=1}^{\infty} \sum_{n=1}^{\infty} \frac{\sin\dfrac{m\pi\delta}{a} \sin\dfrac{n\pi\varepsilon}{b}}{mn\sqrt{\dfrac{m^2}{a^2} + \dfrac{n^2}{b^2}}}$$

$$\times \frac{\sinh\sqrt{\dfrac{m^2}{a^2} + \dfrac{n^2}{b^2}}\,\pi z}{\cosh\sqrt{\dfrac{m^2}{a^2} + \dfrac{n^2}{b^2}}\,\pi c} \cos\frac{m\pi x}{a} \cos\frac{n\pi y}{b} + \text{const},$$

where J is the current and σ the conductivity.

*196. A cylindrical pipe $a \leqslant r \leqslant b$ made from material of magnetic permeability μ is placed in a homogeneous magnetic field H_0. Find the resulting distribution of magnetic potential. Plot the lines of force for the values $\mu = 5$ and $b/a = 1.5$.

Ans.

$$u_1 = H_0 x \left[1 - \frac{(\mu^2 - 1)(b^2 - a^2)}{b^2(\mu + 1)^2 - a^2(\mu - 1)^2} \left(\frac{b}{r}\right)^2 \right] + \text{const}, \qquad b \leqslant r < \infty,$$

$$u_2 = H_0 x \frac{2b^2[(\mu + 1) + (\mu - 1)(a/r)^2]}{b^2(\mu + 1)^2 - a^2(\mu - 1)^2} + \text{const}, \qquad a \leqslant r \leqslant b,$$

$$u_3 = H_0 x \frac{4\mu b^2}{b^2(\mu + 1)^2 - a^2(\mu - 1)^2} + \text{const}, \qquad 0 \leqslant r \leqslant a.$$

The lines of force are plotted in Figure 42.

197. Find the magnetic field due to a current J flowing in a wire placed inside a cylindrical hole of radius a drilled in iron of magnetic permeability μ, if the wire is at distance b from the axis of the hole. Plot the lines of force for the values $\mu = 3$, $b/a = 0.5$.

Ans.

$$A_1 = -\frac{2J}{c} \ln R - \frac{2J}{c} \frac{\mu - 1}{\mu + 1} \ln \bar{R} + \text{const}, \qquad 0 \leqslant r < a,$$

$$A_2 = -\frac{2J\mu}{c} \frac{\mu - 1}{\mu + 1} \ln r - \frac{2J\mu}{c} \frac{2}{\mu + 1} \ln R + \text{const}, \qquad a < r < \infty,$$

where A_1 and A_2 are the values of the z-component of the vector potential of the magnetic field in the air and in the iron, and

$$R = \sqrt{r^2 + b^2 - 2br \cos \varphi}, \quad \bar{R} = \sqrt{r^2 + \bar{b}^2 - 2\bar{b}r \cos \varphi}, \qquad b\bar{b} = a^2.$$

The lines of force are shown in Figure 43.

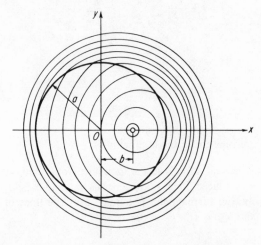

FIGURE 43

198. Solve the preceding problem for the limiting case $\mu = \infty$. Find the equation of the lines of force in the air and in the iron.

Ans.

$$A_1 = -\frac{2J}{c} \ln R\bar{R} + \text{const},$$

$$\frac{1}{\mu} A_2 = -\frac{2J}{c} \ln r + \text{const}.$$

The lines of force are ovals of Cassini

$$R\bar{R} = \text{const}$$

in the air, and circles

$$r = \text{const}$$

in the iron.

199. A sphere of radius a made from material of magnetic permeability μ is introduced into a homogeneous magnetic field with components $H_x = H_y = 0$, $H_z = -H_0$. Show that the field inside the sphere is homogeneous, and find its value.

Ans.

$$H = -\frac{3H_0}{\mu + 2}.$$

200. A hollow sphere $a \leqslant r \leqslant b$ of magnetic permeability μ is placed in a homogeneous magnetic field $H_x = H_y = 0$, $H_z = -H_0$. Solve the corresponding problem of magnetostatics.

Ans.

$$u = \begin{cases} \dfrac{9b^3\mu H_0 z}{b^3(\mu + 2)(2\mu + 1) - 2a^3(\mu - 1)^2} + \text{const}, & 0 \leqslant r \leqslant a, \\[3mm] \dfrac{3b^3[(2\mu + 1)r^3 + (\mu - 1)a^3]H_0 z}{r^3[b^3(\mu + 2)(2\mu + 1) - 2a^3(\mu - 1)^2]} + \text{const}, & a \leqslant r \leqslant b, \\[3mm] H_0 z + \dfrac{b^3(a^3 - b^3)(\mu - 1)(2\mu + 1)H_0 z}{r^3[b^3(\mu + 2)(2\mu + 1) - 2a^3(\mu - 1)^2]} + \text{const}, & b \leqslant r < \infty. \end{cases}$$

*201. Find the magnetic field due to a d-c current J flowing in a circular loop of radius r_0 inside a hollow spherical shield made from material of magnetic permeability μ (see Figure 44).

Ans. The components of the vector potential of the magnetic field are

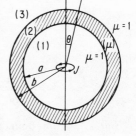

FIGURE 44

$$A_r = A_\theta = 0$$

$$A_\varphi = A(r, \theta) = \frac{2\pi J\mu}{c} \sum_{n=0}^{\infty} \frac{(4n + 3)^2}{(2n + 1)(2n + 2)} P^1_{2n+1}(0)$$

$$\times \frac{(r_0/r)^{2n+2} P^1_{2n+1}(\cos\theta)}{[(2n + 1)\mu + (2n + 2)][(2n + 2)\mu + (2n + 1)]}$$

$$- (a/b)^{4n+3}(2n + 1)(2n + 2)(\mu - 1)^2$$

$$r \geqslant b,$$

in terms of the associated Legendre functions $P^1_{2n+1}(x)$. Note that

$$P^1_{2n+1}(0) = (-1)^{n+1} \frac{1 \cdot 3 \cdot 5 \cdots (2n + 1)}{2 \cdot 4 \cdot 6 \cdots 2n}.$$

202. A lossless open-ended transmission line of length l with parameters L and C is charged to a constant potential E (cf. Prob. 23). Determine the current distribution along the line, assuming that a coil of self-inductance L_0 is connected across the end $x = l$ at the time $t = 0$.

Ans.

$$I(x, t) = \frac{2\alpha E}{Z} \sum_{n=1}^{\infty} \frac{\sin(\gamma_n vt/l) \sin(\gamma_n x/l)}{\cos \gamma_n[\gamma_n^2 + \alpha(1 + \alpha)]},$$

where the γ_n are consecutive positive roots of the equation

$$\tan \gamma = \frac{\alpha}{\gamma},$$

$\alpha = Ll/L_0$, $v = 1/\sqrt{LC}$ is the velocity of wave propagation along the line, and $Z = Lv$ is the wave resistance of the line.

203. A transmission line with parameters L, C and R is short-circuited at one end $x = l$ and connected at the other end $x = 0$ to a source of constant e.m.f. E. Find the voltage distribution along the line, for the case of zero initial conditions.

Ans.

$$u(x, t) = E\left(1 - \frac{x}{l}\right) - \frac{2E}{\pi} e^{-Rt/2L} \sum_{n=1}^{\infty}$$

$$\times \left(\cos \frac{n\pi v^* t}{l} + \frac{Rl}{2n\pi Z^*} \sin \frac{n\pi v^* t}{l}\right) \frac{\sin(n\pi x/l)}{n}$$

where

$$v^* = \frac{1}{\sqrt{LC}} \sqrt{1 - \frac{R^2 C l^2}{4n^2 \pi^2 L}}, \qquad Z^* = Lv^*.$$

Hint. Make the boundary conditions homogeneous by subtracting out a particular solution of the differential equation depending only on the coordinate x.

204. A plane electromagnetic wave with electric field components $E_x = E_y = 0$, $E_z = E_0 e^{i(\omega t - kx)}$ (where $k = \omega/c$ is the wave number) is incident on an infinite perfectly conducting cylinder of radius a. Find the resulting diffracted electric field.

Ans.

$$E_z = E_0 \left[e^{-ikr\cos\varphi} - \frac{J_0(ka)}{H_0^{(2)}(ka)} H_0^{(2)}(kr) \right.$$

$$\left. - 2\sum_{n=1}^{\infty} e^{-in\pi/2} \frac{J_n(ka)}{H_n^{(2)}(ka)} H_n^{(2)}(kr) \cos n\varphi \right] e^{i\omega t},$$

where $J_n(x)$ and $H_n^{(2)}(x)$ are Bessel functions of the first and third kinds.

205. Solve the preceding problem, assuming that the cylinder is made from material of conductivity σ and dielectric constant ε.

Ans.

$$E_z = E_0 \left[e^{-ik_1 r\cos\varphi} + \frac{k_1 J_0(k_2 a)J_0'(k_1 a) - k_2 J_0(k_1 a)J_0'(k_2 a)}{k_2 H_0^{(2)}(k_1 a)J_0'(k_2 a) - k_1 J_0(k_2 a)H_0^{(2)\prime}(k_1 a)} H_0^{(2)}(k_1 r) \right.$$

$$\left. + 2\sum_{n=1}^{\infty} e^{-in\pi/2} \frac{k_1 J_n(k_2 a)J_n'(k_1 a) - k_2 J_n(k_1 a)J_n'(k_2 a)}{k_2 H_n^{(2)}(k_1 a)J_n'(k_2 a) - k_1 J_n(k_2 a)H_n^{(2)\prime}(k_1 a)} H_n^{(2)}(k_1 r) \cos n\varphi \right] e^{i\omega t},$$

$$r \geqslant a,$$

$$E_z = E_0 \frac{2i}{\pi a} \left[\frac{J_0(k_2 r)}{k_1 J_0(k_2 a)H_0^{(2)\prime}(k_1 a) - k_2 H_0^{(2)}(k_1 a)J_0'(k_2 a)} \right.$$

$$\left. + 2\sum_{n=1}^{\infty} e^{-in\pi/2} \frac{J_n(k_2 r)\cos n\varphi}{k_1 J_n(k_2 a)H_n^{(2)\prime}(k_1 a) - k_2 H_n^{(2)}(k_1 a)J_n'(k_2 a)} \right] e^{i\omega t},$$

$$0 \leqslant r \leqslant a,$$

$$k_1 = \frac{\omega}{c}, \qquad k_2 = \sqrt{\frac{\omega^2 \varepsilon - 4\pi i\sigma\omega}{c^2}}.$$

*206. Find the electromagnetic oscillations in a spherical resonator of radius a excited by a dipole of moment P located at the center of the sphere, assuming that the direction of the dipole coincides with the direction of the z-axis.

Ans. The complex amplitudes of the field components are

$$H_r = H_\theta = 0,$$

$$H_\varphi = H(r, \theta) = \frac{P}{cr} \left[\frac{1 + ikr}{r} e^{-ikr} \right.$$

$$\left. + ke^{-ika} \frac{1 + ika - k^2 a^2}{ka\cos ka + (k^2 a^2 - 1)\sin ka} \left(\frac{\sin kr}{kr} - \cos kr \right) \right] \sin\theta,$$

$$E_r = -\frac{i}{kr\sin\theta} \frac{\partial}{\partial\theta}(H_\varphi \sin\theta),$$

$$E_\theta = \frac{i}{kr} \frac{\partial}{\partial r}(rH_\varphi),$$

$$E_\varphi = 0,$$

in terms of the spherical coordinates r, θ and φ, where ω is the frequency of the oscillations and $k = \omega/c$ is the wave number.

References

Bateman (B2), Frank and von Mises (F6), Franklin (F7), Gray and Mathews (G2), Grinberg (G5), Jackson (J1), Jeffreys and Jeffreys (J4), Lebedev (L9, Chaps. 6 and 8), McLachlan (M5), Morse and Feshbach (M9), Tikhonov and Samarski (T1), Tolstov (T7), Webster (W5). For further problems, see Budak, Samarski and Tikhonov (B6), Gyunter and Kuzmin (G7, Chap. 15), Smirnov (S5).

5

THE EIGENFUNCTION METHOD FOR
SOLVING INHOMOGENEOUS PROBLEMS

In this chapter we study various inhomogeneous problems of mathematical physics leading to integration of the equation

$$\frac{1}{r(x)}\left\{\frac{\partial}{\partial x}\left[p(x)\frac{\partial u}{\partial x}\right] - q(x)u\right\} + M_y u = F(x, y) \qquad (a < x < b, c < y < d),$$

$$(1)$$

which is the same as equation (1) of Chap. 4, except for the presence of the given function $F(x, y)$ in the right-hand side.[1] This time we require that the solution of (1) satisfy *inhomogeneous* boundary conditions

$$\alpha_a \frac{\partial u}{\partial x}\bigg|_{x=a} + \beta_a u\big|_{x=a} = f_a(x), \qquad \alpha_b \frac{\partial u}{\partial x}\bigg|_{x=b} + \beta_b u\big|_{x=b} = f_b(y), \qquad (2)$$

where α_a, α_b, β_a, β_b are constants and $f_a(y)$, $f_b(y)$ are given functions. In the elliptic case,

$$\gamma_c \frac{\partial u}{\partial y}\bigg|_{y=c} + \delta_c u\big|_{y=c} = g_c(x), \qquad \gamma_d \frac{\partial u}{\partial y}\bigg|_{y=d} + \delta_d u\big|_{y=d} = g_d(x), \qquad (3)$$

where again γ_c, γ_d, δ_c, δ_d are constants and $g_c(x)$, $g_d(x)$ are given functions. In the hyperbolic and parabolic cases, the boundary conditions (3) are replaced by the conditions (4′) and (4″), p. 57.

It is sometimes possible to find a particular solution u^* of equation (1)

---

[1] In particular, the functions $p(x)$, $q(x)$, $r(x)$ and the differential operator M_y have the same meaning as on p. 56.

satisfying the conditions (2), and then the substitution $u = u^* + v$ reduces the present problem to the homogeneous problem which can be solved by the Fourier method. The problem can also be solved easily in the case where only the differential equation (1) is inhomogeneous, but not the boundary conditions (2), so that $f_a = f_b = 0$. Then we can look for a solution in the form of an expansion

$$u = \sum_{n=1}^{\infty} u_n(y) X_n(x), \qquad a < x < b \qquad (4)$$

with respect to the eigenfunctions $X_n(x)$ of the homogeneous problem, i.e., the nontrivial solutions of the equation

$$(pX')' + (\lambda r - q)X = 0 \qquad (5)$$

satisfying the homogeneous boundary conditions

$$\alpha_a X'(a) + \beta_a X(a) = 0, \qquad \alpha_b X'(b) + \beta_b X(b) = 0. \qquad (6)$$

Suppose the right-hand side of (1) can be expanded in a series with respect to the functions $X_n(x)$, so that

$$F(x, y) = \sum_{n=1}^{\infty} F_n(y) X_n(x), \qquad a < x < b,$$

where

$$F_n(y) = \frac{\int_a^b rF(x, y)X_n(x)\, dx}{\int_a^b rX_n^2(x)\, dx}.$$

Then, after substituting (4) into (1), the problem reduces to the integration of the ordinary differential equation

$$M_y u_n - \lambda_n u_n = F_n(y),$$

where the λ_n are the eigenvalues of the homogeneous problem. To determine the resulting constants of integration, we substitute (4) into (3) [or into equations (4'), (4''), p. 57], expand the functions on the right in terms of the eigenfunctions $X_n(x)$, and then equate corresponding coefficients of the functions $X_n(x)$.

The general case of inhomogeneous boundary conditions can be reduced to the problem just considered (an inhomogeneous differential equation and homogeneous boundary conditions) by looking for a solution of the form $u = u^* + v$, where u^* is a sufficiently smooth function which satisfies the boundary conditions (2) but, unlike the case mentioned above, is not necessarily a solution of the differential equation. For example, if the boundary conditions are of the first kind, i.e.,

$$u\big|_{x=a} = f_a(y), \qquad u\big|_{x=b} = f_b(y),$$

we can choose u^* to be the following linear function of x:

$$u^* = \frac{x-a}{b-a} f_b(y) + \frac{b-x}{b-a} f_a(y). \tag{7}$$

Similarly, if the boundary conditions are of the second kind, i.e.,

$$\frac{\partial u}{\partial x}\bigg|_{x=a} = f_a(y), \qquad \frac{\partial u}{\partial x}\bigg|_{x=b} = f_b(y),$$

we can choose

$$u^* = \frac{1}{2}\frac{(x-a)^2}{b-a} f_b(y) - \frac{1}{2}\frac{(b-x)^2}{b-a} f_a(y), \tag{8}$$

and so on. However, it should be noted that this method, involving as it does a function u^* which is to a large extent arbitrary, is not always successful (for example, in cases where the boundary conditions are discontinuous). In fact, improper choice of u^* [even such simple functions as (7) and (8)] can lead to great complication in later stages of the calculations.

A more adequate method of solving inhomogeneous problems has been proposed by Grinberg (G4),[2] and is free from the need to choose the function u^* in each particular case (which sometimes requires great ingenuity). In Grinberg's method, we try to solve the inhomogeneous problem by again representing the solution as a series of the form (4), whose coefficients are given by the formula

$$u_n(y) = \frac{\int_a^b ru X_n(x)\,dx}{\int_a^b r X_n^2(x)\,dx} = \frac{\bar{u}_n}{\int_a^b r X_n^2(x)\,dx}, \tag{9}$$

in keeping with the general theory of expansion in series of orthogonal functions. Thus, to obtain a formal solution of the problem, we need only find the value of the integral $\bar{u}_n$. This can be done by the following device: First we multiply equation (1) by $X_n(x)$ and integrate the result from a to b. Then we integrate by parts twice, obtaining

$$\left(p\frac{\partial u}{\partial x}X_n - pX_n'u\right)\bigg|_a^b + \int_a^b [(pX_n')' - qX_n]u\,dx$$
$$+ M_y\int_a^b ru X_n\,dx = \int_a^b rFX_n\,dx. \tag{10}$$

---

[2] In cases where the boundary conditions are homogeneous and only the differential equation is inhomogeneous, Grinberg's method gives the same result as the classical method of solution.

Taking account of equation (5) and the boundary conditions (2) and (6), we can write (10) in the form[3]

$$M_y \bar{u}_n - \lambda_n \bar{u}_n = \bar{F}_n - \frac{p(b)}{\alpha_b} X_n(b) f_b(y) + \frac{p(a)}{\alpha_a} X_n(a) f_a(y) \qquad (11)$$

in terms of the eigenvalues λ_n, where

$$\bar{F}_n = \int_a^b rF X_n \, dx.$$

Equation (11) serves to determine $\bar{u}_n$, since its right-hand side involves only known functions. The resulting constants of integration are found from the equations which result when the same method [i.e., multiplication by $rX_n(x)$, followed by integration from a to b] is applied to equation (3) [or to equations (4′), (4″), p. 57].

The method just described can also be applied to problems of mathematical physics involving the Sturm-Liouville problem with singular end points (see p. 59), provided that the eigenvalue spectrum is discrete. Moreover, the method can be extended to certain problems involving higher-order equations (see Probs. 236–241), or to problems where the solution depends on a larger number of variables.

It should be pointed out that for inhomogeneous boundary conditions of the first kind, the series representing the solution will not be uniformly convergent near the end points of the interval (a, b).[4] To improve the convergence, we can apply the methods ordinarily used in such cases.[5] In the simplest problems, we can improve the convergence by separating out the slowly converging part of the series and summing it by using the tables given in Sec. 2 of the Mathematical Appendix (see p. 381).

The problems in this chapter, as in the preceding one, are grouped into five sections, two on mechanics, two on heat conduction (including a problem on diffusion), and one on electricity and magnetism. Problems involving coordinate systems more complicated than rectangular or polar coordinates (both cylindrical and spherical) will be deferred until Chap. 7. Problems with concentrated sources are usually regarded as limiting cases

---

[3] In the case of boundary conditions of the first kind ($\alpha_a = \alpha_b = 0$), the right-hand side of (11) should be replaced by

$$F_n + \frac{p(b)}{\beta_b} X_n'(b) f_b(y) - \frac{p(a)}{\beta_a} X_n'(a) f_a(y).$$

[4] If the boundary conditions are inhomogeneous only at one end point $x = a$, this statement applies only at $x = a$. In the case of boundary conditions of the second kind, the series representing the derivative $\partial u / \partial x$ exhibits similar behavior.

[5] See K1, Chap. 1, Sec. 5. Another method, of a completely general character, is given by Grinberg (G5, Chap. 12).

of the corresponding problems with distributed sources; this greatly simplifies the calculations, allowing us to write the solutions in compact and symmetric form. For example, the field due to a linear oscillator inside a cylindrical resonator can easily be solved in this way (see Prob. 256), whereas the usual method of solution (which involves subtracting out the singularity) leads to very complicated calculations.

In the case of problems with inhomogeneous boundary conditions, the choice of a method of solution is left to the reader, although we are of the opinion that in such cases, Grinberg's method has indisputable methodological advantages. Of course, by proper choice of u^*, certain problems can be solved quite easily, without recourse to this method.

As a rule, the answers are given in the form of series, obtained after improving convergence, or in closed form. In some cases, the solution is given in two forms, corresponding to expansions in functions of each of the two independent variables.

I. Mechanics: Vibrating Systems

207. A string of length l with fastened ends vibrates under the action of a uniformly distributed pulsating load $q \sin \omega t$. Describe the vibrations, assuming that the string is at rest at the time $t = 0$.

Ans.

$$u(x, t) = \frac{4qvl}{\pi^2 \omega T} \sum_{n=0}^{\infty} \frac{\sin \dfrac{(2n+1)\pi vt}{l} - \dfrac{(2n+1)\pi v}{\omega l} \sin \omega t}{1 - \left[\dfrac{(2n+1)\pi v}{\omega l}\right]^2} \frac{\sin \dfrac{(2n+1)\pi x}{l}}{(2n+1)^2},$$

$$0 \leqslant x \leqslant l,$$

where $v = \sqrt{T/\rho}$, T is the tension and ρ is the linear density of the string.

208. Solve the preceding problem, assuming that the pulsating load acts only on the section $a < x < b$ of the string.

Ans.

$$u(x, t) = \frac{2qvl}{\pi^2 \omega T} \sum_{n=1}^{\infty} \left(\cos \frac{n\pi a}{l} - \cos \frac{n\pi b}{l}\right) \frac{\sin \dfrac{n\pi vt}{l} - \dfrac{n\pi v}{\omega l} \sin \omega t}{1 - \left(\dfrac{n\pi v}{\omega l}\right)^2} \frac{\sin \dfrac{n\pi x}{l}}{n^2}.$$

209. Study the vibrations of a string due to a concentrated pulsating load $A \sin \omega t$ applied at the time $t = 0$ to an arbitrary point $x = c$ of the string.

Ans.

$$u(x, t) = \frac{2Av}{\pi\omega T} \sum_{n=1}^{\infty} \frac{\sin\dfrac{n\pi vt}{l} - \dfrac{n\pi v}{\omega l}\sin\omega l}{1 - \left(\dfrac{n\pi v}{\omega l}\right)^2} \frac{\sin\dfrac{n\pi c}{l}\sin\dfrac{n\pi x}{l}}{n}$$

Hint. Pass to the limit in the solution of Prob. 208.

*210. Find the general solution of the problem of a vibrating string under the action of an external load $q(x, t)$, assuming that the string is at rest at the time $t = 0$.

Ans.

$$u(x, t) = \frac{2v}{\pi T} \sum_{n=1}^{\infty} \frac{1}{n} \sin\frac{n\pi x}{l} \int_0^t \sin\frac{n\pi v(t - \tau)}{l}\, d\tau \int_0^l q(\xi, \tau) \sin\frac{n\pi\xi}{l}\, d\xi,$$
$$0 < x < l.$$

211. Solve Prob. 106 on the longitudinal oscillations of a rod, which was solved by another method in Chap. 4.

Ans.[6]

$$u(x, t) = \frac{2Al}{ES} \sum_{n=0}^{\infty} (-1)^n \frac{\sin\omega t - \dfrac{2\omega l}{(2n + 1)\pi v}\sin\dfrac{(2n + 1)\pi vt}{2l}}{\left[\dfrac{(2n + 1)\pi}{2}\right]^2 - \left(\dfrac{\omega l}{v}\right)^2} \sin\frac{(2n + 1)\pi x}{2l}.$$

212. Investigate the vertical longitudinal oscillations of a rod of length l suspended from the end $x = 0$ under the action of its own weight, subject to zero initial conditions.

Ans.

$$u(x, t) = \frac{16gl^2}{\pi^3 v^2} \sum_{n=0}^{\infty} \frac{1 - \cos\left[(2n + 1)\pi vt/2l\right]}{(2n + 1)^3} \sin\frac{(2n + 1)\pi x}{2l}.$$

where g is the acceleration of gravity, E is Young's modulus, ρ is the density, and $v = \sqrt{E/\rho}$.

213. Investigate the longitudinal oscillations of the pyramid-shaped cantilever of square cross section shown in Figure 45, due to a force $A\sin\omega t$ applied at the time $t = 0$ to its free end.

---

[6] To verify that the two forms of the solution given in the answers to Probs. 106 and 211 coincide, use the expansion

$$\sum_{n=0}^{\infty} (-1)^n \frac{\sin(2n + 1)\xi}{(2n + 1)^2 - \alpha^2} = \frac{\pi\sin\alpha\xi}{4\alpha\cos\pi(\alpha/2)}, \qquad |\xi| \leqslant \pi/2.$$

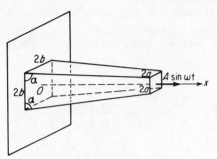

FIGURE 45

Ans.

$$u(x, t) = \frac{Av}{2a\omega E(b - x\tan\alpha)} \sum_{n=1}^{\infty} \frac{\sin\frac{\gamma_n vt}{l} - \frac{\gamma_n v}{\omega l}\sin\omega t}{1 - \left(\frac{\gamma_n v}{\omega l}\right)^2} \frac{\sin\gamma_n}{\gamma_n} \frac{\sin\frac{\gamma_n x}{l}}{1 - \frac{\sin 2\gamma_n}{2\gamma_n}},$$

where the γ_n are consecutive positive roots of the equation

$$\tan\gamma = -\gamma\frac{a}{b - a}.$$

214. An inhomogeneous rod consisting of two sections made from different materials is clamped at one end and is initially at rest. Find the longitudinal oscillations which result if a constant force P is applied to the free end of the rod (see Figure 46).

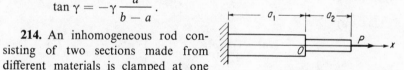

FIGURE 46

Ans.

$$u(x, t) = 2P\frac{a_1^2}{v_1^2} \sum_{n=1}^{\infty} \frac{\sin\gamma_n}{\gamma_n^2} \frac{(1 - \cos\gamma_n t)X_n(x)}{\rho_1 S_1 a_1 \cos^2(\gamma_n a_2 v_1/a_1 v_2) + \rho_2 S_2 a_2 \sin^2\gamma_n},$$

$$X_n(x) = \begin{cases} \cos\dfrac{\gamma_n a_2 v_1}{a_1 v_2}\sin\dfrac{\gamma_n(a_1 + x)}{a_1}, & -a_1 \leqslant x \leqslant 0, \\[2ex] \sin\gamma_n\cos\dfrac{\gamma_n v_1(a_2 - x)}{a_1 v_2}, & 0 \leqslant x \leqslant a_2, \end{cases}$$

where the γ_n are consecutive positive roots of the equation

$$v_1 E_2 S_2 \tan\gamma = v_2 E_1 S_1 \cot\frac{\gamma a_2 v_1}{a_1 v_2},$$

the two sections have Young's moduli E_i, cross-sectional areas S_i and densities ρ_i $(i = 1, 2)$, and $v_i = \sqrt{E_i/\rho_i}$.

215. A beam of length l, simply supported at its ends, is originally in a state of equilibrium. Investigate the transverse oscillations of the beam after applying an arbitrary load, uniformly distributed over the section $x_1 < x < x_2$.

Ans.

$$u(x, t) = \frac{2l^2 a^2}{\pi^3 EJ} \sum_{n=1}^{\infty} \left(\cos \frac{n\pi x_1}{l} - \cos \frac{n\pi x_2}{l} \right) \frac{\sin (n\pi x/l)}{n^3}$$

$$\times \int_0^t q(\tau) \sin \frac{n^2 \pi^2 a^2 (t - \tau)}{l^2} \, d\tau, \qquad 0 \leqslant x \leqslant l,$$

where $a^2 = \sqrt{EJ/\rho S}$, E is Young's modulus, J the moment of inertia of a cross section, ρ the density and S the cross-sectional area of the beam.

216. Solve the preceding problem, assuming that a) the load is uniformly distributed over the whole length of the beam and is a periodic function of time $q(t) = q \sin \omega t$; b) a concentrated pulsating force $A \sin \omega t$ is applied to the point $x = c$ of the beam.

Ans. a)

$$u(x, t) = \frac{4l^2 a^2 q}{\pi^3 \omega EJ} \sum_{n=0}^{\infty} \frac{\sin \dfrac{(2n+1)^2 \pi^2 a^2 t}{l^2} - \dfrac{(2n+1)^2 \pi^2 a^2}{\omega l^2} \sin \omega t}{1 - \left[\dfrac{(2n+1)^2 \pi^2 a^2}{\omega l^2} \right]^2} \frac{\sin \dfrac{(2n+1)\pi x}{l}}{(2n+1)^3};$$

b) $$u(x, t) = \frac{2A a^2 l}{\pi^2 \omega EJ} \sum_{n=0}^{\infty} \frac{\sin \dfrac{n^2 \pi^2 a^2 t}{l^2} - \dfrac{n^2 \pi^2 a^2}{\omega l^2} \sin \omega t}{1 - \left(\dfrac{n^2 \pi^2 a^2}{\omega l^2} \right)^2} \frac{\sin \dfrac{n\pi c}{l} \sin \dfrac{n\pi x}{l}}{n^2}.$$

*217. Find the transverse oscillations of a beam $-l \leqslant x \leqslant l$ with clamped ends under the action of a pulsating force $q \sin \omega t$, uniformly distributed over the whole length of the beam, assuming that the beam is at rest before the load is applied.

Ans.

$$u(x, t) = \frac{2q l^2 a^2}{\omega EJ} \sum_{n=1}^{\infty} \frac{\tan \gamma_n}{\gamma_n^3 \cosh \gamma_n \cos \gamma_n} \frac{\sin \dfrac{\gamma_n^2 a^2 t}{l^2} - \dfrac{\gamma_n^2 a^2}{\omega l^2} \sin \omega t}{1 - \left(\dfrac{\gamma_n^2 a^2}{\omega l^2} \right)^2} X_n(x),$$

where $a^2 = \sqrt{EJ/\rho S}$, in the notation of Prob. 215,

$$X_n(x) = \cosh \gamma_n \cos \frac{\gamma_n x}{l} - \cos \gamma_n \cosh \frac{\gamma_n x}{l},$$

and the γ_n are consecutive positive roots of the equation $\tan \gamma + \tanh \gamma = 0$.

218. Solve the preceding problem for the case where the external load is a concentrated force $A \sin \omega t$ applied to the center of the beam.

Ans.

$$u(x, t) = \frac{A l a^2}{2\omega E J} \sum_{n=1}^{\infty} \frac{\cosh \gamma_n - \cos \gamma_n}{\gamma_n^2 \cos^2 \gamma_n \cosh^2 \gamma_n} \frac{\sin \dfrac{\gamma_n^2 a^2 t}{l^2} - \dfrac{\gamma_n^2 a^2}{\omega l^2} \sin \omega t}{1 - \left(\dfrac{\gamma_n^2 a^2}{\omega l^2}\right)^2} X_n(x).$$

Hint. First replace the concentrated load by a load uniformly distributed over the section $-\varepsilon < x < \varepsilon$ of the beam, and then take the limit as $\varepsilon \to 0$.

219. Solve Prob. 217 for a beam $0 \leqslant x \leqslant l$ if the end $x = 0$ is simply supported, while the end $x = l$ is clamped.

Ans.

$$u(x, t) = \frac{q l^2 a^2}{\omega E J} \sum_{n=1}^{\infty} \frac{\sinh \gamma_n - 2 \cosh \gamma_n \sin \gamma_n + \sin \gamma_n}{\gamma_n^3 \sinh^2 \gamma_n \sin^2 \gamma_n}$$

$$\times \frac{\sin \dfrac{\gamma_n^2 a^2 t}{l^2} - \dfrac{\gamma_n^2 a^2}{\omega l^2} \sin \omega t}{1 - \left(\dfrac{\gamma_n^2 a^2}{2}\right)^2} X_n(x),$$

where

$$X_n(x) = \sinh \gamma_n \sin \frac{\gamma_n x}{l} - \sin \gamma_n \sinh \frac{\gamma_n x}{l},$$

and the γ_n are consecutive positive roots of the equation $\tan \gamma = \tanh \gamma$.

220. A concentrated force P is applied to the free end of a cantilever initially in equilibrium (see Figure 47). Investigate the resulting transverse oscillations, assuming that the force does not change subsequently.

Ans.

$$u(x, t) = \frac{2P l^3}{EJ} \sum_{n=1}^{\infty} \frac{1 - \cos (\gamma_n^2 a^2 t/l^2)}{\gamma_n^4 (\sinh \gamma_n + \sin \gamma_n)} X_n(x),$$

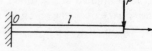

FIGURE 47

where

$$X_n(x) = (\cosh \gamma_n + \cos \gamma_n)\left(\sinh \frac{\gamma_n x}{l} - \sin \frac{\gamma_n x}{l}\right)$$

$$- (\sinh \gamma_n + \sin \gamma_n)\left(\cosh \frac{\gamma_n x}{l} - \cos \frac{\gamma_n x}{l}\right),$$

and the γ_n are consecutive positive roots of the equation $\cosh \gamma \cos \gamma = -1$.

221. Solve the preceding problem for the case where the force is a periodic function of time $P = A \sin \omega t$.

Ans.

$$u(x, t) = \frac{2Ala^2}{\omega EJ} \sum_{n=1}^{\infty} \frac{X_n(x)}{\gamma_n^2(\sinh \gamma_n + \sin \gamma_n)} \frac{\sin \dfrac{\gamma_n^2 a^2 t}{l^2} - \dfrac{\gamma_n^2 a^2}{\omega l^2} \sin \omega t}{1 - \left(\dfrac{\gamma_n^2 a^2}{\omega l^2}\right)^2}.$$

*222. Solve Prob. 220 for the case where the force $P = P(t)$ is arbitrary.

$$u(x, t) = \frac{2la^2}{EJ} \sum_{n=1}^{\infty} \frac{X_n(x)}{\gamma_n^2(\sinh \gamma_n + \sin \gamma_n)} \int_0^t P(\tau) \sin \frac{\gamma_n^2 a^2(t - \tau)}{l^2}\, d\tau.$$

*223. Investigate the transverse oscillations of a beam of mass M clamped at the points $x = 0$ and $x = l$, due to a concentrated pulsating load $A \sin \omega t$ moving along the beam with constant velocity v. Assume that at the time $t = 0$, the beam is at rest and the moving load is at the point $x = 0$.[7]

Ans.

$$u(x, t) = \frac{A}{M\omega^2} \sum_{n=1}^{\infty} \left[\frac{\cos \dfrac{n^2\pi^2 a^2 t}{l^2} - \cos\left(1 + \dfrac{n\pi v}{\omega l}\right)\omega t}{\left(\dfrac{n^2\pi^2 a^2}{\omega l^2}\right)^2 - \left(1 + \dfrac{n\pi v}{\omega l}\right)^2} \right.$$

$$\left. - \frac{\cos \dfrac{n^2\pi^2 a^2 t}{l^2} - \cos\left(1 - \dfrac{n\pi v}{\omega l}\right)\omega t}{\left(\dfrac{n^2\pi^2 a^2}{\omega l^2}\right)^2 - \left(1 - \dfrac{n\pi v}{\omega l}\right)^2} \right] \sin \frac{n\pi x}{l}.$$

224. Investigate the vibrations of a circular membrane of radius a due to a load applied at the time $t = 0$, if the load is uniformly distributed with density $q(t)$ over the circular ring $r_1 < r < r_2$. Consider the special case $q(t) = q \sin \omega t$.

---

[7] This is the problem of a locomotive moving along a railway bridge (see T2, Sec. 59).

Ans.

$$u(r, t) = \frac{2v}{T} \sum_{n=1}^{\infty} \frac{r_2 J_1(\gamma_n r_2/a) - r_1 J_1(\gamma_n r_1/a)}{\gamma_n^2 J_1^2(\gamma_n)} J_0\left(\frac{\gamma_n r}{a}\right) \int_0^t q(\tau) \sin \frac{\gamma_n v(t - \tau)}{a} \, d\tau,$$

where the γ_n are consecutive positive roots of the equation $J_0(\gamma) = 0$, ρ is the surface density, T is the tension per unit length of the boundary of the membrane, and $v = \sqrt{T/\rho}$. In the special case,

$$u(r, t) = \frac{2qv}{\omega T} \sum_{n=1}^{\infty} \frac{\sin \frac{\gamma_n vt}{a} - \frac{\gamma_n v}{\omega a} \sin \omega t}{1 - \left(\frac{\gamma_n a}{\omega a}\right)^2} \frac{r_2 J_1\left(\frac{\gamma_n r_2}{a}\right) - r_1 J_1\left(\frac{\gamma_n r_1}{a}\right)}{\gamma_n^2 J_1^2(\gamma_n)} J_0\left(\frac{\gamma_n r}{a}\right).$$

*225. Investigate the vibrations of a circular membrane of radius a due to a pulsating load $p \sin \omega t$ applied at the time $t = 0$ along the circumference of a circle of radius $b < a$.

Ans.

$$u(r, t) = \frac{2vpb}{\omega a T} \sum_{n=1}^{\infty} \frac{\sin \frac{\gamma_n vt}{a} - \frac{\gamma_n v}{\omega a} \sin \omega t}{1 - \left(\frac{\gamma_n v}{\omega a}\right)^2} \frac{J_0\left(\frac{\gamma_n b}{a}\right) J_0\left(\frac{\gamma_n r}{a}\right)}{\gamma_n J_1^2(\gamma_n)}.$$

Hint. Replace the load by a load distributed with constant density over the area of the ring $b - \varepsilon < r < b + \varepsilon$, and then take the limit as $\varepsilon \to 0$.

226. A circular elastic plate of radius a, clamped along its boundary, begins to oscillate under the action of a suddenly applied pulsating load $q \sin \omega t$, uniformly distributed over the area of the plate. Find the resulting transverse oscillations.

Ans.

$$u(r, t) = \frac{2qa^2 b^2}{\omega D} \sum_{n=1}^{\infty} \frac{\sin \frac{\gamma_n^2 b^2 t}{a^2} - \frac{\gamma_n^2 b^2}{\omega a^2} \sin \omega t}{1 - \left(\frac{\gamma_n^2 b^2}{\omega a^2}\right)^2} \frac{J_1(\gamma_n) R_{\gamma_n}(r)}{\gamma_n^3 J_0^2(\gamma_n) I_0(\gamma_n)},$$

where

$$R_\gamma(r) = I_0(\gamma) J_0\left(\frac{\gamma r}{a}\right) - J_0(\gamma) I_0\left(\frac{\gamma r}{a}\right)$$

is a linear combination of cylinder functions, the γ_n are consecutive positive roots of the equation $R'_y(a) = 0$, D is the flexural rigidity, h the thickness and ρ the density of the plate, and $b^2 = \sqrt{D/\rho h}$.

*227. Solve the preceding problem, assuming that the oscillations are due to a concentrated pulsating force $A \sin \omega t$ applied at the center of the plate (oscillations of the diaphragm of a loudspeaker).

Ans.

$$u(r, t) = \frac{Ab^2}{2\pi\omega D} \sum_{n=1}^{\infty} \frac{\sin\dfrac{\gamma_n^2 b^2 t}{a^2} - \dfrac{\gamma_n^2 b^2}{\omega a^2}\sin\omega t}{1 - \left(\dfrac{\gamma_n^2 b^2}{\omega a^2}\right)^2} \frac{[I_0(\gamma_n) - J_0(\gamma_n)]R_{\gamma_n}(r)}{\gamma_n^2 J_0^2(\gamma_n) I_0^2(\gamma_n)}.$$

Hint. Replace the concentrated load by a load distributed over a disk of small radius ε, and then take the limit $\varepsilon \to 0$.

2. Mechanics: Statics of Deformable Media

228. Find the deflection of a rectangular membrane $-a \leqslant x \leqslant a$, $-b \leqslant y \leqslant b$ due to a load uniformly distributed with density q over the rectangle $-c \leqslant x \leqslant c$, $-d \leqslant y \leqslant d$ forming part of the membrane.

Ans.

$$u(x, y)\big|_{|y|<d} = \frac{16qa^2}{\pi^3 T} \sum_{n=0}^{\infty} \frac{\sin\left[(2n+1)\pi c/2a\right]}{(2n+1)^3}$$

$$\times \left[1 - \cosh\frac{(2n+1)\pi(b-d)}{2a} \frac{\cosh\dfrac{(2n+1)\pi y}{2a}}{\cosh\dfrac{(2n+1)\pi b}{2a}}\right]\cos\frac{(2n+1)\pi x}{2a},$$

$$u(x, y)\big|_{|y|>d} = \frac{16qa^2}{\pi^3 T} \sum_{n=0}^{\infty} \frac{\sin\left[(2n+1)\pi c/2a\right]\sinh\left[(2n+1)\pi d/2a\right]}{(2n+1)^3}$$

$$\times \frac{\sinh\dfrac{(2n+1)\pi(b-|y|)}{2a}}{\cosh\dfrac{(2n+1)\pi b}{2a}}\cos\frac{(2n+1)\pi x}{2a},$$

where T is the tension per unit length of the boundary of the membrane.

229. Find the deflection of a uniformly loaded rectangular membrane (this is a special case of the preceding problem), and compare the answer with that found earlier in Prob. 131.[8]

Ans.

$$u(x, y) = \frac{16qa^2}{\pi^3 T} \sum_{n=0}^{\infty} \frac{(-1)^n}{(2n + 1)^3} \left\{ 1 - \frac{\cosh [(2n + 1)\pi y/2a]}{\cosh [(2n + 1)\pi b/2a]} \right\} \cos \frac{(2n + 1)\pi x}{2a}.$$

*230. Find the static deflection of a rectangular membrane under the action of a line load p uniformly distributed along an axis of symmetry (see Figure 48).

Ans.

$$u(x, y) = \frac{4pa}{\pi^2 T} \sum_{n=0}^{\infty} \left\{ 1 - \frac{\cosh [(2n+1)\pi y/2a]}{\cosh [(2n+1)\pi b/2a]} \right\}$$

$$\times \frac{\cos [(2n + 1)\pi x/2a]}{(2n + 1)^2}. \qquad (12)$$

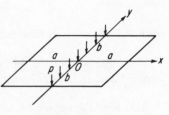

FIGURE 48

Another form of the answer is

$$u(x, y) = \frac{4pb}{\pi^2 T} \sum_{n=0}^{\infty} \frac{(-1)^n}{(2n + 1)^2} \frac{\sinh \dfrac{(2n + 1)\pi(a - |x|)}{2b}}{\cosh \dfrac{(2n + 1)\pi a}{2b}} \cos \frac{(2n + 1)\pi y}{2b}. \qquad (13)$$

231. Find the deflection of a circular membrane of radius a due to the action of a line load p uniformly distributed along a diameter.

Ans.

$$u(r, \varphi) = \frac{pa}{2\pi T} \left[1 - \frac{r^2}{a^2} - \frac{2r^2}{a^2} \cos \varphi \ln \frac{r}{a} - 2 \sum_{n=2}^{\infty} \frac{(r/a)^{2n} - (r/a)^2}{n^2 - 1} \cos 2n\varphi \right],$$

where the series can be summed easily.

Hint. To solve the problem, replace the line load by a load uniformly distributed over the sector $-\varepsilon < \varphi < \varepsilon$, $\pi - \varepsilon < \varphi < \pi + \varepsilon$, and then take the limit as $\varepsilon \to 0$.

232. Investigate the twisting of a rod whose cross section is a semicircle of radius a. Calculate the tangential stresses τ on the surface of the rod.

---

[8] To compare the two answers, use formula 16, p. 385.

Ans.

$$\tau|_{r=a} = 2a\theta G\left[\sin\varphi - \frac{2}{\pi}\left(1 - \cos\varphi\ln\cot\frac{\varphi}{2}\right)\right]\sin\varphi,$$

$$\tau|_{\varphi=0} = \tau|_{\varphi=\pi} = \frac{a\theta G}{\pi}\left[1 - \frac{a^2}{r^2} + \frac{2 - \dfrac{r^2}{a^2} - \dfrac{a^2}{r^2}}{1 - \dfrac{r^2}{a^2}} + \frac{a}{r}\left(\frac{a^2}{r^2} - \frac{r^2}{a^2}\right)\ln\frac{a+r}{a-r}\right],$$

where θ is the angle of twist per unit length and G is the shear modulus.

Hint. The sum of the series needed to represent the solution in closed form is found in the solution to Prob. 132.

233. Find the torsion function $u(r, \varphi)$ for the twisting of a circular shaft of radius a weakened by a radial crack going from the surface of the shaft to its axis. Calculate the torsional rigidity C of the shaft.

$$u(r, \varphi) = \frac{32a^2}{\pi}\sum_{n=0}^{\infty}\frac{\left(\dfrac{r}{a}\right)^{(2n+1)/2} - \left(\dfrac{r}{a}\right)^2}{(2n+1)[16 - (2n+1)^2]}\sin\frac{(2n+1)\varphi}{2},$$

$$C = Ga^4\left\{\frac{512}{\pi}\sum_{n=0}^{\infty}\frac{1}{(2n+1)^2(2n+5)[16 - (2n+1)^2]} - \frac{\pi}{2}\right\} = 0.878Ga^4.$$

234. Investigate the twisting of a rod whose cross section is a circular sector of radius a and vertex angle α.

Ans. The torsion function is

$$u(r, \varphi) = \frac{8a^2}{\pi}\sum_{n=0}^{\infty}\frac{\left(\dfrac{r}{a}\right)^2 - \left(\dfrac{r}{a}\right)^{(2n+1)\pi/\alpha}}{(2n+1)\left[\dfrac{(2n+1)^2\pi^2}{\alpha^2} - 4\right]}\sin\frac{(2n+1)\pi\varphi}{\alpha}$$

$$(0 \leqslant r \leqslant a, 0 \leqslant \varphi \leqslant \alpha).$$

235. Solve the preceding problem for a rod whose cross section is a "curvilinear rectangle" $a \leqslant r \leqslant b, 0 \leqslant \varphi \leqslant \alpha$.

Ans.

$$u(r, \varphi) = \frac{8b^2}{\pi} \sum_{n=0}^{\infty} \frac{\sin \dfrac{(2n+1)\pi\varphi}{\alpha}}{(2n+1)\left[\dfrac{(2n+1)^2\pi^2}{\alpha^2} - 4\right]}$$

$$\times \left\{ \left(\frac{r}{b}\right)^2 - \frac{\left(\dfrac{a}{b}\right)^2\left[\left(\dfrac{r}{b}\right)^{(2n+1)\pi/\alpha} - \left(\dfrac{b}{r}\right)^{(2n+1)\pi/\alpha}\right] - \left[\left(\dfrac{r}{a}\right)^{(2n+1)\pi/\alpha} - \left(\dfrac{a}{r}\right)^{(2n+1)\pi/\alpha}\right]}{\left(\dfrac{a}{b}\right)^{(2n+1)\pi/\alpha} - \left(\dfrac{b}{a}\right)^{(2n+1)\pi/\alpha}} \right\}.$$

236. A rectangular elastic plate $0 \leqslant x \leqslant a$, $-b/2 \leqslant y \leqslant b/2$ is simply supported along its boundary and loaded by a concentrated force P applied at the center of the plate. Find the deflection along the midline $y = 0$.

Ans.

$$u\big|_{y=0} = \frac{Pa^2}{4\pi^3 D} \sum_{n=0}^{\infty} \frac{(-1)^n}{(2n+1)^3} \frac{\sinh \dfrac{(2n+1)\pi b}{a} - \dfrac{(2n+1)\pi b}{a}}{\cosh^2 \dfrac{(2n+1)\pi b}{2a}} \sin \frac{(2n+1)\pi x}{a},$$

where D is the flexural rigidity of the plate.

Hint. Replace the concentrated load by a load uniformly distributed over the rectangle

$$\frac{a}{2} - \delta < x < \frac{a}{2} + \delta, \qquad -\varepsilon < y < \varepsilon,$$

and then take the limit as $\delta, \varepsilon \to 0$.

237. Solve Prob. 134, using the method of this chapter.

Ans.[9]

$$u(0, 0) = \frac{64qa^4}{\pi^5 D} \sum_{n=0}^{\infty} \frac{(-1)^n}{(2n+1)^5} \left[1 - \frac{2 + \dfrac{(2n+1)\pi b}{2a} \tanh \dfrac{(2n+1)\pi b}{2a}}{2 \cosh \dfrac{(2n+1)\pi b}{2a}}\right].$$

238. A rectangular elastic plate with sides a and b is simply supported along the edges $x = 0$ and $x = a$ and clamped along the edges $y = \pm b/2$.

---

[9] To reduce the solution to the form given in Prob. 134, use the formula

$$\sum_{n=0}^{\infty} \frac{(-1)^n}{(2n+1)^5} = \frac{5\pi^5}{1536}.$$

Find the deflection of the plate under the action of a load p applied along the midline $x = a/2$.

Ans.

$$u(x, y) = \frac{2pa^3}{\pi^4 D} \sum_{n=0}^{\infty} (-1)^n \frac{\sin\left[(2n+1)\pi x/a\right]}{(2n+1)^4}$$

$$\times \left\{ 1 - 2\frac{\left[\sinh\frac{(2n+1)\pi b}{2a} + \frac{(2n+1)\pi b}{2a}\cosh\frac{(2n+1)\pi b}{2a}\right]\cosh\frac{(2n+1)\pi y}{a}}{\sinh\frac{(2n+1)\pi b}{a} + \frac{(2n+1)\pi b}{a}} \right.$$

$$\left. + 2\frac{\frac{(2n+1)\pi y}{a}\sinh\frac{(2n+1)\pi b}{2a}\sinh\frac{(2n+1)\pi y}{a}}{\sinh\frac{(2n+1)\pi b}{a} + \frac{(2n+1)\pi b}{a}} \right\}.$$

239. A rectangular elastic plate, simply supported along its boundary, is acted upon by bending moments m uniformly distributed along two opposite edges (see Figure 49). Find the deflection of an arbitrary point of the plate.

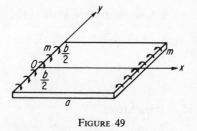

FIGURE 49

$$u(x, y) = \frac{4ma^2}{\pi^3 D}\left\{ \frac{\pi^3 x}{8a}\left(1 - \frac{x}{a}\right) \right.$$

$$-\frac{1}{2}\sum_{n=0}^{\infty}\left(\frac{\left[2\cosh\frac{(2n+1)\pi b}{2a} + \frac{(2n+1)\pi b}{2a}\sinh\frac{(2n+1)\pi b}{2a}\right]\cosh\frac{(2n+1)\pi y}{a}}{(2n+1)^3\cosh^2\frac{(2n+1)\pi b}{2a}}\right.$$

$$\left.\left. - \frac{\frac{(2n+1)\pi y}{a}\cosh\frac{(2n+1)\pi b}{2a}\sinh\frac{(2n+1)\pi y}{a}}{(2n+1)^3\cosh^2\frac{(2n+1)\pi b}{2a}}\right)\sin\frac{(2n+1)\pi x}{a} \right\}.$$

*240. Solve the preceding problem, assuming that the edges $y = \pm b/2$ are clamped.

$$u(x, y) = \frac{4ma^2}{\pi^3 D} \left\{ \frac{\pi^3 x}{8a}\left(1 - \frac{x}{a}\right) \right.$$

$$-2\sum_{n=0}^{\infty} \left(\frac{\left[\sinh\dfrac{(2n+1)\pi b}{2a} + \dfrac{(2n+1)\pi b}{2a}\cosh\dfrac{(2n+1)\pi b}{2a}\right]\cosh\dfrac{(2n+1)\pi y}{a}}{\left[\sinh\dfrac{(2n+1)\pi b}{a} + \dfrac{(2n+1)\pi b}{a}\right](2n+1)^3} \right.$$

$$\left. \left. - \frac{\dfrac{(2n+1)\pi y}{a}\sinh\dfrac{(2n+1)\pi b}{2a}\sinh\dfrac{(2n+1)\pi y}{a}}{\left[\sinh\dfrac{(2n+1)\pi b}{a} + \dfrac{(2n+1)\pi b}{a}\right](2n+1)^3} \right) \sin\frac{(2n+1)\pi x}{a} \right\}.$$

*241. Find the deflection of the center of a circular plate of radius a with a clamped boundary under the action of a line load p uniformly distributed along one of its radii.

Ans.

$$u\big|_{r=0} = \frac{pa^3}{64\pi D}.$$

3. Heat Conduction: Nonstationary Problems

*242. A slab is heated by a thermal current of constant density q flowing through the face $x = 0$ starting from the time $t = 0$, while the face $x = a$ is held at temperature T_0. Find the subsequent temperature distribution in the slab, assuming that the initial temperature of the slab is zero.

Ans.

$$T(x, t) = T_0 + \frac{qa}{k}\left(1 - \frac{x}{a}\right)$$

$$- \frac{8a}{\pi^2}\sum_{n=0}^{\infty}\left[\frac{q}{k} + (-1)^n\frac{2n+1}{2a}\pi T_0\right]e^{-(2n+1)^2\pi^2\tau/4a^2}\frac{\cos\left[(2n+1)\pi x/2a\right]}{(2n+1)^2}.$$

where k is the thermal conductivity, c the specific heat and ρ the density of the slab, and $\tau = kt/c\rho$.

243. Solve the preceding problem, assuming that the face $x = 0$ is held at temperature $T = f(\tau)$, while the other face $x = a$ is held at temperature zero. Consider the special case $f(\tau) = A\tau$.

Ans.

$$T(x, t) = \frac{2\pi}{a^2} \sum_{n=1}^{\infty} n \sin \frac{n\pi x}{a} \int_0^\tau f(s)e^{-n^2\pi^2(\tau-s)/a^2}\, ds.$$

In the special case.

$$T(x, t) = A\left[\left(1 - \frac{x}{a}\right)\tau - \frac{2a^2}{\pi^3} \sum_{n=1}^{\infty} \frac{1 - e^{-n^2\pi^2\tau/a^2}}{n^3} \sin \frac{n\pi x}{a} \right].$$

244. Find the temperature distribution in a slab if the face $x = 0$ radiates heat into the surrounding medium according to Newton's law, while the other face $x = a$ is held at the temperature T_0 equal to the initial temperature of the slab.

Ans.

$$T(x, t) = T_0\left\{ \frac{1 + hx}{1 + ha} - 2 \sum_{n=1}^{\infty} \frac{\cos \gamma_n}{\gamma_n[1 - (\sin 2\gamma_n/2\gamma_n)]} e^{-\gamma_n^2\tau/a^2} \sin \frac{\gamma_n(a - x)}{a} \right\},$$

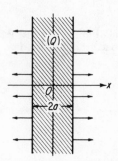

FIGURE 50

where h is the heat exchange coefficient figuring in Newton's law, and the γ_n are consecutive positive roots of the equation

$$\tan \gamma = - \frac{\gamma}{ah}.$$

245. Find the temperature distribution in a conductor with the cross section shown in Figure 50, heated from the time $t = 0$ by a d-c current producing Joule heat with density Q. It is assumed that the initial temperature is zero, and that the loss of heat into the surrounding medium is described by Newton's law.

Ans.

$$T(x, t) = \frac{2Qa^2}{k}\left\{ \frac{1}{4}\left(1 - \frac{x^2}{a^2}\right) + \frac{1}{2ah} - \sum_{n=1}^{\infty} \frac{\sin \gamma_n e^{-\gamma_n^2\tau/a^2}}{\gamma_n^3[1 + (\sin 2\gamma_n/2\gamma_n)]} \cos \frac{\gamma_n x}{a} \right\},$$

where the γ_n are consecutive positive roots of the equation

$$\tan \gamma = \frac{ah}{\gamma}.$$

Hint. Unless a particular solution of the inhomogeneous equation is subtracted out first, the expansion

$$\frac{1}{4}\left(1 - \frac{x^2}{a^2}\right) + \frac{1}{2ah} = \sum_{n=1}^{\infty} \frac{\sin \gamma_n}{\gamma_n^3[1 + (\sin 2\gamma_n/2\gamma_n)]} \cos \frac{\gamma_n x}{a}, \qquad -a < x < a$$

must be used to reduce the solution to the form given in the answer.

246. Find the temperature distribution $T(r, t)$ in a cylinder of radius a whose surface temperature varies according to the law

$$T\big|_{r=a} = f(\tau),$$

where $\tau = kt/c\rho$, assuming that the initial temperature of the cylinder is zero. Consider the special cases a) $f(\tau) = A\tau$; b) $f(\tau) = A \sin \omega\tau$.

Ans.

$$T(r, t) = \frac{2}{a^2} \sum_{n=1}^{\infty} \frac{\gamma_n J_0(\gamma_n r/a)}{J_1(\gamma_n)} \int_0^{\tau} f(s) e^{-\gamma_n^2(\tau-s)/a^2}\, ds,$$

where the γ_n are consecutive positive roots of the equation $J_0(\gamma) = 0$. In the special cases,

a) $$T(r, t) = A\left[\tau - 2a^2 \sum_{n=1}^{\infty} \frac{1 - e^{-\gamma_n^2 \tau/a^2}}{\gamma_n^3 J_1(\gamma_n)} J_0\left(\frac{\gamma_n r}{a}\right)\right];$$

b) $$T(r, t) = A\left\{\sin \omega\tau + 2\omega a^2 \sum_{n=1}^{\infty} \frac{\gamma_n}{(\gamma_n^4 + a^4\omega^2) J_1(\gamma_n)}\right.$$
$$\left. \times \left[e^{-\gamma_n^2 \tau/a^2} - \cos \omega\tau - \frac{a^2\omega}{\gamma_n^2} \sin \omega\tau\right] J_0\left(\frac{\gamma_n r}{a}\right)\right\}.$$

Hint. Use formula 17, p. 385.

*247. Find the temperature distribution in a cylindrical conductor of radius a heated from the time $t = 0$ by a d-c current producing Joule heat with density Q. It is assumed that the initial temperature distribution is zero and that the loss of heat from the surface of the cylinder is described by Newton's law.

Ans.

$$T(r, t) = \frac{2Qa^2}{k}\left\{\frac{1}{8}\left(1 - \frac{r^2}{a^2}\right) + \frac{1}{4ah} - ah \sum_{n=1}^{\infty} \frac{e^{-\gamma_n^2 \tau/a^2} J_0(\gamma_n r/a)}{\gamma_n^4[1 + (ah/\gamma_n)^2] J_0(\gamma_n)}\right\},$$

where the γ_n are consecutive positive roots of the equation

$$\gamma J_1(\gamma) = ah J_0(\gamma).$$

248. Solve Prob. 150, using the method of this chapter.

Ans.[10]

$$T(r, t) = \frac{2Qa^2}{k} \sum_{n=1}^{\infty} \frac{1 - e^{-\gamma_n^2 \tau/a^2}}{\gamma_n^3 J_1(\gamma_n)} J_0\left(\frac{\gamma_n r}{a}\right), \qquad J_0(\gamma_n) = 0.$$

249. Find the temperature distribution in a cylinder of radius a in which heat is produced with volume density Q, assuming that the initial temperature

---

[10] To reduce the answer to the form given in Prob. 150, use formula 18, p. 385.

of the cylinder is zero and that heat flows out of the cylinder with surface density q.

Ans.

$$T(r, t) = \left(\frac{Q}{k} - \frac{2q}{ka}\right)\tau + \frac{qa}{4k}\left(1 - 2\frac{r^2}{a^2}\right) + \frac{2qa}{k}\sum_{n=1}^{\infty} \frac{e^{-\gamma_n^2 \tau/a^2}}{\gamma_n^2 J_0(\gamma_n)} J_0\left(\frac{\gamma_n r}{a}\right),$$

where the γ_n are consecutive positive roots of the equation $J_1(\gamma) = 0$.

Hint. Use formula 19, p. 385.

250. The outer surface of a cylindrical pipe $a \leqslant r \leqslant b$ is held at temperature $T|_{r=b} = f(\tau)$, while the inner surface is held at temperature zero. Find the temperature distribution, assuming that the initial temperature is zero.

Ans.

$$T(r, t) = \frac{\ln(r/a)}{\ln(b/a)} f(\tau)$$

$$+ \pi \sum_{n=1}^{\infty} \frac{J_0(\gamma_n)J_0(\gamma_n b/a)}{J_0^2(\gamma_n b/a) - J_0^2(\gamma_n)} R_{\gamma_n}(r)\left[f(\tau) - \frac{\gamma_n^2}{a^2}\int_0^\tau f(s)e^{-\gamma_n^2(\tau-s)/a^2} ds\right],$$

where

$$R_\gamma(r) = Y_0(\gamma)J_0\left(\frac{\gamma r}{a}\right) - J_0(\gamma)Y_0\left(\frac{\gamma r}{a}\right)$$

is a linear combination of Bessel functions of the first and second kinds, and the γ_n are consecutive positive roots of the equation $R_\gamma(b) = 0$.

251. Find the temperature distribution in a cylinder $0 \leqslant r \leqslant a, 0 \leqslant z \leqslant l$, assuming that the initial temperature is zero, and that starting from the time $t = 0$, the face $z = l$ of the cylinder is held at temperature T_0, while the rest of the surface is held at temperature zero.

Ans.

$$T(r, z, t) = T_0\left[\frac{z}{l} + \frac{2}{\pi}\sum_{m=1}^{\infty} \frac{(-1)^m}{m} \frac{I_0(m\pi r/l)}{I_0(m\pi a/l)} \sin\frac{m\pi z}{l}\right.$$

$$+ 4\pi\sum_{m=1}^{\infty}\sum_{n=1}^{\infty} \frac{(-1)^m m}{(m\pi)^2 + (\gamma_n l/a)^2}$$

$$\left.\times \exp\left\{-\left[(m\pi)^2 + \left(\frac{\gamma_n l}{a}\right)^2\right]\frac{\tau}{l^2}\right\} \frac{J_0(\gamma_n r/a)}{\gamma_n J_1(\gamma_n)} \sin\frac{m\pi z}{l}\right],$$

where $J_0(x)$, $J_1(x)$ and $I_0(x)$ are Bessel functions, and the γ_n are consecutive positive roots of the equation $J_0(\gamma) = 0$.

Hint. Make the boundary conditions homogeneous by setting

$$T = T_0\frac{z}{l} + u.$$

252. Solve the preceding problem, assuming that the surface of the cylinder is held at temperature zero and that heat is produced inside the cylinder with density Q.

Ans.

$$T(r, z, t) = \frac{8Qa^2}{\pi k} \sum_{m=1,3,5,\ldots}^{\infty} \sum_{n=1}^{\infty} \frac{J_0(\gamma_n r/a)}{\gamma_n J_1(\gamma_n)} \frac{1 - e^{-[(\gamma_n/a)^2 + (m\pi/l)^2]\tau}}{\gamma_n^2 + (m\pi a/l)^2} \frac{\sin(m\pi z/l)}{m},$$

where the γ_n are consecutive positive roots of the equation $J_0(\gamma) = 0$.

253. Find the temperature distribution in a sphere of radius a inside which heat is produced with density Q, starting from the time $t = 0$. It is assumed that the sphere is initially at temperature zero and that its surface is held at constant temperature zero.

Ans.

$$T(r, t) = T_0 + \frac{Q}{6k}(a^2 - r^2) + \frac{2a}{\pi r} \sum_{n=1}^{\infty} \frac{(-1)^n}{n} \left(T_0 + \frac{Qa^2}{kn^2\pi^2} \right) e^{-n^2\pi^2\tau/a^2} \sin\frac{n\pi r}{a}.$$

254. Solve the preceding problem if a) heat flows out of the sphere with surface density q; b) heat is radiated into the surrounding medium according to Newton's law.

Ans.

a) $$T(r, t) = \left(\frac{Q}{k} - \frac{3q}{ka} \right)\tau - \frac{2qa^2}{kr} \sum_{n=1}^{\infty} \frac{1 - e^{-\gamma_n^2\tau/a^2}}{\gamma_n^2 \sin\gamma_n} \sin\frac{\gamma_n r}{a},$$

where the γ_n are consecutive positive roots of the equation $\tan\gamma = \gamma$;

b) $$T(r, t) = \frac{2Qha^4}{kr(1 - ah)} \sum_{n=1}^{\infty} \frac{\cos\gamma_n}{\gamma_n^3[1 - (\sin 2\gamma_n/2\gamma_n)]} (1 - e^{-\gamma_n^2\tau/a^2}) \sin\frac{\gamma_n r}{a},$$

where h is the heat exchange coefficient and the γ_n are consecutive positive roots of the equation

$$\tan\gamma = \frac{\gamma}{1 - ah}.$$

255. A diffusing substance enters a thin tube of length l with impermeable walls. Find the concentration distribution in the tube if the density with which the substance flows into the end $x = 0$ is a given function of time $q(t)$. It is assumed that the initial concentration in the tube is zero and that the other end of the tube is joined to a vessel in which a given concentration C_l is maintained.

Ans.

$$C(x, t) = C_l - 2 \sum_{n=0}^{\infty} \left[\frac{2(-1)^n C_l}{\pi(2n+1)} e^{-D(2n+1)^2 \pi^2 t/4l^2} \right.$$

$$\left. - \frac{1}{l} \int_0^l e^{-D(2n+1)^2 \pi^2 (t-\tau)/4l^2} q(\tau) \, d\tau \right] \cos \frac{(2n+1)\pi x}{2l} ,$$

where D is the diffusion coefficient.

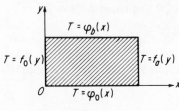

FIGURE 51

4. Heat Conduction: Stationary Problems

256. Find the stationary temperature distribution in a bar of rectangular cross section, given the temperature distribution on its faces (see Figure 51).

Ans.

$$T(x, y) = \frac{2}{a} \sum_{n=1}^{\infty} \left\{ \sinh \frac{n\pi y}{a} \int_0^a \varphi_b(\xi) \sin \frac{n\pi\xi}{a} \, d\xi \right.$$

$$+ \sinh \frac{n\pi(b-y)}{a} \int_0^a \varphi_0(\xi) \sin \frac{n\pi\xi}{a} \, d\xi$$

$$\left. + \int_0^b [f_0(\eta) - (-1)^n f_a(\eta)] G_n(\eta, y) \, d\eta \right\} \frac{\sin (n\pi x/a)}{\sinh (n\pi b/a)} ,$$

where

$$G_n(\eta, y) = \begin{cases} \sinh \dfrac{n\pi\eta}{a} \sinh \dfrac{n\pi(b-y)}{a} , & \eta \leqslant y, \\[2mm] \sinh \dfrac{n\pi y}{a} \sinh \dfrac{n\pi(b-\eta)}{a} , & \eta \geqslant y. \end{cases}$$

257. Study the special case of the preceding problem corresponding to the boundary conditions

$$f_0(y) = 0, \qquad f_a(y) = T_0, \qquad \varphi_0(x) = T_0, \qquad \varphi_b(x) = 0.$$

Ans.

$$T(x, y) = T_0 \left[\frac{x}{a} + \frac{2}{\pi} \sum_{n=1}^{\infty} \frac{\sinh \dfrac{n\pi(b-y)}{a} + (-1)^n \sinh \dfrac{n\pi y}{a}}{n \sinh \dfrac{n\pi b}{a}} \sin \frac{n\pi x}{a} \right].$$

Hint. Use formula 2, p. 384.

258. A heat current Q flows into a bar of rectangular cross section through two opposite faces and leaves the bar through the other two faces (see Figure 52). Find the stationary temperature distribution in the bar, assuming that both the incoming and the outgoing currents are uniformly distributed over the faces.

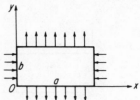

Ans.

$$T(x, y) = \frac{Q}{2abk} \left[y(b - y) - x(a - x) \right],$$

where k is the thermal conductivity.

Hint. To obtain the solution in closed form, use formula 9, p. 385.

FIGURE 52

259. Solve the preceding problem for an arbitrary distribution of current density on the face, i.e.,

$$q_x|_{x=0} = f_0(y), \qquad q_x|_{x=a} = f_a(y), \qquad q_y|_{y=0} = \varphi_0(x), \qquad q_y|_{y=b} = \varphi_b(x),$$

where the functions on the right satisfy the condition

$$\int_0^a [\varphi_b(x) - \varphi_0(x)]\, dx + \int_0^b [f_a(y) - f_0(y)]\, dy = 0$$

for the solvability of the Neumann problem.

Ans.

$$
\begin{aligned}
T(x, y) = {} & \frac{1}{ak} \int_0^y (y - \eta)[f_a(\eta) - f_0(\eta)]\, d\eta - \frac{y}{ak} \int_0^a \varphi_0(\xi)\, d\xi \\
& + \frac{2}{\pi k} \sum_{n=1}^{\infty} \left\{ \cosh \frac{n\pi(b - y)}{a} \int_0^a \varphi_0(\xi) \cos \frac{n\pi\xi}{a}\, d\xi \right. \\
& \quad - \cosh \frac{n\pi y}{a} \int_0^a \varphi_b(\xi) \cos \frac{n\pi\xi}{a}\, d\xi \\
& \quad \left. - \int_0^b [(-1)^n f_a(\eta) - f_0(\eta)] G_n(\eta, y)\, d\eta \right\} \frac{\cos (n\pi x/a)}{n \sinh (n\pi b/a)} + \text{const,}
\end{aligned}
$$

where

$$
G_n(\eta, y) =
\begin{cases}
\cosh \dfrac{n\pi\eta}{a} \cosh \dfrac{n\pi(b - y)}{a}, & \eta \leqslant y, \\[2ex]
\cosh \dfrac{n\pi y}{a} \cosh \dfrac{n\pi(b - \eta)}{a}, & \eta \geqslant y.
\end{cases}
$$

260. Two faces of a rectangular bar are thermally insulated, and the other two are held at temperature zero (see Figure 53). Find the stationary temperature distribution, assuming that heat is produced with density Q inside the bar.

Ans.

$$T(x, y) = \frac{16Qa^2}{\pi^3 k} \sum_{n=0}^{\infty} \frac{(-1)^n}{(2n + 1)^3} \left\{ 1 - \frac{\cosh [(2n + 1)\pi y/2a]}{\cosh [(2n + 1)\pi b/2a]} \right\} \cos \frac{(2n + 1)\pi x}{2a}.$$

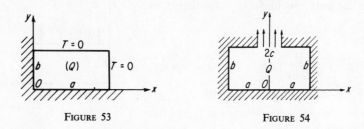

FIGURE 53   FIGURE 54

*261. Heat is produced with density Q in the bar shown in Figure 54. Find the stationary temperature distribution, assuming that a heat current of constant density q leaves the bar through the section $|x| < c$ of the upper face, while the rest of the surface of the bar is thermally insulated.

Ans.

$$T(x, y) = -\frac{Q}{k} \left[\frac{y^2}{2} + \frac{2a^2 b}{\pi^2 c} \sum_{n=1}^{\infty} \frac{\sin (n\pi c/a)}{n^2} \frac{\cosh (n\pi y/a)}{\sinh (n\pi b/a)} \cos \frac{n\pi x}{a} \right]$$

$$+ \text{ const, (14)}$$

where k is the thermal conductivity. Another form of the solution, suitable for $a/b \gg 1$, is

$$T(x, y) = \frac{Q}{kc} \left\{ \frac{a - c}{2} x^2 - \frac{a}{2} y^2 \right.$$

$$\left. + \frac{2ab^2}{\pi^2} \sum_{n=1}^{\infty} \frac{(-1)^n \sinh [n\pi(a - c)/b]}{n^2 \sinh (n\pi a/b)} \cosh \frac{n\pi x}{b} \cos \frac{n\pi y}{b} \right\} + \text{const},$$

$$|x| < c,$$

$$T(x, y) = \frac{Q}{k} \left\{ a |x| - \frac{x^2}{2} \right.$$

$$\left. - \frac{2ab^2}{\pi^2 c} \sum_{n=1}^{\infty} \frac{(-1)^n \sinh (n\pi c/b)}{n^2 \sinh (n\pi a/b)} \cosh \frac{n\pi(a - |x|)}{b} \cos \frac{n\pi y}{b} \right\} + \text{const},$$

$$|x| > c.$$

262. Find the stationary temperature distribution in a conductor of rectangular cross section heated by a d-c current producing heat with density Q, if the surface of the conductor gives off heat according to Newton's law.

Ans.

$$T(x, y) = \frac{4Qa^2}{k} \sum_{n=1}^{\infty} \frac{\sin \gamma_n}{\gamma_n^2 (2\gamma_n + \sin 2\gamma_n)}$$

$$\times \left[1 - \frac{ah}{\gamma_n} \frac{1}{\tanh (\gamma_n b/a) + (ah/\gamma_n)} \frac{\cosh (\gamma_n y/a)}{\cosh (\gamma_n b/a)} \right] \cos \frac{\gamma_n x}{a},$$

where h is the heat exchange coefficient, and the γ_n are consecutive positive roots of the equation

$$\tan \gamma = \frac{ah}{\gamma}.$$

263. Find the stationary temperature distribution in a rectangular parallelepiped $0 \leqslant x \leqslant a$, $0 \leqslant y \leqslant b$, $0 \leqslant z \leqslant c$, if the faces $x = 0$, $y = 0$, $z = 0$ are held at temperature zero, while the other faces have the temperature distribution

$$T|_{x=a} = f_a(y, z), \qquad T|_{y=b} = f_b(x, z), \qquad T|_{z=c} = f_c(x, y).$$

Ans.

$$T(x, y, z) = \frac{4}{ab} \sum_{m=1}^{\infty} \sum_{n=1}^{\infty} \left\{ \frac{\sinh \gamma_{mn} z}{\sinh \gamma_{mn} c} \int_0^a \int_0^b f_c(\xi, \eta) \sin \frac{m\pi\xi}{a} \sin \frac{n\pi\eta}{b} \, d\xi \, d\eta \right.$$

$$- \frac{1}{\gamma_{mn}} \int_0^c G_{mn}(\zeta, z) \left[(-1)^m \frac{m\pi}{a} \int_0^b f_a(\eta, \zeta) \sin \frac{n\pi\eta}{b} \, d\eta \right.$$

$$\left. + (-1)^n \frac{n\pi}{b} \int_0^a f_b(\xi, \zeta) \sin \frac{m\pi\xi}{a} \, d\xi \right] d\zeta \left. \right\} \sin \frac{m\pi x}{a} \sin \frac{n\pi y}{b},$$

where

$$\gamma_{mn} = \pi \sqrt{\frac{m^2}{a} + \frac{n^2}{b^2}},$$

$$G_{mn}(\zeta, z) = \frac{1}{\sinh \gamma_{mn} c} \times \begin{cases} \sinh \gamma_{mn}\zeta \, \sinh \gamma_{mn}(c - z), & \zeta \leqslant z, \\ \sinh \gamma_{mn} z \, \sinh \gamma_{mn}(c - \zeta), & \zeta \geqslant z. \end{cases}$$

264. A heat current Q enters a bar of semicircular cross section through its plane face and leaves through the curved face (see Figure 55). Find the stationary temperature distribution in the bar, assuming that the incoming and outgoing currents have constant density.

Ans.

$$T(r, \varphi) = \frac{Qr}{\pi ka} \left[1 - 2 \sum_{n=1}^{\infty} (-1)^n \right.$$

$$\times \left. \frac{1 - (1/2n)(r/a)^{2n-1}}{4n^2 - 1} \cos 2n\varphi \right] + \text{const},$$

where k is the thermal conductivity.

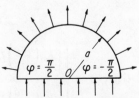

FIGURE 55

265. Find the temperature distribution in a bar whose cross section is the "curvilinear rectangle" $a \leqslant r \leqslant b$, $0 \leqslant \varphi \leqslant a$, given the following temperature distribution on the faces of the bar:

$$T|_{r=a} = 0, \qquad T|_{r=b} = T_0, \qquad T|_{\varphi=0} = 0, \qquad T|_{\varphi=a} = T_0.$$

Ans.

$$T(r, \varphi) = T_0 \left[\frac{\ln (r/a)}{\ln (b/a)} + \frac{2}{\pi} \right.$$

$$\times \sum_{n=1}^{\infty} \frac{(-1)^n \sinh \dfrac{n\pi(\alpha - \varphi)}{\ln (b/a)} + \sinh \dfrac{n\pi\varphi}{\ln (b/a)}}{n \sinh \dfrac{n\pi\alpha}{\ln (b/a)}} \left. \sin \frac{n\pi \ln (r/a)}{\ln (b/a)} \right].$$

Another form of the solution is

$$T(r, \varphi) = T_0 \left[\frac{\varphi}{\alpha} + \frac{2}{\pi} \right.$$

$$\times \sum_{n=1}^{\infty} \frac{\left[1 - (-1)^n \left(\dfrac{a}{b}\right)^{n\pi/\alpha}\right] \left(\dfrac{r}{a}\right)^{n\pi/\alpha} - \left[1 - (-1)^n \left(\dfrac{b}{a}\right)^{n\pi/\alpha}\right] \left(\dfrac{a}{r}\right)^{n\pi/\alpha}}{n \left[\left(\dfrac{b}{a}\right)^{n\pi/\alpha} - \left(\dfrac{a}{b}\right)^{n\pi/\alpha} \right]} \left. \sin \frac{n\pi\varphi}{\alpha} \right].$$

266. Find the stationary temperature distribution $T(r, z)$ in a cylinder $0 \leqslant r \leqslant a$, $0 \leqslant z \leqslant l$ with an arbitrary axially symmetric temperature distribution along its surface:

$$T|_{z=0} = f_0(r), \qquad T|_{z=l} = f_l(r), \qquad T_{r=a} = \varphi(z).$$

Ans.

$$T(r, z) = \frac{2}{l} \sum_{n=1}^{\infty} \frac{1}{I_0(n\pi a/l)} \left\{ I_0\left(\frac{n\pi r}{l}\right) \int_0^l \varphi(\zeta) \sin \frac{n\pi\zeta}{l} \, d\zeta \right.$$

$$\left. + \frac{n\pi}{l} \int_0^a [(-1)^n f_l(\rho) - f_0(\rho)] \rho G_n(r, \rho) \, d\rho \right\} \sin \frac{n\pi z}{l},$$

$$G_n(r, \rho) = \begin{cases} \left[K_0\left(\dfrac{n\pi a}{l}\right) I_0\left(\dfrac{n\pi r}{l}\right) - I_0\left(\dfrac{n\pi a}{l}\right) K_0\left(\dfrac{n\pi r}{l}\right) \right] I_0\left(\dfrac{n\pi\rho}{l}\right), & \rho \leqslant r, \\[12pt] \left[K_0\left(\dfrac{n\pi a}{l}\right) I_0\left(\dfrac{n\pi\rho}{l}\right) - I_0\left(\dfrac{n\pi a}{l}\right) K_0\left(\dfrac{n\pi\rho}{l}\right) \right] I_0\left(\dfrac{n\pi r}{l}\right), & \rho \geqslant r, \end{cases}$$

where $I_0(x)$ and $K_0(x)$ are cylinder functions of imaginary argument. Another form of the solution is

$$T(r, z) = \frac{2}{a^2} \sum_{n=1}^{\infty} \frac{1}{\sinh(\gamma_n l/a) J_1^2(\gamma_n)} \left\{ \sinh \frac{\gamma_n z}{a} \int_0^a f_l(\rho) \rho J_0\left(\frac{\gamma_n \rho}{a}\right) d\rho \right.$$

$$+ \sinh \frac{\gamma_n(l-z)}{a} \int_0^a f_0(\rho) \rho J_0\left(\frac{\gamma_n \rho}{a}\right) d\rho$$

$$\left. + a J_1(\gamma_n) \int_0^l \varphi(\zeta) G_n(z, \zeta) \, d\zeta \right\} J_0\left(\frac{\gamma_n r}{a}\right),$$

$$G_n(z, \zeta) = \begin{cases} \sinh \dfrac{\gamma_n \zeta}{a} \sinh \dfrac{\gamma_n(l-z)}{a}, & \zeta \leqslant z, \\[2mm] \sinh \dfrac{\gamma_n z}{a} \sinh \dfrac{\gamma_n(1-\zeta)}{a}, & \zeta \geqslant z, \end{cases}$$

where the γ_n are consecutive positive roots of the equation $J_0(\gamma) = 0$.

267. A heat current Q enters a cylinder $0 \leqslant r \leqslant a$, $0 \leqslant z \leqslant l$ through its ends and leaves through the lateral surface. Find the temperature distribution in the cylinder, assuming that the incoming and outgoing currents have constant density.

Ans.

$$T(r, z) = \frac{Ql}{2\pi a^2 k}\left[\frac{z^2}{l^2} - \frac{z}{l} - \frac{r^2}{2l^2}\right] + \text{const},$$

where k is the thermal conductivity.

268. Find the temperature distribution in a cylinder $0 \leqslant r \leqslant a$, $-l \leqslant z \leqslant l$ inside which heat is produced with density Q, if the surface radiates heat into the surrounding medium according to Newton's law.

Ans.

$$T(r, z) = \frac{2Ql^2}{k} \sum_{n=1}^{\infty} \frac{\sin \gamma_n}{\gamma_n^3[1 + (\sin 2\gamma_n/2\gamma_n)]}$$

$$\times \left[1 - \frac{\dfrac{hl}{\gamma_n}}{\dfrac{I_1(\gamma_n a/l)}{I_0(\gamma_n a/l)} + \dfrac{hl}{\gamma_n}} \frac{I_0(\gamma_n r/l)}{I_0(\gamma_n a/l)} \right] \cos \frac{\gamma_n z}{l},$$

where h is the heat exchange coefficient, $I_0(x)$ and $I_1(x)$ are Bessel functions of imaginary argument, and the γ_n are consecutive positive roots of the equation

$$\tan \gamma = \frac{hl}{\gamma}.$$

Another form of the solution is

$$T(r, z) = \frac{2Qha^3}{k} \sum_{n=1}^{\infty} \frac{1 - \dfrac{(ah/\gamma_n)\cosh(\gamma_n z/a)}{[\tanh(\gamma_n l/a) + (ah/\gamma_n)]\cosh(\gamma_n l/a)}}{[1 + (ah/\gamma_n)^2]\gamma_n^4 J_0(\gamma_n)} J_0\left(\frac{\gamma_n r}{a}\right),$$

where the γ_n are consecutive positive roots of the equation

$$\gamma J_1(\gamma) = ah J_0(\gamma).$$

*269. A thin wire heated by a d-c current producing Joule heat Q per unit length is placed inside a cylindrical object (see Figure 56). Find the temperature distribution in the object, assuming that the lateral surface of the cylinder is held at temperature zero, while the ends radiate heat into the surrounding medium according to Newton's law.

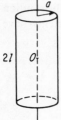

FIGURE 56

Ans.

$$T(r, z) = \frac{Q}{\pi k} \sum_{n=1}^{\infty} \frac{1}{\gamma_n^2 J_1^2(\gamma_n)}$$

$$\times \left[1 - \frac{ah\cosh(\gamma_n z/a)}{\gamma_n \sinh(\gamma_n l/a) + ah\cosh(\gamma_n l/a)} \right] J_0\left(\frac{\gamma_n r}{a}\right),$$

where the γ_n are consecutive positive roots of the equation $J_0(\gamma) = 0$.

Hint. Replace the line source by a source distributed over a cylinder of small radius ε, and then take the limit as $\varepsilon \to 0$.

270. Find the stationary temperature distribution $T(r, \theta)$ in a sphere of radius a, assuming that heat is produced with density Q inside the sphere, while the boundary condition

$$\left(\frac{\partial T}{\partial r} + hT \right) \bigg|_{r=a} = f(\theta),$$

involving a given function $f(\theta)$, is satisfied on the surface of the sphere.

Ans.

$$T(r, \theta) = \frac{Q}{6k}(a^2 - r^2) + \frac{Qa}{3kh} + \frac{1}{2h}\int_0^\pi f(\theta)\sin\theta\,d\theta$$

$$+ \frac{a}{2}\sum_{n=1}^\infty \frac{2n+1}{ah+n}\left(\frac{r}{a}\right)^n P_n(\cos\theta)\int_0^\pi f(\theta)P_n(\cos\theta)\sin\theta\,d\theta.$$

5. Electricity and Magnetism

271. Calculate the two-dimensional electrostatic field due to the electrodes shown in Figures 57(a) and 57(b).

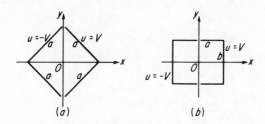

FIGURE 57

Ans.

a) $u(x, y) = \dfrac{\sqrt{2}\,V}{a}(x + y)$

$$- \frac{4V}{\pi}\sum_{n=1}^\infty \left\{(-1)^n \sinh\frac{n\pi[a - \sqrt{2}(y - x)]}{2a}\right.$$

$$\left.+ \sinh\frac{n\pi[a + \sqrt{2}(y - x)]}{2a}\right\} \frac{\sin\dfrac{n\pi[a - \sqrt{2}(y + x)]}{2a}}{n\sinh n\pi}$$

b) $u(x, y) = V\left[\dfrac{x}{a} + \dfrac{2}{\pi}\sum_{n=1}^\infty \dfrac{\cosh(n\pi y/a)}{\cosh(n\pi b/a)}\dfrac{\sin(n\pi x/a)}{n}\right],$

where $u(x, y)$ is the electrostatic potential.

*272. Find the electrostatic field in the electron-optical device shown in Figure 58.[11] What is the distribution of potential along the axis of symmetry?

---

[11] By an *electron-optical device* (for example, a lens), we mean a system of conductors at given potentials producing an electrostatic field used to govern the trajectories of charged particles.

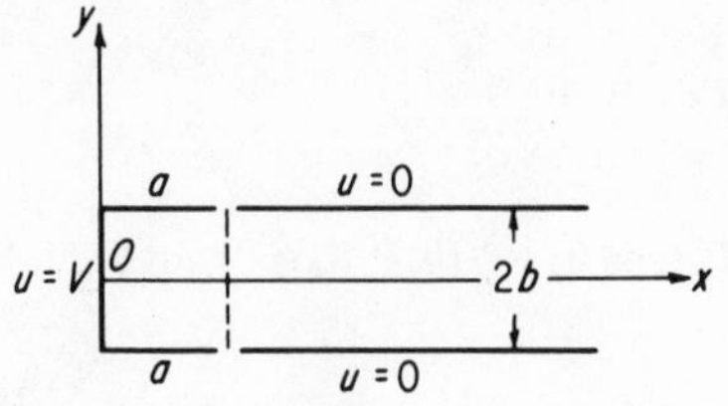

Ans.

FIGURE 58

$$u\big|_{y=0,x<a} = V\left[1 + \frac{4}{\pi}\sum_{n=0}^{\infty}\frac{(-1)^{n+1}}{2n+1}\right.$$
$$\left. \times e^{-(2n+1)\pi a/2b}\sinh\frac{(2n+1)\pi x}{2b}\right],$$

$$u\big|_{y=0,x>a} = \frac{4V}{\pi}\sum_{n=0}^{\infty}\frac{(-1)^n}{2n+1}$$
$$\times \cosh\frac{(2n+1)\pi a}{2b}\,e^{-(2n+1)\pi a/2b}.$$

273. Find the electrostatic potential $u(x, y)$ inside a box of rectangular cross section $0 \leqslant x \leqslant a$, $0 \leqslant y \leqslant b$ with grounded walls, due to a charged wire along the line $x = x_0$, $y = y_0$.

Ans.

$$u(x, y) = 8q\sum_{n=1}^{\infty}\frac{\sinh\dfrac{n\pi(a - x_0)}{b}\sin\dfrac{n\pi y_0}{b}}{n\sinh\dfrac{n\pi a}{b}}\sinh\frac{n\pi x}{b}\sin\frac{n\pi y}{b}, \qquad x < x_0,$$

$$u(x, y) = 8q\sum_{n=1}^{\infty}\frac{\sinh\dfrac{n\pi x_0}{b}\sin\dfrac{n\pi y_0}{b}}{n\sinh\dfrac{n\pi a}{b}}\sinh\frac{n\pi(a - x)}{b}\sin\frac{n\pi y}{b}, \qquad x > x_0,$$

where q is the charge per unit length of the wire.

Hint. Solve Poisson's equation, regarding the charge as uniformly distributed over the rectangle $x_0 - \delta < x < x_0 + \delta$, $y_0 - \varepsilon < y < y_0 + \varepsilon$, and then take the limit as $\delta, \varepsilon \to 0$.

274. Find the electrostatic field $u(x, y, z)$ due to a charge at the point x_0, y_0, z_0 inside a rectangular parallelepiped $0 \leqslant x \leqslant a, 0 \leqslant y \leqslant b, 0 \leqslant z \leqslant c$ with grounded walls.

Ans.

$$u(x, y, z) = \frac{16\pi q}{ac}\sum_{m=1}^{\infty}\sum_{n=1}^{\infty}\frac{G_{mn}(y, y_0)}{\gamma_{mn}}\sin\frac{m\pi x_0}{a}\sin\frac{n\pi z_0}{c}\sin\frac{m\pi x}{a}\sin\frac{n\pi z}{c},$$

where

$$\gamma_{mn} = \pi\sqrt{\frac{m^2}{a^2} + \frac{n^2}{c^2}},$$

$$G_{mn}(y, y_0) = \begin{cases} \dfrac{\sinh\gamma_{mn}y_0\sinh\gamma_{mn}(b - y)}{\sinh\gamma_{mn}b}, & y_0 < y, \\[4mm] \dfrac{\sinh\gamma_{mn}y\sinh\gamma_{mn}(b - y_0)}{\sinh\gamma_{mn}b}, & y_0 > y. \end{cases}$$

Hint. First assume that the charge is uniformly distributed over a small volume, and then pass to the limit.

275. A charged wire, with charge q per unit length, is placed inside a grounded metal box whose cross section is a "curvilinear rectangle" $a \leqslant r \leqslant b$, $0 \leqslant \varphi \leqslant \alpha$. Find the electrostatic potential $u(r, \varphi)$ inside the box.

Ans.

$$u(r, \varphi) = 4q \sum_{n=1}^{\infty} \frac{\left(\dfrac{b}{r_0}\right)^{n\pi/\alpha} - \left(\dfrac{r_0}{b}\right)^{n\pi/\alpha}}{\left(\dfrac{b}{a}\right)^{n\pi/\alpha} - \left(\dfrac{a}{b}\right)^{n\pi/\alpha}} \frac{\sin \dfrac{n\pi\varphi_0}{\alpha}}{n} \left[\left(\dfrac{r}{a}\right)^{n\pi/\alpha} - \left(\dfrac{a}{r}\right)^{n\pi/\alpha} \right] \sin \frac{n\pi\varphi}{\alpha},$$

$$r \leqslant r_0,$$

$$u(r, \varphi) = 4q \sum_{n=1}^{\infty} \frac{\left(\dfrac{r_0}{a}\right)^{n\pi/\alpha} - \left(\dfrac{a}{r_0}\right)^{n\pi/\alpha}}{\left(\dfrac{b}{a}\right)^{n\pi/\alpha} - \left(\dfrac{a}{b}\right)^{n\pi/\alpha}} \frac{\sin \dfrac{n\pi\varphi_0}{\alpha}}{n} \left[\left(\dfrac{b}{r}\right)^{n\pi/\alpha} - \left(\dfrac{r}{b}\right)^{n\pi/\alpha} \right] \sin \frac{n\pi\varphi}{\alpha},$$

$$r \geqslant r_0,$$

where r_0, φ_0 are the polar coordinates of the wire.

276. Examine the following special cases of the preceding problem:

a) $a = 0$, $b = \infty$ (charged wire inside a wedge);
b) $a = 0$, $b = \infty$, $\alpha = 2\pi$ (charged wire near the edge of a conducting half-plane);
c) $b = \infty$, $\alpha = \pi$, $\varphi_0 = \pi/2$ (charged wire over a plane with a semi-cylindrical boss).

Ans.

a) $$u(r, \varphi) = q \ln \frac{1 - 2\left(\dfrac{r}{r_0}\right)^{\pi/\alpha} \cos \dfrac{\pi(\varphi_0 + \varphi)}{\alpha} + \left(\dfrac{r}{r_0}\right)^{2\pi/\alpha}}{1 - 2\left(\dfrac{r}{r_0}\right)^{\pi/\alpha} \cos \dfrac{\pi(\varphi_0 - \varphi)}{\alpha} + \left(\dfrac{r}{r_0}\right)^{2\pi/\alpha}};$$

b) $$u(r, \varphi) = q \ln \frac{1 - 2\sqrt{\dfrac{r}{r_0}} \cos \dfrac{\varphi_0 + \varphi}{2} + \dfrac{r}{r_0}}{1 - 2\sqrt{\dfrac{r}{r_0}} \cos \dfrac{\varphi_0 - \varphi}{2} + \dfrac{r}{r_0}};$$

c) $$u(r, \varphi) = q \ln \left[\frac{1 + 2\dfrac{r}{r_0} \sin \varphi + \left(\dfrac{r}{r_0}\right)^2 \ 1 - 2\dfrac{a^2}{r_0 r} \sin \varphi + \left(\dfrac{a^2}{r_0 r}\right)^2}{1 - 2\dfrac{r}{r_0} \sin \varphi + \left(\dfrac{r}{r_0}\right)^2 \ 1 + 2\dfrac{a^2}{r_0 r} \sin \varphi + \left(\dfrac{a^2}{r_0 r}\right)^2} \right].$$

*277. Find the electrostatic field inside a grounded cylindrical shell $0 \leqslant r \leqslant a, 0 \leqslant z \leqslant l$ due to a charge q at the point $r = 0, z = c$.

Ans. The electrostatic potential is

$$u(r, z) = \frac{4q}{a} \sum_{n=1}^{\infty} \frac{\sinh\left[\gamma_n(l - c)/a\right]}{\sinh\left(\gamma_n l/a\right)} \sinh \frac{\gamma_n z}{a} \frac{J_0(\gamma_n r/a)}{\gamma_n J_1^2(\gamma_n)}, \qquad z < c,$$

$$u(r, z) = \frac{4q}{a} \sum_{n=1}^{\infty} \frac{\sinh\left(\gamma_n c/a\right)}{\sinh\left(\gamma_n l/a\right)} \sinh \frac{\gamma_n(l - z)}{a} \frac{J_0(\gamma_n r/a)}{\gamma_n J_1^2(\gamma_n)}, \qquad z > c,$$

where the γ_n are consecutive positive roots of the equation $J_0(\gamma) = 0$. Another form of the solution is

$$u(r, z) = \frac{4q}{l} \sum_{n=1}^{\infty} \frac{I_0(n\pi a/l)K_0(n\pi r/l) - K_0(n\pi a/l)I_0(n\pi r/l)}{I_0(n\pi a/l)} \sin \frac{n\pi c}{l} \sin \frac{n\pi z}{l}.$$

278. Find the electrostatic field inside a cylindrical shell $0 \leqslant r \leqslant a$, $0 \leqslant z \leqslant l$ whose ends and lateral surface are at the potentials V_0, V_l and V, respectively.

Ans. The electrostatic potential is

$$u(r, z) = V_0 \left(1 - \frac{z}{l}\right) + V_l \frac{z}{l}$$

$$+ \frac{2}{\pi} \sum_{n=1}^{\infty} \frac{[1 - (-1)^n]V + (-1)^n V_l - V_0}{n} \frac{I_0(n\pi r/l)}{I_0(n\pi a/l)} \sin \frac{n\pi z}{l}.$$

Another form of the solution is

$$u(r, z) = V + 2(V_l - V) \sum_{n=1}^{\infty} \frac{\sinh\left(\gamma_n z/a\right)}{\sinh\left(\gamma_n l/a\right)} \frac{J_0(\gamma_n r/a)}{\gamma_n J_1(\gamma_n)}$$

$$+ 2(V_0 - V) \sum_{n=1}^{\infty} \frac{\sinh\left[\gamma_n(l - z)/a\right]}{\sinh\left(\gamma_n l/a\right)} \frac{J_0(\gamma_n r/a)}{\gamma_n J_1(\gamma_n)},$$

where $J_0(x)$, $J_1(x)$ and $I_0(x)$ are cylinder functions, and the γ_n are consecutive positive roots of the equation $J_0(\gamma) = 0$.

279. Find the electrostatic potential along the axis of a cylindrical shell $0 \leqslant r \leqslant a, 0 \leqslant z \leqslant l$ if the lateral surface is held at a given potential

$$u\big|_{r=a} = f(z),$$

while the ends are held at potentials $V_0 = 0$ and $V_l = V$.

Ans.

$$u\big|_{r=0} = \frac{Vz}{l} + \frac{2}{l} \sum_{n=1}^{\infty} \left[\frac{Vl}{\pi} \frac{(-1)^n}{n} + \int_0^l f(\zeta) \sin \frac{n\pi\zeta}{l} \, d\zeta\right] \frac{\sin\left(n\pi z/l\right)}{I_0(n\pi a/l)},$$

where $I_0(x)$ is the Bessel function of imaginary argument.

280. Examine the special cases of the preceding problem which correspond to the following potential distributions on the lateral surface of the cylinder:

a) $f(z) = V_a$;

b) $f(z) = V_k$ for $\dfrac{(k-1)}{N} < z < \dfrac{k}{N}$ $(k = 1, 2, \ldots, N)$;

c) $f(z) = \begin{cases} 0, & 0 \leqslant z < c, \\ V, & c < z \leqslant l. \end{cases}$

Ans.

a) $u|_{r=0} = \dfrac{Vz}{l} + \dfrac{2}{\pi} \displaystyle\sum_{n=1}^{\infty} [(-1)^n (V - V_a) + V_a] \dfrac{\sin(n\pi z/l)}{n I_0(n\pi a/l)}$;

b) $u|_{r=0} = \dfrac{Vz}{l} + \dfrac{2}{\pi}$

$\times \displaystyle\sum_{n=1}^{\infty} \left[(-1)^n V + 2 \sin \dfrac{n\pi}{2N} \sum_{k=1}^{N} V_k \sin \dfrac{(2k-1)n\pi}{2N} \right] \dfrac{\sin(n\pi z/l)}{n I_0(n\pi a/l)}$;

c) $u|_{r=0} = V \left[\dfrac{z}{l} + \dfrac{2}{\pi} \displaystyle\sum_{n=1}^{\infty} \dfrac{\cos(n\pi c/l)}{n} \dfrac{\sin(n\pi z/l)}{I_0(n\pi a/l)} \right]$.

Comment. Case b corresponds to a piecewise constant potential, produced in electronic practice by the use of a voltage divider. Case c is the problem of the distribution of electrostatic potential between two conducting cylindrical caps separated by a negligibly small space.

281. What potential distribution must be maintained along the lateral surface of the cylinder of Prob. 279 in order to obtain the distribution

$$u|_{r=0} = V \left[\frac{z}{l} + \sum_{n=1}^{N} a_n \sin \frac{n\pi z}{l} \right]$$

along the axis, where the a_n are any given numbers?

Ans.

$$u|_{r=a} = V \left[\frac{z}{l} + \sum_{n=1}^{N} a_n I_0 \left(\frac{n\pi a}{l} \right) \sin \frac{n\pi z}{l} \right].$$

*282. Determine the electric field on the axis of the electron-optical lens shown in Figure 59, consisting of two cylinders at potentials V, if the potential distribution in the space between the cylinders is given approximately by the formula

$$u|_{r=a, -\delta \leqslant z \leqslant \delta} = V \sin \frac{\pi z}{2\delta} .$$

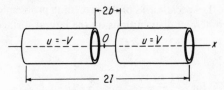

FIGURE 59

Ans.

$$E\big|_{r=0} = -\frac{V}{l}\left[1 + 2\sum_{n=1}^{\infty} \frac{\cos(n\pi\delta/l)}{1 - (2n\delta/l)^2}\frac{\cos(n\pi z/l)}{I_0(n\pi a/l)}\right].$$

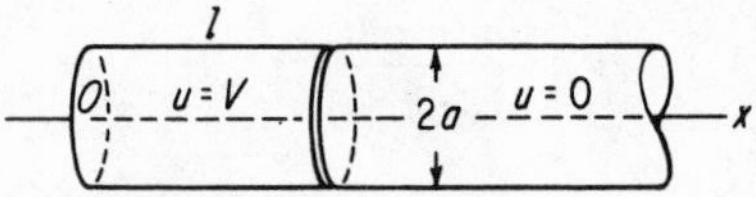

283. Find the potential distribution in the electron-optical device shown in Figure 60.

FIGURE 60

Ans.

$$u(r, z) = V\left[1 - 2\sum_{n=1}^{\infty} e^{-\gamma_n l/a} \sinh\frac{\gamma_n z}{a}\frac{J_0(\gamma_n r/a)}{\gamma_n J_1(\gamma_n)}\right], \qquad z < l,$$

$$u(r, z) = 2V\sum_{n=1}^{\infty}\cosh\frac{\gamma_n l}{a}\, e^{-\gamma_n z/a}\frac{J_0(\gamma_n r/a)}{\gamma_n J_1(\gamma_n)}, \qquad z > l,$$

where the γ_n are consecutive positive roots of the equation $J_0(\gamma) = 0$.

Hint. Use formula 17, p. 385.

284. Find the potential distribution in the electron-optical device shown in Figure 61, consisting of two semicylinders (with closed ends) at potentials $u = 0$ and $u = V$, separated by a negligibly small space.

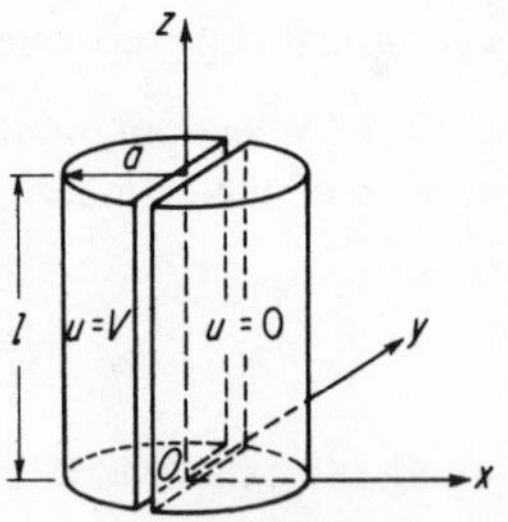

Ans.

$$u(r, \varphi, z) = \frac{V}{2} + \frac{4V}{\pi}\sum_{m=1}^{\infty}\sum_{n=1}^{\infty}\frac{\sin(m\pi/a)\cos m\varphi}{m}$$

$$\times\left[1 - \frac{m^2\int_0^a J_m\left(\dfrac{\gamma_{mn}r}{a}\right)\dfrac{dr}{r}}{\gamma_{mn}J'_m(\gamma_{mn})}\frac{\cosh\dfrac{\gamma_{mn}(2z-l)}{2a}}{\cosh\dfrac{\gamma_{mn}l}{2a}}\right]\frac{J'_m(\gamma_{mn})J_m(\gamma_{mn}r/a)}{J^2_{m+1}(\gamma_{mn})},$$

FIGURE 61

where $J_m(x)$ is the Bessel function of order m, a is the radius and l the length of the semicylinders, and the γ_{mn} are consecutive positive roots of the equation $J_m(\gamma) = 0$.

285. Find the distribution of d-c current in a thin conducting sheet, if a current J enters and leaves via point electrodes applied at the points $(\pm c, 0)$ [see Figure 62].[12]

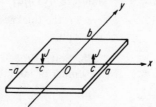

FIGURE 62

Ans. The potential of the current distribution is

$$u(x, y) = -\frac{2J}{\pi\sigma h} \sum_{n=0}^{\infty} \frac{\sin\left[(2n+1)\pi c/2a\right]}{(2n+1)\sinh\left[(2n+1)\pi b/2a\right]}$$
$$\times \cosh\frac{(2n+1)\pi(b-|y|)}{2a} \sin\frac{(2n+1)\pi x}{2a} + \text{const},$$

where h is the thickness and σ the conductivity of the sheet.

Hint. Regard the current as distributed over two small rectangles, and then pass to the limit.

286. Find the distribution of d-c current in a thin conducting disk of radius a, if a current J enters and leaves via point electrodes applied at the points $r = b, \varphi = 0$ and $r = b, \varphi = \pi$ $(b < a)$.

Ans.

$$u(r, \varphi) = \frac{J}{4\pi\sigma h} \ln \frac{\left[1 - 2\dfrac{r}{b}\cos\varphi + \dfrac{r^2}{b^2}\right]\left[1 - 2\dfrac{br}{a^2}\cos\varphi + \left(\dfrac{br}{a^2}\right)^2\right]}{\left[1 + 2\dfrac{r}{b}\cos\varphi + \dfrac{r^2}{b^2}\right]\left[1 + 2\dfrac{br}{a^2}\cos\varphi + \left(\dfrac{br}{a^2}\right)^2\right]}.$$

Hint. To represent the solution in closed form, use the expansion

$$-\frac{1}{2}\ln(1 - 2\rho\cos\varphi + \rho^2) = \sum_{n=1}^{\infty}\frac{\rho^n}{n}\cos n\varphi, \qquad 0 \leqslant \rho < 1.$$

287. Find the distribution of d-c current in a thin cylindrical shell of radius a, if a current J enters and leaves via point electrodes applied at the points $(a, -\pi/2, 0)$ and $(a, \pi/2, 0)$ [see Figure 63].

---

[12] The differential equation for the potential of the current distribution in a thin conducting shell is given in Prob. 21.

Ans.

$$u(\varphi, z) = -\frac{2J}{\pi \sigma h} \sum_{n=0}^{\infty} \frac{(-1)^n}{2n+1} \frac{\cosh\left[(2n+1)(l-|z|)/a\right]}{\sinh\left[(2n+1)l/a\right]} \sin(2n+1)\varphi + \text{const},$$

where h is the thickness and σ the conductivity of the shell.

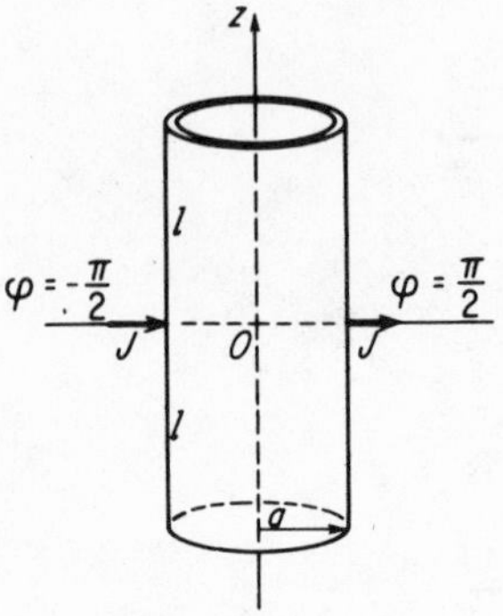

288. Solve Prob. 287 for the limiting case of a cylinder of infinite length.

Ans.

$$u(\varphi, z) = \frac{J}{2\pi \sigma h} \ln \frac{\cosh(z/a) - \sin \varphi}{\cosh(z/a) + \sin \varphi}.$$

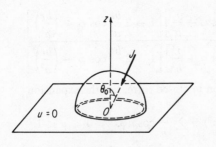

FIGURE 63

*289. A thin conducting shell of hemispherical shape lies on a plane base, made of a good conductor (see Figure 64). Find the distribution of d-c current in the shell, assuming that a current J enters the shell by an electrode applied to the hemisphere at the point $r = a$, $\theta = \theta_0$, $\varphi = 0$, while the current leaves through the rim of the hemisphere (in contact with the plane).

Ans.

$$u(\theta, \varphi) = \frac{J}{4\pi \sigma h} \ln \frac{1 - 2 \tan \frac{\theta_0}{2} \tan \frac{\theta}{2} \cos \varphi + \tan^2 \frac{\theta_0}{2} \tan^2 \frac{\theta}{2}}{\tan^2 \frac{\theta}{2} - 2 \tan \frac{\theta_0}{2} \tan \frac{\theta}{2} \cos \varphi + \tan^2 \frac{\theta_0}{2}}.$$

Hint. Introduce $\tan(\theta/2)$ as a new independent variable.

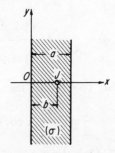

FIGURE 64

FIGURE 65

290. Suppose an infinite slab of conductivity σ contains a line current source (see Figure 65), from which a current J per unit length flows into the slab. Find the distribution of current in the slab, assuming that the slab is surrounded by a nonconducting medium.

Ans. The potential of the current field is

$$u(x, y) = -\frac{J}{4\pi\sigma} \ln \left\{ \left[\cosh \frac{\pi y}{a} - \cos \frac{\pi(x + b)}{a} \right] \right.$$

$$\left. \times \left[\cosh \frac{\pi y}{a} - \cos \frac{\pi(x - b)}{a} \right] \right\} + \text{const.}$$

291. Find the voltage distribution in a lossless transmission line of length l, if the end $x = 0$ is connected at the time $t = 0$ to a source of variable e.m.f. $Ee^{-\alpha t}$ and the end $x = l$ is kept open. It is assumed that the current and voltage in the line are initially zero.

Ans.

$$u(x, t) = \frac{2Ev}{\alpha l} \sum_{n=0}^{\infty} \frac{\sin \dfrac{(2n + 1)\pi vt}{2l} + \dfrac{(2n + 1)\pi v}{2l\alpha} \left[e^{-\alpha t} - \cos \dfrac{(2n + 1)\pi vt}{2l} \right]}{\left[1 + \dfrac{(2n + 1)\pi v}{2l\alpha} \right]^2}$$

$$\times \sin \frac{(2n + 1)\pi x}{2l},$$

where L and C are the self-inductance and capacitance of the line per unit length, and $v = 1/\sqrt{LC}$ is the velocity of wave propagation along the line.

292. One end $x = 0$ of a transmission line of length l with parameters L, C and R is connected to a source of constant e.m.f. E, while the other end $x = l$ is connected to a resistance R_0. Find the voltage in the line if the load R_0 is suddenly disconnected.

Ans.

$$u(x, t) = E - \frac{8E}{\pi^2(1 + \alpha)} e^{-Rt/2L}$$

$$\times \sum_{n=0}^{\infty} \left[\cos \frac{(2n + 1)\pi v_n t}{2l} - \frac{Rl}{(2n + 1)\pi L v_n} \sin \frac{(2n + 1)\pi v_n t}{2l} \right]$$

$$\times \frac{(-1)^n \sin \dfrac{(2n + 1)\pi x}{2l}}{(2n + 1)^2},$$

where

$$\alpha = \frac{R_0}{lR}, \qquad v_n = \frac{1}{\sqrt{LC}} \sqrt{1 - \frac{R^2 C l^2}{(2n + 1)^2 \pi^2 L}}.$$

293. Find the steady-state electromagnetic oscillations in a perfectly conducting waveguide whose cross section is a rectangle $0 \leqslant x \leqslant a$,

$0 \leqslant y \leqslant b$, assuming that the oscillations are excited by an infinite line current source $J = J_0 \sin \omega t$ passing through the point (x_0, y_0).

Ans. The complex amplitude of the vector potential of the electromagnetic field is

$$A = A_z(x, y) = \frac{8\pi J_0}{ca} \sum_{n=1}^{\infty} \frac{\sin \dfrac{n\pi x_0}{a} \sinh \sqrt{\left(\dfrac{n\pi}{a}\right)^2 - k^2}\,(b - y_0)}{\sqrt{\left(\dfrac{n\pi}{a}\right)^2 - k^2} \sinh \sqrt{\left(\dfrac{n\pi}{a}\right)^2 - k^2}\,b}$$

$$\times \sinh \sqrt{\left(\frac{n\pi}{a}\right)^2 - k^2}\, y \sin \frac{n\pi x}{a}, \qquad 0 \leqslant y \leqslant y_0,$$

$$A = A_z(x, y) = \frac{8\pi J_0}{ca} \sum_{n=1}^{\infty} \frac{\sin \dfrac{n\pi x_0}{a} \sinh \sqrt{\left(\dfrac{n\pi}{a}\right)^2 - k^2}\, y_0}{\sqrt{\left(\dfrac{n\pi}{a}\right)^2 - k^2} \sinh \sqrt{\left(\dfrac{n\pi}{a}\right)^2 - k^2}\,b}$$

$$\times \sinh \sqrt{\left(\frac{n\pi}{a}\right)^2 - k^2}\,(b - y) \sin \frac{n\pi x}{a}, \qquad y_0 \leqslant y \leqslant b,$$

where $0 \leqslant x \leqslant a$ and $k = \omega/c$ is the wave number.

Hint. Integrate the inhomogeneous wave equation for A, assuming that the current J is uniformly distributed over a rectangle whose dimensions are then made to approach zero.

294. Find the electromagnetic field due to an infinite linear current source $J = J_0 \sin \omega t$ placed between two parallel perfectly conducting planes (see Figure 66).

Ans. The complex amplitude of the vector potential of the electromagnetic field is

$$A = A_z(x, y) = \frac{2\pi J}{ca} \sum_{n=1}^{\infty} \left[\cos \frac{n\pi x}{a} - \cos \frac{n\pi(x + 2b)}{a} \right]$$

$$\times \frac{\exp\left\{ -\sqrt{\left(\dfrac{n\pi}{a}\right)^2 - \left(\dfrac{\omega}{c}\right)^2}\,|y| \right\}}{\sqrt{\left(\dfrac{n\pi}{a}\right)^2 - \left(\dfrac{\omega}{c}\right)^2}}.$$

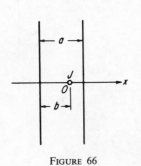

FIGURE 66

295. Find the steady-state electromagnetic oscillations in a perfectly conducting waveguide whose cross section is a circular sector $0 \leqslant r \leqslant a$, $0 \leqslant \varphi \leqslant \alpha$, assuming that the oscillations are excited by an infinite line current source $J = J_0 \sin \omega t$ passing through the point $r = r_0$, $\varphi = \varphi_0$.

Ans. The vector potential of the electromagnetic field is $A = \text{Im}\{A_0 e^{i\omega t}\}$, where

$$A_0(r, \varphi) = \frac{2\pi^2 J_0 i}{c\alpha} \sum_{n=1}^{\infty} [H_n^{(2)}(ka)H_n^{(1)}(kr_0) - H_n^{(2)}(kr_0)H_n^{(1)}(ka)]$$

$$\times \frac{J_n(kr)}{J_n(ka)} \sin \frac{n\pi\varphi_0}{\alpha} \sin \frac{n\pi\varphi}{\alpha}, \qquad r < r_0,$$

$$A_0(r, \varphi) = \frac{2\pi^2 J_0 i}{c\alpha} \sum_{n=1}^{\infty} [H_n^{(2)}(ka)H_n^{(1)}(kr) - H_n^{(2)}(kr)H_n^{(1)}(ka)]$$

$$\times \frac{J_n(kr_0)}{J_n(ka)} \sin \frac{n\pi\varphi_0}{\alpha} \sin \frac{n\pi\varphi}{\alpha}, \qquad r > r_0,$$

$k = \omega/c$ is the wave number, and $J_n(x)$, $H_n^{(1)}(x)$, $H_n^{(2)}(x)$ are cylinder functions.

*296. Find the electromagnetic oscillations in a cylindrical resonator $0 \leqslant r \leqslant a$, $-l \leqslant z \leqslant l$ excited by a dipole of moment P located at the origin of coordinates and directed along the z-axis.

Ans. The complex amplitude of the z-component of the vector potential is given by the Fourier expansion

$$A(r, z) = \frac{\pi P}{lc}\left[a_0(r) + 2\sum_{n=1}^{\infty} a_n(r) \cos \frac{n\pi z}{l}\right],$$

where the coefficients a_n have the values

$$a_n = \frac{1}{I_0(\alpha_n a)} [I_0(\alpha_n a)K_0(\alpha_n r) - K_0(\alpha_n a)I_0(\alpha_n r)], \qquad \alpha_n = \sqrt{\left(\frac{n\pi}{l}\right)^2 - k^2},$$

k is the wave number, and $I_0(x)$, $K_0(x)$ are Bessel functions of imaginary argument.

297. Find the electromagnetic field in an infinite cylindrical waveguide with perfectly conducting walls, assuming that the source of the oscillations is a current $J \sin \omega t$ in a coil of given dimensions, with a single uniformly wound layer (see Figure 67).

Ans. The complex amplitudes of the components of the electromagnetic field are

$$E_r = E_z = 0,$$

$$E_\varphi = -\frac{8\pi i\omega b J N}{c^2 a^2 h} \sum_{n=1}^{\infty} \frac{J_1(\gamma_n b/a)}{\alpha_n^2 J_0^2(\gamma_n)} J_1\left(\frac{\gamma_n r}{a}\right)$$

$$\times \begin{cases} 1 - e^{-\alpha_n h/2} \cosh \alpha_n z, & 0 < x < h/2, \\ \sinh \dfrac{\alpha_n h}{2} e^{-\alpha_n z}, & z > h/2, \end{cases}$$

$$H_r = \frac{c}{i\omega} \frac{\partial E_\varphi}{\partial z}, \quad H_z = -\frac{c}{i\omega} \frac{1}{r} \frac{\partial}{\partial r}(rE_\varphi), \quad H_\varphi = 0,$$

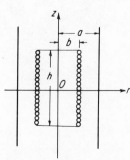

FIGURE 67

where

$$\alpha_n = \sqrt{\frac{\gamma_n^2}{a^2} - k^2},$$

$J_0(x)$ and $J_1(x)$ are Bessel functions, k is the wave number, c the velocity of light, N the number of turns in the coil, and the γ_n are consecutive positive roots of the equation $J_1(\gamma) = 0$.

References

Grinberg (G4, G5, G6), Morse and Feshbach (M9), Tikhonov and Samarski (T1).

6

INTEGRAL TRANSFORMS

If application of the Fourier method to a given problem leads to a set of particular solutions depending continuously on some real or complex parameter, we say that the problem has a *continuous spectrum*.[1] Characteristically, the solution of a problem of this kind is constructed from appropriate particular solutions by integrating with respect to the parameter, i.e., the solution takes the form of an integral expansion involving the eigenfunctions (the continuous analogue of the series expansions considered in Chaps. 4 and 5).[2] Problems with continuous spectra are encountered in all branches of mathematical physics, and can often be solved by the method of integral transforms, to which the present chapter is devoted. We begin by reminding the reader of the necessary background information.

By an *integral transform* of a function $f(x)$, defined in an interval (a, ∞), we mean an expression of the form

$$\tilde{f}(\tau) = \int_a^\infty f(x)K(x, \tau)\, dx, \qquad c \leqslant \tau < \infty, \tag{1}$$

where a and c are real numbers (the value $-\infty$ is allowed), and K is a function called the *kernel* of the transform. More generally, we allow K to depend on a complex parameter $p = \sigma + i\tau$ varying over some region D of the complex

---

[1] As a rule, such problems involve unbounded domains.

[2] The theory of integral expansions has undergone considerable development in recent years (see e.g., A1, L13, L14, L15, T6, T7 and S6, Vol. V). We also mention the classic paper by Weyl (W7).

plane. Then (1) is replaced by

$$\bar{f}(p) = \int_a^\infty f(x)K(x, p) \, dx, \qquad p \in D. \tag{2}$$

Examples of transformations of type (1):

1. The Fourier transform

$$K(x, \tau) = \frac{1}{\sqrt{2\pi}} e^{i\tau x}, \qquad a = -\infty, \quad c = -\infty.$$

2. The Fourier cosine transform

$$K(x, \tau) = \sqrt{\frac{2}{\pi}} \cos \tau x, \qquad a = 0, \quad c = 0.$$

3. The Fourier sine transform

$$K(x, \tau) = \sqrt{\frac{2}{\pi}} \sin \tau x, \qquad a = 0, \quad c = 0.$$

4. The Hankel transform

$$K(x, \tau) = x J_\nu(\tau x), \qquad a = 0, \quad c = 0,$$

where $J_\nu(x)$ is the Bessel function of the first kind of order $\nu > -\frac{1}{2}$.

5. The transform

$$K(x, \tau) = K_{i\tau}(x) \quad \text{or} \quad \frac{K_{i\tau}(x)}{\sqrt{x}}, \qquad a = 0, \quad c = 0,$$

where $K_\nu(x)$ is Macdonald's cylinder function.[3]

6. The Mehler-Fock transform

$$K(x, \tau) = P_{-\frac{1}{2}+i\tau}(x), \qquad a = 1, \quad c = 0,$$

where $P_\nu(x)$ is the Legendre function of the first kind.

Examples of transformations of type (2):

7. The Laplace transform

$$K(x, p) = e^{-px}, \qquad a = 0,$$

where D is the half-plane lying to the right of some line $\sigma = \sigma_1$ parallel to the imaginary axis.

8. The Mellin transform

$$K(x, p) = x^{p-1}, \qquad a = 0,$$

where D is the strip between the parallel lines $\sigma = \sigma_1$ and $\sigma = \sigma_2$.

---

[3] The second expression for $K(x, \tau)$ leads to a more symmetric inversion formula [see formula (21), p. 195].

Provided that the function $f(x)$ belongs to an appropriate class (depending on the integral transform in question), we can express $f(x)$ in terms of its integral transform by using a suitable *inversion formula*, which for transforms of type (1) takes the form

$$f(x) = \int_c^\infty \bar{f}(\tau)M(x, \tau)\, d\tau. \tag{3}$$

Here $M(x, \tau)$ is a suitable function defined in the region $a < x < \infty$, $c < \tau < \infty$ and called the *kernel of the inverse transform*. In the case of the transforms 1–6 just enumerated, we have

1. $\quad M(x, \tau) = \dfrac{1}{\sqrt{2\pi}}\, e^{-i\tau x}$,

2. $\quad M(x, \tau) = \sqrt{\dfrac{2}{\pi}}\, \cos \tau x$,

3. $\quad M(x, \tau) = \sqrt{\dfrac{2}{\pi}}\, \sin \tau x$,

4. $\quad M(x, \tau) = \tau J_\nu(\tau x)$,

5. $\quad M(x, \tau) = \dfrac{2\tau \sinh \pi\tau}{\pi^2}\, \dfrac{K_{i\tau}(x)}{x}$ or $\dfrac{2\tau \sinh \pi\tau}{\pi^2}\, \dfrac{K_{i\tau}(x)}{\sqrt{x}}$,

6. $\quad M(x, \tau) = \tau \tanh \pi\tau P_{-\frac{1}{2}+i\tau}(x)$.

In the case of transforms of type (2), the inversion formula takes the form

$$f(x) = \frac{1}{2\pi i} \int_\Gamma \bar{f}(p)M(x, p)\, dp, \tag{4}$$

where $M(x, p)$ is the kernel of the inverse transform, defined for all x in the interval (a, ∞) and p in the region D, while Γ is a suitable path of integration contained in D. For example,

$$M(x, p) = e^{px}$$

for the Laplace transform, while

$$M(x, p) = x^{-p}$$

for the Mellin transform. In both cases, Γ is a straight line parallel to the imaginary axis and lying in the region D.

We now turn to the integral transform method for solving partial differential equations. The basic idea is to look for some integral transform $\bar{u}$ of the solution, rather than for u itself, deferring the calculation of u until the end of the problem. In many cases, we can choose the kernel K in such a way that the original equation for u is transformed into a simpler equation for $\bar{u}$, with one less independent variable. Of course, the extra conditions on the function u are transformed into corresponding conditions on its integral transform, but the conditions involving the behavior of u as $x \to a$, $x \to \infty$ are automatically taken into account when transforming the original equation for u. The integral transform method has many advantages, e.g., it is applicable to both homogeneous and inhomogeneous problems, it simplifies calculations and singles out the purely computational part of the solution, it allows us to construct an operational calculus for a given kernel by using tables of direct and inverse transforms of the functions most commonly encountered in the applications, and so on.[4]

The present chapter is devoted to the solution of problems with continuous spectra by writing the solutions as integral expansions involving suitable functions or by using the method of integral transforms.[5] The problems are not classified by physical content, but rather by the particular transform used. There are five sections, the first on the Fourier integral and the Fourier transform, the second on Hankel's expansion and the related transform, the third and fourth on the Laplace and Mellin transforms, and the fifth on expansions with respect to cylinder functions of imaginary argument.[6] Many of the more difficult problems are equipped with solutions.

1. The Fourier Transform

Given a real function $f(x)$, defined in the interval $(-\infty, \infty)$, suppose that

1. $f(x)$ is piecewise continuous and of bounded variation in every finite subinterval $[a, b]$, where $-\infty < a < b < \infty$;[7]

2. The integral

$$\int_{-\infty}^{\infty} |f(x)|\, dx$$

is finite.

---

[4] Although the literature on the application of integral transforms to physical problems emphasizes the Laplace and Fourier transforms, a number of works have appeared in recent years on the application of various other integral transforms (see e.g., G5, H3, K3, L8, L10, S8, S9, S10, T8).

[5] As already noted, every integral expansion of the form (1) or (2) is accompanied by an inversion formula of the form (3) or (4), and conversely, and hence the distinction between the method of integral *expansions* and that of integral *transforms* is purely formal. Thus problems on the Fourier integral will be grouped with those on Fourier transforms, problems on Hankel's integral formula with those on Hankel transforms, and so on.

[6] Other integral expansions and transforms will be found in Chap. 7, which is concerned with the method of curvilinear coordinates.

[7] In particular, this condition is satisfied if $f(x)$ is piecewise smooth in $[a, b]$, or if $f(x)$ satisfies so-called *Dirichlet conditions* in $[a, b]$ (see W8, p. 161).

Then $f(x)$ satisfies the *Fourier integral theorem*

$$f(x) = \frac{1}{2\pi} \int_{-\infty}^{\infty} e^{-i\lambda x} \int_{-\infty}^{\infty} f(\xi) e^{i\lambda\xi} \, d\xi, \qquad -\infty < x < \infty, \tag{5}$$

where, if $f(x)$ has a jump discontinuity at the point $x = c$, the left-hand side should be replaced by the sum

$$\tfrac{1}{2}[f(c - 0) + f(c + 0)]$$

(see T5, p. 13). Formula (5) is valid under other conditions (see T5, Chap. 1), and can be written in the alternative form

$$f(x) = \frac{1}{\pi} \int_0^{\infty} \left[\cos \lambda x \int_{-\infty}^{\infty} f(\xi) \cos \lambda\xi \, d\xi + \sin \lambda x \int_{-\infty}^{\infty} f(\xi) \sin \lambda\xi \, d\xi \right] d\lambda,$$
$$-\infty < x < \infty.$$

The *Fourier transform* of a function satisfying the above conditions is defined as

$$\tilde{f}(\lambda) = \frac{1}{\sqrt{2\pi}} \int_{-\infty}^{\infty} f(x) e^{i\lambda x} \, dx, \qquad -\infty < \lambda < \infty. \tag{6}$$

Then, according to formula (5), the inverse of (6) is given by

$$f(x) = \frac{1}{\sqrt{2\pi}} \int_{-\infty}^{\infty} \tilde{f}(\lambda) e^{-i\lambda x} \, d\lambda, \qquad -\infty < x < \infty. \tag{7}$$

Formulas (5)–(7) play an important role in solving a wide variety of physical problems, in particular, boundary value problems for the Laplace and Helmholtz equations involving infinite strips, infinite cylinders, etc. In general, the application of these formulas is called for in problems leading to integration of the equation

$$\frac{\partial^2 u}{\partial x^2} + Lu = f(x, \ldots) \qquad -\infty < x < \infty,$$

where L is a linear differential operator which does not contain x, and $f(x, \ldots)$ is a given function.

Besides the formulas already written, many problems of mathematical physics involve the application of the Fourier sine and cosine integrals

$$f(x) = \frac{2}{\pi} \int_0^{\infty} \sin \lambda x \, d\lambda \int_0^{\infty} f(\xi) \sin \lambda\xi \, d\xi, \qquad 0 < x < \infty, \tag{8}$$

$$f(x) = \frac{2}{\pi} \int_0^{\infty} \cos \lambda x \, d\lambda \int_0^{\infty} f(\xi) \cos \lambda\xi \, d\xi, \qquad 0 < x < \infty, \tag{9}$$

valid for functions obeying the obvious analogues of the above conditions, i.e., such that

1. $f(x)$ is piecewise continuous and of bounded variation in every finite subinterval $[a, b]$, where $0 < a < b < \infty$;

2. The integral

$$\int_0^\infty |f(x)| \, dx$$

is finite.

The analogues of (6) and (7) are then

$$\bar{f}_s(\lambda) = \sqrt{\frac{2}{\pi}} \int_0^\infty f(x) \sin \lambda x \, dx, \qquad f(x) = \sqrt{\frac{2}{\pi}} \int_0^\infty \bar{f}_s(\lambda) \sin \lambda x \, d\lambda, \quad (10)$$

$$\bar{f}_c(\lambda) = \sqrt{\frac{2}{\pi}} \int_0^\infty f(x) \cos \lambda x \, dx, \qquad f(x) = \sqrt{\frac{2}{\pi}} \int_0^\infty \bar{f}_c(\lambda) \cos \lambda x \, d\lambda. \quad (11)$$

Formulas (8)–(11) are encountered in solving boundary value problems for the Laplace and Helmholtz equations involving half-strips, semi-infinite cylinders, etc.

The problems which follow are taken from various branches of physics, and are all susceptible to solution by using expansions or transforms like formulas (5)–(11).

298. Solve the problem of the temperature distribution in an infinite rod, with the following special initial temperature distributions $T\big|_{t=0} = f(x)$:

a)
$$f(x) = \begin{cases} T_0, & |x| < x_0, \\ 0, & |x| > x_0; \end{cases}$$

b)
$$f(x) = T_0 e^{-\alpha^2 x^2}.$$

Ans.

a)
$$T(x, t) = \frac{T_0}{\sqrt{2}} \left[\Phi\left(\frac{x_0 + x}{2\sqrt{\tau}}\right) + \Phi\left(\frac{x_0 - x}{2\sqrt{\tau}}\right) \right],$$

where $\Phi(x)$ is the probability integral;

b)
$$T(x, t) = \frac{T_0}{\sqrt{1 + 4\alpha^2\tau}} e^{-\alpha^2 x^2/(1 + 4\alpha^2 \tau)}.$$

Here k is the thermal conductivity, c the specific heat and ρ the density of the rod, and $\tau = kt/c\rho$.

299. A semi-infinite body bounded by the plane $x = 0$ has a given initial temperature distribution

$$T\big|_{t=0} = f(x), \qquad 0 \leqslant x < \infty.$$

Find the subsequent temperature distribution in the body, assuming that its boundary is held at temperature zero starting from the time $t = 0$. Apply the general result to the special case $f(x) = T_0$.

Ans.

$$T(x, t) = \frac{1}{\sqrt{\pi}}\left[\int_{-\infty}^{\frac{x}{2\sqrt{\tau}}} e^{-s^2}f(x - 2\sqrt{\tau}\,s)\,ds - \int_{\frac{x}{2\sqrt{\tau}}}^{\infty} e^{-s^2}f(-x + 2\sqrt{\tau}\,s)\,ds\right],$$

where

$$T(x, t) = T_0\Phi\left(\frac{x}{2\sqrt{\tau}}\right)$$

in the special case.

300. Find the stationary temperature distribution $T(x, y)$ in a semi-infinite body bounded by the plane $y = 0$, if the part $|x| < a$ is held at temperature T_0, while the other part $|x| > a$ is held at temperature zero (see Figure 68).

Ans.

$$T(x, y) = \frac{T_0}{\pi}\,\psi,$$

where ψ is the angle subtended by the segment $-a \leqslant x \leqslant a$, $y = 0$ at the point $P = (x, y)$.

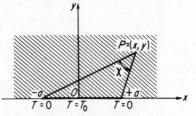

FIGURE 68

301. Find the stationary temperature distribution $T(x, y)$ in a semi-infinite slab $0 \leqslant x < \infty$, $0 \leqslant y \leqslant b$ if the face $y = b$ is held at temperature T_0, while the other two faces are held at temperature zero.

Ans.

$$T(x, y) = \frac{2T_0}{\pi}\,\text{arc tan}\left(\tanh\frac{\pi x}{2b}\tan\frac{\pi y}{2b}\right).$$

Hint. Use the formula

$$1 = \frac{2}{\pi}\int_0^\infty \frac{\sin\lambda x}{\lambda}\,d\lambda, \qquad x > 0.$$

302. A heat current Q enters a semi-infinite body through the section $|x| < a$ of its plane boundary (see Figure 69). Find the stationary temperature distribution in the body, assuming that the current is uniformly distributed and that the surface of the body radiates heat into the surrounding medium according to Newton's law.

Ans.

$$T(x, y) = \frac{Q}{\pi a k}\int_0^\infty \frac{\sin\lambda a}{\lambda(\lambda + h)}\,e^{-\lambda y}\cos\lambda x\,d\lambda,$$

where k is the thermal conductivity of the body and h is its heat exchange coefficient. In particular, the temperature of the part of the surface $|x| < a$ has the representation in closed form

$$T|_{y=0,|x|<a} = \frac{Q}{2kah}\left[1 - \cos ah \cos xh + \frac{1}{\pi}\left\{\mathrm{Si}\left[(a+x)h\right]\cos(a+x)h\right.\right.$$
$$+ \mathrm{Si}\left[(a-x)h\right]\cos(a-x)h$$
$$\left.\left.- \mathrm{Ci}\left[(a+x)h\right]\sin(a+x)h - \mathrm{Ci}\left[(a-x)h\right]\sin(a-x)h\right\}\right],$$

where $\mathrm{Si}(x)$ and $\mathrm{Ci}(x)$ are the sine and cosine integrals.

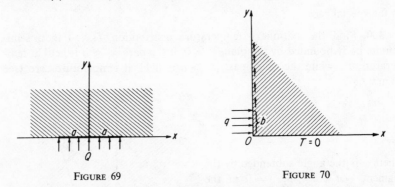

FIGURE 69   FIGURE 70

*303. Solve the two-dimensional stationary heat conduction problem for a quadrant of thermal conductivity k (see Figure 70), if the face $y = 0$ is held at temperature zero, while the other face is covered by a thermal insulator except for the section $0 < y < b$ through which heat flows with constant density q. Find the distribution of heat current through the face $y = 0$.

Ans.

$$T(x, y) = \frac{2q}{\pi k}\int_0^\infty \frac{1 - \cos \lambda b}{\lambda^2}e^{-\lambda x}\sin \lambda y\, d\lambda, \qquad q(x, 0) = \frac{q}{\pi}\ln\left(1 + \frac{b^2}{x^2}\right).$$

304. Find the stationary temperature distribution in the quadrant $x \geqslant 0$, $y \geqslant 0$ if the face $y = 0$ is held at temperature T_0, while the face $x = 0$ radiates heat into the surrounding medium according to Newton's law. Find the temperature distribution along the radiating face.

Ans.

$$T(x, y) = T_0\left[1 - \frac{2h}{\pi}\int_0^\infty \frac{e^{-\lambda x}\sin \lambda y}{\lambda(\lambda + h)}\right]d\lambda,$$

$$T|_{x=0} = \frac{2T_0}{\pi}\left[\left(\frac{\pi}{2} - \mathrm{Si}(yh)\right)\cos yh + \mathrm{Ci}(yh)\sin yh\right],$$

where $\mathrm{Si}(x)$ and $\mathrm{Ci}(x)$ are the sine and cosine integrals.

Hint. Take the Fourier sine transform of the required function $T(x, y)$, i.e., multiply the relevant differential equation by $\sin \lambda y$ and integrate with respect to y from 0 to ∞.

305. The end of a semi-infinite cylinder $0 \leqslant r \leqslant a$, $0 \leqslant z \leqslant \infty$ is held at constant temperature T_0, while the lateral surface is held at temperature zero. Find the stationary temperature distribution in the cylinder, by expanding the required function in a Fourier sine integral with respect to z.

Ans.

$$T = T_0 \left[1 - \frac{2}{\pi} \int_0^\infty \frac{I_0(\lambda r)}{I_0(\lambda a)} \frac{\sin \lambda z}{\lambda} \, d\lambda \right],$$

where $I_0(x)$ is the Bessel function of imaginary argument.

Hint. Introduce a new unknown function $u = T - T_0$, and use the integral

$$1 = \frac{2}{\pi} \int_0^\infty \frac{\sin \lambda x}{\lambda} \, d\lambda, \qquad x > 0.$$

306. Solve the preceding problem, assuming that a given temperature distribution $T|_{z=0} = f(r)$ is maintained on the end of the cylinder, while the lateral surface radiates heat into the surrounding medium according to Newton's law.

Ans.

$$T(r, z) = \frac{2}{\pi} \int_0^\infty \lambda \sin \lambda z \left[\int_0^a G_\lambda(\rho, r) \rho f(\rho) \, d\rho \right] d\lambda,$$

where

$$G_\lambda(\rho, r) = \begin{cases} \{ [\lambda K_1(\lambda a) - h K_0(\lambda a)] I_0(\lambda r) \\ \qquad + [\lambda I_1(\lambda a) + h I_0(\lambda a)] K_0(\lambda r) \} \dfrac{I_0(\lambda \rho)}{D(\lambda)}, & \rho \leqslant r, \\ \{ [\lambda K_1(\lambda a) - h K_0(\lambda a)] I_0(\lambda \rho) \\ \qquad + [\lambda I_1(\lambda a) + h I_0(\lambda a)] K_0(\lambda \rho) \} \dfrac{I_0(\lambda r)}{D(\lambda)}, & \rho \geqslant r, \end{cases}$$

$$D(\lambda) = \lambda I_1(\lambda a) + h I_0(\lambda a),$$

$I_n(z)$ and $K_n(z)$ are Bessel functions of imaginary argument, and h is the heat exchange coefficient.

307. Find the stationary temperature distribution in a semi-infinite cylinder $0 \leqslant r \leqslant a$, $0 \leqslant z < \infty$ if the lateral surface is maintained at the temperature $T|_{r=a} = f(z)$, while the end radiates heat into the surrounding medium.

Ans.

$$T(r, z) = \frac{2}{\pi} \int_0^\infty \varphi_\lambda(z) \frac{I_0(\lambda r)}{I_0(\lambda a)} \frac{d\lambda}{h^2 + \lambda^2} \int_0^\infty f(\zeta) \varphi_\lambda(\zeta) \, d\zeta,$$

$$\varphi_\lambda(z) = \lambda \cos \lambda z + h \sin \lambda z.$$

Hint. To solve the problem, use the following generalization of the Fourier integral theorem (see L13, p. 79):

$$f(x) = \frac{2}{\pi} \int_0^\infty \frac{\varphi_\lambda(x)}{h^2 + \lambda^2} \, d\lambda \int_0^\infty f(\xi)\varphi_\lambda(\xi) \, d\xi, \qquad 0 \leqslant x < \infty.$$

308. Find the two-dimensional electrostatic potential in the half-space $-\infty < x < \infty, y \geqslant 0$, if the potential distribution $u|_{y=0} = f(x)$ is maintained on the plane $y = 0$.

Ans.

$$u(x, y) = \frac{y}{\pi} \int_{-\infty}^{+\infty} \frac{f(\xi)}{(\xi - x)^2 + y^2} \, d\xi = \frac{1}{\pi} \int_{-\pi/2}^{\pi/2} f(x + y \tan \theta) \, d\theta.$$

Hint. To reduce the solution to final form, use the integral

$$\int_0^\infty e^{-ax} \cos bx \, dx = \frac{a}{a^2 + b^2}, \qquad a > 0.$$

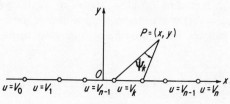

FIGURE 71

309. Examine the special case of the preceding problem corresponding to the piecewise constant potential distribution in the plane $y = 0$ shown in Figure 71.

Ans.

$$u(x, y) = \frac{1}{\pi} \sum_{k=0}^n V_k \psi_k,$$

where V_k is the value of the potential in the interval (x_{k-1}, x_k) and ψ_k is the angle subtended by (x_{k-1}, x_k) at the point $P = (x, y)$.

310. Find the distribution of electrostatic potential in the planar electron-optical lens shown in Figure 72 (cf. Prob. 282).[8]

Ans.

$$u(x, y) = \frac{V_2 + V_1}{2} + \frac{V_2 - V_1}{\pi} \int_0^\infty \frac{\cosh \lambda y}{\cosh \lambda h} \frac{\sin \lambda x}{\lambda} \, d\lambda.$$

---

[8] Note that the integrals representing the solutions of Probs. 310–311 can be expressed in terms of elementary functions.

Hint. Subtract out the particular solution $\frac{1}{2}(V_2 + V_1)$ of Laplace's equation, and then use the expansion

$$1 = \frac{2}{\pi} \int_0^\infty \frac{\sin \lambda x}{\lambda} \, d\lambda \qquad x > 0.$$

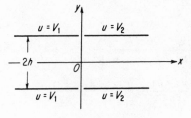

FIGURE 72

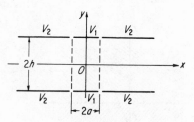

FIGURE 73

311. Find the distribution of electrostatic potential on the axis of the planar electron-optical lens shown in Figure 73.

Ans.

$$u(x, y) = V_2 + 2 \frac{V_1 - V_2}{\pi} \int_0^\infty \frac{\sin \lambda a}{\lambda} \frac{\cosh \lambda y}{\cosh \lambda h} \cos \lambda x \, d\lambda.$$

312. A thin charged wire of charge q per unit length is placed between two parallel conducting planes (see Figure 74). Find the resulting distribution of electrostatic potential, and also the density of charge on the planes $y = 0$ and $y = h$.

Ans. The potential distribution is

$$u(x, y) = 4q \int_0^\infty \frac{\sinh \lambda(h - a)}{\lambda \sinh \lambda h}$$
$$\times \sinh \lambda y \cos \lambda x \, d\lambda, \quad y < a,$$

where the corresponding formula for $y > a$ is obtained by permuting y and a. The charge density on the planes is

$$\sigma(x) = -\frac{q}{2h} \sin \frac{\pi a}{h} \times \begin{cases} \dfrac{1}{\cosh(\pi x/h) - \cos(\pi a/h)}, & y = 0, \\[2ex] \dfrac{1}{\cosh(\pi x/h) + \cos(\pi a/h)}, & y = h. \end{cases}$$

Hint. First assume that the charge is uniformly distributed over the rectangle $-\delta < x < \delta$, $a - \varepsilon < y < a + \varepsilon$, and then take the limit as δ, $\varepsilon \to 0$. To solve the corresponding Poisson equation, take the Fourier cosine transform of the unknown function, by multiplying the equation by $\cos \lambda x$ and integrating with respect to x from 0 to ∞.

*313. Find the electrostatic field of a thin charged wire of charge q per unit length located near the plane interface between two dielectric slabs (see Figure 75).

Ans.

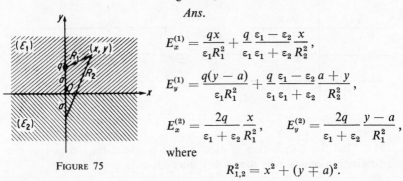

$$E_x^{(1)} = \frac{qx}{\varepsilon_1 R_1^2} + \frac{q}{\varepsilon_1} \frac{\varepsilon_1 - \varepsilon_2}{\varepsilon_1 + \varepsilon_2} \frac{x}{R_2^2},$$

$$E_y^{(1)} = \frac{q(y-a)}{\varepsilon_1 R_1^2} + \frac{q}{\varepsilon_1} \frac{\varepsilon_1 - \varepsilon_2}{\varepsilon_1 + \varepsilon_2} \frac{a+y}{R_2^2},$$

$$E_x^{(2)} = \frac{2q}{\varepsilon_1 + \varepsilon_2} \frac{x}{R_1^2}, \qquad E_y^{(2)} = \frac{2q}{\varepsilon_1 + \varepsilon_2} \frac{y-a}{R_1^2},$$

where

FIGURE 75

$$R_{1,2}^2 = x^2 + (y \mp a)^2.$$

Hint. To avoid any difficulties associated with the behavior of the logarithmic potential at infinity, set up a system of equations for the components of the electrostatic field.

314. Find the potential distribution in the electron-optical lens shown in Figure 76.

Ans.

$$u(r, z) = \frac{V_1 + V_2}{2} + \frac{V_2 - V_1}{\pi} \int_0^\infty \frac{I_0(\lambda r)}{I_0(\lambda a)} \frac{\sin \lambda z}{\lambda} \, d\lambda,$$

where $I_0(x)$ is the Bessel function of imaginary argument.

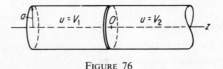

FIGURE 76

Hint. Reduce the problem to one with boundary conditions which are odd in the variable z, and then make a sine expansion, using the formula

$$1 = \frac{2}{\pi} \int_0^\infty \frac{\sin \lambda z}{\lambda} \, d\lambda, \qquad z > 0.$$

315. Find the potential of the electrostatic field due to a point charge q placed on the axis of an infinite conducting cylinder of radius a.

Ans.

$$u(r, z) = \frac{q}{\sqrt{r^2 + z^2}} - \frac{2q}{\pi} \int_0^\infty \frac{K_0(\lambda a)}{I_0(\lambda a)} I_0(\lambda r) \cos \lambda z \, d\lambda,$$

where $I_0(x)$ and $K_0(x)$ are Bessel functions of imaginary argument.

Hint. In the course of the calculations, use the following integral representation of Macdonald's function:

$$K_0(\lambda a) = \int_0^\infty \frac{\cos \lambda z}{\sqrt{a^2 + z^2}} \, dz.$$

316. Find the distribution of electrostatic potential inside a conducting cone $0 \leqslant r < \infty$, $0 \leqslant \theta \leqslant \theta_0$ due to a point charge q on its axis (see Figure 77).

Ans.

$$u(r, \theta) = \frac{q}{\sqrt{a^2 - 2ar \cos \theta + r^2}}$$

$$- \frac{q}{\sqrt{ra}} \int_0^\infty \frac{P_{-\frac{1}{2}+i\tau}(-\cos \theta_0)}{P_{-\frac{1}{2}+i\tau}(\cos \theta_0)} P_{-\frac{1}{2}+i\tau}(\cos \theta) \cos \left(\tau \ln \frac{r}{a} \right) \frac{d\tau}{\cosh \pi\tau},$$

where $P_\nu(x)$ is the Legendre function of the first kind.

Hint. Introduce new variables

$$x = \ln \frac{r}{a}, \qquad u = r^{-1/2} v.$$

To expand the source, use the following integral representation of the Legendre function:

$$P_{-\frac{1}{2}+i\tau}(\cos \alpha) = \frac{2}{\pi} \cosh \pi\tau \int_0^\infty \frac{\cos \tau x}{\sqrt{2 \cosh x - 2 \cos \alpha}} \, dx.$$

FIGURE 77

317. A point current source is placed on the axis of a cylindrical tube filled with a medium of conductivity σ_1 and surrounded by a medium of conductivity σ_2. Find the potential of the current field in each medium.[9]

Ans.

$$u_1(r, z) = \frac{J}{4\pi\sigma_1} \left[\frac{1}{\sqrt{r^2 + z^2}} + \frac{2}{\pi} (\sigma_1 - \sigma_2) \int_0^\infty \frac{K_0(\lambda a) K_1(\lambda a) I_0(\lambda r) \cos \lambda z \, d\lambda}{\sigma_1 K_0(\lambda a) I_1(\lambda a) + \sigma_2 I_0(\lambda a) K_1(\lambda a)} \right],$$

$$u_2(r, z) = \frac{J}{2\pi^2 a} \int_0^\infty \frac{K_0(\lambda r) \cos \lambda z}{\sigma_1 K_0(\lambda a) I_1(\lambda a) + \sigma_2 I_0(\lambda a) K_1(\lambda a)} \frac{d\lambda}{\lambda},$$

---

[9] This is the problem of "electrical coring" (see Fock's paper F2).

where $I_n(x)$ and $K_n(x)$ are Bessel functions of imaginary argument, J is the current emanating from the electrode, and a is the radius of the tube.

318. A line current J is placed between the boundary planes of two massive bodies made from iron of magnetic permeability μ (see Figure 78). Find the magnetic field in the air space.

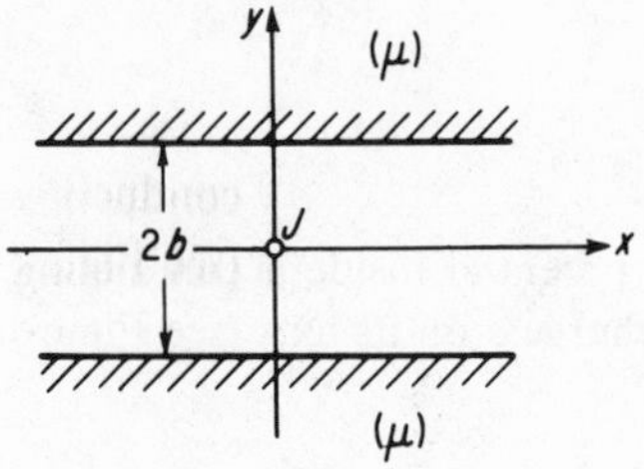

Ans.

$$H_x = -\frac{2J}{c}\left[\frac{y}{x^2 + y^2} - (\mu - 1) \right.$$
$$\left. \times \int_0^\infty \frac{e^{-\lambda b}\sinh \lambda y}{\cosh \lambda b + \mu \sinh \lambda b}\cos \lambda x \, d\lambda\right],$$

$$H_y = \frac{2J}{c}\left[\frac{x}{x^2 + y^2} + (\mu - 1) \right.$$
$$\left. \times \int_0^\infty \frac{e^{-\lambda b}\cosh \lambda y}{\cosh \lambda b + \mu \sinh \lambda b}\sin \lambda x \, d\lambda\right].$$

FIGURE 78

Hint. Take the Fourier transform of the equations for the components of the magnetic field.

319. Solve the preceding problem, assuming that the iron has infinite magnetic permeability.

Ans.

$$H_x = -\frac{2J}{c}\left[\frac{y}{x^2 + y^2} - \int_0^\infty e^{-\lambda b}\frac{\sinh \lambda y}{\sinh \lambda b}\cos \lambda x \, d\lambda\right],$$

$$H_y = \frac{2J}{c}\left[\frac{x}{x^2 + y^2} + \int_0^\infty e^{-\lambda b}\frac{\cosh \lambda y}{\sinh \lambda b}\sin \lambda x \, d\lambda\right].$$

320. A current J flows in a circular loop placed on a cylindrical core made from material of magnetic permeability μ (see Figure 79). Find the distribution of magnetic field on the axis of the core (see Lebedev's paper L3).

Ans.

$$H\big|_{r=0} = -\frac{4\mu J}{c}\int_0^\infty \frac{K_1(\lambda a)\cos \lambda z}{I_0(\lambda a)K_1(\lambda a) + \mu I_1(\lambda a)K_0(\lambda a)}\, d\lambda,$$

where $I_n(x)$ and $K_n(x)$ are Bessel functions of imaginary argument, and c is the velocity of light.

*321. Find the electromagnetic field radiated by a line current $J_0 e^{i\omega t}$ placed inside an ideally conducting shield of rectangular cross section (see Figure 80). Investigate the limiting case $b \to \infty$.

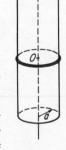

FIGURE 79

Ans.

$$E_z(x, y) = -\frac{4ikJ}{c} \int_0^\infty \frac{\sin \lambda a}{\sqrt{\lambda^2 - k^2}} \frac{\sinh [\sqrt{\lambda^2 - k^2}(b - |y|)]}{\cosh \sqrt{\lambda^2 - k^2} \, b} \sin \lambda x \, d\lambda,$$

where $k = \omega/c$ is the wave number and E_z is the complex amplitude. For $b \to \infty$, we have

$$E_z(x, y) = \frac{\pi kJ}{c} [H_0^{(2)}(k\sqrt{(x + a)^2 + y^2})$$
$$- H_0^{(2)}(k\sqrt{(x - a)^2 + y^2})]$$

in terms of the Hankel function $H_0^{(2)}(x)$, which gives the familiar law for reflection by a conducting plane of the radiation due to a line source.

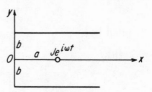

FIGURE 80

322. Find the electromagnetic field produced in a cylindrical waveguide by a dipole of moment P placed at the origin and directed along the axis of the cylinder. Find an expression for the longitudinal component of the electric field.

Ans. The complex amplitude of the z-component of the electric field is

$$E = \frac{2iP}{\pi\omega} \int_0^\infty (\lambda^2 - k^2) \left[\frac{I_0(\sqrt{\lambda^2 - k^2}\, a)K_0(\sqrt{\lambda^2 - k^2}\, r)}{I_0(\sqrt{\lambda^2 - k^2}\, a)} \right.$$
$$\left. - \frac{K_0(\sqrt{\lambda^2 - k^2}\, a)I_0(\sqrt{\lambda^2 - k^2}\, r)}{I_0(\sqrt{\lambda^2 - k^2}\, a)} \right] \cos \lambda z \, d\lambda,$$

where a is the radius of the cylinder, and $I_0(x)$, $K_0(x)$ are Bessel functions of imaginary argument.

323. Find the steady-state oscillations produced by a point source of sound of frequency ω placed on the axis of an infinite cylindrical tube with ideally reflecting walls.

Ans. The velocity potential is

$$u(r, z, t) = A \frac{\sin (\omega t - k\sqrt{r^2 + z^2})}{\sqrt{r^2 + z^2}}$$
$$+ \mathrm{Im} \left[\frac{2A}{\pi} e^{i\omega t} \int_0^\infty \frac{K_1(\sqrt{\lambda^2 - k^2}\, a)}{I_1(\sqrt{\lambda^2 - k^2}\, a)} I_0(\sqrt{\lambda^2 - k^2}\, r) \cos \lambda z \, d\lambda \right],$$

where $I_0(x)$, $I_1(x)$ and $K_1(x)$ are Bessel functions of imaginary argument.

Hint. Concerning the character of the singularity at the source point, see Prob. 85.

*324. Study the stress distribution in an elastic half-plane due to arbitrary stresses

$$\sigma_y\big|_{y=0} = f(x), \qquad \tau_{xy}\big|_{y=0} = g(x)$$

applied to its boundary.

Ans.

$$\sigma_x = \frac{2}{\pi} \int_{-\infty}^{\infty} \frac{yf(\xi) + (x - \xi)g(\xi)}{[y^2 + (x - \xi)^2]^2} (x - \xi)^2 \, d\xi,$$

$$\tau_{xy} = \frac{2y}{\pi} \int_{-\infty}^{\infty} \frac{yf(\xi) + (x - \xi)g(\xi)}{[y^2 + (x - \xi)^2]^2} (x - \xi) \, d\xi,$$

$$\sigma_y = \frac{2y^2}{\pi} \int_{-\infty}^{\infty} \frac{yf(\xi) + (x - \xi)g(\xi)}{[y^2 + (x - \xi)^2]^2} \, d\xi.$$

Hint. Take the Fourier transform of the system of equations

$$\frac{\partial \sigma_x}{\partial x} + \frac{\partial \tau_{xy}}{\partial y} = 0, \quad \frac{\partial \tau_{xy}}{\partial x} + \frac{\partial \sigma_y}{\partial y} = 0,$$

$$\frac{\partial^2 \sigma_x}{\partial y^2} + \frac{\partial^2 \sigma_y}{\partial x^2} - 2 \frac{\partial^2 \tau_{xy}}{\partial x \, \partial y} = 0$$

from two-dimensional elasticity theory.

325. Examine the special case of the preceding problem obtained when a concentrated force **P** with components $P_x = 0$ and $P_y = P$ is applied at the origin.[10]

Ans.

$$\sigma_x = -\frac{2Px^2y}{\pi(x^2 + y^2)^2}, \qquad \sigma_y = -\frac{2Py^3}{\pi(x^2 + y^2)^2},$$

$$\tau_{xy} = -\frac{2Pxy^3}{\pi(x^2 + y^2)^2}.$$

326. Study the stress distribution in an elastic half-plane $y \geqslant 0$ due to a concentrated force P applied at the point $x = 0$, $y = a$ and directed along the y-axis.[11] Find an expression for the shear stress τ_{xy}.

Ans.

$$\tau_{xy} = \frac{Px}{4\pi}\left\{(1 - \nu)\left[\frac{1}{R_1^2} - \frac{1}{R_2^2}\right] + 2(1 + \nu)(a - y)\left[\frac{y + a}{R_1^4} + \frac{y - a}{R_2^4}\right]\right.$$

$$\left. - 4(1 - \nu)y\,\frac{y + a}{R_1^4} - 4a(1 + \nu)y\,\frac{x^2 - (y + a)^2}{R_1^6}\right\},$$

---

[10] This is Flamant's problem, solved by inspection (without recourse to Fourier transforms) in courses on elasticity.

[11] Another way of solving this problem is given in M6.

where

$$R_{1,2} = \sqrt{x^2 + (y \pm a)^2}$$

and ν is Poisson's ratio.

Hint. Regard the force P as a distributed body force with components X and Y, and use the equations

$$\frac{\partial \sigma_x}{\partial x} + \frac{\partial \tau_{xy}}{\partial y} + X = 0, \qquad \frac{\partial \tau_{xy}}{\partial x} + \frac{\partial \sigma_y}{\partial y} + Y = 0,$$

$$\frac{\partial^2}{\partial y^2}(\sigma_x - \nu\sigma_y) + \frac{\partial^2}{\partial x^2}(\sigma_y - \nu\sigma_x) = 2(1 + \nu)\frac{\partial^2 \tau_{xy}}{\partial x \, \partial y}$$

from two-dimensional elasticity theory.

327. Study the two-dimensional stress distribution in an elastic strip compressed by two concentrated forces P applied at the points $x = 0$, $y = \pm b$ (see Figure 81). Find the normal stress σ_y along the axis of symmetry.

Ans.

$$\sigma_y\big|_{y=0} = -\frac{2P}{\pi}\int_0^\infty \frac{\sinh \lambda b + \lambda b \cosh \lambda b}{2\lambda b + \sinh 2\lambda b} \cos \lambda x \, d\lambda.$$

Hint. See Prob. 324.

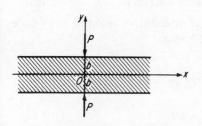

FIGURE 81

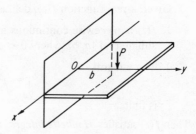

FIGURE 82

***328.** A semi-infinite thin elastic plate, clamped along the edge $y = 0$, is loaded by a concentrated force P at the point $(0, b)$. Find the bending moment M and the shear force N along the clamped edge (see Figure 82).

Ans.

$$M\big|_{y=0} = -\frac{P}{\pi}\frac{b^2}{b^2 + x^2}, \qquad N\big|_{y=0} = \frac{2P}{\pi}\frac{b^2}{(b^2 + x^2)^2}.$$

Hint. Replace the concentrated force by a force uniformly distributed over the rectangle $-\delta < x < \delta$, $b - \varepsilon < y < b + \varepsilon$, and take the Fourier cosine transform with respect to the variable x of the differential equation for deflection of the plate. Then pass to the limit $\delta, \varepsilon \to 0$.

329. A thin elastic plate, in the form of an infinite strip $-\infty < x < \infty$, $0 \leqslant y \leqslant b$ of width b, is clamped along its edges and loaded by a concentrated force P at the point $(0, a)$. Find the bending moment along the edge $y = 0$.

Ans.

$$-M|_{y=0}$$
$$= \frac{P}{\pi} \int_0^\infty [a \sinh \lambda b \sinh \lambda(b - a) - (b - a)\lambda b \sinh \lambda a] \frac{\cos \lambda x \, d\lambda}{\sinh^2 \lambda b - \lambda^2 b^2}.$$

330. Solve the preceding problem, assuming that the edges of the strip are simply supported and that the force is applied at the point $(0, b/2)$. Find the deflection of the center of the strip due to the force.

Ans.

$$u(0, b/2) = \frac{Pb^2}{8\pi D} \int_0^\infty \frac{\sinh \mu - \mu}{\mu^3 \cosh^2 (\mu/2)} \, d\mu,$$

where D is the flexural rigidity of the plate.

2. The Hankel Transform

Given a real function $f(r)$, defined in the interval $(0, \infty)$, suppose that

1. $f(r)$ is piecewise continuous and of bounded variation in every finite subinterval $[a, b]$, where $0 < a < b < \infty$;

2. The integral

$$\int_0^\infty \sqrt{r} \, |f(r)| \, dr$$

is finite.

Then $f(r)$ satisfies *Hankel's integral theorem*[12]

$$f(r) = \int_0^\infty J_\nu(\lambda r)\lambda \, d\lambda \int_0^\infty f(\rho)J_\nu(\lambda\rho)\rho \, d\rho, \qquad 0 < r < \infty, \qquad (12)$$

where $J_\nu(x)$ is the Bessel function of the first kind of order $\nu > -\frac{1}{2}$. If $f(x)$ has a jump discontinuity at the point $r = c$, the left-hand side should be replaced by the sum

$$\tfrac{1}{2}[f(c - 0) + f(c + 0)]$$

(see W4, p. 456 ff.). Formula (12) is one of the most important integral expansions encountered in mathematical physics.

The *Hankel transform* of a function satisfying the above conditions is defined as

$$\bar{f}(\lambda) = \int_0^\infty f(r)J_\nu(\lambda r)r \, dr, \qquad 0 < \lambda < \infty. \qquad (13)$$

---

[12] Sometimes called the *Fourier-Bessel integral*.

Then, according to formula (12), the inverse of (13) is given by

$$f(r) = \int_0^\infty \bar{f}(\lambda)J_\nu(\lambda r)\lambda \, d\lambda, \qquad 0 < r < \infty. \tag{14}$$

There is a generalization of formula (12), known as *Weber's integral* (see T6, p. 75)

$$f(r) = \int_0^\infty \frac{\varphi_\lambda(r)\lambda \, d\lambda}{J_\nu^2(\lambda a) + Y_\nu^2(\lambda a)} \int_a^\infty f(\rho)\varphi_\lambda(\rho)\rho \, d\rho, \qquad a < r < \infty, \tag{15}$$

involving the linear combination

$$\varphi_\lambda(r) = J_\nu(\lambda a)Y_\nu(\lambda r) - Y_\nu(\lambda a)J_\nu(\lambda r)$$

of Bessel functions of the first and second kinds ($\nu > -\frac{1}{2}$). A sufficient condition for validity of (15) is that $f(r)$ be piecewise continuous and of bounded variation in every finite subinterval $[\alpha, \beta]$, where $a < \alpha < \beta < \infty$, and that the integral

$$\int_a^\infty \sqrt{r}\,|f(r)| \, dr$$

be finite. It should be noted that Weber's integral reduces to Hankel's integral as $a \to 0$.

Hankel's integral expansion and the Hankel transform can be used to solve a number of problems of mathematical physics, e.g., boundary value problems for the Laplace and Helmholtz equations involving half-spaces and regions bounded by parallel planes, certain problems of elasticity theory, etc.[13] The problems that follow can be solved quite readily, as soon as one has acquired the necessary experience in handling Bessel functions.

331. Find the stationary temperature distribution in the half-space $z \geqslant 0$, if a given temperature distribution $T|_{z=0} = f(r)$ is maintained on the boundary $z = 0$. Examine the special case

$$f(r) = \begin{cases} T_0, & r < a, \\ 0, & r > a. \end{cases}$$

Ans.

$$T(r, z) = \int_0^\infty e^{-\lambda z} J_0(\lambda r)\lambda \, d\lambda \int_0^\infty f(\rho)J_0(\lambda \rho)\rho \, d\rho.$$

---

[13] In general, application of these formulas is called for in problems leading to integration of the equation

$$\frac{1}{r}\frac{\partial}{\partial r}\left(r\frac{\partial u}{\partial r}\right) - \frac{\nu^2}{r^2}u + Lu = f(r, \ldots), \qquad 0 < r < \infty,$$

where L is a linear operator which does not contain r, and $f(r, \ldots)$ is a given function. Weber's expansion plays the same role for the interval $a < r < \infty$ (see Probs. 335–337).

In the special case,

$$T(r, z) = T_0 a \int_0^\infty e^{-\lambda z} J_1(\lambda a) J_0(\lambda r)\, d\lambda,$$

where $J_0(x)$ and $J_1(x)$ are Bessel functions.

Hint. To evaluate the integral

$$\int_0^a J_0(\lambda \rho)\rho\, d\rho,$$

use the differential equation for the Bessel function.

332. Solve the preceding problem, assuming that the half-space is heated by a thermal current of constant density q, incident on the disk of radius a with center at the origin, while the rest of the boundary exchanges heat with the surrounding medium according to Newton's law.

Ans.

$$T(r, z) = \frac{qa}{k} \int_0^\infty \frac{e^{-\lambda z}}{\lambda + h}\, J_1(\lambda a) J_0(\lambda r)\, d\lambda,$$

where h is the heat exchange coefficient.

333. A cylindrical rod of radius a, heated to temperature T_0, is introduced into an unbounded medium whose initial temperature is zero. Find the temperature distribution $T(r, t)$, assuming that the medium and the rod have the same thermal conductivity k, specific heat c and density ρ.

Ans.

$$T = T_0 a \int_0^\infty e^{-\lambda^2 \tau} J_1(\lambda a) J_0(\lambda r)\, d\lambda,$$

where $\tau = kt/c\rho$.

*334. Examine the process of temperature equalization (in unbounded space) of an arbitrary axially symmetric initial temperature distribution

$$T\big|_{t=0} = f(r), \qquad 0 \leqslant r < \infty.$$

Ans.

$$T(r, t) = \frac{1}{2\tau} \int_0^\infty e^{-(r^2 + s^2)/4\tau} I_0\left(\frac{rs}{2\tau}\right) f(s) s\, ds,$$

where $I_0(x)$ is the Bessel function of imaginary argument.

Hint. To calculate the coefficient in the Fourier-Bessel integral, use the formula

$$\int_0^\infty e^{-\lambda^2 \tau} J_0(\lambda s) J_0(\lambda r) \lambda\, d\lambda = \frac{1}{2\tau}\, e^{-(r^2 + s^2)/4\tau} I_0\left(\frac{rs}{2\tau}\right).$$

(see W4, p. 395).

*335. A cylindrical hole of radius a is drilled in an infinite body, and the walls of the hole are maintained at temperature T_0 starting from the time $t = 0$. Examine the evolution of the temperature distribution in the body, assuming that its initial temperature is zero.

Ans.

$$T(r, t) = \frac{2T_0}{\pi} \int_0^\infty \frac{\varphi_\lambda(r)[1 - e^{-\lambda^2\tau}]}{J_0^2(\lambda a) + Y_0^2(\lambda a)} \frac{d\lambda}{\lambda},$$

where

$$\varphi_\lambda(r) = J_0(\lambda a)Y_0(\lambda r) - Y_0(\lambda a)J_0(\lambda r),$$

and $J_0(x)$ and $Y_0(x)$ are Bessel functions of the first and second kinds.

Hint. Set $\nu = 0$ in formula (15).

336. A cylindrical conductor of radius a heated by a d-c current passes through an infinite slab of width $2h$ (see Figure 83). Find the stationary temperature distribution in the slab, assuming that the surface temperature of the conductor is T_0, while the faces of the slab have temperature zero.

Ans.

$$T(r, z) = T_0\left[1 - \frac{2}{\pi} \int_0^\infty \frac{\varphi_\lambda(r)}{J_0^2(\lambda a) + Y_0^2(\lambda a)} \frac{\cosh \lambda z}{\cosh \lambda h} \frac{d\lambda}{\lambda}\right],$$

where $\varphi_\lambda(r)$ has the same meaning as in the preceding problem.

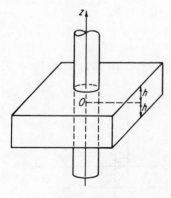

FIGURE 83

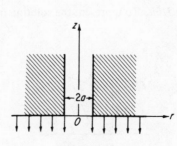

FIGURE 84

337. The walls of a cylindrical hole terminating at the plane surface of an infinite body (see Figure 84) are held at a given temperature T_0. Find the stationary temperature distribution in the body, assuming that it radiates heat from its surface into the surrounding medium according to Newton's law.

Ans.

$$T(r, z) = T_0\left[1 - \frac{2h}{\pi}\int_0^\infty \frac{\varphi_\lambda(r)e^{-\lambda z}\,d\lambda}{\lambda(\lambda + h)[J_0^2(\lambda a) + Y_0^2(\lambda a)]}\right],$$

where

$$\varphi_\lambda(r) = J_0(\lambda a)Y_0(\lambda r) - Y_0(\lambda a)J_0(\lambda r),$$

and h is the heat exchange coefficient.

338. Find the distribution of electrostatic potential in the space between two grounded plane electrodes $z = a$, due to a point charge q at the point $r = 0$, $z = 0$.

Ans.

$$u(r, z) = \frac{q}{\sqrt{r^2 + z^2}} - q\int_0^\infty e^{-\lambda a}\frac{\cosh \lambda z}{\cosh \lambda a}\,J_0(\lambda r)\,d\lambda,$$

in terms of the Bessel function $J_0(x)$.

Hint. Use the formula

$$\frac{1}{\sqrt{r^2 + z^2}} = \int_0^\infty e^{-\lambda z}J_0(\lambda r)\,d\lambda, \qquad z > 0.$$

339. Find the electrostatic field due to a point charge q located near the plane interface between two media with different dielectric constants (see Figure 85).

Ans.

$$u_1(r, z) = \frac{q}{\varepsilon_1}\frac{1}{R_1} + \frac{q}{\varepsilon_1}\frac{\varepsilon_1 - \varepsilon_2}{\varepsilon_1 + \varepsilon_2}\frac{1}{R_2},$$

$$u_2(r, z) = \frac{2q}{\varepsilon_1 + \varepsilon_2}\frac{1}{R_1}, \qquad R_{1,2} = \sqrt{r^2 + (z \mp a)^2}.$$

Hint. To represent the solution in closed form, use the hint to Prob. 338.

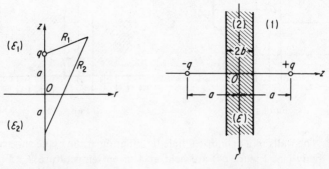

FIGURE 85   FIGURE 86

PROB. 343 — INTEGRAL TRANSFORMS 165

340. Find the electrostatic field produced by two point charges $+q$ and $-q$, between which there is a slab of material of dielectric constant ε (see Figure 86). Calculate the field on the line joining the charges.

Ans.

$$E_z\big|_{r=0} = \begin{cases} -2q\displaystyle\int_0^\infty \frac{\lambda e^{-\lambda(a-b)}\cosh\lambda z}{\sinh\lambda b + \varepsilon\cosh\lambda b}\,d\lambda, & |z| < b, \\[3mm] \dfrac{q}{(|z|-a)^2} + q\displaystyle\int_0^\infty \frac{\sinh\lambda b - \varepsilon\cosh\lambda b}{\sinh\lambda b + \varepsilon\cosh\lambda b}e^{-\lambda(a-2b+|z|)}\lambda\,d\lambda, & |z| > b. \end{cases}$$

341. A d-c current J enters the ground through an electrode making contact with the earth's surface $(z = 0)$ over the area of a disk of radius a. Find the current distribution in the earth, and examine the limiting case of a point contact.

Ans. The potential of the current field is

$$u(r, z) = \frac{J}{\pi a\sigma}\int_0^\infty e^{-\lambda z}J_1(\lambda a)J_0(\lambda r)\frac{d\lambda}{\lambda}, \qquad z > 0,$$

where $J_0(x)$ and $J_1(x)$ are Bessel functions, and σ is the conductivity of the earth. In the limiting case,

$$u(r, z) = \frac{J}{2\pi\sigma\sqrt{r^2 + z^2}}.$$

342. A point electrode carrying current J is placed on terrain consisting of two layers of different conductivities (see Figure 87). Calculate the potential of the current field on the earth's surface.

Ans.

$$u\big|_{z=0} = \frac{J}{2\pi\sigma_1 r} + \frac{J}{2\pi\sigma_1}(\sigma_1 - \sigma_2)\int_0^\infty \frac{e^{-\lambda a}J_0(\lambda r)}{\sigma_1\sinh\lambda a + \sigma_2\cosh\lambda a}\,d\lambda.$$

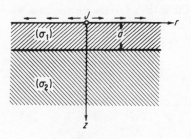

FIGURE 87

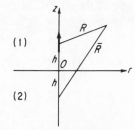

FIGURE 88

343. Determine the electromagnetic field of a vertical radiator (antenna) placed at height h over the plane surface of the earth, assumed to be perfectly conducting (see Figure 88).

Ans. The z-component of the vector potential of the electromagnetic field is

$$A(r, z) = \frac{P}{c}\left(\frac{e^{-ikR}}{R} + \frac{e^{-ik\bar{R}}}{\bar{R}}\right),$$

(all other components vanish), where

$$R = \sqrt{(z-h)^2 + r^2}, \qquad \bar{R} = \sqrt{(z+h)^2 + r^2},$$

ω is the frequency of the oscillations, c is the velocity of light, P is the moment of the radiating dipole, and $k = \omega/c$.

Hint. Use the expansion

$$\frac{e^{-ik\sqrt{r^2+z^2}}}{\sqrt{r^2 + z^2}} = \int_0^\infty \frac{e^{-\sqrt{\lambda^2-k^2}\,z}J_0(\lambda r)\lambda}{\sqrt{\lambda^2 - k^2}}\,d\lambda.$$

344. Solve the preceding problem, assuming that the earth has finite conductivity.

Ans. The vector potential of the electromagnetic field is

$$A^{(1)} = \frac{P}{c}\left\{\frac{e^{-ik_1R}}{R} + \int_0^\infty \frac{\lambda}{\sqrt{\lambda^2 - k_1^2}}\right.$$
$$\times \frac{k_2^2\sqrt{\lambda^2 - k_1^2} - k_1^2\sqrt{\lambda^2 - k_2^2}}{k_2^2\sqrt{\lambda^2 - k_1^2} + k_1^2\sqrt{\lambda^2 - k_2^2}}\,e^{-(h+z)\sqrt{\lambda^2-k_1^2}}J_0(\lambda r)\,d\lambda\bigg\},$$

$$A^{(2)} = \frac{2Pk_2^2}{c}\int_0^\infty \frac{\lambda e^{-h\sqrt{\lambda^2-k_1^2}+z\sqrt{\lambda^2-k_1^2}}}{k_2^2\sqrt{\lambda^2 - k_1^2} + k_1^2\sqrt{\lambda^2 - k_2^2}}\,J_0(\lambda r)\,d\lambda,$$

where

$$k_1 = \frac{\omega}{c}, \qquad k_2 = \frac{\sqrt{(\varepsilon\omega - 4\pi\sigma i)\omega}}{c},$$

in terms of the earth's dielectric constant ε and conductivity σ.[14]

345. Using the solution of Prob. 344, find an expression for the normal component of the electric field on the earth's surface for the case where the dipole is placed directly on the surface itself $(h \to 0)$.

Ans.

$$E_z\big|_{z=0} = \frac{2P\omega i}{c^2 r}\left(\frac{k_2}{k_1}\right)^2 \frac{k_1 k_2}{k_1^4 - k_2^4}\left\{\left(\frac{k_1 k_2^3}{k_1^2 + k_2^2} - \frac{ik_2}{r} - \frac{k_2}{k_1}\frac{1}{r^2}\right)e^{-ik_1r}\right.$$
$$- \left(\frac{k_1^3 k_2}{k_1^2 + k_2^2} - \frac{ik_1}{r} - \frac{k_1}{k_2}\frac{1}{r^2}\right)e^{-ik_2r}$$
$$\left.- ir\frac{k_1^3 k_2^3}{(k_1^2 + k_2^2)^{3/2}}\int_{\sqrt{k_1^2+k_2^2}/k_2}^{\sqrt{k_1^2+k_2^2}/k_1} e^{-irk_1k_2s/\sqrt{k_1^2+k_2^2}}\frac{ds}{\sqrt{s^2-1}}\right\}.$$

---

[14] Details on the transformation of these expressions into a form suitable for calculation as well as an analysis of the corresponding physical picture of wave propagation, can be found in the specialized literature (see e.g., F6, Chap. 23, Sec. 1 and S14, Secs. 31–32).

Hint. Use the Van der Pol substitution

$$\frac{1}{k_2^2\sqrt{\lambda^2 - k_1^2} + k_1^2\sqrt{\lambda^2 - k_2^2}} = \frac{k_1 k_2}{k_1^4 - k_2^4} \int_{\sqrt{k_1^2 + k_2^2}/k_2}^{\sqrt{k_1^2 + k_2^2}/k_1} \frac{d\,\dfrac{1}{\sqrt{s^2 - 1}}}{\sqrt{\lambda^2 - \dfrac{k_1^2 k_2^2}{k_1^2 + k_2^2}s^2}}.$$

346. Determine the electromagnetic field of a horizontal radiator located at height h above the plane surface of the earth, assumed to be a perfect conductor (see Figure 89).

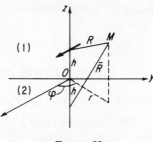

Ans. The vector potential of the electromagnetic field has the components

$$A_x = A(r, z) = \frac{P}{c}\left(\frac{e^{-ikR}}{R} - \frac{e^{-ik\bar{R}}}{\bar{R}}\right),$$

$$A_y = A_z = 0.$$

FIGURE 89

347. Solve the preceding problem, assuming that the earth has finite conductivity.

Ans.

$$A_x^{(1)} = \frac{P}{c}\left\{\frac{e^{-ik_1 R}}{R}\right.$$

$$\left. + \int_0^\infty \frac{\lambda}{\sqrt{\lambda^2 - k_1^2}}\frac{\sqrt{\lambda^2 - k_1^2} - \sqrt{\lambda^2 - k_2^2}}{\sqrt{\lambda^2 - k_1^2} + \sqrt{\lambda^2 - k_2^2}} e^{-\sqrt{\lambda^2 - k_1^2}(z+h)}J_0(\lambda r)\,d\lambda\right\},$$

$$A_z^{(1)} = \frac{2P}{c}(k_1^2 - k_2^2)\cos\varphi$$

$$\times \int_0^\infty \frac{\lambda^2 e^{-\sqrt{\lambda^2 - k_1^2}(z+h)}J_1(\lambda r)}{(k_2^2\sqrt{\lambda^2 - k_1^2} + k_1^2\sqrt{\lambda^2 - k_2^2})(\sqrt{\lambda^2 - k_1^2} + \sqrt{\lambda^2 - k_2^2})}\,d\lambda,$$

$$A_x^{(2)} = \frac{2P}{c}\int_0^\infty \frac{\lambda e^{-\sqrt{\lambda^2 - k_1^2}h + \sqrt{\lambda^2 - k_2^2}z}J_0(\lambda r)}{\sqrt{\lambda^2 - k_1^2} + \sqrt{\lambda^2 - k_2^2}}\,d\lambda,$$

$$A_z^{(2)} = \frac{2P}{c}(k_1^2 - k_2^2)\cos\varphi$$

$$\times \int_0^\infty \frac{\lambda^2 e^{-\sqrt{\lambda^2 - k_1^2}h + \sqrt{\lambda^2 - k_2^2}z}J_1(\lambda r)}{(k_2^2\sqrt{\lambda^2 - k_1^2} + k_1^2\sqrt{\lambda^2 - k_2^2})(\sqrt{\lambda^2 - k_1^2} + \sqrt{\lambda^2 - k_2^2})}\,d\lambda.$$

For further details, see F6, Chap. 23, Sec. 2 and S14, Sec. 33.

348. Find the magnetic field of a horizontal radiator lying on the plane surface of the earth, assumed to have dielectric constant ε and conductivity σ.

Show that the magnetic field on the earth's surface can be expressed in terms of elementary functions.

Ans.

$$H_z\big|_{z=0} = \frac{2P}{c} \frac{\sin \varphi}{k_2^2 - k_1^2} \frac{d}{dr}\left[\frac{1}{r}\frac{d}{dr}\left(\frac{1}{r}e^{-ik_2r} - \frac{1}{r}e^{-ik_1r}\right)\right].$$

349. Find the steady-state acoustic vibrations due to a disk-shaped piston of radius a inserted in an infinite screen and vibrating with velocity $v_0 \sin \omega t$.

Ans. The velocity potential is

$$u(r, z, t) = \operatorname{Im}\left\{v_0 a e^{i\omega t}\int_0^\infty \frac{J_1(\lambda a)}{\sqrt{\lambda^2 - k^2}}e^{-\sqrt{\lambda^2-k^2}z}J_0(\lambda r)\,d\lambda\right\},$$

where $k = \omega/c$ and $J_0(x)$, $J_1(x)$ are Bessel functions.

350. A concentrated normal force P is applied to the plane surface of a semi-infinite elastic body $z \geqslant 0$. Study the resulting stress distribution in the body, and find expressions for σ_z and τ_{rz}.

Ans.

$$\sigma_z = -\frac{3P}{2\pi}z^3(r^2 + z^2)^{-5/2}, \qquad \tau_{rz} = -\frac{3P}{2\pi}rz^2(r^2 + z^2)^{-5/2},$$

where the force P is assumed to be applied at the point $r = z = 0$.

Hint. Use the formulas

$$\sigma_z = \frac{\partial}{\partial z}\left[(2 - \nu)\,\Delta u - \frac{\partial^2 u}{\partial z^2}\right], \qquad \tau_{rz} = \frac{\partial}{\partial r}\left[(1 - \nu)\,\Delta u - \frac{\partial^2 u}{\partial z^2}\right]$$

expressing the stresses σ_z and τ_{rz} in terms of the biharmonic stress function (ν is Poisson's ratio). Then expand the quantities Δu and $\partial^2 u/\partial z^2$ in Hankel integrals (see also the solution of Prob. 351). In the boundary conditions, first replace the concentrated force P by a force uniformly distributed over a small disk of radius ε, and then take the limit as $\varepsilon \to 0$.

*351. Generalize the preceding problem to the case of a concentrated force **P** with components $P_x = P_y = 0$, $P_z = P$, applied at an arbitrary interior point of the body (with coordinates $r = 0, z = a$). Find an expression for the stress σ_z.

Ans.

$$\sigma_z = -\frac{P}{8\pi(1 - \nu)}\left\{(1 - 2\nu)(z - a)\left(\frac{1}{R_1^3} - \frac{1}{R_2^3}\right) + \frac{3(z - a)^3}{R_1^5}\right.$$
$$\left. + \frac{3(z + a)^2}{R_2^5}[a + (3 - 4\nu)z] + 2az(z + a)\left[\frac{15(z + a)^2}{R_2^7} - \frac{9}{R_2^5}\right]\right\}.$$

where $R_{1,2} = \sqrt{r^2 + (z \mp a)^2}$.

Hint. Use the stress function

$$\frac{P}{8\pi(1 - \nu)} \sqrt{r^2 + z^2},$$

corresponding to a concentrated force P (with components $P_x = P_y = 0$, $P_z = P$) applied to the point $r = z = 0$ of an infinite elastic body (see T4, p. 355).

352. Study the transverse oscillations of an infinite elastic plate due to a concentrated force $P(t)$ applied at the point $r = z = 0$ starting from the time $t = 0$.

Ans.

$$u(r, t) = \frac{1}{4\pi\sqrt{D\rho}} \int_0^t P(s) \left\{ \frac{\pi}{2} - \text{Si} \left[\frac{r^2}{4\sqrt{D/\rho(t - s)}} \right] \right\} ds,$$

where D is the flexural rigidity and ρ the density of the plate, and $\text{Si}(x)$ is the sine integral.

3. The Laplace Transform

The Laplace transform is acknowledged to be the most effective tool for dealing with the nonstationary problems of mathematical physics. Since the whole subject has been thoroughly treated in the literature,[15] we shall confine ourselves to a few brief remarks, mainly for reference purposes.

Let $f(t)$ be a real function defined in the interval $(0, \infty)$ such that

1. $f(t)$ is piecewise continuous in every finite subinterval $[a, T]$, where $0 < a < T < \infty$;

2. The product $f(t)e^{-\sigma_1 t}$ is absolutely integrable on $(0, \infty)$ for some suitable $\sigma_1 \geqslant 0$.

Then the *Laplace transform* of $f(t)$ is defined by the formula

$$f(p) = \int_0^\infty f(t)e^{-pt} \, dt, \tag{16}$$

where $p = \sigma + i\tau$ is any complex number in the half-plane $\text{Re } p > \sigma_1$.[16] If it is also assumed that $f(t)$ is of bounded variation in every finite subinterval

---

[15] See the relevant books cited at the end of this chapter (p. 202).

[16] Laplace transforms can also be defined for functions satisfying weaker conditions. Note that the function f is an analytic function of p in the domain $\text{Re } p > \sigma_1$. The values of f in the rest of the complex plane can be determined by analytic continuation.

[a, T],[17] then formula (16) can be inverted by using the *Fourier-Mellin theorem*

$$f(t) = \frac{1}{2\pi i} \int_\Gamma \bar{f}(p)e^{pt}\,dp, \tag{17}$$

where Γ is a straight line parallel to the imaginary axis lying to the right of the line $\text{Re}\,p = \sigma_1$ (see Figure 90). Conversely, (17) implies (16) if $\bar{f}(p)$ satisfies appropriate conditions.

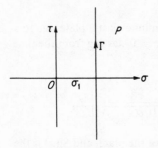

FIGURE 90

The application of the Laplace transform method is called for in nonstationary problems leading to integration of the equation

$$Lu - \frac{1}{a^2}\frac{\partial^2 u}{\partial t^2} - b\frac{\partial u}{\partial t} = f(t, \ldots),$$

where L is a linear differential operator which does not contain t, a and b are given constants, and $f(t, \ldots)$ is a given function. Its use allows us to eliminate the time t, thereby reducing the problem to the determination of a function $\bar{u}$ satisfying a simpler equation. In particular, if the unknown function u depends only on one spatial variable (in addition to the time), the equation for $\bar{u}$ will be an ordinary differential equation.

After finding $\bar{u}$, the problem can be solved by using the inversion formula (17), where the path of integration Γ must be chosen in such a way that all the singular points of $\bar{u}$ lie to the left of Γ. The actual calculation of the complex integral (17) can be carried out by various methods, the most important of which involve the use of Cauchy's theorem and residue theory, expansion in series, application of the convolution theorem, use of appropriate tables,[18] etc. The variety of available methods makes it possible to obtain the solution of the problem quickly, in the form most suitable for understanding the physics of the situation and making subsequent numerical calculations. This constitutes the great advantage of the Laplace transform method, which is particularly suitable for studying wave propagation along transmission lines, physical problems with boundary conditions involving time derivatives (see Probs. 365, 367, 370), and so on.

This section contains a variety of nonstationary problems, dealing first with heat conduction, then with electricity and magnetism, and finally with mechanics. Because of the abundance of specialized literature on Laplace transforms, we have omitted the simplest problems belonging to these categories. At the end of the section, we give a few problems of a more

---

[17] In particular, this condition is satisfied if $f(t)$ is piecewise smooth in $[a, T]$, or if $f(t)$ satisfies Dirichlet conditions in $[a, T]$.

[18] The tables in E3 are particularly complete.

complicated nature (e.g., Probs. 391, 405, 406), to be solved by combining the Laplace transform with some other integral transform (e.g., the Fourier transform or the Hankel transform).

353. Starting from the time $t = 0$, the plane boundary of a semi-infinite body of thermal conductivity k, specific heat c and density ρ is maintained at the temperature $T\big|_{x=0} = f(\tau)$, where $\tau = kt/c\rho$. Find the subsequent temperature distribution in the body, assuming that the initial temperature is zero.

Ans.

$$T(x, t) = \frac{2}{\sqrt{\pi}} \int_{\frac{x}{2\sqrt{\tau}}}^{\infty} e^{-u^2} f\left(\tau - \frac{x^2}{4u^2}\right) du, \qquad x > 0.$$

354. Consider the following special cases of the preceding problem:

    a) $f(\tau) = T_0$;   b) $f(\tau) = A\tau$;

    c) $f(\tau) = \begin{cases} T_0, & 0 \leqslant \tau < \tau_0, \\ 0, & \tau > \tau_0; \end{cases}$ d) $f(\tau) = T_0 \sin \omega\tau$.

Ans.

    a) $T(x, t) = T_0\left[1 - \Phi\left(\dfrac{x}{2\sqrt{\tau}}\right)\right]$;

    b) $T(x, t) = A\tau\left\{\left(1 + \dfrac{x^2}{2\tau}\right)\left[1 - \Phi\left(\dfrac{x}{2\sqrt{\tau}}\right)\right] - \dfrac{x}{\sqrt{\pi\tau}} e^{-x^2/4\tau}\right\}$;

    c) $T(x, t) = T_0\begin{cases} 1 - \Phi\left(\dfrac{x}{2\sqrt{\tau}}\right), & 0 \leqslant \tau \leqslant \tau_0, \\[2mm] \Phi\left(\dfrac{x}{2\sqrt{\tau - \tau_0}}\right) - \Phi\left(\dfrac{x}{2\sqrt{\tau}}\right), & \tau > \tau_0; \end{cases}$

    d) $T(x, t) = \dfrac{2T_0}{\sqrt{\pi}} \int_{\frac{x}{2\sqrt{\tau}}}^{\infty} \sin\left[\omega\left(\tau - \dfrac{x^2}{4u^2}\right)\right] e^{-u^2} du,$

where $\Phi(x)$ is the probability integral.

*355. Solve Prob. 353 for a given density q of heat current incident on the boundary (instead of a given surface temperature distribution). Examine the special cases[19]

$$\text{a) } q = q_0; \qquad \text{b) } q = q_0 \sin \omega\tau.$$

Ans.

$$T(x, t) = \frac{q_0}{k}\left\{\frac{2\sqrt{\tau}}{\sqrt{\pi}} e^{-x^2/4\tau} - x\left[1 - \Phi\left(\frac{x}{2\sqrt{\tau}}\right)\right]\right\},$$

$$T\big|_{x=0} = \frac{\sqrt{2}q_0}{k\sqrt{\omega}}\left[C\left(\sqrt{\frac{2}{\pi}\omega\tau}\right)\sin \omega\tau - S\left(\sqrt{\frac{2}{\pi}\omega\tau}\right)\cos \omega\tau\right],$$

---

[19] In Case b, consider only the surface temperature.

where $C(x)$ and $S(x)$ are the Fresnel integrals, and k is the thermal conductivity.

356. Find the evolution in time of the temperature on the plane boundary of a semi-infinite body, if the density of heat current incident on the body is a given function of time $q = q(\tau)$. Consider the special case $q = $ const.

Ans.

$$T\big|_{x=0} = \frac{1}{k\sqrt{\pi}} \int_0^\tau q(s) \frac{ds}{\sqrt{\tau - s}} .$$

In the special case,

$$T\big|_{x=0} = \frac{2q}{k\sqrt{\pi}} \sqrt{\tau}.$$

*357. A semi-infinite body, heated to the initial temperature T_0, radiates heat from its plane boundary $x = 0$. Find the distribution of temperature in the body, assuming that the radiation obeys Newton's law and that the temperature of the surrounding medium is zero.

Ans.

$$T(x, t) = T_0\left\{\Phi\left(\frac{x}{2\sqrt{\tau}}\right) + e^{h(x+h\tau)}\left[1 - \Phi\left(h\sqrt{\tau} + \frac{x}{2\sqrt{\tau}}\right)\right]\right\},$$

where h is the heat exchange coefficient.

Hint. To simplify the calculations, substitute the integral representation

$$\frac{1}{\sqrt{p} + h} = \int_0^\infty e^{-s(\sqrt{p} + h)} \, ds$$

into the inversion formula, and then reverse the order of integration.

358. Starting from the time $t = 0$, a train of heat current pulses $q = f(\tau)$ such that

$$f(\tau) = \begin{cases} q_0, & 0 < \tau < \tau_0, \\ 0, & \tau_0 < \tau < \tau^*, \end{cases} \quad f(\tau + \tau^*) = f(\tau)$$

flows through the plane boundary of a semi-infinite body. Find the temperature distribution in the body after a large number of cycles, assuming that the initial temperature is zero and neglecting heat exchange between the surface of the body and the surrounding medium.

Ans. For finite x,

$$T \approx \frac{2q_0\tau_0}{k\sqrt{\pi\tau^*}} \sqrt{\tau} \qquad \text{as} \quad \tau \to \infty.$$

359. Two semi-infinite bodies made from different materials, one heated to temperature T_0 and the other held at temperature zero, are put into contact starting from the time $t = 0$ (see Figure 91). Describe the subsequent equalization of temperature.

Ans.

$$T(x, t) = \frac{T_0\alpha}{1 + \alpha}\left[1 + \frac{1}{\alpha}\Phi\left(\frac{x\sqrt{b_1}}{2\sqrt{t}}\right)\right], \quad x > 0,$$

$$T(x, t) = \frac{T_0\alpha}{1 + \alpha}\left[1 - \Phi\left(-\frac{x\sqrt{b_2}}{2\sqrt{t}}\right)\right], \quad x < 0,$$

where $\Phi(x)$ is the probability integral, and

$$b_i = \frac{c_i\rho_i}{k_i} \quad (i = 1, 2), \qquad \alpha = \sqrt{\frac{c_1\rho_1 k_1}{c_2\rho_2 k_2}}.$$

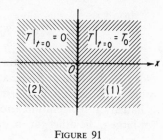

FIGURE 91

360. The temperature distribution

$$T|_{x=0} = f(y)$$

is maintained on the plane boundary of the half-space $0 \leqslant x < \infty$, $-\infty < y < \infty$, starting from the time $t = 0$. Solve the corresponding problem of heat conduction, assuming that the initial temperature equals zero.

Ans.

$$T(x, y, t) = \frac{x}{\pi} \int_{-\infty}^{\infty} \frac{f(s)}{x^2 + (y - s)^2} e^{-[x^2 + (y-s)^2]/4\tau} ds, \qquad \tau = \frac{kt}{c\rho}.$$

Hint. Take Laplace and Fourier transforms in succession.

361. Find the temperature distribution inside a body shaped like a quadrant ($x \geqslant 0, y \geqslant 0$), whose surface is held at temperature T_0 starting from the time $t = 0$ (the initial temperature is assumed to be zero). Plot the corresponding isotherms.

Ans.

$$T(x, y, t) = T_0\left[1 - \Phi\left(\frac{x}{2\sqrt{\tau}}\right)\Phi\left(\frac{y}{2\sqrt{\tau}}\right)\right].$$

The result of the calculations is shown in Figure 92.

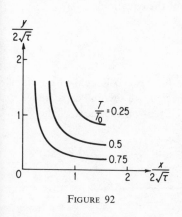

FIGURE 92

Hint. Look for a solution of the form $T = T_0[1 + u(x, t)v(y, t)]$, and then reduce the problem to Prob. 354, Case a.

362. Find the temperature distribution inside a body shaped like an octant ($x \geqslant 0, y \geqslant 0, z \geqslant 0$), whose surface is held at temperature T_0

starting from the time $t = 0$ (the initial temperature is assumed to be zero).

Ans.

$$T(x, y, z, t) = T_0\left[1 - \Phi\left(\frac{x}{2\sqrt{\tau}}\right)\Phi\left(\frac{y}{2\sqrt{\tau}}\right)\Phi\left(\frac{z}{2\sqrt{\tau}}\right)\right].$$

363. Find the temperature $T(x, t)$ in a slab of finite thickness, if one face $x = 0$ is held at temperature T_0 starting from the time $t = 0$, while the other $x = a$ is held at temperature zero. It is assumed that the whole slab is initially at temperature zero. Give two forms of the solution, one suitable for large t, the other for small t.

Ans.

$$T(x, t) = T_0\left[1 - \frac{x}{a} - \frac{2}{\pi}\sum_{n=1}^{\infty}\frac{\sin(n\pi x/a)}{n}e^{-n^2\pi^2\tau/a^2}\right],$$

$$T(x, t) = T_0\sum_{n=0}^{\infty}\left[\Phi\left(\frac{2a - x + 2na}{2\sqrt{\tau}}\right) - \Phi\left(\frac{x + 2na}{2\sqrt{\tau}}\right)\right].$$

Hint. To obtain the second form of the solution, expand the Laplace transform of the desired function in ascending powers of the quantity $e^{-x\sqrt{p}}$.

364. Solve the preceding problem, assuming that a thermal current of constant density q is incident on the face $x = a$, while the face $x = 0$ radiates heat according to Newton's law.

Ans.

$$T(x, t) = \frac{qa}{k}\left\{\frac{1}{ah} + \frac{x}{a} - 2(ah)^2\sum_{n=1}^{\infty}\frac{\cos[\gamma_n(a - x)/a]\,e^{-\gamma_n^2\tau/a^2}}{[ah(1 + ah) + \gamma_n^2]\gamma_n^2\sin^2\gamma_n}\right\},$$

where the γ_n are consecutive positive roots of the equation

$$\cot\gamma = \frac{\gamma}{ah},$$

and h is the heat exchange coefficient.

365. Solve Prob. 363 assuming that the face $x = 0$ is held at constant temperature T_0, while the face $x = a$ is connected to a thermal capacitance.[20] Derive expressions for the density of heat current on the faces of the slab.

Ans.

$$q\big|_{x=0} = \frac{2kT_0}{a}\sum_{n=1}^{\infty}\frac{1 + \alpha^2\gamma_n^2}{1 + \alpha + \alpha^2\gamma_n^2}e^{-\gamma_n^2\tau/a^2},$$

$$q\big|_{x=a} = \frac{2akT_0}{a}\sum_{n=1}^{\infty}\frac{\gamma_n}{(1 + \alpha + \alpha^2\gamma_n^2)\sin\gamma_n}e^{-\gamma_n^2\tau/a^2},$$

---

[20] By a "thermal capacitance" we mean a body in which any temperature drop can be neglected. In Probs. 365, 367, etc., C_0 denotes the amount of heat needed to raise the temperature of the body by 1 degree, referred to unit area, unit length, etc.

where the γ_n are consecutive positive roots of the equation

$$\cot \gamma = \alpha\gamma, \qquad \alpha = \frac{C_0}{c\rho a}.$$

Hint. The boundary condition at $x = a$ has the form

$$C_0 \frac{\partial T}{\partial t}\bigg|_{x=a} = -k \frac{\partial T}{\partial x}\bigg|_{x=a},$$

where C_0 is the thermal capacitance per unit area and k is the thermal conductivity.

366. Find the temperature distribution in a slab $-a \leqslant x \leqslant a$ in which, starting from the time $t = 0$, there is a process periodically producing heat according to the law shown in Figure 93. The temperature of the faces of the slab and the initial temperature are assumed to be zero.

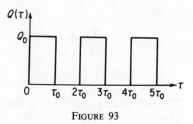

FIGURE 93

Ans.

$$T(x, t) = \frac{Q_0}{k} \left\{ \frac{a^2 - x^2}{4} - \frac{16a^2}{\pi^3} \sum_{n=0}^{\infty} \frac{(-1)^n \cos(\tau_0\lambda_n^2 x/a)}{(2n+1)^3[1 + e^{\tau_0^3\lambda_n^4/a^2}]} e^{-\tau\tau_0^2\lambda_n^4/a^2} \right.$$

$$- 2\frac{\tau_0}{\pi^2} \sum_{n=0}^{\infty} \frac{1}{(2n+1)^2}$$

$$\times \left[\left(1 - 4\frac{\cosh\lambda_n(x+a)\cos\lambda_n(x-a) + \cosh\lambda_n(x-a)\cos\lambda_n(x+a)}{\cosh 2\lambda_n a + \cos 2\lambda_n a}\right)\cos 2\lambda_n^2\tau \right.$$

$$\left. \left. + 4\frac{\sinh\lambda_n(x-a)\sin\lambda_n(x+a) + \sinh\lambda_n(x+a)\sin\lambda_n(x-a)}{\cosh 2\lambda_n a + \cos 2\lambda_n a} \sin 2\lambda_n^2\tau \right] \right\},$$

where

$$\lambda_n = \sqrt{\frac{(2n+1)\pi}{2\tau_0}}.$$

367. A thin cylindrical rod (probe), heated to temperature T_0, is inserted into the ground, in order to measure the ground's thermal properties. Describe how the temperature of the probe varies with time, assuming that the temperature drop inside the rod can be neglected (see T10).

Ans.

$$T\big|_{r=a} = \frac{4T_0\alpha}{\pi^2} \int_0^\infty \frac{e^{-\lambda^2\tau/a^2}}{[\lambda\alpha J_0(\lambda) - J_1(\lambda)]^2 + [\lambda\alpha Y_0(\lambda) - Y_1(\lambda)]^2} \frac{d\lambda}{\lambda},$$

where $J_n(x)$ and $Y_n(x)$ are Bessel functions of the first and second kinds, $\alpha = C_0/2\pi a^2 c\rho$, C_0 is the thermal capacitance of the probe per unit length, a is the radius of the probe, k is the thermal conductivity, c the specific heat and ρ the density of the ground, and $\tau = kt/c\rho$.

Hint. Solve the heat conduction problem for the domain $r > a$ with the boundary condition

$$T|_{r=a,t=0} = T_0, \qquad C_0 \frac{\partial T}{\partial t}\bigg|_{r=a} = 2\pi ak \frac{\partial T}{\partial r}\bigg|_{r=a}, \quad t > 0.$$

368. Use the Laplace transform to solve Prob. 335, and then show that the two answers are equivalent.

Ans.

$$T(r, t) = T_0 \left[1 - \frac{2}{\pi} \int_0^\infty \frac{\varphi_\lambda(r) e^{-\lambda^2 \tau}}{J_0^2(\lambda a) + Y_0^2(\lambda a)} \frac{d\lambda}{\lambda}\right].$$

where

$$\varphi_\lambda(r) = J_0(\lambda a) Y_0(\lambda r) - Y_0(\lambda a) J_0(\lambda r).$$

The equivalence of this result and the answer to Prob. 335 follows from the expansion

$$1 = \frac{2}{\pi} \int_0^\infty \frac{\varphi_\lambda(r)}{J_0^2(\lambda a) + Y_0^2(\lambda a)} \frac{d\lambda}{\lambda}.$$

369. Investigate the heating of a cylindrical cable if, starting from the time $t = 0$, heat is produced with density Q in the core of the cable, while its outer surface is held at temperature T_0 (see Figure 94). Find the temperature of the core, neglecting any temperature drop inside the core.

Ans.

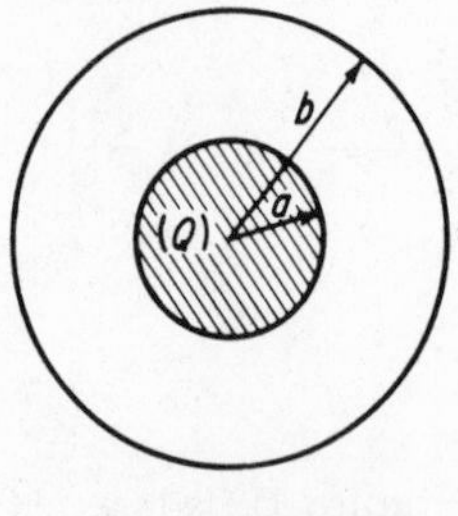

FIGURE 94

$$T|_{r=a} = \frac{Q}{2\pi k}$$

$$\times \left\{\ln \frac{b}{a} - \frac{\pi b}{a} \sum_{n=1}^\infty \frac{J_0(\gamma_n) J_1(\gamma_n a/b)}{\gamma_n[J_0^2(\gamma_n) - J_1^2(\gamma_n a/b)]} \right.$$

$$\left. \times R_{\gamma_n}(a) e^{-\gamma_n^2 \tau/b^2}\right\},$$

where

$$R_\gamma(r) = Y_0(\gamma) J_0(\gamma r/b) - J_0(\gamma) Y_0(\gamma r/b)$$

is a linear combination of Bessel functions, and the γ_n are consecutive positive roots of the equation $R_\gamma'(a) = 0$.

370. Starting from the time $t = 0$, heat is produced with density Q in a cylindrical conductor of radius a. Find the temperature along the axis of the conductor, assuming that heat leaves its surface by way of a thermal capacitance and the initial temperature is zero.

Ans.

$$T\big|_{r=0} = \frac{Q}{k}\left\{\frac{\tau}{2\alpha + 1} + \frac{a^2\alpha(1 + 4\alpha)}{4(2\alpha + 1)^2} + 2a^2\alpha\sum_{n=1}^{\infty}\frac{e^{-\gamma_n^2\tau/a^2}}{\gamma_n^2(\alpha^2\gamma_n^2 + 2\alpha + 1)J_0(\gamma_n)}\right\},$$

where the γ_n are consecutive positive roots of the equation

$$J_1(\gamma) + \alpha\gamma J_0(\gamma) = 0,$$

C_0 is the thermal capacitance per unit length, and $\alpha = C_0/2\pi a^2 c\rho$.

*371. At the time $t = 0$, a cold cylinder of radius a is encased in a thin heated cylindrical sleeve covered on the outside by a thermally insulating layer (see Figure 95). Find the temperature distribution in the cylinder, assuming that the initial temperatures of the cylinder and the sleeve are 0 and T_0, respectively, and neglecting any temperature drop inside the sleeve.

Ans.

$$T(r, t)$$

$$= T_0\left[\frac{1}{1 + \dfrac{1}{2\alpha}} + \sum_{n=1}^{\infty}\frac{J_0\left(\dfrac{\gamma_n r}{a}\right)e^{-\gamma_n^2\tau/a^2}}{J_0(\gamma_n)\left(1 + \dfrac{1}{2\alpha} + \dfrac{\alpha\gamma_n^2}{2}\right)}\right],$$

FIGURE 95

where the γ_n are consecutive positive roots of the equation

$$J_1(\gamma) + \alpha\gamma J_0(\gamma) = 0,$$

C_0 is the thermal capacitance of the sleeve per unit length, and $\alpha = C_0/2\pi a^2 c\rho$.

372. A diffusing substance is distributed in the half-space $x > 0$ with a given initial concentration

$$C\big|_{t=0} = f(x) = \begin{cases} C_0, & 0 < x < a, \\ 0, & x > a. \end{cases}$$

Find the density of the substance through the boundary $x = 0$, assuming that the concentration on the boundary is maintained at zero starting from the time $t = 0$.

Ans.

$$q\big|_{x=0} = \frac{\sqrt{D}\,C_0}{\sqrt{\pi t}}(1 - e^{-a^2/4Dt}),$$

where D is the diffusion coefficient.

373. Find the concentration distribution of a diffusing substance in the half-space $x > 0$ bounded by an impermeable wall, assuming that the initial concentration equals zero and that the substance is released with constant density Q in the layer $0 < x < a$ during a finite time interval T. Derive an expression for the concentration of the substance on the wall $x = 0$.

Ans.

$$C\big|_{x=0,\,t<T} = \frac{Qa^2}{D}\left\{\frac{Dt}{a^2}\,\Phi\left(\frac{a}{2\sqrt{Dt}}\right) - \frac{1}{2}\left[1 - \Phi\left(\frac{a}{2\sqrt{Dt}}\right)\right] + \frac{1}{a}\sqrt{\frac{Dt}{\pi}}\,e^{-a^2/4Dt}\right\},$$

$$C\big|_{x=0,\,t>T} = \frac{Qa^2}{D}\left\{\left(\frac{1}{2} + \frac{Dt}{a^2}\right)\left[\Phi\left(\frac{a}{2\sqrt{Dt}}\right) - \Phi\left(\frac{a}{2\sqrt{D(t-T)}}\right)\right]\right.$$

$$\left. + \frac{DT}{a^2}\,\Phi\left(\frac{a}{2\sqrt{D(t-T)}}\right) + \frac{1}{a}\sqrt{\frac{D}{\pi}}[\sqrt{t}\,e^{-a^2/4Dt} - \sqrt{t-T}\,e^{-a^2/4D(t-T)}]\right\}.$$

374. A substance diffuses outward through the lateral surface of an infinite cylinder of radius a into the surrounding medium, where the concentration of the substance equals zero at the time $t = 0$. Find the subsequent concentration distribution, assuming that the substance flows out of the cylinder with constant density q. Derive a formula for the concentration of the substance on the surface of the cylinder.

Ans.

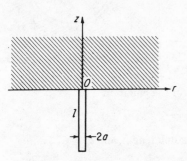

FIGURE 96

$$C\big|_{r=a} = -\frac{4}{\pi^2}\frac{qa}{D}\int_0^\infty \frac{1}{J_1^2(x) + Y_1^2(x)}$$

$$\times\,(1 - e^{-x^2 Dt/a^2})\,\frac{dx}{x^3},$$

in terms of the Bessel functions $J_n(x)$ and $Y_n(x)$.

*375. A diffusing substance emanates from a thin cylindrical tube of length l closed at one end, and enters the half-space $z > 0$ through an opening in the impermeable wall $z = 0$ (see Figure 96). Find the amount of substance inside the tube as a function of time, assuming that the flow of current is constant over a cross section of the tube and that the initial values of the concentration of the substance in the tube and in the half-space equal C_0 (per unit length) and 0, respectively.

Ans.

$$M(t) = \frac{2M_0}{\pi}\int_0^\infty \frac{\left(1 - \cos\frac{ax}{l}\right)e^{-Dtx^2/l^2}}{1 + \sin^2 x - 2\sin x \sin\left(1 + \frac{a}{l}\right)x}\left(\frac{\sin x}{x}\right)^2 dx,$$

where a is the radius of the tube and M_0 is the initial amount of substance inside the tube ($M_0 = C_0 l$).

376. The end $x = 0$ of an infinite transmission line, with self-inductance L and capacitance C per unit length, is joined at the time $t = 0$ to a source of e.m.f. $E = f(t)$. Find the voltage $u(x, t)$ at every point of the line.

Ans.

$$u(x, t) = \begin{cases} 0, & t < \dfrac{x}{v}, \\[2ex] f\left(t - \dfrac{x}{v}\right), & t > \dfrac{x}{v}. \end{cases}$$

where $v = 1/\sqrt{LC}$ is the velocity of wave propagation along the line.

377. Solve the preceding problem for the case of self-inductance L, capacitance C, resistance R and leakage conductance G per unit length, chosen to satisfy the relation $RC = LG$ (a distortionless line).

Ans.

$$u(x, t) = \begin{cases} 0, & t < \dfrac{x}{v}, \\[2ex] e^{-x\sqrt{RG}} f\left(t - \dfrac{x}{v}\right), & t > \dfrac{x}{v}. \end{cases}$$

378. A condenser of capacitance C_0, charged to the potential V, is discharged at the time $t = 0$ into an infinite line with parameters L and C. Find the distribution of current $I(x, t)$ in the line.

Ans.

$$I(x, t) = \begin{cases} 0, & t < \dfrac{x}{v}, \\[2ex] \dfrac{V}{Z} e^{-\alpha[t-(x/v)]}, & t > \dfrac{x}{v}, \end{cases}$$

where $Z = \sqrt{L/C}$ is the wave resistance of the line, and $\alpha = 1/C_0 Z$.

379. The end $x = 0$ of an infinite line with self-inductance L, capacitance C and resistance R per unit length is connected at the time $t = 0$ to a source of constant e.m.f. Study the resulting process of propagation of a voltage wave along the line (see C2, p. 202).

Ans.

$$u(x, t) = \begin{cases} 0, & t < \dfrac{x}{v}, \\[2ex] E\left[e^{-\alpha x/v} + \dfrac{\alpha x}{v} \int_{x/v}^{t} e^{-\alpha \tau} \dfrac{I_1(\alpha\sqrt{\tau^2 - (x/v)^2}}{\sqrt{\tau^2 - (x/v)^2}} \, d\tau \right], & t > \dfrac{x}{v}, \end{cases}$$

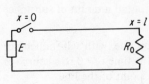

FIGURE 97

where E is the size of the applied e.m.f., $\alpha = R/2L$ and $I_1(x)$ is the Bessel function of imaginary argument.

380. A line of length l with parameters L and C is terminated at the end $x = l$ by a resistance R_0 (see Figure 97). Find the subsequent voltage in the load R_0, assuming that the end $x = 0$ is suddenly connected at the time $t = 0$ to a source of constant e.m.f. E. Under what conditions is there no reflection of waves from the end of the line?

Ans.

$$u\big|_{x=l} = \begin{cases} 0, & 0 < t < T, \\ E\left[1 - \left(\dfrac{Z - R_0}{Z + R_0}\right)^n\right], & (2n - 1)T < t < (2n + 1)T, \\ & n = 1, 2, \ldots, \end{cases}$$

where $Z = \sqrt{L/C}$ is the wave resistance of the line, and $T = l/v$ is the time it takes a wave to go from one end of the line to the other. There is no wave reflection if the resistance R_0 equals the wave resistance Z.

381. Solve the preceding problem, assuming that the line is terminated by a lumped capacitance C_0 rather than by a resistance R_0. Derive an expression for the voltage across the capacitance in two forms: a) as a trigonometric series; b) in closed form for the first few reflections.

Ans.

a) $$u\big|_{x=l} = E\left[1 - 4\sum_{n=1}^{\infty} \frac{\sin \gamma_n}{2\gamma_n + \sin 2\gamma_n} \cos \frac{\gamma_n t}{T}\right],$$

where the γ_n are consecutive positive roots of the equation

$$\cot \gamma = \frac{\gamma}{\alpha}, \qquad \alpha = \frac{lC}{C_0};$$

b)

$$\frac{1}{2E} u\big|_{x=l} = \begin{cases} 0, & 0 < t < T, \\ 1 - e^{-\alpha[(t/T)-1]}, & T < t < 3T, \\ -e^{-\alpha[(t/T)-1]} + \left[1 + 6\alpha\left(\dfrac{t}{3T} - 1\right)\right]e^{-3\alpha[(t/3T)-1]}, & 3T < t < 5T, \end{cases}$$

and so on.

382. Solve Prob. 381 for the case where the line is terminated by a lumped inductance L_0.

Ans.

a) $$u\big|_{x=l} = E\left[\frac{1}{1 + \alpha} - 4\sum_{n=1}^{\infty} \frac{\sin \gamma_n}{2\gamma_n - \sin 2\gamma_n} \cos \frac{\gamma_n t}{T}\right],$$

where the γ_n are consecutive positive roots of the equation

$$\tan \gamma = -\frac{\gamma}{\alpha}, \qquad \alpha = \frac{lL}{L_0};$$

b)

$$\frac{1}{2E} u \big|_{x=l} = \begin{cases} 0, & 0 < t < T, \\ e^{-\alpha[(t/T)-1]}, & T < t < 3T, \\ e^{-\alpha[(t/T)-1]} - \left[1 - 2\alpha\left(\frac{t}{T} - 3\right)\right]e^{-\alpha[(t/T)-3]}, & 3T < t < 5T, \end{cases}$$

and so on.

383. Write the general expression for the reflected waves in Prob. 382.[21]

Ans.

$$u\big|_{x=l} = 2E \sum_{n=1}^{N} (-1)^{n-1} e^{-\alpha[(t/T)-(2n-1)]} L_{n-1}\left\{2\alpha\left(\frac{t}{T} - (2n-1)\right)\right\},$$

where $(2N-1)T < t < (2N+1)T$, $N = 1, 2, 3, \ldots$, and $L_n(x)$ is the Laguerre polynomial, defined by

$$L_n(x) = \frac{e^x}{n!} \frac{d^n}{dx^n} (e^{-x} x^n).$$

Hint. Note that the Laplace transform of the Laguerre polynomial is

$$\overline{L}_n(x) = \frac{1}{p}\left(1 - \frac{1}{p}\right)^n.$$

384. Using the residue theorem, give the solution of Prob. 380 in the form of a Fourier series.

Ans.

$$\frac{1}{E} u\big|_{x=l} = 1 - \frac{R_0 e^{-\alpha vt/l}}{\alpha\sqrt{Z^2 - R_0^2}}\left[1 - 2\alpha \sum_{n=1}^{\infty} (-1)^n \frac{n\pi \sin(n\pi vt/l) - \alpha\cos(n\pi vt/l)}{\alpha^2 + n^2\pi^2}\right],$$

$$R_0 < Z;$$

$$\frac{1}{E} u\big|_{x=l} = 1 - \frac{2R_0 e^{-\beta vt/l}}{\sqrt{R_0^2 - Z^2}}$$

$$\times \sum_{n=1}^{\infty} (-1)^n \left(\frac{(n-\frac{1}{2})\pi \cos[(n-\frac{1}{2})\pi vt/l]}{\beta^2 + (n-\frac{1}{2})^2\pi^2} + \frac{\beta \sin[(n-\frac{1}{2})\pi vt/l]}{\beta^2 + (n-\frac{1}{2})^2\pi^2}\right), \qquad R_0 > Z,$$

---

[21] The details are given in L9, Sec. 4.25.

where

$$\alpha = \frac{1}{2}\ln\frac{Z+R_0}{Z-R_0}, \quad \beta = \frac{1}{2}\ln\frac{R_0+Z}{R_0-Z}, \quad v = \frac{1}{\sqrt{LC}}, \quad Z = Lv.$$

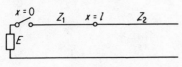

385. Study the propagation of waves along an inhomogeneous transmission line consisting of a finite section of length l with wave resistance Z_1, followed by an infinite section with wave resistance Z_2 (see Figure 98). Find the reflected and refracted waves appearing at the junction, assuming that at the time $t=0$ an arbitrary e.m.f. $E=f(t)$ is applied at the end $x=0$.

FIGURE 98

Ans.

$$u_1(x,t) = \begin{cases} 0, & 0 < t < \dfrac{x}{v_1}, \\[2mm] f\left(t-\dfrac{x}{v_1}\right), & \dfrac{x}{v_1} < t < \dfrac{2l-x}{v_1}, \\[2mm] f\left(t-\dfrac{x}{v_1}\right) - \dfrac{Z_1-Z_2}{Z_1+Z_2}f\left(t-\dfrac{2l-x}{v_1}\right), & \dfrac{2l-x}{v_1} < t < \dfrac{2l+x}{v_1}, \\[2mm] \cdots\cdots\cdots\cdots\cdots\cdots\cdots\cdots\cdots\cdots\cdots\cdots\cdots\cdots, \end{cases}$$

$$u_2(x,t) = \begin{cases} 0, & 0 < t < T + \dfrac{x-l}{v_2}, \\[2mm] \dfrac{2Z_2}{Z_1+Z_2}f\left(t-T-\dfrac{x-l}{v_2}\right), & T + \dfrac{x-l}{v_2} < t < 3T + \dfrac{x-l}{v_2}, \\[2mm] \dfrac{2Z_2}{Z_1+Z_2}\left[f\left(t-T-\dfrac{x-l}{v_2}\right) + \dfrac{Z_1-Z_2}{Z_1+Z_2}f\left(t-3T-\dfrac{x-l}{v_2}\right)\right], \\[2mm] \qquad\qquad 3T + \dfrac{x-l}{v_2} < t < 5T + \dfrac{x-l}{v_2}, \\[2mm] \cdots\cdots\cdots\cdots\cdots\cdots\cdots\cdots\cdots\cdots\cdots\cdots\cdots\cdots, \end{cases}$$

where $v_1 = 1/\sqrt{L_1 C_1}$ and $v_2 = 1/\sqrt{L_2 C_2}$ are the velocities of wave propagation along the two parts of the line, and $T = l/v_1$.

*386.** A voltage wave $E = E_0 e^{-\alpha t}$ produced by a lightning discharge at the end $x=0$ of a transmission line activates a lightning rod at the point $x=l$. Find the voltage in the section of the line after the lightning rod, assuming that the rod behaves like an ohmic resistance R_0 during its time of operation.

Ans.

$$0, \qquad 0 < t < \frac{x}{v},$$

$$u(x, t) = \frac{2E_0}{2 + \dfrac{Z}{R_0}} \frac{1 - \left(\dfrac{Z}{2R_0 + Z}\right)^k e^{2\alpha k l/v}}{1 - \dfrac{Z}{2R_0 + Z} e^{2\alpha l/v}} \, e^{-\alpha[t - (x/v)]},$$

$$\frac{x + (2k - 2)l}{v} < t < \frac{x + 2kl}{v} \qquad\qquad (k = 1, 2, \ldots),$$

where v is the velocity of wave propagation along the line, and Z is the wave resistance.

387. A constant e.m.f. E is applied at the time $t = 0$ to the end $x = 0$ of a semi-infinite cable (a line with parameters R and C). Find the voltage at every point of the cable.

Ans.

$$u(x, t) = E\left[1 - \Phi\left(\frac{x\sqrt{RC}}{2\sqrt{t}}\right)\right],$$

where $\Phi(x)$ is the probability integral.

388. Find the voltage in a cable of length l if a source of constant e.m.f. E is applied to the end $x = 0$, while the end $x = l$ is terminated by an ohmic load R_0.

Ans.

$$u(x, t) = E\left\{\frac{1 + \alpha[1 - (x/l)]}{1 + \alpha} + 2\sum_{n=1}^{\infty} \frac{\sin(\gamma_n x/l)\, e^{-\gamma_n^2 t/RCl^2}}{[1 + \alpha + (\gamma_n^2/\alpha)]\sin\gamma_n \cos\gamma_n}\right\},$$

where the γ_n are consecutive positive roots of the equation

$$\tan \gamma = -\frac{\gamma}{\alpha},$$

C and R are the capacitance and resistance of the cable per unit length, and $\alpha = Rl/R_0$.

389. Solve Prob. 387 for a cable with leakage conductance G per unit length.

Ans.

$$u(x, t) = \frac{E}{2}\left\{e^{\sqrt{RG}x}\left[1 - \Phi\left(\frac{x}{2}\sqrt{\frac{RC}{t}} + \sqrt{\frac{Gt}{C}}\right)\right]\right.$$

$$\left. + e^{-\sqrt{RG}x}\left[1 - \Phi\left(\frac{x}{2}\sqrt{\frac{RC}{t}} - \sqrt{\frac{Gt}{C}}\right)\right]\right\}.$$

390. Study the propagation of voltage waves in the compound line with equivalent circuit shown in Figure 99, caused by switching on a constant e.m.f. E at the end $x = 0$.

Ans.

$$u(x, t) = E\left[1 - \frac{2}{\pi} \int_0^{\pi/2} \sin(\alpha x \tan \varphi) \cos(\beta t \sin \varphi) \cot \varphi \, d\varphi\right],$$

where L, C and K are the self-inductance and capacitances per unit length of the line, $\alpha = \sqrt{C/K}$ and $\beta = 1/\sqrt{LK}$.

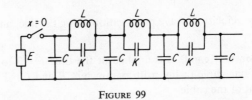

FIGURE 99

Hint. The equations governing the current and voltage in the line are

$$-\frac{\partial u}{\partial x} = L\frac{\partial I_L}{\partial t} = \frac{1}{K}\int_0^t I_K \, dt,$$

$$-\frac{\partial I}{\partial x} = C\frac{\partial u}{\partial t}, \qquad I = I_L + I_K,$$

in terms of the voltage $u(x, t)$, the total current $I(x, t)$, and the currents $I_L(x, t)$ and $I_K(x, t)$ flowing through the self-inductance and capacitances L and K.

391. Near the plane interface between two slabs of material with dielectric constants ε_1 and ε_2, there is a source of electromagnetic oscillations radiating a spherical wave whose Hertz vector has components

$$\Pi_x = \Pi_y = 0,$$

$$\Pi_z = \frac{1}{R} f\left(t - \frac{R}{v_1}\right)$$

where $f(\xi) = 0$ if $\xi < 0$, R is the distance from the source to the observation point, and $v_1 = c/\sqrt{\varepsilon_1}$ is the velocity of propagation of electromagnetic waves. Find the Hertz vector on the interface for the limiting case where the source is located on the interface itself.

Ans.

$$\Pi|_{z=0} = \Pi_1 - \Pi_2,$$

where

$$\Pi_1 = \begin{cases} 0, & t < \dfrac{r}{v_1}, \\[2ex] \dfrac{2v_1^3 v_2}{(v_1^4 - v_2^4)r} \displaystyle\int_{(v_1 t/r)\sqrt{1+(v_2/v_1)^2}}^{\sqrt{1+(v_2/v_1)^2}} f\left(t - \dfrac{rs}{\sqrt{v_1^2 + v_2^2}}\right) d\,\dfrac{1}{\sqrt{s^2 - 1}}, & t > \dfrac{r}{v_1}, \end{cases}$$

$$\Pi_2 = \begin{cases} 0, & t < \dfrac{r}{v_2}, \\[2ex] \dfrac{2v_1^3 v_2}{(v_1^4 - v_2^4)r} \displaystyle\int_{(v_2 t/r)\sqrt{1+(v_1/v_2)^2}}^{\sqrt{1+(v_1/v_2)^2}} f\left(t - \dfrac{rs}{\sqrt{v_1^2 + v_2^2}}\right) d\,\dfrac{1}{\sqrt{s^2 - 1}}, & t > \dfrac{r}{v_2}, \end{cases}$$

and $v_i = c/\sqrt{\varepsilon_i}$ $(i = 1, 2)$ are the velocities of propagation of electromagnetic waves in the two media.

Hint. Take Laplace and Hankel transforms in succession.

392. Consider the special case of the preceding problem corresponding to a wave with the steep front described by the function

$$f(\xi) = \begin{cases} 0, & \xi < 0, \\ 1, & \xi > 0. \end{cases}$$

Ans.

$$\Pi_1 = \begin{cases} 0, & t < \dfrac{r}{v_1}, \\[2ex] \dfrac{2}{1 - (v_2/v_1)^4}\left[\dfrac{1}{r} - \dfrac{v_2}{v_1}\dfrac{1}{\sqrt{t^2(v_1^2 + v_2^2) - r^2}}\right], & t > \dfrac{r}{v_1}, \end{cases}$$

$$\Pi_2 = \begin{cases} 0, & t < \dfrac{r}{v_2}, \\[2ex] \dfrac{2(v_2/v_1)^2}{1 - (v_2/v_1)^4}\left[\dfrac{1}{r} - \dfrac{v_1}{v_2}\dfrac{1}{\sqrt{t^2(v_1^2 + v_2^2) - r^2}}\right], & t > \dfrac{r}{v_2}. \end{cases}$$

393. A force $F(t)$ is applied at the time $t = 0$ to the end $x = 0$ of a semi-infinite rod. Study the resulting propagation of elastic waves in the rod.

Ans. The displacement of an arbitrary point of the rod is

$$u(x, t) = \begin{cases} 0, & 0 < t < \dfrac{x}{v}, \\[2ex] \dfrac{v}{ES} \displaystyle\int_0^{t-(x/v)} F(\tau)\, d\tau, & \dfrac{x}{v} < t < \infty, \end{cases}$$

where E is Young's modulus, ρ the density and S the cross-sectional area of the rod, and $v = \sqrt{E/\rho}$.

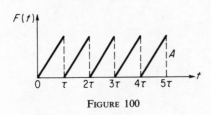

FIGURE 100

394. Suppose one end $x = 0$ of a rod of length l is clamped, while a compressive force $F(t)$ with the saw-tooth wave form shown in Figure 100 is applied to the other end $x = l$, starting from the time $t = 0$. Investigate the resulting longitudinal oscillations, and find the reaction at the fastened end, assuming that the period τ equals the time $T = l/v$ it takes an elastic wave to traverse the rod (v is the velocity of wave propagation).

Ans.

$$\frac{1}{2A} R\big|_{x=0} = \begin{cases} 0, & 0 < t < T, \ (2n+1)T < t < (2n+3)T, \quad n = 1, 3, 5, \ldots, \\ \dfrac{t}{T} - n, & nT < t < (n+1)T, \quad n = 1, 2, 5, 6, 9, 10, \ldots. \end{cases}$$

395. A cantilever clamped at the end $x = 0$ begins to oscillate under the action of an impulse delivered to a concentrated mass M_0 fastened to the free end $x = l$. Find the dynamic reaction at the clamped end, assuming that a velocity v_0 is imparted to the mass M_0 by the impulse.

Ans.

$$R\big|_{x=0} = \frac{2v_0}{v} ESf(t),$$

where

$$f(t) = \begin{cases} 0, & 0 < t < T, \\ e^{-\alpha(t-T)}, & T < t < 3T, \\ e^{-\alpha(t-T)} + [1 - 2\alpha(t-3T)]e^{-\alpha(t-3T)}, & 3T < t < 5T, \\ \cdots\cdots\cdots\cdots\cdots\cdots\cdots\cdots\cdots \end{cases}$$

in terms of the velocity of wave propagation v and the constants $\alpha = ES/M_0v$, $T = l/v$.

Hint. The boundary conditions for the displacement $u(x, t)$ at the end $x = l$ are

$$M_0 \frac{\partial^2 u}{\partial t^2} = -ES \frac{\partial u}{\partial x}, \qquad t > 0,$$

$$\frac{\partial u}{\partial t} = -v_0, \qquad t = 0.$$

396. Use the Laplace transform to solve Prob. 106.

Ans.

$$u(x, t) = \frac{Av}{\omega ES} \begin{cases} \sin(\omega x/v) \\ \cos(\omega l/v) \end{cases} \sin \omega t$$

$$+ \frac{4}{\pi} \sum_{n=0}^{\infty} (-1)^n \frac{\sin[(2n+1)\pi x/2l]}{2n+1} \frac{\sin[(2n+1)\pi vt/2l]}{1 - [(2n+1)\pi v/2l\omega]^2} \Bigg\}.$$

397. A constant force Q is applied to the end $x = 0$ of a semi-infinite beam, starting from the time $t = 0$. Find the deflection at any point of the beam, assuming that the beam is initially at rest.[22]

Ans.

$$u(x, t) = \frac{Qa^2}{2EJ} \, xtf\left(\frac{x}{2a\sqrt{t}}\right),$$

where

$$f(x) = \left(1 + \frac{2x^2}{3}\right)\left[\frac{1}{2} - C\left(\sqrt{\frac{2}{\pi}}x\right)\right] - \left(1 - \frac{2x^2}{3}\right)\left[\frac{1}{2} - S\left(\sqrt{\frac{2}{\pi}}x\right)\right]$$
$$+ \frac{2}{3\sqrt{2\pi}}\left[(1 + x^2)\frac{\sin x^2}{x} + (1 - x^2)\frac{\cos x^2}{x}\right],$$

E is Young's modulus, J the moment of inertia of a cross section, ρ the density and S the cross-sectional area, $a^2 = \sqrt{EJ/\rho S}$, and $C(s)$, $S(x)$ are the Fresnel integrals.

398. Solve the preceding problem, assuming that a constant bending moment (rather than a constant force) is applied to the end $x = 0$. Find the bending moment along the beam at any time t.

Ans.

$$m(x, t) = \frac{M}{2}\left[1 - C\left(\frac{x}{a\sqrt{2\pi t}}\right) - S\left(\frac{x}{a\sqrt{2\pi t}}\right)\right].$$

399. Find the displacement of the end $x = 0$ of a semi-infinite beam struck at the time $t = 0$ by a mass M_0 moving with velocity v_0.

Ans.

$$u\big|_{x=0} = \frac{v_0}{\alpha^2}\left\{\frac{2\alpha\sqrt{t}}{\sqrt{\pi}} - 1 + e^{\alpha^2 t}[1 - \Phi(\alpha\sqrt{t})]\right\},$$

where $\alpha = 2\sqrt{2}\,\rho aS/M$ and $\Phi(x)$ is the probability integral.

400. Find the transverse oscillations of a beam $-l \leqslant x \leqslant l$, simply supported at both ends, due to an impulse P acting at the center of the beam. Write an expression for the deflection of the center of the beam.

Ans.

$$u\big|_{x=0} = \frac{4Pla^2}{\pi^2 EJ}\sum_{n=0}^{\infty}\frac{\sin\,[(2n+1)^2\pi^2 a^2 t/4l^2]}{(2n+1)^2}.$$

401. Find the deflection of an infinite elastic plate, if at the time $t = 0$ a constant force Q is applied to the point $x = y = 0$ (see L16, p. 424).

Ans.

$$u(r, t) = \frac{Q\tau}{4\pi D}\left[\frac{\pi}{2} - \text{Si}\left(\frac{r^2}{4\tau}\right) - \sin\frac{r^2}{4\tau} + \frac{r^2}{4\tau}\,\text{Ci}\left(\frac{r^2}{4\tau}\right)\right],$$

---

[22] Problems 397–399 are treated in Lurye's book L16.

where D is the flexural rigidity and ρ the surface density of the plate, $\tau = t\sqrt{D/\rho}$ and Si(x), Ci(x) are the sine and cosine integrals.

Hint. At the point where the force is applied, $u|_{r=0}$ must be bounded, and moreover

$$r \frac{\partial}{\partial r} \Delta u\Big|_{r=0} = \frac{Q}{2\pi D}.$$

*402. Solve the preceding problem, assuming that an impulse P (rather than a force Q) is applied to the point $x = y = 0$.

Ans.

$$u(r, t) = \frac{P}{4\pi\sqrt{\rho D}}\left[\frac{\pi}{2} - \text{Si}\left(\frac{r^2}{4\tau}\right)\right].$$

403. At the time $t = 0$ an impulse with components $P_x = 0$, $P_y = P$ is applied to the point $x = y = 0$ of an infinite elastic plate. Describe the resulting process of wave propagation.

Ans. The elastic potentials are given by the formulas

$$\varphi(r, t) = \begin{cases} 0, & t < \dfrac{r}{a}, \\[2mm] \dfrac{Py}{2\pi\rho r^2}\sqrt{t^2 - \dfrac{r^2}{a^2}}, & t > \dfrac{r}{a}, \end{cases}$$

$$\psi(r, t) = \begin{cases} 0, & t < \dfrac{r}{b}, \\[2mm] -\dfrac{Px}{2\pi\rho r^2}\sqrt{t^2 - \dfrac{r^2}{b^2}}, & t > \dfrac{r}{b}, \end{cases}$$

where

$$a = \sqrt{\frac{\lambda + 2\mu}{\rho}}, \qquad b = \sqrt{\frac{\mu}{\rho}}$$

are the velocities of propagation of the longitudinal and transverse oscillations, λ and μ are Lamé's constants, and ρ is the density.

404. Solve the preceding problem, assuming that the source of the oscillations is a concentrated force with components $Q_x = 0$, $Q_y = Q$.

Ans.

$$\varphi(r, t)\big|_{t > r/a} = \frac{Qy}{4\pi\rho r^2}\left[t\sqrt{t^2 - \frac{r^2}{a^2}} - \frac{r^2}{a^2}\ln\left(\frac{at}{r} + \sqrt{\left(\frac{at}{r}\right)^2 - 1}\right)\right],$$

$$\psi(r, t)\big|_{t > r/b} = -\frac{Qx}{4\pi\rho r^2}\left[t\sqrt{t^2 - \frac{r^2}{b^2}} - \frac{r^2}{b^2}\ln\left(\frac{bt}{r} + \sqrt{\left(\frac{bt}{r}\right)^2 - 1}\right)\right],$$

where the elastic potentials are zero for smaller values of the time.

405. Show that the solution of the two-dimensional wave equation

$$\frac{\partial^2 u}{\partial x^2} + \frac{\partial^2 u}{\partial y^2} = \frac{1}{v^2}\frac{\partial^2 u}{\partial t^2}$$

in the domain $x > 0$, subject to zero initial conditions and the boundary condition $u|_{x=0} = f(y)$, can be written in the form

$$u(x, y, t) = \begin{cases} 0, & t < \dfrac{x}{v}, \\ \dfrac{xvt}{\pi}\displaystyle\int_{y-\sqrt{v^2t^2-x^2}}^{y+\sqrt{v^2t^2-x^2}} \dfrac{f(\eta)}{x^2+(y-\eta)^2}\dfrac{d\eta}{\sqrt{v^2t^2-x^2-(y-\eta)^2}}, & t > \dfrac{x}{v}. \end{cases}$$

Hint. Take Laplace and Fourier transforms in succession.

*406. Show that the solution of the wave equation

$$\frac{\partial^2 u}{\partial x^2} + \frac{\partial^2 u}{\partial y^2} = \frac{1}{v^2}\frac{\partial^2 u}{\partial t^2} + b\frac{\partial u}{\partial t} \qquad (0 \leqslant x < \infty, -\infty < y < \infty)$$

for a medium with attenuation, subject to zero initial conditions and the boundary condition $u|_{x=0} = f(y)$, can be written in the form

$u(x, y, t)$

$$= \begin{cases} 0, & t < \dfrac{x}{v}, \\ \dfrac{xe^{-v^2bt/2}}{\pi}\displaystyle\int_{y-\sqrt{v^2t^2-x^2}}^{y+\sqrt{v^2t^2-x^2}} \left[\sinh\dfrac{vbr}{2} + \dfrac{vt\cosh(vbr/2)}{r}\right]\dfrac{f(\eta)\,d\eta}{x^2+(y-\eta)^2}, & t > \dfrac{x}{v}, \end{cases}$$

where

$$r = \sqrt{v^2t^2 - x^2 - (y-\eta)^2}.$$

Deduce the solutions of Probs. 308, 360 and 405 as special cases.

4. The Mellin Transform

Let $f(r)$ be a real function defined in the interval $(0, \infty)$ such that

1. $f(r)$ is piecewise continuous and of bounded variation in every finite subinterval $[a, b]$, where $0 < a < b < \infty$;

2. Both integrals

$$\int_0^1 r^{\sigma_1 - 1}|f(r)|\,dr, \qquad \int_1^\infty r^{\sigma_2 - 1}|f(r)|\,dr \tag{18}$$

are finite for suitably chosen real numbers σ_1 and σ_2.

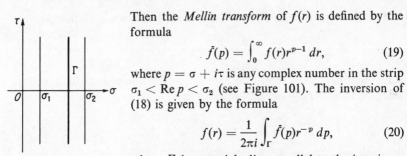

FIGURE 101

Then the *Mellin transform* of $f(r)$ is defined by the formula

$$\tilde{f}(p) = \int_0^\infty f(r) r^{p-1} \, dr, \qquad (19)$$

where $p = \sigma + i\tau$ is any complex number in the strip $\sigma_1 < \operatorname{Re} p < \sigma_2$ (see Figure 101). The inversion of (18) is given by the formula

$$f(r) = \frac{1}{2\pi i} \int_\Gamma \tilde{f}(p) r^{-p} \, dp, \qquad (20)$$

where Γ is a straight line parallel to the imaginary axis lying inside the strip.[23]

The Mellin transform is related to the Laplace and Fourier transforms, and is the appropriate tool to use for solving problems of two-dimensional elasticity theory and potential theory involving angular regions. The required technique can easily be acquired by working through the following small set of problems.[24]

*407. Find the stationary temperature distribution inside the dihedral angle $0 \leqslant r < \infty$, $0 \leqslant \varphi \leqslant \alpha < \pi$, if one boundary is held at temperature zero, while the temperature distribution

$$T\big|_{\varphi=\alpha} = \begin{cases} T_0, & 0 \leqslant r < a, \\ 0, & r > a \end{cases}$$

is maintained on the other boundary.

Ans.

$$T(r, \varphi) = \frac{T_0}{\pi} \arctan \frac{\left(\dfrac{a}{r}\right)^{\pi/\alpha} \sin \dfrac{\pi\varphi}{\alpha}}{1 + \left(\dfrac{a}{r}\right)^{\pi/\alpha} \cos \dfrac{\pi\varphi}{\alpha}}.$$

408. Solve the preceding problem, assuming that a given distribution of heat current

$$q_\varphi\big|_{\varphi=\alpha} = \begin{cases} q_0, & a - \varepsilon < r < a + \varepsilon, \\ 0 & \text{otherwise} \end{cases}$$

[23] See e.g., T5, Secs. 1.5 and 1.29. The conditions imposed on $f(r)$ can be weakened.

[24] A few remarks are in order concerning the choice of the path of integration Γ in the inversion formula (20). If the behavior of the function f as $r \to 0$ and $r \to \infty$ is known in advance (e.g., from physical considerations), then the boundaries of the strip (σ_1, σ_2) can be found from the requirement that both integrals (18) be finite. If the behavior of the function f is known only at one end point of the interval $(0, \infty)$, say as $r \to 0$, we can first determine the left-hand boundary σ_1, and the line Γ must then lie to the right of Γ and to the left of the nearest singular point of the function $\tilde{f}$ figuring in the integral (20).

is maintained on the boundary $\varphi = \alpha$. Consider the case of a concentrated current Q entering the boundary along the line $r = a$, $\varphi = \alpha$.

Ans.

$$T(r, \varphi) = \frac{Q}{2\pi k \cosh \dfrac{\pi \ln (r/a)}{2\alpha}} \ln \frac{\cosh \dfrac{\pi \ln (r/a)}{2\alpha} + \sin \dfrac{\pi\varphi}{2\alpha}}{\cosh \dfrac{\pi \ln (r/a)}{2\alpha} - \sin \dfrac{\pi\varphi}{2\alpha}},$$

where k is the thermal conductivity.

409. Use the Mellin transform to solve Prob. 303.

Ans.

$$T(r, \varphi) = \frac{qb}{\pi k} \int_0^\infty \frac{\cos [\tau \ln (r/b)] + \tau \sin [\tau \ln (r/b)]}{\tau (1 + \tau^2)} \frac{\sinh \tau\varphi}{\cosh (\tau\pi/2)} d\tau.$$

410. A thin charged wire, with charge q per unit length, is placed along the line $r = r_0$, $\varphi = \varphi_0$ inside the dihedral angle $0 < r < \infty$, $0 \leqslant \varphi \leqslant \alpha$, whose boundaries are held at potential zero. Find the potential of the resulting electrostatic field.

Ans.

$$u(r, \varphi)\big|_{\varphi \leqslant \varphi_0} = \frac{2q}{i} \int_{\sigma - i\infty}^{\sigma + i\infty} \frac{\sin p(\alpha - \varphi_0) \sin p\varphi}{p \sin p\alpha} \left(\frac{r_0}{r}\right)^p dp,$$

$$u(r, \varphi)\big|_{\varphi \geqslant \varphi_0} = \frac{2q}{i} \int_{\sigma - i\infty}^{\sigma + i\infty} \frac{\sin p\varphi_0 \sin p(\alpha - \varphi)}{p \sin p\alpha} \left(\frac{r_0}{r}\right)^p dp,$$

where $|\sigma| < \pi/\alpha$. In particular, the imaginary axis can be chosen as the path of integration.

411. Calculate the following special cases of the preceding problem:
a) $\alpha = 2\pi, r_0 = a, \varphi_0 = \pi$ (line charge opposite the edge of a conducting half plane);
b) $\alpha = 3\pi/2, r_0 = a, \varphi_0 = \pi$ (line charge near a conducting right-angular corner);
c) $\alpha = \pi/2$ (line charge inside a dihedral angle).

Ans.

$$\text{a)} \qquad u(r, \varphi) = q \ln \frac{1 + 2\sqrt{\dfrac{r}{a}} \sin \dfrac{\varphi}{2} + \dfrac{r}{a}}{1 - 2\sqrt{\dfrac{r}{a}} \sin \dfrac{\varphi}{2} + \dfrac{r}{a}};$$

b)
$$u(r, \varphi) = q \ln \frac{1 - 2\left(\dfrac{r}{a}\right)^{2/3} \cos \dfrac{2(\pi + \varphi)}{3} + \left(\dfrac{r}{a}\right)^{4/3}}{1 - 2\left(\dfrac{r}{a}\right)^{2/3} \cos \dfrac{2(\pi - \varphi)}{3} + \left(\dfrac{r}{a}\right)^{4/3}};$$

c)
$$u(r, \varphi) = q \ln \frac{1 - 2\left(\dfrac{r}{r_0}\right)^{2} \cos 2(\varphi + \varphi_0) + \left(\dfrac{r}{r_0}\right)^{4}}{1 - 2\left(\dfrac{r}{r_0}\right)^{2} \cos 2(\varphi - \varphi_0) + \left(\dfrac{r}{r_0}\right)^{4}}.$$

412. The common boundary of two media of dielectric constants ε_1 and ε_2 consists of two planes intersecting at the angle 2α (see Figure 102). Find the electrostatic field due to a charged wire lying in the plane of symmetry.

Ans.

$$E_r^{(1)} = \frac{2q}{\varepsilon_1} \left\{ \frac{r + a \cos \varphi}{R^2} + \frac{\beta}{2ir} \right.$$
$$\left. \times \int_{1-i\infty}^{1+i\infty} \frac{\sin 2\alpha(p - 1) \cos (\pi - \varphi)(p - 1)}{\sin \pi(p - 1)[\sin\pi(p - 1) - \beta \sin (\pi - 2\alpha)(p - 1)]} \left(\frac{a}{r}\right)^{p-1} dp \right\},$$

$$E_r^{(2)} = \frac{2qi}{\varepsilon_2 + \varepsilon_1} \int_{1-i\infty}^{1+i\infty} \frac{\cos \varphi(p - 1)}{\sin \pi(p - 1) - \beta \sin (\pi - 2\alpha)(p - 1)} \left(\frac{a}{r}\right)^{p-1} dp,$$

$$E_\varphi^{(1)} = -\frac{2q}{\varepsilon_1} \left\{ \frac{a \sin \varphi}{R^2} + \frac{\beta}{2ir} \right.$$
$$\left. \times \int_{1-i\infty}^{1+i\infty} \frac{\sin 2\alpha(p - 1) \sin (\pi - \varphi)(p - 1)}{\sin \pi(p - 1)[\sin \pi(p - 1) - \beta \sin (\pi - 2\alpha)(p - 1)]} \left(\frac{a}{r}\right)^{p-1} dp \right\},$$

$$E_\varphi^{(2)} = \frac{2qi}{\varepsilon_2 + \varepsilon_1} \int_{1-i\infty}^{1+i\infty} \frac{\sin \varphi(p - 1)}{\sin \pi(p - 1) - \beta \sin (\pi - 2\alpha)(p - 1)} \left(\frac{a}{r}\right)^{p-1} dp,$$

where
$$\beta = \frac{\varepsilon_2 - \varepsilon_1}{\varepsilon_2 + \varepsilon_1}$$

and R is the distance from the charge to the observation point (see G5, Chap. 14).

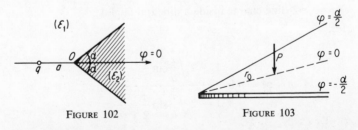

FIGURE 102   FIGURE 103

413. Investigate the bending of a thin wedge-shaped elastic plate with simply supported edges, loaded by a concentrated force P applied at an arbitrary point of the axis of symmetry (see Figure 103).[25]

Ans. The deflection of an arbitrary point of the plate is given by the formula

$$u(r, \varphi) = \frac{Pr_0 r}{4\pi D} \int_0^\infty \{\cos \varphi \sinh (\alpha - \varphi)\tau - \cos (\alpha - \varphi) \sinh \varphi\tau$$

$$+ \tau[\sin \varphi \cosh (\alpha - \varphi)\tau - \sin (\alpha - \varphi) \cosh \varphi\tau]\} \frac{\cos [\tau \ln (r/r_0)]}{\cosh \alpha\tau + \cos \alpha \, \tau(\tau^2 + 1)},$$

$$\varphi > 0,$$

where D is the flexural rigidity of the plate.

414. Let $\alpha = \pi/2$ in the preceding problem. Show that the deflection of the points on the axis of symmetry of the plate is given by the formula

$$u(r, 0) = \frac{Pr_0 r}{4\pi D} \left[\frac{1}{2}\left(\frac{r}{r_0} + \frac{r_0}{r}\right) \ln \frac{\left|1 - \dfrac{r_0^2}{r^2}\right|}{1 + \dfrac{r_0^2}{r^2}} + \ln \frac{1 + \dfrac{r_0}{r}}{\left|1 - \dfrac{r_0}{r}\right|} \right].$$

*415. A thin elastic plate $0 \leqslant r < \infty$, $0 \leqslant \varphi \leqslant \alpha$ is clamped along its edges and loaded at the point (r_0, φ_0) by a concentrated force P. Find the bending moment and shear force along the edge $\varphi = 0$.[26]

Ans.

$$M\big|_{\varphi=0} = \frac{Pr_0}{\pi r} \int_0^\infty \left[\frac{\sin \varphi_0 \sinh \alpha\tau \sinh (\alpha - \varphi_0)\tau}{\sinh^2 \alpha\tau - \tau^2 \sin^2 \alpha} \right.$$

$$\left. - \frac{\sin \alpha \sin (\alpha - \varphi_0) \cdot \tau \sinh \varphi_0\tau}{\sinh^2 \alpha\tau - \tau^2 \sin^2 \alpha} \right] \cos \left(\tau \ln \frac{r_0}{r}\right) d\tau,$$

$$N\big|_{\varphi=0} = \frac{Pr_0}{\pi r^2} \int_0^\infty \{\cos \varphi_0 \sinh \alpha\tau \sinh (\alpha - \varphi_0)\tau$$

$$+ \tau[\sin \alpha \cos (\alpha - \varphi_0) \cosh \alpha\tau \sinh (\alpha - \varphi_0)\tau - \cos \alpha \sin (\alpha - \varphi_0) \sinh \alpha\tau$$

$$\times \cosh (\alpha - \varphi_0)\tau] - \tau^2 \sin \alpha \sin (\alpha - \varphi_0) \cosh \varphi_0\tau\} \frac{\cos [\tau \ln (r_0/r)]}{\sinh^2 \alpha\tau - \tau^2 \sin^2 \alpha} d\tau.$$

---

[25] Problems 413, 414 and 417 are treated in Uflyand's paper U3.
[26] Another way of solving this problem is due to Sakharov (S2).

416. Use the Mellin transform to solve Prob. 328 (a special case of the preceding problem).

Ans.

$$M\big|_{\varphi=0} = \frac{P}{\pi} \frac{r_0^2 \sin^2 \varphi_0}{r^2 + r_0^2 - 2r_0 r \cos \varphi_0}, \qquad N\big|_{\varphi=0} = \frac{2P}{\pi} \frac{r_0^3 \sin^3 \varphi_0}{(r^2 + r_0^2 - 2r_0 r \cos \varphi_0)^2}.$$

Hint. Use the formula

$$\int_0^\infty \frac{\sinh \tau\varphi}{\sinh \tau\pi} \cos \tau\psi \, d\tau = \frac{1}{2} \frac{\sin \varphi}{\cosh \psi + \cos \varphi}.$$

417. Solve Prob. 415 assuming that one of the edges of the plate ($\varphi = \alpha$) is supported. Consider the special cases a) $\alpha = \pi/2$ and b) $\alpha = \pi$ (the quadrant and the half-plane).

Ans.

$$M\big|_{\varphi=0} = \frac{Pr_0}{\pi r} \int_0^\infty [\sin \varphi_0 \sinh (2\alpha - \varphi_0)\tau - \sin (2\alpha - \varphi_0) \sinh \varphi_0\tau]$$

$$\times \frac{\cos [\tau \ln (r/r_0)]}{\sinh 2\alpha\tau - \tau \sin 2\alpha} \, d\tau.$$

In the special cases, we have

a)
$$M\big|_{\varphi=0} = -\frac{2Pr_0}{\pi r} \frac{\sin \varphi_0 \sin 2\varphi_0}{\dfrac{r^2}{r_0^2} + \dfrac{r_0^2}{r^2} - 2 \cos 2\varphi_0} \, ;$$

b)
$$M\big|_{\varphi=0} = -\frac{P}{\pi} \frac{\left(\sqrt{\dfrac{r}{r_0}} + \sqrt{\dfrac{r_0}{r}}\right) \sin \varphi_0 \sin \dfrac{\varphi_0}{2}}{1 + \dfrac{r^2}{r_0^2} - \dfrac{2r}{r_0} \cos \varphi_0}.$$

5. Integral Transforms Involving Cylinder Functions of Imaginary Order

Let $f(x)$ be a real function defined in the interval $(0, \infty)$ such that

1. $f(x)$ is piecewise continuous and of bounded variation in every finite subinterval $[a, b]$, where $0 < a < b < \infty$;

2. Both integrals

$$\int_0^{1/2} |f(x)| \, x^{-1/2} \ln \frac{1}{x} \, dx, \qquad \int_{1/2}^\infty |f(x)| \, dx$$

are finite.

Then $f(x)$ satisfies the formula

$$f(x) = \frac{2}{\pi^2} \int_0^\infty \tau \sinh \pi\tau \, \frac{K_{i\tau}(x)}{\sqrt{x}} \, d\tau \int_0^\infty f(\xi) \frac{K_{i\tau}(\xi)}{\sqrt{\xi}} \, d\xi, \qquad 0 < x \leq \infty, \quad (21)$$

where $K_\nu(x)$ is Macdonald's function. If we write

$$\bar{f}(\tau) = \int_0^\infty f(x) \frac{K_{i\tau}(x)}{\sqrt{x}} \, dx, \qquad 0 \leq \tau < \infty, \quad (22)$$

it follows from (21) that

$$f(x) = \frac{2}{\pi^2} \int_0^\infty \bar{f}(\tau) \frac{K_{i\tau}(x)}{\sqrt{x}} \, \tau \sinh \pi\tau \, d\tau, \qquad 0 < x < \infty. \quad (23)$$

Besides the more familiar transforms considered so far, formula (21), proved by one of the authors of this book (see L6, L8), plays a role in certain physical problems.

If we use the formulas

$$x = \lambda r, \qquad \xi = \lambda\rho, \qquad f(x)\sqrt{x} = g(r) \qquad (\lambda > 0)$$

to introduce new variables, (21) takes the form

$$g(r) = \frac{2}{\pi^2} \int_0^\infty K_{i\tau}(\lambda r)\tau \sinh \pi\tau \, d\tau \int_0^\infty g(\rho) \frac{K_{i\tau}(\lambda\rho)}{\rho} \, d\rho, \qquad 0 < r < \infty, \quad (24)$$

which, although less symmetric than (21), is more suitable for solving the problems encountered in mathematical physics. Formula (24) holds provided the integrals

$$\int_0^{1/2} |g(r)| \, r^{-1} \ln \frac{1}{r} \, dr, \qquad \int_{1/2}^\infty |g(r)| \, r^{-1/2} \, dr \quad (25)$$

are finite. The following expansion of this type is useful in the applications:[27]

$$e^{-\lambda r} = \frac{2}{\pi} \int_0^\infty K_{i\tau}(\lambda r) \, d\tau, \qquad 0 < r < \infty. \quad (26)$$

In addition to the above formulas involving Macdonald's function, there is an analogous expansion in Hankel functions and a corresponding inversion formula, which play a role in certain applications. These relations can be deduced formally from (24) by setting λ equal to a pure imaginary ($\lambda = ik$),

---

[27] However, note that the first of the integrals (25) is not finite for $g(r) = e^{-\lambda r}$.

and then using the relation between the functions $K_\nu(z)$ and $H_\nu^{(2)}(z)$. In this way, we find the formulas[28]

$$\overline{\varphi}(\tau) = \int_0^\infty \varphi(r) e^{\pi\tau/2} \frac{H_{i\tau}^{(2)}(kr)}{r} \, dr, \qquad 0 \leqslant \tau < \infty, \qquad (27)$$

$$\varphi(r) = -\frac{1}{2} \int_0^\infty \overline{\varphi}(\tau) e^{\pi\tau/2} H_{i\tau}^{(2)}(kr) \, \tau \sinh \pi\tau \, d\tau, \qquad 0 < r < \infty. \qquad (28)$$

The integral expansion (24) can be used to solve the Dirichlet problem and other problems of potential theory for regions bounded by two intersecting planes (the three-dimensional problem), for wedge-shaped regions bounded by two parallel planes and two intersecting planes (perpendicular to the parallel planes), and so on. Formulas of the type (27) and (28) are encountered in solving problems involving the diffraction of acoustic and electromagnetic waves by an obstacle in the shape of a dihedral angle or a cone. The following problems illustrate various physical applications of the above expansions.

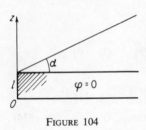

*418. Find the stationary temperature distribution in a wedge-shaped body of thickness l (see Figure 104), if the temperature distribution

FIGURE 104

$$T\big|_{\varphi=\alpha} = f(r) \sin \frac{n\pi z}{l}, \qquad n = 1, 2, \ldots$$

is maintained on the boundary $\varphi = \alpha$, while the other boundaries are held at temperature zero.

Ans.

$$T(r, \varphi, z) = \frac{2}{\pi} \sin \frac{n\pi z}{l} \int_0^\infty \left\{ f(0) + \frac{\tau}{\pi} \sinh \pi\tau \right.$$

$$\left. \times \int_0^\infty [f(\rho) - e^{-n\pi\rho/l} f(0)] K_{i\tau}\left(\frac{n\pi\rho}{l}\right) \frac{d\rho}{\rho} \right\} \frac{\sinh \varphi\tau}{\sinh \alpha\tau} K_{i\tau}\left(\frac{n\pi r}{l}\right) d\tau,$$

where $K_\nu(z)$ is Macdonald's function.

419. Solve the preceding problem for an arbitrary temperature distribution

$$T\big|_{\varphi=\alpha} = f(r, z)$$

on the face $\varphi = \alpha$.

---

[28] For conditions under which (28) implies (27), see the paper K3.

Ans.

$$T(r, \varphi, z) = \frac{2}{l} \sum_{n=1}^{\infty} T_n(r, \varphi) \sin \frac{n\pi z}{l},$$

$$T_n(r, \varphi) = \frac{2}{\pi} \int_0^{\infty} \left\{ f_n(0) + \frac{\tau}{\pi} \sinh \pi\tau \int_0^{\infty} [f_n(\rho) - e^{-n\pi\rho/l} f_n(0)] K_{i\tau} \left(\frac{n\pi\rho}{l} \right) \frac{d\rho}{\rho} \right\}$$

$$\times \frac{\sinh \varphi\tau}{\sinh \alpha\tau} K_{i\tau} \left(\frac{n\pi r}{l} \right) d\tau,$$

$$f_n(r) = \int_0^l f(r, z) \sin \frac{n\pi z}{l} \, dz.$$

Hint. Expand the function $f(r, z)$ in a Fourier series with respect to $\sin (n\pi z/l)$, and then use the result of Prob. 418.

420. Find the stationary temperature distribution in the "quadrant-shaped" slab $0 \leqslant x < \infty$, $0 \leqslant y < \infty$, $0 \leqslant z \leqslant l$, if the boundary $x = 0$ is held at constant temperature T_0, while the other boundaries are held at temperature zero.

Ans.

$$T = \frac{8T_0}{\pi^2} \sum_{n=0}^{\infty} \frac{\sin [(2n+1)\pi z/l]}{2n+1} \int_0^{\infty} \cosh \frac{\pi\tau}{2} \frac{\sinh \varphi\tau}{\sinh (\pi\tau/2)} K_{i\tau} [(2n+1)\pi r/l] \, d\tau.$$

By using the representation

$$K_{i\tau}(x) = \frac{1}{\cosh (\pi\tau/2)} \int_0^{\infty} \cos (x \sinh t) \cos \tau t \, dt$$

of Macdonald's function, this result can be brought into simpler form:

$$T = \frac{8T_0}{\pi^2} \sin 2\varphi \sum_{n=0}^{\infty} \frac{\sinh [(2n+1)\pi z/l]}{2n+1} \int_0^{\infty} \frac{\cos [(2n+1)\pi r \sinh t/l]}{\cosh 2t + \cos 2\varphi} \, dt.$$

421. Find the distribution of the electric charge density induced by a point charge q placed near the edge of a thin conducting half-plane (see Figure 105).

Ans.

$$\sigma = -\frac{q\sqrt{a}}{2\pi^2 \sqrt{x}[(x+a)^2 + z^2]}.$$

*422. Solve the preceding problem, assuming that the charge q is located at an arbitrary point $r = r_0$, $\varphi = \varphi_0$, $z = 0$. Find the distribution of electrostatic potential.

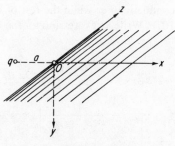

FIGURE 105

Ans.

$$u(r,\ \varphi,\ z) = \frac{2q}{\pi}\left[\frac{1}{R_1}\ \text{arc tan}\ \psi_1 - \frac{1}{R_2}\ \text{arc tan}\ \psi_2\right],$$

where

$$R_{1,2}^2 = r^2 + z^2 + r_0^2 - 2r_0 r\cos(\varphi \mp \varphi_0),$$

$$\psi_{1,2} = \sqrt[\mp]{\frac{\cosh \tfrac{1}{2}\alpha + \cos \tfrac{1}{2}(\varphi \mp \varphi_0)}{\cosh \tfrac{1}{2}\alpha - \cos \tfrac{1}{2}(\varphi \mp \varphi_0)}},\qquad \cosh \alpha = \frac{r^2 + r_0^2 + z^2}{2r_0 r}.$$

423. A point charge q is placed near the edge of a conductor of rectangular shape held at potential $u = 0$ (see Figure 106). Find the distribution of charge density on the boundaries of the conductor.[29]

Ans.

$$\sigma(r,\ z) = \frac{q}{4\pi\sqrt{2ar}}\left[\frac{2\cosh \lambda + \sqrt{2} - 1}{r(2\cosh \lambda + \sqrt{2})^{3/2}} \right.$$
$$\left. - \frac{16}{3\pi r}\int_{\cosh \frac{1}{3}\lambda}^{\infty} \frac{x\ dx}{(2x^2 - 1)^2\sqrt{4x^3 - 3x - \cosh \lambda}}\right],$$

where

$$\cosh \lambda = \frac{r^2 + z^2 + a^2}{2ar}.$$

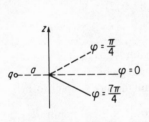

FIGURE 106

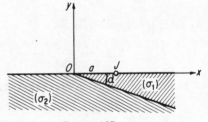

FIGURE 107

424. Find the current distribution produced in the ground by a point electrode located near a wedge-shaped layer, assuming that the layer has conductivity σ_1 while the rest of the ground has conductivity σ_2 (see Figure 107). Write an expression for the potential distribution on the earth's surface.

Ans.

$$u\big|_{\varphi=0} = -\frac{2J\beta}{\pi^3\sigma_1}\int_0^\infty \cos \lambda z\ d\lambda \int_0^\infty \frac{K_{i\tau}(\lambda a)K_{i\tau}(\lambda r)\sinh 2(\pi - \alpha)\tau}{\sinh \pi\tau + \beta \sinh(\pi - 2\alpha)\tau}\ d\tau,$$

$$u\big|_{\varphi=\pi} = \frac{4J}{\pi^3(\sigma_1 + \sigma_2)}\int_0^\infty \cos \lambda z\ d\lambda \int_0^\infty \frac{K_{i\tau}(\lambda a)K_{i\tau}(\lambda r)\sinh \pi\tau}{\sinh \pi\tau + \beta \sinh(\pi - 2\alpha)\tau}\ d\tau,$$

[29] This problem was first solved by Macdonald (M1).

where

$$\beta = \frac{\sigma_2 - \sigma_1}{\sigma_2 + \sigma_1},$$

J is the current and $K_\nu(x)$ is Macdonald's function (see S4).

425. Show that the solution of the preceding problem can be reduced to the form

$$u\big|_{\varphi=0} = -\frac{J\beta}{2\pi\sigma_1}\left\{\frac{2}{\sqrt{r^2 + a^2 + z^2}} - \frac{\beta}{\sqrt{(r+a)^2 + z^2}}\right.$$

$$\left. -\frac{1 - \beta^2}{\sin\delta}\int_0^\infty \frac{\sinh\delta\xi\,d\xi}{\sinh\pi\xi\,\sqrt{(r+a)^2 + z^2 + 4ar\sinh^2\frac{1}{4}\pi\xi}}\right\},$$

$$u\big|_{\varphi=\pi} = \frac{J}{\pi(\sigma_1 + \sigma_2)}\left\{\frac{1}{\sqrt{(r+a)^2 + z^2}}\right.$$

$$\left. -\frac{\beta}{\sin\delta}\int_0^\infty \frac{\sinh\delta\xi\,d\xi}{\sinh\pi\xi\,\sqrt{(r+a)^2 + z^2 + 4ar\sinh^2\frac{1}{4}\pi\xi}}\right\},$$

$$\delta = \arccos\frac{\beta}{2}$$

if $\alpha = \pi/4$, and to the form

$$u\big|_{\varphi=0} = -\frac{J\beta}{\pi\sigma_1}\left\{\frac{1}{\sqrt{r^2 + a^2 + z^2 + ar}}\right.$$

$$\left. -\frac{1 + \beta}{2\cos\frac{1}{2}\delta}\int_0^\infty \frac{\cosh\delta\xi\,d\xi}{\cosh\pi\xi\,\sqrt{(r+a)^2 + z^2 + 4ar\sinh^2\frac{1}{3}\xi\pi}}\right\},$$

$$u\big|_{\varphi=\pi} = \frac{3J}{\pi(\sigma_1 + \sigma_2)}\left\{\frac{1}{\sqrt{(r+a)^2 + z^2}}\right.$$

$$\left. -\frac{\beta}{\sin\delta}\int_0^\infty \frac{\sinh\delta\xi\,d\xi}{\sinh\pi\xi\,\sqrt{(r+a)^2 + z^2 + 4ar\sinh^2\frac{1}{3}\pi\xi}}\right\},$$

$$\delta = \arccos\frac{1 + \beta}{2}$$

if $\alpha = \pi/3$.

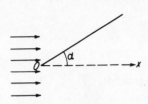

FIGURE 108

426. Solve the problem of diffraction of a plane electromagnetic wave

$$E_x = E_y = 0, \qquad E_z = E_0 e^{i(\omega t - kx)}$$

(where $k = \omega/c$ is the wave number) incident on a thin perfectly conducting sheet (see Figure 108) making an angle α with the direction of wave propagation (Sommerfeld's problem).

Ans. The complex amplitude of the z-component of the total field is

$$E = E_0 e^{-ikr\cos\varphi} \left[\frac{1}{2} + \frac{e^{i\pi/4}}{\sqrt{\pi}} \int_0^{\sqrt{2kr}\sin\frac{1}{2}\varphi} e^{-is^2}\, ds \right]$$

$$+ E_0 e^{-ikr\cos(\varphi-2\alpha)} \left[-\frac{1}{2} + \frac{e^{i\pi/4}}{\sqrt{\pi}} \int_0^{\sqrt{2kr}\sin\frac{1}{2}(\varphi-2\alpha)} e^{-is^2}\, ds \right]$$

(see K3).

427. Using the result of the preceding problem, find the current distribution on each side of the sheet. Consider the special cases where a) $\alpha = 0$; b) $\alpha = \pi/2$.

Ans. The required densities are determined by the system of linear equations

$$j_1 + j_2 = j = \frac{E_0 c}{2\pi} \sqrt{\frac{2}{\pi}} \left[\frac{e^{-ikr}}{\sqrt{ikr}} \cos\frac{\alpha}{2} + \frac{2e^{i\pi/4}}{\sqrt{\pi}} \sin\alpha\, e^{-ikr\cos\alpha} \int_0^{\sqrt{2kr}\sin\frac{1}{2}\alpha} e^{-is^2}\, ds \right],$$

$$j_1 - j_2 = \frac{E_0 c}{2\pi} \sin\alpha\, e^{-ikr\cos\alpha},$$

where j_1 and j_2 are the densities on the upper and lower sides of the sheet, respectively. In the special cases,

a) $$j_1 = j_2 = \frac{E_0 c}{2\pi\sqrt{2\pi}} \frac{e^{-ikr}}{\sqrt{ikr}}\ ;$$

b) $$j_{1,2} = \frac{E_0 c}{4\pi} \left[\frac{1}{\sqrt{\pi}} \frac{e^{-ikr}}{\sqrt{ikr}} + \frac{2e^{i\pi/4}}{\sqrt{\pi}} \int_0^{\sqrt{kr}} e^{-is^2}\, ds \pm 1 \right].$$

428. A line source of a-c current $J = J_0 e^{i\omega t}$ is placed parallel to the edge of a thin conducting sheet $0 \leqslant x < \infty$, $-\infty < y < \infty$. Find the distribution of induced currents if the source lies in the plane of the sheet at a distance a from its edge.

Ans. The complex amplitude of the current density is

$$j = \frac{J_0}{2\pi} \sqrt{\frac{a}{x}} \frac{e^{-ik(x+a)}}{x+a}.$$

429. Find the electromagnetic field of a dipole of moment P located on the axis of a perfectly conducting conical reflector of vertex angle 2α, if the dipole lies at a distance a from the vertex of the cone (see L10).

Ans. If $r < a$, the complex amplitude of the magnetic field is given by the series

$$H(r, \theta) = \frac{P\pi^2}{ica} \sum_{n=1}^{\infty} (\nu_n + \tfrac{1}{2}) \frac{H^{(2)}_{\nu_n+\frac{1}{2}}(ka)}{\sqrt{a}} \frac{J_{\nu_n+\frac{1}{2}}(kr)}{\sqrt{r}} \frac{P_{\nu_n}(-\cos\alpha)P^1_{\nu_n}(\cos\theta)}{\sin\pi\nu_n \left[\dfrac{\partial P_\nu(\cos\alpha)}{\partial\nu}\right]_{\nu=\nu_n}},$$

where the ν_n are consecutive positive roots of the equation $P_\nu(\cos\alpha) = 0$ involving the Legendre function $P_\nu(x)$, and $J_\nu(x)$, $H^{(2)}_\nu$ are cylinder functions. The corresponding formula for $r > a$ is obtained by permuting the symbols r and a in the general term of the series.

430. A plane acoustic wave $u_0 e^{i(\omega t - kx)}$ is incident on a screen in the form of a half-plane $r \geqslant 0$, $\varphi = \alpha$. Find the wave reflected from the screen.

Ans. The complex amplitude of the velocity potential at an arbitrary point is

$$u = u_0 e^{-ikr\cos\varphi} \left[\frac{1}{2} + \frac{e^{i\pi/4}}{\sqrt{\pi}} \int_0^{\sqrt{2kr}\sin\frac{1}{2}\varphi} e^{-is^2}\,ds\right]$$

$$- u_0 e^{-ikr\cos(\varphi-2\alpha)} \left[-\frac{1}{2} + \frac{e^{i\pi/4}}{\sqrt{\pi}} \int_0^{\sqrt{2kr}\sin\frac{1}{2}(\varphi-2\alpha)} e^{-is^2}\,ds\right].$$

431. A point source of sound, radiating a spherical wave

$$u = u_0 \frac{\sin(\omega t - kR)}{R},$$

is placed on the axis of a conical resonator $0 \leqslant \theta \leqslant \alpha$ with perfectly reflecting walls. Find the velocity potential inside the cone.

Ans. The complex amplitude of the velocity potential is

$$u\big|_{r<a} = u_0 \pi^2 i \sum_{n=1}^{\infty} (\nu_n + \tfrac{1}{2}) \frac{H^{(2)}_{\nu_n+\frac{1}{2}}(ka)}{\sqrt{a}} \frac{J_{\nu_n+\frac{1}{2}}(kr)}{\sqrt{r}} \frac{P'_{\nu_n}(-\cos\alpha)P_{\nu_n}(\cos\theta)}{\sin\pi\nu_n \left[\dfrac{\partial P'_\nu(\cos\alpha)}{\partial\nu}\right]_{\nu=\nu_n}},$$

where the ν_n are consecutive positive roots of the equation $P'_\nu(\cos\alpha) = 0$, and a is the distance from the source to the vertex of the cone. The corresponding formula for $r > a$ is obtained by permuting the symbols r and a.

References

Books: Campbell and Foster (C1), Carslaw and Jaeger (C2), Churchill (C4), Doetsch (D1), Fuchs and Levin (F9), Gray and Mathews (G2), Grinberg (G5), Lebedev (L8, L9), Levitan (L13), Lurye (L16), Morse and Feshbach (M9), Sneddon (S10), Titchmarsh (T5, T6), Tranter (T8), Van der Pol and Bremmer (V1), Watson (W4), Widder (W9).

Papers: Fock (F4), Kontorovich and Lebedev (K3), Lebedev (L6), Lebedev and Kontorovich (L10), Sneddon (S8, S9), Weyl (W7).

7

CURVILINEAR COORDINATES

A physical problem can often be greatly simplified by the introduction of a suitable system of orthogonal curvilinear coordinates, facilitating the formulation of the boundary conditions and making it possible to solve the problem by using the techniques of Chaps. 4–6. These earlier chapters contain an abundance of examples illustrating the simplest systems of curvilinear coordinates, i.e., polar, cylindrical and spherical coordinates. We now turn to more complicated coordinate systems, whose effective use will allow the reader to solve a much larger class of problems.

Perhaps the most important use of curvilinear coordinates is to solve boundary value problems for the Laplace and Helmholtz equations. However, neither the three-dimensional Laplace equation nor the Helmholtz equation permits separation of variables when written in *arbitrary* orthogonal curvilinear coordinates, a fact which prevents us from applying the methods developed in the preceding three chapters. Therefore a problem of great theoretical and practical interest is to find all coordinate systems which actually lead to separation of variables in these equations. Some special results pertaining to this problem, which has not yet been solved completely, will be found in concise form in Sec. 8, p. 247.[1]

The material given here is organized as follows: All problems involving a given coordinate system are grouped together, regardless of their physical content or spectral character (the latter determines whether the solution takes the form of a series or an integral). In the case of two-dimensional systems, considered in Secs. 1–3, all the necessary preliminary material is presented in problem form. However, in the case of three-dimensional

---

[1] See also the papers cited at the end of the chapter (p. 252).

systems, considered in Secs. 4–7, more background information on differential equations, special functions, etc. is needed, and this material is summarized at the beginning of each section.

Besides problems of the simpler kind, this chapter contains some relatively difficult problems, whose solution requires the use of various integral transforms, knowledge of the properties of certain special functions, and so on (see e.g., Probs. 483, 497, 502–504). These problems are intended for the adequately prepared reader, and can serve as practice material for those trying to deepen their understanding of the methods of mathematical physics. Finally, it should be kept in mind that some of the problems can be solved more simply by using other methods (e.g., conformal mapping or inversion).

I. Elliptic Coordinates

432. Study the system of elliptic coordinates α, β related to the rectangular coordinates x, y by the formula

$$x + iy = c \cosh (\alpha + i\beta) \qquad (0 \leqslant \alpha < \infty, -\pi < \beta < \pi). \qquad (1)$$

Show that the curves $\alpha = $ const, $\beta = $ const form an orthogonal system of confocal ellipses and hyperbolas (see Figure 109). What is the appropriate expression for ds^2, the square of the element of arc length? Write Laplace's equation in elliptic coordinates.

Ans.

$$ds^2 = c^2(\cosh^2 \alpha - \cos^2 \beta)(d\alpha^2 + d\beta^2),$$

$$\Delta u = \frac{1}{c^2(\cosh^2 \alpha - \cos^2 \beta)}\left(\frac{\partial^2 u}{\partial \alpha^2} + \frac{\partial^2 u}{\partial \beta^2}\right) = 0.$$

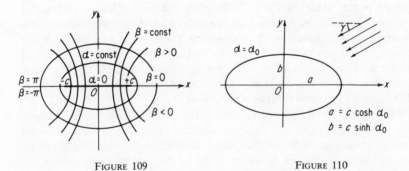

FIGURE 109   FIGURE 110

433. A conducting elliptic cylinder with semiaxes a and b is placed in a homogeneous electric field (see Figure 110). Find the distribution of electric charge density on the surface of the cylinder.

Ans.

$$\sigma = \frac{E_0}{4\pi}(a+b)\frac{\cos(\beta-\gamma)}{\sqrt{a^2\sin^2\beta + b^2\cos^2\beta}},$$

where E_0 is the external field.

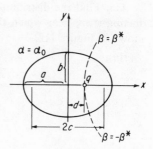

Hint. Introduce elliptic coordinates α, β (see the preceding problem), where the parameter c equals the eccentricity of the given ellipse.

434. A wire with charge q per unit length is placed inside a hollow conducting elliptic cylinder with semiaxes a and b. Find the poten-

FIGURE 111

tial distribution inside the cylinder, assuming that the wire is parallel to the axis of the cylinder (see Figure 111).

Ans.

$$u(\alpha,\beta) = 2q\left[\alpha_0 - \alpha + 2\sum_{n=1}^{\infty}\frac{\sinh n(\alpha_0-\alpha)}{n\cosh n\alpha_0}\cos n\beta^*\cos n\beta\right],$$

in terms of the elliptic coordinates α and β, where α_0 and β^* are the parameters defined by the relations

$$\tanh\alpha_0 = \frac{b}{a}, \quad \cos\beta^* = \frac{d}{c} \qquad (d < c).$$

Hint. Regard the charge q as uniformly distributed over the "curvilinear rectangle"

$$0 < \alpha < \delta, \qquad \beta^* - \frac{\varepsilon}{2} < |\beta| < \beta^* + \frac{\varepsilon}{2},$$

and then take the limit as $\delta, \varepsilon \to 0$.

435. Solve the preceding problem, assuming that the charged wire is placed outside the cylinder at the point $x = d$ $(d > a)$, $y = 0$.

Ans.

$$u(\alpha,\beta) = q\ln\frac{\cosh(\alpha^* + \alpha - 2\alpha_0) - \cos\beta}{\cosh(\alpha^* - \alpha) - \cos\beta}, \quad \cosh\alpha^* = \frac{d}{c}, \quad c = \sqrt{a^2 - b^2}.$$

Hint. To sum the series, use the formula

$$\ln(2\cosh y - 2\cos x) = y - 2\sum_{n=1}^{\infty}\frac{e^{-ny}\cos nx}{n}, \qquad y > 0.$$

436. Find the distribution of induced charge on an infinitely long conducting strip, near which there is a wire with charge q per unit length, as shown in Figure 112.

Ans.

$$\sigma = -\frac{q}{2\pi}\frac{\sqrt{d^2 - a^2}}{(d - x)\sqrt{a^2 - x^2}}.$$

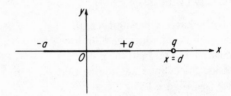

FIGURE 112

*437. An elliptic cylinder of given dimensions, made from material of magnetic permeability μ, is introduced into a homogeneous magnetic field making angle γ with the major axis (see Figure 110). Find the magnetic potential, and show that the field outside the cylinder is homogeneous.

Ans.

$$u = H_0(x \cos \gamma + y \sin \gamma)$$
$$+ H_0(1 - \mu)ab\,\frac{a + b}{\sqrt{a^2 - b^2}}\left(\frac{\cos \gamma \cos \beta}{a + \mu b} + \frac{\sin \gamma \sin \beta}{b + \mu a}\right)e^{-\alpha}$$
$$+ \text{const outside the cylinder,}$$

$$u = H_0(a + b)\left(\frac{\cos \gamma}{a + \mu b}\,x + \frac{\sin \gamma}{b + \mu a}\,y\right) + \text{const inside the cylinder,}$$

where H_0 is the external field, and the ellipse has semiaxes a and b.

Hint. The choice of the particular solutions for the region outside the cylinder is dictated by the requirement that grad u be bounded.

438. A hollow elliptic cylinder, made from material of magnetic permeability μ, has a cross section bounded by the confocal ellipses

$$\frac{x^2}{a_1^2} + \frac{y^2}{b_1^2} = 1, \quad \frac{x^2}{a_2^2} + \frac{y^2}{b_2^2} = 1 \quad (\sqrt{a_1^2 - b_1^2} = \sqrt{a_2^2 - b_2^2} = c).$$

Suppose the cylinder is introduced into a homogeneous magnetic field with components

$$H_x = -H_0, \qquad H_y = H_z = 0.$$

Find the distribution of potential in the body of the cylinder.

Ans.

$$u(\alpha, \beta)$$

$$= H_0 c \frac{\sinh \alpha_1 \sinh (\alpha - \alpha_1) + \mu \cosh \alpha_1 \cosh (\alpha - \alpha_1)}{\sinh (\alpha_2 - \alpha_1)(\sinh \alpha_1 + \mu^2 \cosh \alpha_1) + \mu e^{\alpha_1} \cosh (\alpha_2 - \alpha_1)} e^{\alpha_2} \cos \beta,$$

where

$$\tanh \alpha_i = \frac{b_i}{a_i} \qquad (i = 1, 2).$$

439. A cavity in the shape of an elliptic cylinder with semiaxes a and b is hollowed out of iron of magnetic permeability μ, and contains a line current J whose direction is parallel to the axis of the cylinder. Find the vector potential of the magnetic field, assuming that the current passes through the point $x_0 < \sqrt{a^2 - b^2}$, $y_0 = 0$ of the semimajor axis.

Ans.

$$A_1 = -\frac{2J}{c} \ln R + \frac{4J}{c} (\mu - 1)$$

$$\times \sum_{n=1}^{\infty} \frac{e^{-n\alpha_0} \cos n\beta_0}{n(\cosh n\alpha_0 + \mu \sinh n\alpha_0)} \cosh n\alpha \cos n\beta + \text{const}, \qquad 0 \leqslant \alpha < \alpha_0,$$

$$A_2 = -\frac{2J\mu}{c} \ln R - \frac{4J\mu}{c} (\mu - 1)$$

$$\times \sum_{n=1}^{\infty} \frac{\sinh n\alpha_0 \cos n\beta_0}{n(\cosh n\alpha_0 + \mu \sinh n\alpha_0)} e^{-n\alpha} \cos n\beta + \text{const}, \qquad \alpha > \alpha_0,$$

where A_1 and A_2 are the values of the z-component of the vector potential in the air and in the iron, respectively, R is the distance from the point $(x_0, 0)$ to the point (x, y), α and β are elliptic coordinates, c is the velocity of light, and

$$\tanh \alpha_0 = \frac{b}{a}, \qquad \cos \beta_0 = \frac{x_0}{\sqrt{a^2 - b^2}}.$$

440. Solve the preceding problem for the limiting case $\mu = \infty$. Find the tangential component of the magnetic field on the interface between the air and the iron.

Ans.

$$A_1 = -\frac{2J}{c} \ln R + \frac{4J}{c} \sum_{n=1}^{\infty} \frac{e^{-n\alpha_0} \cos n\beta_0}{n \sinh n\alpha_0} \cosh n\alpha \cos n\beta + \text{const},$$

$$\frac{1}{\mu} A_2 = -\frac{2J\alpha}{c} + \text{const}.$$

The tangential component of the magnetic field on the interface is

$$H_\beta\big|_{\alpha=\alpha_0} = \frac{2J}{c} \frac{1}{\sqrt{a^2 - b^2} \sqrt{\cosh^2 \alpha_0 - \cos^2 \beta}} = \frac{2J}{c} \frac{p}{ab},$$

where p is the length of the perpendicular dropped from the origin of coordinates onto the tangent to the ellipse at the point $M = (\alpha_0, \beta)$.

441. A d-c current flows in a conductor whose cross section is an ellipse with semiaxes a and b, producing heat with volume density Q. Find the temperature distribution inside the conductor, assuming that its surface is held at temperature zero.

Ans.

$$T(\alpha, \beta) = \frac{Q}{8k}(a^2 - b^2)\left(1 - \frac{\cosh 2\alpha}{\cosh 2\alpha_0}\right)(\cosh 2\alpha_0 - \cos 2\beta),$$

where k is the thermal conductivity and

$$\tanh \alpha_0 = \frac{b}{a}.$$

Hint. Subtract out a particular solution $u = P(x, y)$ of the inhomogeneous heat conduction equation, where $P(x, y)$ is a polynomial in x and y.

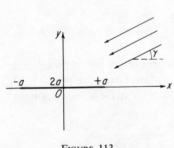

FIGURE 113

442. A thin sheet of width $2a$ is placed in a plane-parallel flow of an ideal fluid. Find the velocity potential, assuming that the direction of the flow makes angle γ with the plane of the sheet (see Figure 113).

Ans.

$$u = v_\infty(x \cos \gamma + y \sin \gamma + a \sin \gamma e^{-\alpha} \sin \beta),$$

where v_∞ is the velocity of the flow far from the sheet.

*443. Solve the problem of the twisting of a rod of elliptical cross section with two cuts extending to its foci, as shown in Figure 114. Calculate the torsional rigidity C numerically for the cases where the ratio of the semiaxes is $\frac{1}{4}$, $\frac{1}{2}$ and $\frac{3}{4}$.

Ans. The torsion function is

$$u(\alpha, \beta) = -c^2 \sinh^2 \alpha \sin^2 \beta + \frac{8b^2}{\pi} \sum_{n=0}^{\infty} \frac{\cosh (2n+1)\alpha}{\cosh (2n+1)\alpha_0} \frac{\sin (2n+1)\beta}{(1 - 4n^2)(2n+3)}.$$

The torsional rigidity is

$$\frac{C}{C_0} = \frac{1 + (b^2/a^2)}{b/a}\left[\frac{32}{\pi^2}\left(1 - \frac{b^2}{a^2}\right)S - \frac{b}{2a}\right],$$

$$S = \sum_{n=0}^{\infty}\left[\frac{(2n - 1)\sinh(2n + 3)\alpha_0 + (2n + 3)\sinh(2n - 1)\alpha_0}{2(2n + 1)}\right.$$

$$\left. - \sinh(2n + 1)\alpha_0\right]\frac{1}{(1 - 4n^2)(2n - 1)(2n + 3)^2\cosh(2n + 1)\alpha_0},$$

where $\tanh\alpha_0 = b/a$, and C_0 is the torsional rigidity of the ellipse without the cut. The result of the numerical calculations are

$$\left.\frac{C}{C_0}\right|_{b/a=1/4} = 0.997, \quad \left.\frac{C}{C_0}\right|_{b/a=1/2} = 0.970,$$

$$\left.\frac{C}{C_0}\right|_{b/a=3/4} = 0.826.$$

Hint. Subtract out the particular solution $-y^2$.

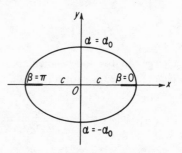

FIGURE 114

444. Find the torsion function of a rod of semielliptic cross section.

Ans.

$$u(\alpha, \beta) = -c^2\sinh^2\alpha\sin^2\beta - \frac{8b^2}{\pi}\sum_{n=0}^{\infty}\frac{\sinh(2n + 1)\alpha}{\sinh(2n + 1)\alpha_0}\frac{\sin(2n + 1)\beta}{(4n^2 - 1)(2n + 3)},$$

where a and b are the semiaxes of the ellipse, and

$$\tanh\alpha_0 = \frac{b}{a}, \qquad c^2 = a^2 - b^2.$$

445. Find the stationary temperature distribution in a body whose surface is the hyperbolic cylinder

$$\frac{x^2}{a^2} - \frac{y^2}{b^2} = 1, \qquad x > 0,$$

given the temperature distribution on the surface.

Ans.

$$T(\alpha, \beta) = \frac{1}{\pi}\int_0^{\infty}\left[f_c\frac{\cosh\lambda\beta}{\cosh\lambda\beta_0}\cos\lambda\alpha + f_s\frac{\sinh\lambda\beta}{\sinh\lambda\beta_0}\sin\lambda\alpha\right]d\lambda,$$

where

$$f_c = \int_{-\infty}^{\infty} f(\alpha) \cos \lambda\alpha \, d\alpha, \qquad f_s = \int_{-\infty}^{\infty} f(\alpha) \sin \lambda\alpha \, d\alpha$$

are the Fourier cosine and sine transforms of the function $f(\alpha)$ figuring in the boundary condition

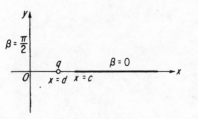

FIGURE 115

$$T|_{\beta=\beta_0} = f(\alpha) \qquad \left(\tan \beta_0 = \frac{b}{a}\right).$$

Hint. In Probs. 445–447 use elliptic coordinates defined by

$$x + iy = c \cosh(\alpha + i\beta)$$
$$(-\infty < \alpha < \infty, \, 0 \le \beta \le \pi),$$

instead of by formula (1), p. 204.

446. Find the density of electric charge on two perpendicular grounded planes between which there is a charged wire, as shown in Figure 115.

Ans.

$$\sigma|_{\beta=0} = -\frac{qd}{\pi\sqrt{(x/c)^2 - 1}} \frac{\sqrt{1 - (d/c)^2}}{x^2 - d^2}, \qquad c < x < \infty,$$

$$\sigma|_{\beta=\pi/2} = -\frac{qd}{\pi\sqrt{(y/c)^2 + 1}} \frac{\sqrt{1 - (d/c)^2}}{y^2 + d^2}, \qquad -\infty < y < \infty.$$

447. A charged wire with charge q per unit length is placed on the axis of symmetry of a slot of width $2a$ cut in a grounded conducting metal plane. Find the resulting electrostatic potential u. What is the charge density on the two parts of the plane?

Ans.

$$u(\alpha, \beta) = q \ln \frac{\cosh \alpha + \sin \beta}{\cosh \alpha - \sin \beta}.$$

The charge density is

$$\sigma = \frac{q}{2\pi x} \frac{\cdot a}{\sqrt{x^2 - a^2}}, \qquad x > a.$$

Hint. In elliptic coordinates, the two parts of the plane have equations $\beta = 0$ and $\beta = \pi$.

2. Parabolic Coordinates

448. Study the system of parabolic coordinates α, β, related to the rectangular coordinates x, y by the formula

$$x + iy = \frac{c}{2}(\alpha + i\beta)^2 \qquad (-\infty < \alpha < \infty, \, 0 \le \beta < \infty). \qquad (2)$$

Show that the curves $\alpha = \text{const}$, $\beta = \text{const}$ form two orthogonal families of parabolas (see Figure 116). What is the appropriate expression for the square of the element of arc length? Write Laplace's equation in parabolic coordinates.

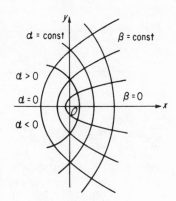

Ans.

$$ds^2 = c^2(\alpha^2 + \beta^2)(d\alpha^2 + d\beta^2),$$

$$\Delta u = \frac{1}{c^2(\alpha^2 + \beta^2)}\left(\frac{\partial^2 u}{\partial \alpha^2} + \frac{\partial^2 u}{\partial \beta^2}\right) = 0.$$

*449. A charged wire with charge q per unit length is placed at the focus of a conducting screen in the form of a parabolic cylinder. Find the resulting electrostatic field.

FIGURE 116

Ans. The electrostatic potential is

$$u(\alpha, \beta) = 4q \int_0^\infty \frac{\sinh \lambda(\beta_0 - \beta)}{\lambda \cosh \lambda \beta_0} \cos \lambda\alpha \, d\lambda.$$

in terms of β_0, the value of the coordinate β on the surface of the cylinder, given by

$$\beta_0 = \sqrt{p/c},$$

where p is the focal distance of the parabola and c is the scale factor figuring in formula (2). Using formula 13, p. 385, we can write the solution in closed form:

$$u(\alpha, \beta) = 2q \ln \frac{\cosh \dfrac{\pi\alpha}{2\beta_0} + \cos \dfrac{\pi\beta}{2\beta_0}}{\cosh \dfrac{\pi\alpha}{2\beta_0} - \cos \dfrac{\pi\beta}{2\beta_0}}. \tag{3}$$

450. Write a solution of the preceding problem in the form of a series of functions depending on the variable β.

Ans.

$$u(\alpha, \beta) = 8q \sum_{n=0}^\infty \frac{e^{-(2n+1)\pi|\alpha|/2\beta_0} \cos \left[(2n + 1)\pi\beta/2\beta_0\right]}{2n + 1}.$$

Using the formula

$$\sum_{n=0}^\infty \frac{p^{2n+1} \cos (2n + 1)x}{2n + 1} = \frac{1}{4} \ln \frac{1 + 2p \cos x + p^2}{1 - 2p \cos x + p^2}, \qquad p^2 \leqslant 1$$

to sum the series, we arrive at formula (3).

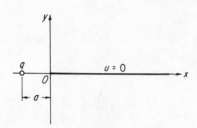

FIGURE 117

451. A charged wire with charge q per unit length is placed parallel to the edge of a thin conducting half-plane (see Figure 117). Find the resulting charge distribution on the half-plane.

Ans.

$$\sigma = \frac{q}{2\pi(a + x)}\sqrt{\frac{a}{x}}.$$

Hint. The equation of the half-plane in parabolic coordinates is $\beta = 0$. In solving the problem, regard the charge as uniformly distributed over the area of a curvilinear rectangle bounded by appropriate curves $\alpha = $ const, $\beta = $ const, and then make the dimensions of the rectangle go to zero.

3. Two-Dimensional Bipolar Coordinates

452. Study the system of two-dimensional bipolar coordinates α, β, related to the rectangular coordinates x, y by the formula

$$x + iy = c \tanh \frac{\alpha + i\beta}{2} \qquad (-\infty < \alpha < \infty, -\pi < \beta < \pi). \qquad (4)$$

Show that the curves $\beta = $ const are circles

$$x^2 + (y - c \cot \beta)^2 = \frac{c^2}{\sin^2 \beta}$$

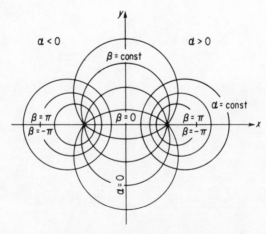

FIGURE 118

going through the points $x = \pm c$, while the curves $\alpha = $ const are the orthogonal circles

$$(x - c \coth \alpha)^2 + y^2 = \frac{c^2}{\sinh^2 \alpha}$$

(see Figure 118). What is the appropriate expression for the square of the element of arc length? Write Laplace's equation in two-dimensional bipolar coordinates.

Ans.

$$ds^2 = \frac{c^2}{(\cosh \alpha + \cos \beta)^2} (d\alpha^2 + d\beta^2),$$

$$\Delta u = \frac{(\cosh \alpha + \cos \beta)^2}{c^2} \left(\frac{\partial^2 u}{\partial \alpha^2} + \frac{\partial^2 u}{\partial \beta^2} \right) = 0.$$

453. Find the electrostatic potential in the region between two parallel cylinders of radius a, held at potentials $\pm V$, respectively, if the axes of the cylinders are a distance $2l$ apart (see Figure 119). Calculate the capacitance per unit length between the pair of cylinders.

Ans. In terms of bipolar coordinates α, β with parameter $c = \sqrt{l^2 - a^2}$,

$$u = V\frac{\alpha}{\alpha_0}, \qquad C = \frac{1}{4\alpha_0},$$

where

$$\cosh \alpha_0 = \frac{l}{a}$$

(the two cylinders have equations $\alpha = \pm\alpha_0$).

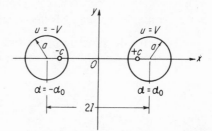

FIGURE 119

454. A cylindrical pipe of radius a is buried in the ground at depth b (see Figure 120). Find the stationary temperature distribution in the region surrounding the pipe, if the temperature of the ground is zero while a heat current Q, uniformly distributed with respect to angle, leaves the pipe's surface.

Ans.

$$T(\alpha, \beta) = \frac{Q}{k\pi \sinh \alpha_0} \left[\frac{\alpha}{2} + \sum_{n=1}^{\infty} \frac{(-1)^n}{n} \frac{e^{-n\alpha_0}}{\cosh n\alpha_0} \sinh n\alpha \cos n\beta \right],$$

where $\cosh \alpha_0 = b/a$ and k is the thermal conductivity.

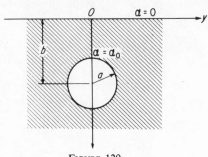

FIGURE 120

455. Two parallel cylinders of radius a with axes a distance $2l$ apart are placed in a plane-parallel flow of an ideal fluid, making angle α with the line joining the centers of the cylinders. Find the resulting velocity potential.

Ans.

$$u(\alpha, \beta) = v_\infty \sqrt{l^2 - a^2}$$

$$\times \left\{ \cos \gamma \left[\frac{\sinh \alpha}{\cosh \alpha + \cos \beta} + 2 \sum_{n=1}^{\infty} (-1)^n \frac{e^{-n\alpha_0}}{\cosh n\alpha_0} \sinh n\alpha \cos n\beta \right] \right.$$

$$\left. + \sin \gamma \left[\frac{\sin \beta}{\cosh \alpha + \cos \beta} + 2 \sum_{n=1}^{\infty} (-1)^n \frac{e^{-n\alpha_0}}{\sinh n\alpha_0} \cosh n\alpha \sin n\beta \right] \right\},$$

where $\cosh \alpha_0 = l/a$ and v_∞ is the velocity of the flow far from the cylinders.

456. Solve the problem of the twisting of a circular shaft weakened by an eccentrically drilled hole, as shown in Figure 121 (see W6).

Ans. The torsion function is

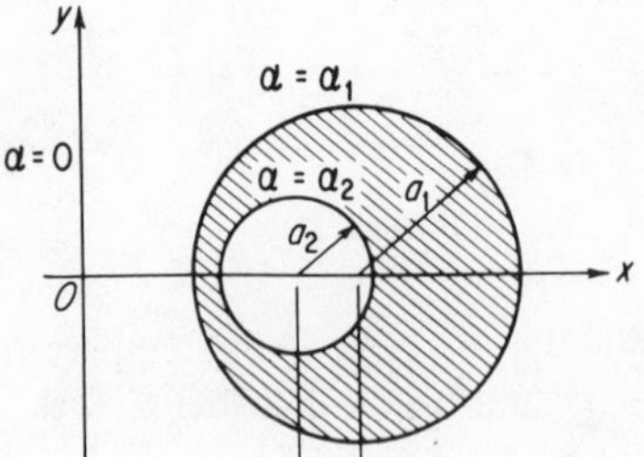

$$u(\alpha, \beta) = a_1^2 \sinh^2 \alpha_1 \left\{ \coth \alpha_1 + \coth \alpha_2 \right.$$

$$- \frac{\cosh \alpha}{\cosh \alpha + \cos \beta}$$

$$+ 2 \sum_{n=1}^{\infty} \frac{(-1)^n}{\sinh n(\alpha_2 - \alpha_1)}$$

$$\times [e^{-n\alpha_1} \coth \alpha_1 \sinh n(\alpha_2 - \alpha)$$

$$\left. + e^{-n\alpha_2} \coth \alpha_2 \sinh n(\alpha - \alpha_1)] \cos n\beta \right\},$$

FIGURE 121

where α_1 and α_2 are determined from the relations

$$\cosh \alpha_1 = \frac{a_1^2 - a_2^2 + d^2}{2da_1}, \qquad \cosh \alpha_2 = \frac{a_1^2 - a_2^2 - d^2}{2da_2}.$$

*457. A narrow slot is cut in a circular shaft subject to twisting (see Figure 122). Find the torsion function and calculate the torsional rigidity. Calculate the rigidity numerically for the case $h/a = \frac{1}{2}$.

Ans.

The torsion function is

$$u(\alpha, \beta) = a^2 \sinh^2 \alpha_0 \left\{ -\frac{\sin^2 \beta}{(\cosh \alpha + \cos \beta)^2} \right.$$

$$\left. + \frac{1}{\pi} \sum_{n=0}^{\infty} a_n e^{-(n+1/2)(\alpha - \alpha_0)} \sin (n + \tfrac{1}{2})\beta \right\},$$

where

$$a_n = \int_0^{2\pi} \frac{\sin^2 \beta \sin (n + \tfrac{1}{2})\beta}{(\cosh \alpha_0 + \cos \beta)^2} \, d\beta,$$

$$\sinh \alpha_0 = \frac{h(2a - h)}{2a(a - h)} \qquad (h < a).$$

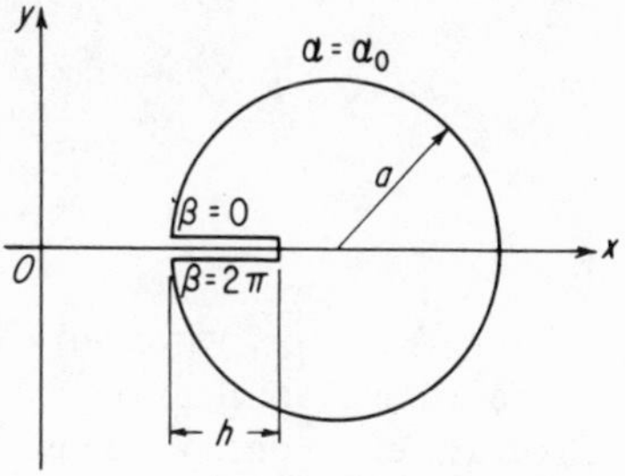

FIGURE 122

The rigidity is

$$C = Ga^4 \sinh^4 \alpha_0 \left[\frac{2}{\pi} \sinh \alpha_0 \sum_{n=0}^{\infty} a_n b_n - \frac{1}{2\pi} \sum_{n=0}^{\infty} (2n + 1)a_n^2 - \frac{\pi}{2 \sinh^4 \alpha_0} \right],$$

where

$$b_n = \int_0^{2\pi} \frac{\sin^2 \beta \sin (n + \tfrac{1}{2})\beta}{(\cosh \alpha + \cos \beta)^3} \, d\beta.$$

In the case $h/a = \frac{1}{2}$, it is found that $C = 1.28 Ga^4$ (compare with the result of Prob. 233).

Hint. Subtract out the particular solution $-y^2$. To calculate the rigidity, use the formula

$$\iint_S (u \, \Delta \psi - \psi \, \Delta u) d\sigma + \int_0^{2\pi} \left[u \frac{\partial \psi}{\partial \alpha} - \psi \frac{\partial u}{\partial \alpha} \right]_{\alpha = \alpha_0} d\beta = 0.$$

In the numerical calculation of the coefficients a_n and b_n, use the relations

$$a_n = \frac{2n - 1}{2} A_{n-1} - \frac{2n + 3}{2} A_{n+1},$$

$$A_n = \frac{1}{\sinh \alpha_0} \left[\frac{2}{2n + 1} + (2n + 1) \sum_{m=1}^{\infty} (-1)^m e^{-m\alpha_0} \frac{1}{(n + \tfrac{1}{2})^2 - m^2} \right]$$

$$b_n = \frac{1}{2} \left[\frac{2n - 1}{2} B_{n-1} - \frac{2n + 3}{2} B_{n+1} \right],$$

$$B_n = \frac{1}{\sinh^2 \alpha_0} \left[A_n \cosh \alpha_0 + (2n + 1) \sum_{m=1}^{\infty} m(-1)^m e^{-m\alpha_0} \frac{1}{(n + \tfrac{1}{2})^2 - m^2} \right].$$

458. An eccentrically drilled tube, with the cross section shown in Figure 121, is subjected to a pressure uniformly distributed over its interior surface. Find the resulting (two-dimensional) deformation of the tube, if no forces act on its outer surface. Calculate the normal stresses on the inner surface of the tube (see J3).

Ans.

$$\frac{\sigma_\beta}{p} = \frac{2a_1^2}{a_1^2 + a_2^2} \frac{(a_1^2 - d^2)^2 - a_2^2(a_2 + 2d \cos \beta)^2}{[a_1^2 - (a_2 - d)^2][a_1^2 - (a_2 + d)^2]} - 1,$$

where p is the pressure.

459. A charged wire with charge q per unit length is placed at height h inside a long tunnel of semicircular profile. Find the electrostatic field in the plane of symmetry, assuming that the walls of the tunnel constitute an equipotential surface.

Ans.

$$E|_{\alpha=0} = \frac{4q}{a} \frac{\sin 2\beta^*(1 + \cos \beta)}{\cos 2\beta - \cos 2\beta^*},$$

where a is the radius of the semicircle, and β^* is determined from the formula

$$\tan \frac{\beta^*}{2} = \frac{h}{a}.$$

Hint. In a system of bipolar coordinates, the region in question is bounded by the coordinate surfaces $\beta = 0$ and $\beta = \pi/2$. Expand the solution in a Fourier integral with respect to the variable α.

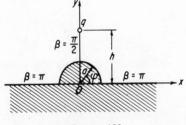

FIGURE 123

460. A conducting plane has a semicylindrical boss of radius a, as shown in Figure 123. Find the distribution of electric charge induced on the surface of the conductor by a wire carrying charge q per unit length placed in the plane of symmetry (see Figure 123). Calculate the maximum value of the electric field on the surface.

Ans. The charge density is

$$\sigma(x) = -\frac{q}{2\pi a} \frac{\sin 2\beta^*(\cosh \alpha - 1)}{\sinh^2 \alpha + \sin^2 \beta^*}, \qquad x = a \coth \frac{\alpha}{2}$$

on the plane, and

$$\sigma(\varphi) = -\frac{q}{2\pi a}\frac{\sin 2\beta^* \cosh \alpha}{\cosh^2 \alpha - \sin^2 \beta^*}, \qquad \sin \varphi = \frac{1}{\cosh \alpha}$$

on the boss, where β^* is determined from the formula

$$\tan \frac{\beta^*}{2} = \frac{h}{a}.$$

The maximum field is

$$E_{\max} = \frac{8qh}{h^2 - a^2}.$$

Hint. Use formula 15, p. 385.

461. Solve the preceding problem, assuming that there is a semicylindrical groove in the plane (rather than a boss).

Ans. The charge density is

$$\sigma(x) = -\frac{q}{3\pi a}\frac{\sin\left(\dfrac{\pi + 2\beta^*}{3}\right)(\cosh \alpha - 1)}{\cosh \dfrac{2\alpha}{3} + \cos\left(\dfrac{\pi + 2\beta^*}{3}\right)}, \qquad x = a \coth \frac{\alpha}{2}$$

on the plane, and

$$\sigma(\varphi) = -\frac{q}{3\pi a}\frac{\sin\dfrac{2(\pi - \beta^*)}{3}\cosh \alpha}{\cosh \dfrac{2\alpha}{3} + \cos\dfrac{2(\pi - \beta^*)}{3}}, \qquad \sin \varphi = \frac{1}{\cosh \alpha}$$

on the surface of the groove.

462. A cylindrical body with cross section in the shape of a symmetrical circular lune is placed in a homogeneous plane-parallel flow of an ideal fluid, with velocity components $v_x = -v_\infty$, $v_y = v_z = 0$ (see Figure 124). Find the resulting velocity potential.

Ans.

$$u(\alpha, \beta)$$

$$= v_\infty \sqrt{a^2 - b^2}\left[\frac{\sinh \alpha}{\cosh \alpha + \cos \beta}\right.$$

$$\left. + 2\int_0^\infty \frac{\sinh \lambda\beta_0 \cosh \lambda(\pi - \beta)}{\sinh \pi\lambda \sinh (\pi - \beta_0)\lambda}\sin \lambda\alpha\, d\lambda\right],$$

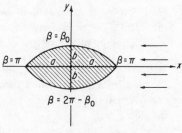

FIGURE 124

where

$$\tan \frac{\beta_0}{2} = \frac{b}{a}$$

and v_∞ is the velocity of the flow far from the body.

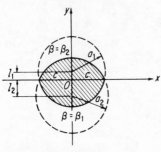

FIGURE 125

Hint. Use the system of bipolar coordinates

$$x + iy = c \tanh \frac{\alpha + i\beta}{2}$$

$$(-\infty < \alpha < \infty, \quad \beta_0 < \beta < 2\pi - \beta_0),$$

instead of the system given by formula (4), p. 212.

463. Find the torsion function for a cylinder whose cross section is a circular lune bounded by the curves $\beta = \beta_1$ and $\beta = \beta_2$ in bipolar coordinates, as shown in Figure 125 (see U1).

Ans.

$$u(\alpha, \beta) = c^2 \left\{ \frac{\cos \beta}{\cosh \alpha + \cos \beta} - 2 \cot \beta_2 \int_0^\infty \frac{\sinh \lambda \beta_2 \sinh \lambda(\beta - \beta_1)}{\sinh \lambda \pi \sinh \lambda(\beta_2 - \beta_1)} \cos \lambda \alpha \, d\lambda \right.$$

$$\left. - 2 \cot \beta_1 \int_0^\infty \frac{\sinh \lambda \beta_1 \sinh \lambda(\beta_2 - \beta)}{\sinh \lambda \pi \sinh \lambda(\beta_2 - \beta_1)} \cos \lambda \alpha \, d\lambda \right\},$$

where the parameters β_1 and β_2 are determined from the relations

$$\left| \frac{\cot \beta_2}{\cot \beta_1} \right| = \frac{l_2}{l_1}, \qquad \left| \frac{\sin \beta_2}{\sin \beta_1} \right| = \frac{a_1}{a_2},$$

$$c = \sqrt{a_1^2 - l_1^2} = \sqrt{a_2^2 - l_2^2}.$$

Hint. To make the problem homogeneous, subtract out the particular solution $\frac{1}{2}(c^2 - x^2 - y^2)$ from the equation for the torsion function, where c is the scale factor of the system of bipolar coordinates.

464. A semicircular elastic plate of radius a is clamped along its edges and loaded by a concentrated force P applied at an arbitrary point of its axis of symmetry. Find the distribution of bending moments along the edges of the plate.[2]

---

[2] Problems 464–466 are treated in Uflyand's book U2.

Ans.

$$M|_{\beta=0} = \frac{P}{2\pi}(\cosh\alpha + 1)\int_0^\infty\left[\sinh\frac{\lambda\pi}{2}\sinh\lambda\left(\frac{\pi}{2} - \beta^*\right)\right.$$

$$\left. - \lambda\cot\beta^*\sinh\lambda\beta^*\right]\frac{\cos\lambda\alpha\,d\lambda}{\sinh^2(\lambda\pi/2) - \lambda^2},$$

$$M|_{\beta=\pi/2} = -\frac{P}{2\pi}\cosh\alpha\int_0^\infty\left[\sinh\frac{\lambda\pi}{2}\sinh\lambda\beta^*\cot\beta^*\right.$$

$$\left. - \lambda\sinh\lambda\left(\frac{\pi}{2} - \beta^*\right)\right]\frac{\cos\lambda\alpha\,d\lambda}{\sinh^2(\lambda\pi/2) - \lambda^2},$$

where β^* is determined from the relation

$$\tan\frac{\beta^*}{2} = \frac{b}{a},$$

b is the distance from the point of application of the force to the rectilinear edge of the plate, and α, β is a system of bipolar coordinates.

465. Solve the preceding problem for the case of a uniformly distributed external load q. Write an expression for the deflection along the axis of symmetry.

Ans.

$$u|_{\alpha=0} = \frac{qa^4}{32D\cos^2(\beta/2)}\left\{\int_0^\infty\left[\sinh\lambda\beta\cos\beta - \lambda\sin\beta\left(\cosh\lambda\beta - \coth\frac{\lambda\pi}{2}\sinh\lambda\beta\right)\right]\right.$$

$$\left.\times\frac{\lambda\,d\lambda}{\sinh^2(\lambda\pi/2) - \lambda^2} - \cos\beta\tan^2\frac{\beta}{2}\right\},$$

where D is the flexural rigidity of the plate.

466. Find the distribution of bending moments along the edges of an elastic plate in the form of a symmetric circular lune $-\beta_0 \leqslant \beta \leqslant \beta_0$, due to a concentrated load P applied at the center of the plate.

Ans.

$$M|_{\beta=\pm\beta_0} = -\frac{P\sin\beta_0}{2\pi}(\cosh\alpha + \cos\beta_0)\int_0^\infty\frac{\sinh\lambda\beta_0\cos\lambda\alpha}{\sinh2\lambda\beta_0 + \lambda\sin2\beta_0}\,d\lambda.$$

4. Spheroidal Coordinates

Turning to three-dimensional coordinate systems, we first consider the case where the region of interest is an ellipsoid. If all three semiaxes of the ellipsoid are different, it is necessary to deal with Lamé functions, whose theory lies beyond the scope of this book.[3] However, in most cases of practical interest, two of the semiaxes of the ellipsoid are equal. Then the ellipsoid reduces to a *spheroid*, i.e., an ellipsoid of revolution, the corresponding coordinate systems are called *spheroidal coordinates*, and the appropriate particular solutions of Laplace's equation can be written in terms of elementary functions and spherical harmonics, whose theory, unlike that of Lamé functions, has been fully developed. Moreover, these particular solutions can be used to solve boundary value problems for the region bounded by a hyperboloid of revolution (see Probs. 483–485).[4]

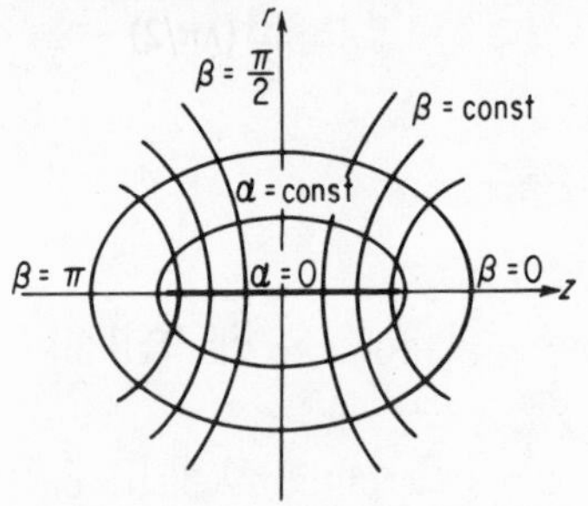

By *prolate spheroidal coordinates* we mean coordinates α, β, φ related to the rectangular coordinates x, y, z by the formulas

$$x = c \sinh \alpha \sin \beta \cos \varphi,$$
$$y = c \sinh \alpha \sin \beta \sin \varphi,$$
$$z = c \cosh \alpha \cos \beta,$$

where

$$0 \leqslant \alpha < \infty, \quad 0 \leqslant \beta \leqslant \pi, \quad -\pi < \varphi \leqslant \pi,$$

FIGURE 126

and $c > 0$ is a scale factor.[5] Then every point of space is characterized by a unique triple of numbers α, β, φ. The corresponding triply orthogonal system of surfaces consists of the prolate spheroids $\alpha = $ const with foci at the points $(0, 0, \pm c)$, the double-sheeted hyperboloids of revolution $\beta = $ const, which are confocal with the spheroids, and the planes $\varphi = $ const passing through the z-axis (see Figure 126). The square of the element of arc length and Laplace's equation take the form

$$ds^2 = c^2(\sinh^2 \alpha + \sin^2 \beta)(d\alpha^2 + d\beta^2) + c^2 \sinh^2 \alpha \sin^2 \beta \, d\varphi^2,$$

$$\Delta u = \frac{1}{c^2(\sinh^2 \alpha + \sin^2 \beta)} \left[\frac{1}{\sinh \alpha} \frac{\partial}{\partial \alpha} \left(\sinh \alpha \frac{\partial u}{\partial \alpha} \right) + \frac{1}{\sin \beta} \frac{\partial}{\partial \beta} \left(\sin \beta \frac{\partial u}{\partial \beta} \right) \right.$$
$$\left. + \left(\frac{1}{\sinh^2 \alpha} + \frac{1}{\sin^2 \beta} \right) \frac{\partial^2 u}{\partial \varphi^2} \right] = 0.$$

If there is no dependence on the angle φ, the appropriate particular solutions

---

[3] For the general theory of ellipsoidal coordinates and Lamé functions, see e.g., H4 and W4. Some problems involving ellipsoidal coordinates, but not requiring knowledge of Lamé functions, are given at the end of this section (see Probs. 486–489).

[4] Spheroidal coordinates can also be used to solve boundary value problems for Helmholtz's equation, but then the particular solutions involve more complicated functions, called *spheroidal wave functions* (see S18, S19).

[5] If a point has cylindrical coordinates r, φ, z, then $z + ir = c \cosh (\alpha + i\beta)$.

of Laplace's equation for dealing with boundary conditions specified on the surface of a prolate spheroid ($\alpha = \alpha_0$) are given by

$$u = u_n = M_n P_n(\cosh \alpha) P_n(\cos \beta), \qquad n = 0, 1, 2, \ldots$$

in the case of the interior problem ($0 \leqslant \alpha < \alpha_0$), and by

$$u = u_n = N_n Q_n(\cosh \alpha) P_n(\cos \beta), \qquad n = 0, 1, 2, \ldots$$

in the case of the exterior problem ($\alpha_0 < \alpha < \infty$).[6] Here $P_n(z)$ is the Legendre polynomial of degree n, $Q_n(z)$ is the Legendre function of the second kind (of degree n), and M_n, N_n are arbitrary constants.[7]

Similarly, if there is no dependence on the angle φ, the use of the superposition method to solve boundary value problems for the region bounded by the hyperboloid of revolution $\beta = \beta_0$ starts from the following particular solutions of Laplace's equation, which depend continuously on the parameter τ:

$$u = u_\tau = M_\tau P_{-\frac{1}{2}+i\tau}(\cosh \alpha) P_{-\frac{1}{2}+i\tau}(\pm \cos \beta), \qquad \tau \geqslant 0. \tag{5}$$

Here $P_\nu(z)$ is the Legendre function of the first kind, and the plus sign pertains to the interior region $0 \leqslant \beta < \beta_0$ and the minus sign to the exterior region $\beta_0 < \beta \leqslant \pi$. The general solution is now constructed by integrating (5) with respect to τ. To determine M_τ, we use the *Mehler-Fock theorem*,[8] instead of the theory of expansions in series of spherical harmonics.

Next we consider *oblate spheroidal coordinates* α, β, φ related to the rectangular coordinates x, y, z by the formulas

$$x = c \cosh \alpha \sin \beta \cos \varphi, \quad y = c \cosh \alpha \sin \beta \sin \varphi, \quad z = c \sinh \alpha \cos \beta,$$

where

$$0 \leqslant \alpha < \infty, \quad 0 \leqslant \beta \leqslant \pi, \quad -\pi < \varphi \leqslant \pi,$$

and $c > 0$ is a scale factor.[9] In this case, the triply orthogonal system of

---

[6] For particular solutions in the more general case of dependence on φ, see e.g., L9, p. 218.

[7] The functions $Q_n(z)$ can be expressed in terms of elementary functions by using the recurrence relation

$$(n + 1)Q_{n+1}(z) - (2n + 1)zQ_n(z) + nQ_{n-1}(z) = 0, \qquad n = 1, 2, \ldots,$$

together with the formulas

$$Q_0(z) = \frac{1}{2} \ln \frac{z+1}{z-1}, \qquad Q_1(z) = \frac{z}{2} \ln \frac{z+1}{z-1} - 1.$$

[8] See L9, Sec. 8.9 and also Probs. 483–485.

[9] If a point has cylindrical coordinates r, φ, z, we now have $z + ir = c \sinh (\alpha + i\beta)$.

surfaces consists of the oblate spheroids $\alpha = $ const, the single-sheeted hyperboloids of revolution $\beta = $ const and the planes $\varphi = $ const (see Figure 127). The square of the element of arc length and Laplace's equation now take the form

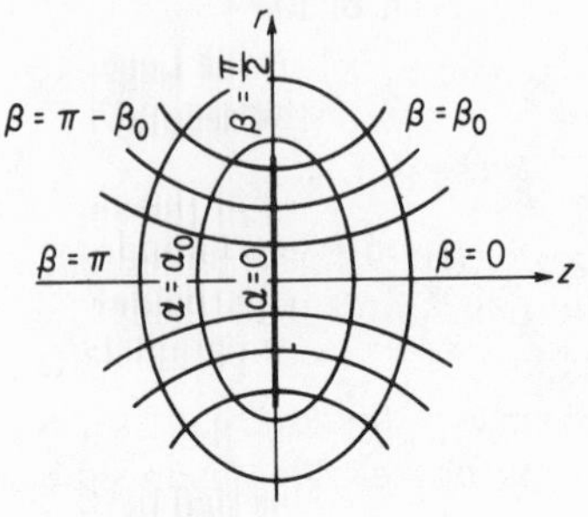

$$ds^2 = c^2(\cosh^2 \alpha - \sin^2 \beta)(d\alpha^2 + d\beta^2) + c^2 \cosh^2 \alpha \sin^2 \beta \, d\varphi^2,$$

$$\Delta u = \frac{1}{c^2(\cosh^2 \alpha - \sin^2 \beta)}\left[\frac{1}{\cosh \alpha}\right.$$
$$\times \frac{\partial}{\partial \alpha}\left(\cosh \alpha \frac{\partial u}{\partial \alpha}\right) + \frac{1}{\sin \beta}\frac{\partial}{\partial \beta}\left(\sin \beta \frac{\partial u}{\partial \beta}\right)$$
$$\left. + \left(\frac{1}{\sin^2 \beta} - \frac{1}{\cosh^2 \alpha}\right)\frac{\partial^2 u}{\partial \varphi^2}\right].$$

FIGURE 127

If there is no dependence on the angle φ, the appropriate particular solutions of Laplace's equation for dealing with boundary conditions specified on the surface of an oblate spheroid ($\alpha = \alpha_0$) are given by

$$u = u_n = M_n P_n(i \sinh \alpha)P_n(\cos \beta), \qquad n = 0, 1, 2, \ldots$$

for the interior problem ($0 \leqslant \alpha < \alpha_0$), and by

$$u = u_n = N_n Q_n(i \sinh \alpha)P_n(\cos \beta)$$

for the exterior problem ($\alpha_0 \leqslant \alpha < \infty$). Here the boundedness of grad u plays a role (see L9, p. 217).

Having made these preliminary remarks, we now give a number of physical problems whose solution involves the use of spheroidal coordinates.

467. Find the charge density on the surface of a conductor in the form of a prolate spheroid with semiaxes a and b, carrying total charge Q. What is the capacitance of the spheroid?

Ans.

$$\sigma = \frac{Q}{4\pi c^2 \sinh \alpha_0}\frac{1}{\sqrt{\cosh^2 \alpha_0 - \cos^2 \beta}} = \frac{Q}{4\pi ab^2}\frac{1}{\sqrt{\dfrac{z^2}{a^4} + \dfrac{r^2}{b^4}}},$$

$$C = \frac{2c}{\ln \dfrac{a + c}{a - c}},$$

where

$$c = \sqrt{a^2 - b^2}, \qquad \tanh \alpha_0 = \frac{b}{a}.$$

Hint. Introduce a system of prolate spheroidal coordinates such that the surface of the ellipsoid has equation $\alpha = \alpha_0$.

468. A point charge q is placed at the center of a hollow conducting shield in the form of a prolate ellipsoid with semiaxes a and b. Find the potential distribution inside the shield, assuming that its surface is at zero potential.

Ans.

$$u(\alpha, \beta) = \frac{q}{\sqrt{a^2 - b^2}}\left[\frac{1}{\sqrt{\sinh^2 \alpha + \cos^2 \beta}}\right.$$
$$\left. - \sum_{n=0}^{\infty}(4n + 1)P_{2n}(0)\frac{Q_{2n}(\cosh \alpha_0)}{P_{2n}(\cosh \alpha_0)}P_{2n}(\cosh \alpha)P_{2n}(\cos \beta)\right],$$

where $P_n(x)$ and $Q_n(x)$ are the Legendre functions of the first and second kind, $\tanh \alpha_0 = b/a$, and

$$P_0(0) = 1, \quad P_{2n}(0) = (-1)^n \frac{1 \cdot 3 \cdot 5 \cdots (2n - 1)}{2 \cdot 4 \cdot 6 \cdots 2n}, \quad n = 1, 2, \ldots$$

Hint. Subtract the potential of the point charge from the solution. To express the solution in series, use the integral

$$\int_{-1}^{1} \frac{P_{2n}(x)\,dx}{\sqrt{\sinh^2 \alpha + x^2}} = 2P_{2n}(0)Q_{2n}(\cosh \alpha).$$

469. Solve Prob. 467 for the case of an oblate spheroid.

Ans.

$$\sigma = \frac{Q}{4\pi c^2 \cosh \alpha_0} \frac{1}{\sqrt{\cosh^2 \alpha_0 - \sin^2 \beta}} = \frac{Q}{4\pi a^2 b}\frac{1}{\sqrt{\dfrac{r^2}{a^4} + \dfrac{z^2}{b^4}}},$$

$$C = \frac{c}{\text{arc sin}\,\dfrac{c}{a}},$$

where

$$c = \sqrt{a^2 - b^2}, \qquad \tanh \alpha_0 = \frac{b}{a}.$$

470. Find the charge density on the surface of a conducting disk of radius a, carrying total charge Q. What is the capacitance of the disk?

Ans.

$$\sigma = \frac{Q}{4\pi a\sqrt{a^2 - r^2}}, \qquad C = \frac{2a}{\pi},$$

where r is the distance from the center of the disk.

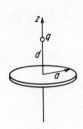

FIGURE 128

*471. Find the surface charge density induced on a disk by a point charge q located at an arbitrary point of its axis of symmetry (see Figure 128).[10]

Ans.

$$\sigma = -\frac{q}{4\pi a^2}\left[\left(\frac{a}{d}\right)^2\left(1 + \frac{a^2\sin^2\beta}{d^2}\right)^{-3/2} + \frac{2i}{\sqrt{\pi}\cos\beta}\right.$$

$$\left. \times \sum_{n=0}^{\infty}(-1)^n\frac{(4n+1)n!}{\Gamma(n+\frac{1}{2})}Q_{2n}\left(\frac{id}{a}\right)P_{2n}(\cos\beta)\right],$$

in terms of the Legendre functions of the first and second kind $P_n(z)$ and $Q_n(z)$, where $\beta = \arcsin(r/a)$, and r is the distance from the center of the disk to an arbitrary point on its surface.

Hint. Use the expansion

$$\frac{1}{\sqrt{\cosh^2\alpha - x^2}} = i\sum_{n=0}^{\infty}(4n+1)P_{2n}(0)Q_{2n}(i\sinh\alpha)P_{2n}(x).$$

472. A grounded plane screen with a circular aperture of radius a is placed in an electric field which is homogeneous at a great distance from the screen. Suppose the field has the value E_1 to the left of the screen and the value E_2 to the right of the screen (see Figure 129). Find the potential in the surrounding space, and calculate the field along the axis of symmetry (a problem of interest in electron optics).

Ans. The potential is

$$u\big|_{z>0} = \frac{a}{\pi}(E_1 - E_2)$$

$$\times \left(1 - \sinh\alpha\arctan\frac{1}{\sinh\alpha}\right)\cos\beta - E_2 z,$$

$$u\big|_{z<0} = \frac{a}{\pi}(E_2 - E_1)$$

$$\times \left(1 - \sinh\alpha\arctan\frac{1}{\sinh\alpha}\right)\cos\beta - E_1 z,$$

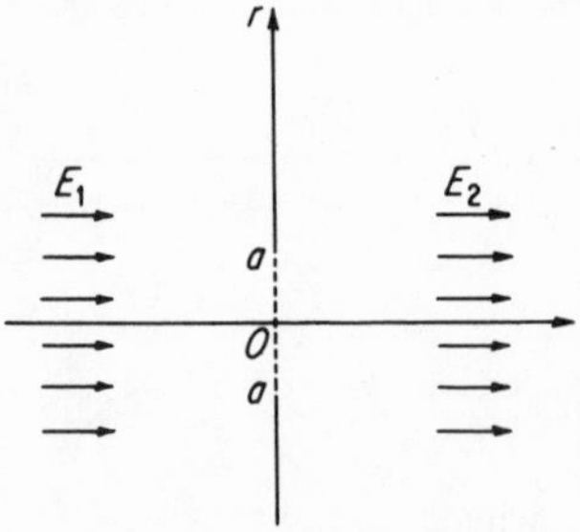

FIGURE 129

while

$$E\big|_{r=0,z>0} = E_2 + \frac{E_1 - E_2}{\pi}\left(\arctan\frac{1}{\sinh\alpha} - \frac{\sinh\alpha}{\cosh^2\alpha}\right),$$

$$E\big|_{r=0,z<0} = E_1 + \frac{E_2 - E_1}{\pi}\left(\arctan\frac{1}{\sinh\alpha} - \frac{\sinh\alpha}{\cosh^2\alpha}\right)$$

is the field along the axis.

---

[10] This problem can be solved more easily by using integral equations (see Prob. 551b).

Hint. Introduce spheroidal coordinates with parameter c equal to the radius of the aperture, and look for a solution of the form

$$u\big|_{z<0} = A_1(\alpha) \cos \beta - E_1 z,$$
$$u\big|_{z>0} = A_2(\alpha) \cos \beta - E_2 z.$$

473. An oblate dielectric spheroid, with semiaxes a and b and dielectric constant ε, is placed in a homogeneous electric field E_0 directed along its axis of symmetry (in the negative z-direction). Solve the resulting problem of electrostatics.

Ans. The potential is

$$u = \frac{E_0 z}{\varepsilon \cosh^2 \alpha_0 - \sinh^2 \alpha_0 - (\varepsilon - 1) \sinh \alpha_0 \cosh^2 \alpha_0 \operatorname{arc\,cot} \sinh \alpha_0} + \text{const}$$

in the dielectric, and

$$u = E_0 z - \frac{E_0 c(\varepsilon - 1) \sinh \alpha_0 \cosh^2 \alpha_0 (1 - \sinh \alpha \operatorname{arc\,cot} \sinh \alpha) \cos \beta}{\varepsilon \cosh^2 \alpha_0 - \sinh^2 \alpha_0 - (\varepsilon - 1) \sinh \alpha_0 \cosh^2 \alpha_0 \operatorname{arc\,cot} \sinh \alpha_0}$$
$$+ \text{const}$$

in the air, where

$$\tanh \alpha_0 = \frac{b}{a}.$$

474. Find the resistance of a grounding rod inserted in ground of conductivity σ (see Figure 130), assuming that the rod is shaped like half of a prolate spheroid with semiaxes a and b, where $a > b$ (see O1).

Ans.

$$R = \frac{1}{4\pi\sigma\sqrt{a^2 - b^2}} \ln \frac{a + \sqrt{a^2 - b^2}}{a - \sqrt{a^2 - b^2}}.$$

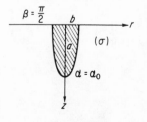

FIGURE 130

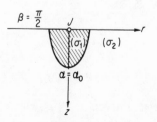

FIGURE 131

475. A constant current J enters the ground through a point contact placed on the earth's surface over a hole filled with material of conductivity σ_1, different from the conductivity σ_2 of the rest of the ground (see Figure 131).

Find the current distribution in the ground, assuming that the boundary between the two media is the prolate spheroid with equation

$$\frac{z^2}{a^2} + \frac{r^2}{b^2} = 1, \quad z > 0.$$

Ans. The potentials of the current field in the two media are given by

$$u_1 = \frac{J}{2\pi\sigma_1 R} + \frac{J(\sigma_2 - \sigma_1)}{2\pi\sigma_1\sqrt{a^2 - b^2}} \sum_{n=0}^{\infty}(4n + 1)P_{2n}(0)$$

$$\times \frac{Q_{2n}(\cosh\alpha_0)Q'_{2n}(\cosh\alpha_0)P_{2n}(\cosh\alpha)P_{2n}(\cos\beta)}{\sigma_1 Q_{2n}(\cosh\alpha_0)P'_{2n}(\cosh\alpha_0) - \sigma_2 P_{2n}(\cosh\alpha_0)Q'_{2n}(\cosh\alpha_0)},$$

$$u_2 = \frac{J}{2\pi\sqrt{a^2 - b^2}\,\sinh^2\alpha_0} \sum_{n=0}^{\infty}(4n + 1)P_{2n}(0)$$

$$\times \frac{Q_{2n}(\cosh\alpha)P_{2n}(\cos\beta)}{\sigma_1 Q_{2n}(\cosh\alpha_0)P'_{2n}(\cosh\alpha_0) - \sigma_2 P_{2n}(\cosh\alpha_0)Q'_{2n}(\cosh\alpha_0)},$$

where R is the distance from the source to the field point, $\tanh\alpha_0 = b/a$, $P_n(x)$ and $Q_n(x)$ are Legendre functions, and

$$P_0(0) = 1, \quad P_{2n}(0) = (-1)^n \frac{1 \cdot 3 \cdot 5 \cdots (2n - 1)}{2 \cdot 4 \cdot 6 \cdots 2n}, \quad n = 1, 2, \ldots$$

476. A d-c current enters ground of conductivity σ through a grounding plate in the form of a disk of radius a (see Figure 132). Find the distribution of current under the plate, and calculate the resistance of the plate.

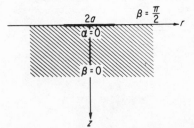

FIGURE 132

Ans. The potential of the current field is

$$u = \frac{2V}{\pi}\,\text{arc cot sinh}\,\alpha,$$

where V is the potential of the plate. The resistance is

$$R = \frac{1}{4\sigma a}.$$

Hint. Introduce a system of spheroidal coordinates ($0 \leqslant \alpha < \infty$, $0 \leqslant \beta \leqslant \pi/2$).

477. A prolate spheroid made from material of magnetic permeability μ is introduced into a homogeneous magnetic field H_0 directed along its axis of symmetry (in the negative z-direction). Solve the resulting problem of magnetostatics, and show that the field inside the spheroid is homogeneous.

Ans. The field inside the spheroid is

$$H_z = - \frac{H_0}{\cosh^2 \alpha_0 - \mu \sinh^2 \alpha_0 + (\mu - 1) \sinh^2 \alpha_0 \cosh \alpha_0 \ln \coth \frac{\alpha_0}{2}}$$

where

$$\tanh \alpha_0 = \frac{b}{a}.$$

Substitution for α_0 leads to the expression

$$H_z = - \frac{H_0}{1 - \frac{b^2}{c^2}(\mu - 1)\left[1 - \frac{a}{2c} \ln \frac{a+c}{a-c}\right]}, \qquad c = \sqrt{a^2 - b^2}.$$

478. Find the stationary temperature distribution in a prolate spheroid $\alpha = \alpha_0$, if a given axially symmetric temperature distribution

$$T(\alpha, \beta)|_{\alpha=\alpha_0} = f(\beta)$$

is maintained on its surface. Consider the special case where one half of the surface of the spheroid ($z < 0$) is held at temperature zero, while the other half ($z > 0$) is held at temperature T_0.

Ans.

$$T(\alpha, \beta) = \frac{1}{2} \sum_{n=0}^{\infty} \frac{2n+1}{P_n(\cosh \alpha_0)} P_n(\cosh \alpha) P_n(\cos \beta) \int_0^{\pi} f(\eta) P_n(\cos \eta) \sin \eta \, d\eta,$$

in terms of the Legendre polynomials $P_n(x)$. In the special case,

$$T(\alpha, \beta) = \frac{T_0}{2}\left[1 + \sum_{n=0}^{\infty} \frac{4n+3}{2(n+1)} \frac{P_{2n}(0)}{P_{2n+1}(\cosh \alpha_0)} P_{2n+1}(\cos \beta) P_{2n+1}(\cosh \alpha)\right].$$

Hint. To calculate the integral

$$\int_0^1 P_n(x) \, dx,$$

use the recurrence relation

$$(2n + 1)P_n(x) = P'_{n+1}(x) - P'_{n-1}(x).$$

479. Find the stationary temperature distribution in a prolate spheroid with semiaxes a and b, whose surface is held at temperature zero, if heat is produced inside the spheroid with constant density Q.

Ans.

$$T(\alpha, \beta) = - \frac{Qc^2}{4k}\left[\sinh^2 \alpha \sin^2 \beta - \frac{2b^2}{3c^2}\right.$$

$$\left. + \frac{b^2}{3(3a^2 - c^2)}(3 \cos^2 \beta - 1)(3 \cosh^2 \alpha - 1)\right],$$

in terms of the spheroidal coordinates α and β, where $c = \sqrt{a^2 - b^2}$ and k is the thermal conductivity.

Hint. Make the problem homogeneous by subtracting out the particular solution $-Qr^2/4k$ of the inhomogeneous heat conduction equation.

480. A body in the shape of a prolate spheroid with semiaxes a and b is placed in a homogeneous flow of an ideal fluid, directed along its axis of symmetry (in the negative z-direction). Find the resulting velocity potential.

Ans.

$$u(\alpha, \beta) = v_\infty c \left[\cosh \alpha + \frac{\cosh \alpha \ln \coth \dfrac{\alpha}{2} - 1}{\dfrac{ac}{b^2} - \dfrac{1}{2} \ln \dfrac{a+c}{a-c}} \right] \cos \beta + \text{const},$$

where $c = \sqrt{a^2 - b^2}$ and v_∞ is the velocity far from the body.

*481. Calculate the gravitational potential due to a homogeneous prolate spheroid with semiaxes a and b, and find an asymptotic expression for the potential in the case of small eccentricity c.

Ans. The potential outside the spheroid is

$$u(\alpha, \beta) = \frac{\pi \rho a b^2}{\sqrt{a^2 - b^2}} \Big\{ \cosh \alpha (3 \cos^2 \beta - 1)$$

$$+ [2(\sin^2 \beta - \sinh^2 \alpha) + 3 \sin^2 \beta \sinh^2 \alpha] \ln \coth \frac{\alpha}{2} \Big\},$$

where ρ is the density, and the gravitational constant is taken to be unity. For small c,

$$u \approx M \left[\frac{1}{R} + \frac{c^2}{5R^3} P_2(\cos \theta) \right],$$

where

$$M = \tfrac{4}{3} \pi \rho a b^2$$

is the mass of the ellipsoid, and

$$R = \sqrt{r^2 + z^2}, \quad \theta = \arctan \frac{r}{z}, \quad P_2(x) = \frac{3x^2 - 1}{2}.$$

Hint. Inside the spheroid, subtract out the particular solution $-\pi \rho r^2$ of the inhomogeneous equation.

482. Solve the preceding problem for the case of an oblate spheroid.

Ans. Outside the spheroid the gravitational potential is

$$u(\alpha, \beta) = \frac{\pi a^2 b \rho}{\sqrt{a^2 - b^2}} \{ [2(\cosh^2 \alpha + \sin^2 \beta)$$

$$- 3 \cosh^2 \alpha \sin^2 \beta] \operatorname{arc cot} \sinh \alpha - \sinh \alpha (3 \cos^2 \beta - 1) \}.$$

For small c,

$$u \approx M\left[\frac{1}{R} - \frac{c^2}{5R^3} P_2(\cos\theta)\right],$$

where

$$M = \tfrac{4}{3}\pi\rho a^2 b.$$

Hint. Inside the spheroid subtract out the particular solution

$$-\pi\rho(a^2 - b^2)\cosh^2\alpha\sin^2\beta$$

of the inhomogeneous equation.

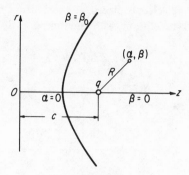

FIGURE 133

*483. A point charge q is placed at the focus of a grounded conducting screen shaped like a hyperboloid of revolution (see Figure 133). Solve the resulting problem of electrostatics.

Ans. The electrostatic potential is

$$u(\alpha,\beta) = \frac{q}{R} - \frac{q\pi}{c}\int_0^\infty \frac{\tau\tanh\pi\tau}{\cosh\pi\tau}\frac{P_{-\frac{1}{2}+i\tau}(-\cos\beta_0)}{P_{-\frac{1}{2}+i\tau}(\cos\beta_0)}P_{-\frac{1}{2}+i\tau}(\cosh\alpha)$$
$$\times\, P_{-\frac{1}{2}+i\tau}(\cos\beta)\,d\tau,\qquad 0\leqslant\beta\leqslant\beta_0,$$

where $P_\nu(x)$ is the Legendre function of the first kind.

Hint. Introduce prolate spheroidal coordinates α, β, φ such that the hyperboloid has equation $\beta = \beta_0$, and make use of the Mehler-Fock theorem

$$f(\alpha) = \int_0^\infty \tau\tanh\pi\tau P_{-\frac{1}{2}+i\tau}(\cosh\alpha)\,d\tau\int_0^\infty f(\xi)P_{-\frac{1}{2}+i\tau}(\cosh\xi)\sinh\xi\,d\xi$$

(see L9, p. 221).

484. A point charge q is placed near the vertex of an electrode shaped like a hyperboloid of revolution. Find the potential in the surrounding space, assuming that the charge lies on the axis of the hyperboloid (see Figure 134).

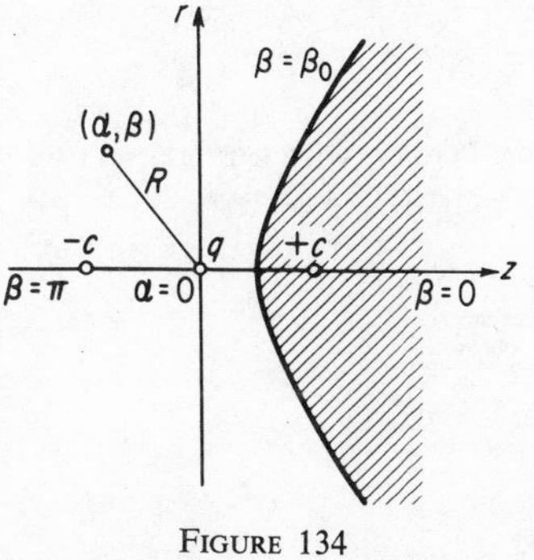

Ans.

$$u = \frac{q}{R} - \frac{q}{2c\sqrt{\pi}}\int_0^\infty \tau\tanh\pi\tau\,\Gamma\!\left(\frac{1}{4} + \frac{i\tau}{2}\right)$$
$$\times\,\Gamma\!\left(\frac{1}{4} - \frac{i\tau}{2}\right)\frac{P_{-\frac{1}{2}+i\tau}(-\cos\beta)}{P_{-\frac{1}{2}+i\tau}(-\cos\beta_0)}$$
$$\times\, P_{-\frac{1}{2}+i\tau}(\cos\beta_0)P_{-\frac{1}{2}+i\tau}(\cosh\alpha)\,d\tau,$$
$$\beta_0\leqslant\beta\leqslant\pi.$$

FIGURE 134

485. A d-c current J flows into ground of conductivity σ through an electrode placed at the bottom of a hollow shaped like a hyperboloid of revolution, with equation $\beta = \beta_0$ in spheroidal coordinates (see Figure 135).

FIGURE 135

Find the current distribution in the ground.

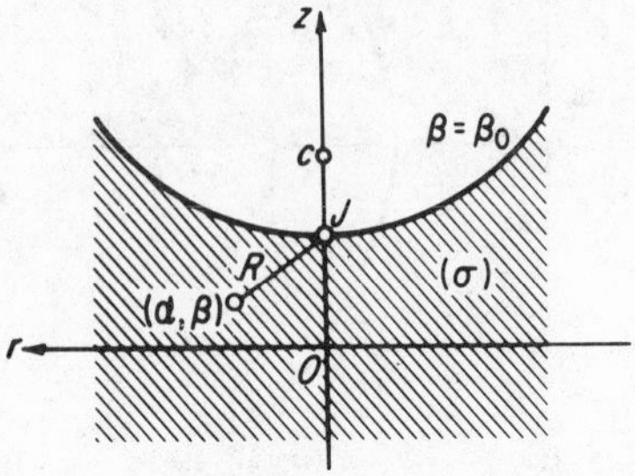

Ans. The potential of the current field is

$$u = \frac{J}{2\pi\sigma R} - \frac{J}{2\pi\sigma c}$$

$$\times \int_0^\infty \frac{\tau \sinh \pi\tau}{\cosh^2 \pi\tau} \frac{P'_{-\frac{1}{2}+i\tau}(\cos \beta_0)}{P'_{-\frac{1}{2}+i\tau}(-\cos \beta_0)}$$

$$\times P_{-\frac{1}{2}+i\tau}(-\cos \beta_0) P_{-\frac{1}{2}+i\tau}(-\cos \beta)$$

$$\times P_{-\frac{1}{2}+i\tau}(\cosh \alpha) \, d\tau,$$

where R is the distance from the source to the field point, c is the eccentricity of the hyperbola $\beta = \beta_0$, and $P_\nu(x)$ is Legendre's function.

486. Find the charge density on the surface of an ellipsoidal conductor with semiaxes a, b and c, carrying total charge Q. What is the capacitance of the ellipsoid?

Ans.

$$\sigma = \frac{Q}{4\pi abc} \frac{1}{\sqrt{\dfrac{x^2}{a^4} + \dfrac{y^2}{b^4} + \dfrac{z^2}{c^4}}},$$

$$C = \frac{2}{\displaystyle\int_0^\infty \frac{ds}{\sqrt{(a^2 + s)(b^2 + s)(c^2 + s)}}}.$$

Hint. Introduce ellipsoidal coordinates α, β, γ, defined as the roots of the cubic equation

$$\frac{x^2}{a^2 + \lambda} + \frac{y^2}{b^2 + \lambda} + \frac{z^2}{c^2 + \lambda} = 1.$$

Then look for a solution depending only on α.

487. Find the charge density on a thin elliptic plate with semiaxes a and b, carrying total charge Q. What is the capacitance of the plate?

Ans.

$$\sigma = \frac{Q}{4\pi ab} \frac{1}{\sqrt{1 - \dfrac{x^2}{a^2} - \dfrac{y^2}{b^2}}},$$

$$C = \frac{a}{K\left(\dfrac{\sqrt{a^2 - b^2}}{a}\right)},$$

where $K(k)$ is the complete elliptic integral of the first kind.

Hint. Take the limit $c \to 0$ in the solution of the preceding problem.

488. An ellipsoid with semiaxes a, b and c, made from material of magnetic permeability μ, is placed in a homogeneous magnetic field H_0 directed along its major axis. Find the resulting magnetic field inside the ellipsoid.

Ans. The direction of the field coincides with that of the external field. The magnitude of the field equals

$$H = \frac{H_0}{1 + \dfrac{abc}{2}(\mu - 1) \displaystyle\int_0^\infty \frac{ds}{(a^2 + s)\sqrt{(a^2 + s)(b^2 + s)(c^2 + s)}}}.$$

489. Calculate the gravitational potential of a homogeneous ellipsoid of density ρ (see S16, p. 161).

Ans.

$$u = \pi \rho abc \int_\lambda^\infty \left[1 - \frac{x^2}{a^2 + s} - \frac{y^2}{b^2 + s} - \frac{z^2}{c^2 + s} \right] \frac{ds}{\sqrt{(a^2 + s)(b^2 + s)(c^2 + s)}},$$

where λ is the positive root of the equation

$$\frac{x^2}{a^2 + \lambda} + \frac{y^2}{b^2 + \lambda} + \frac{z^2}{c^2 + \lambda} - 1 = 0,$$

and the gravitational constant is taken to be unity.

5. Paraboloidal Coordinates

Physical problems involving a region bounded by a paraboloid of revolution can be solved by introducing *paraboloidal coordinates* α, β, φ related to the rectangular coordinates x, y, z by the formulas

$$x = c\alpha\beta \cos \varphi, \quad y = c\alpha\beta \sin \varphi, \quad z = \frac{c}{2}(\alpha^2 - \beta^2),$$

where

$$0 \leqslant \alpha < \infty, \quad 0 \leqslant \beta < \infty, \quad -\pi < \varphi \leqslant \pi,$$

and $c > 0$ is a scale factor.[11] In this case, the triply orthogonal system of coordinate surfaces consists of the two families of paraboloids of revolution

---

[11] If a point has cylindrical coordinates r, φ, z, then

$$z + ir = \frac{c}{2}(\alpha + i\beta)^2.$$

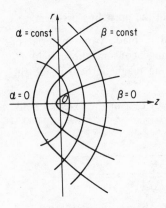

$\alpha = $ const and $\beta = $ const, together with the planes $\varphi = $ const (see Figure 136). The square of the element of arc length and Laplace's equation take the form

$$ds^2 = c^2(\alpha^2 + \beta^2)(d\alpha^2 + d\beta^2) + c^2\alpha^2\beta^2\,d\varphi^2,$$

$$\Delta u = \frac{1}{c^2(\alpha^2 + \beta^2)}\left[\frac{1}{\alpha}\frac{\partial}{\partial\alpha}\left(\alpha\frac{\partial u}{\partial\alpha}\right)\right.$$
$$\left. + \frac{1}{\beta}\frac{\partial}{\partial\beta}\left(\beta\frac{\partial u}{\partial\beta}\right) + \left(\frac{1}{\alpha^2} + \frac{1}{\beta^2}\right)\frac{\partial^2 u}{\partial\varphi^2}\right].$$

If there is no dependence on the angle φ,[12] the use of the superposition method to solve boundary value problems for the region bounded by a paraboloid of revolution

FIGURE 136

$\beta = \beta_0$ starts from the following particular solutions of Laplace's equation, which depend continuously on the parameter λ:[13]

$$u = u_\lambda(\alpha, \beta)J_0(\lambda\alpha)\begin{matrix}I_0(\lambda\beta)\\K_0(\lambda\beta)\end{matrix}, \qquad \lambda \geqslant 0. \tag{6}$$

Here $I_0(x)$, $J_0(x)$ and $K_0(x)$ are cylinder functions, the upper row pertains to the interior region $(0 \leqslant \beta \leqslant \beta_0)$ and the lower row to the exterior region $(\beta_0 \leqslant \beta < \infty)$. The general solution is now constructed by integrating (6) with respect to λ, where, to determine M_λ, we use Hankel's integral theorem [see formula (12), p. 160]. Paraboloidal coordinates can also be used to solve boundary value problems for Helmholtz's equation, but then the particular solutions involve confluent hypergeometric functions (see E2, Vol. 2, Secs. 8.7–8.8).

490. Solve Prob. 483, assuming that the conducting screen is shaped like a paraboloid of revolution, with equation $\beta = \beta_0$ in paraboloidal coordinates.

Ans.

$$u(\alpha, \beta) = \frac{2q}{c(\alpha^2 + \beta^2)} - \frac{2q}{c}\int_0^\infty \frac{K_0(\lambda\beta_0)}{I_0(\lambda\beta_0)}I_0(\lambda\beta)J_0(\lambda\alpha)\,\lambda\,d\lambda,$$

in terms of the Bessel function of the first kind $J_0(x)$ and the Bessel functions of imaginary argument $I_0(x)$ and $K_0(x)$. Note that

$$\beta_0 = \sqrt{\frac{p}{c}},$$

---

[12] See Prob. 492 for the case where dependence on φ is present.

[13] Formula (6) is an abbreviated way of writing two formulas, one involving the function $I_0(\lambda\beta)$, the other $K_0(\lambda\beta)$.

where p is the focal distance and c is the scale factor figuring in the definition of the paraboloidal coordinates.

Hint. Use the integral

$$\int_0^\infty \frac{x J_0(ax)}{x^2 + b^2}\, dx = K_0(ab) \qquad (a > 0,\, b > 0).$$

491. Find the stationary temperature distribution in a body shaped like a paraboloid of revolution $\beta = \beta_0$, if a given axially symmetric temperature distribution

$$T(\alpha, \beta)|_{\beta=\beta_0} = f(\alpha)$$

is maintained on its surface.

Ans.

$$T(\alpha, \beta) = \int_0^\infty \frac{I_0(\lambda\beta)}{I_0(\lambda\beta_0)} J_0(\lambda\alpha)\lambda\, d\lambda \int_0^\infty f(\xi) J_0(\lambda\xi)\xi\, d\xi.$$

492. Solve the Dirichlet problem for the domain bounded by the paraboloid of revolution $\beta = \beta_0$, assuming that the boundary condition is of the form

$$u|_{\beta=\beta_0} = f_n(\alpha) \frac{\cos n\varphi}{\sin n\varphi}, \qquad n = 0, 1, 2, \ldots,$$

where $f_n(\alpha)$ is a given function. Use the result to construct solutions for arbitrary boundary conditions depending on φ.

Ans. Inside the paraboloid,

$$u(\alpha, \beta, \varphi) = \int_0^\infty \bar{f}_n(\lambda) \frac{I_n(\lambda\beta)}{I_n(\lambda\beta_0)} J_n(\lambda\alpha)\lambda\, d\lambda \frac{\cos n\varphi}{\sin n\varphi},$$

where $\bar{f}_n(\lambda)$ is the Hankel transform of $f_n(\alpha)$:[14]

$$\bar{f}_n(\lambda) = \int_0^\infty f_n(\alpha) J_n(\lambda\alpha)\alpha\, d\alpha.$$

6. Toroidal Coordinates

Besides spherical and spheroidal coordinates, there are other coordinate systems whose use is intimately connected with Legendre functions. First we consider *toroidal coordinates* α, β, φ related to the rectangular coordinates x, y, z by the formulas

$$x = \frac{c \sinh \alpha \cos \varphi}{\cosh \alpha - \cos \beta}, \quad y = \frac{c \sinh \alpha \sin \varphi}{\cosh \alpha - \cos \beta}, \quad z = \frac{c \sin \beta}{\cosh \alpha - \cos \beta}, \quad (7)$$

---

[14] Cf. formula (13), p. 160.

where

$$0 \leqslant \alpha < \infty, \quad -\pi < \beta \leqslant \pi, \quad -\pi < \varphi \leqslant \pi,$$

and $c > 0$ is a scale factor.[15,16] The corresponding triply orthogonal system of surfaces consists of the toroidal surfaces $\alpha = \text{const}$, satisfying the equation

$$(r - c \coth \alpha)^2 + z^2 = \left(\frac{c}{\sinh \alpha}\right)^2$$

where $r = \sqrt{x^2 + y^2}$, the spheres $\beta = \text{const}$, satisfying the equation

$$(r - c \cot \beta) + r^2 = \left(\frac{c}{\sin \beta}\right)^2, \tag{8}$$

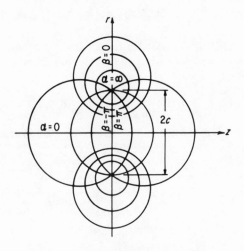

FIGURE 137

and the planes $\varphi = \text{const}$ (see Figure 137). Note that all the spheres (8) intersect in the circle $r = c$, $z = 0$. It is clear from (7) that x, y and z are

---

[15] In the next section, we shall consider a closely related coordinate system, i.e., three-dimensional bipolar coordinates.

[16] If a point has cylindrical coordinates r, φ, z, then

$$r = \frac{c \sinh \alpha}{\cosh \alpha - \cos \beta}, \qquad z = \frac{c \sin \beta}{\cosh \alpha - \cos \beta},$$

or more concisely,

$$z + ir = ic \coth \frac{\alpha + i\beta}{2}.$$

periodic in β and φ, with period 2π. Therefore we can choose $\beta_1 < \beta \leqslant \beta_1 + 2\pi$, $\varphi_1 < \varphi \leqslant \varphi_1 + 2\pi$ instead of $-\pi < \beta \leqslant \pi$, $-\pi < \varphi \leqslant \pi$ (which corresponds to the particular choice $\beta_1 = \varphi_{1.} = -\pi$), and it is sometimes convenient to do so.

In toroidal coordinates, the square of the element of arc length is

$$ds^2 = \frac{c^2}{(\cosh \alpha - \cos \beta)^2} (d\alpha^2 + d\beta^2 + \sinh^2 \alpha \, d\varphi^2),$$

and Laplace's equation takes the form

$$\frac{\partial}{\partial \alpha}\left(\frac{\sinh \alpha}{\cosh \alpha - \cos \beta} \frac{\partial u}{\partial \alpha}\right) + \frac{\partial}{\partial \beta}\left(\frac{\sinh \alpha}{\cosh \alpha - \cos \beta} \frac{\partial u}{\partial \beta}\right)$$
$$+ \frac{1}{\sinh \alpha(\cosh \alpha - \cos \beta)} \frac{\partial^2 u}{\partial \varphi^2} = 0. \quad (9)$$

Unlike the cases considered so far, equation (9) does not permit separation of variables directly. However, if we first introduce a new function v by making the substitution

$$u = \sqrt{2 \cosh \alpha - 2 \cos \beta} \, v,$$

(9) goes into a new equation belonging to the class which permits separation of variables (see L9, p. 223). If there is no dependence on the angle φ, it turns out that Laplace's equation (9) has particular solutions of the form

$$u = u_v = \sqrt{2 \cosh \alpha - 2 \cos \beta} \, [A_v P_{v-\frac{1}{2}}(\cosh \alpha) + B_v Q_{v-\frac{1}{2}}(\cosh \alpha)]$$
$$\times [C_v \cos v\beta + D_v \sin v\beta],$$

in terms of the Legendre functions of the first and second kinds, where v is a parameter and $A_v, \ldots, D_v$ are arbitrary constants. In boundary value problems involving the region bounded by a torus, the parameter v is determined by the requirement that the solution be periodic in β. This leads to the particular solutions

$$u = u_n = \sqrt{2 \cosh \alpha - 2 \cos \beta} \, [M_n \cos n\beta + N_n \sin n\beta] \frac{Q_{n-\frac{1}{2}}(\cosh \alpha)}{P_{n-\frac{1}{2}}(\cosh \alpha)},$$

where the upper row pertains to the interior problem ($\alpha_0 < \alpha \leqslant \infty$) and the lower row to the exterior problem ($0 \leqslant \alpha < \alpha_0$). In problems involving the region bounded by two intersecting spheres $\beta = \beta_1$ and $\beta = \beta_2$, the appropriate particular solutions are obtained by choosing $v = i\tau$ ($\tau \geqslant 0$), and are of the form

$$u = u_\tau = \sqrt{2 \cosh \alpha - 2 \cos \beta} \, [M_\tau \cosh \tau\beta + N_\tau \sinh \tau\beta] P_{-\frac{1}{2}+i\tau}(\cosh \alpha),$$
$$(10)$$

where $\beta_1 < \beta < \beta_2$ for the interior problem and $\beta_2 < \beta < 2\pi + \beta_1$ for the exterior problem. Then the solution of the problem is constructed by integrating (10) with respect to τ, where the factors M_τ and N_τ are determined by using the Mehler-Fock theorem (see L9, Sec. 8.12).

This section contains a number of physical problems which can be solved by using toroidal coordinates. Most of the problems are rather difficult, and are intended for those with the necessary background in the theory of special functions.[17]

493. Find the electrostatic potential due to a charged toroidal conductor at potential V, with the dimensions shown in Figure 138. Calculate the capacitance of the conductor.

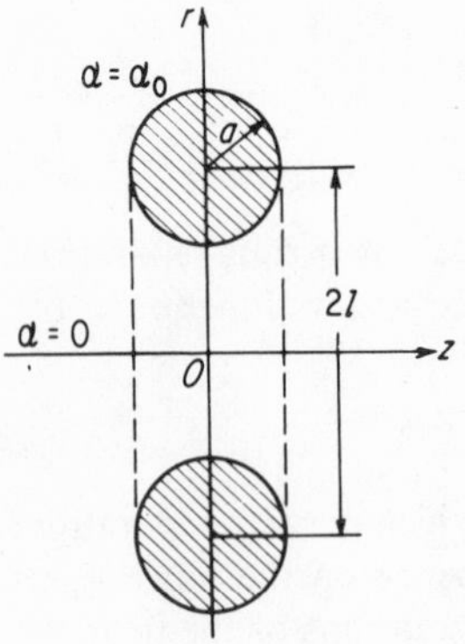

FIGURE 138

Ans. The potential is

$$u(\alpha, \beta) = \frac{V}{\pi} \sqrt{2 \cosh \alpha - 2 \cos \beta}$$

$$\times \left[\frac{Q_{-1/2}(\cosh \alpha_0)}{P_{-1/2}(\cosh \alpha_0)} P_{-1/2}(\cosh \alpha) \right.$$

$$\left. + 2 \sum_{n=1}^{\infty} \frac{Q_{n-\frac{1}{2}}(\cosh \alpha_0)}{P_{n-\frac{1}{2}}(\cosh \alpha_0)} P_{n-\frac{1}{2}}(\cosh \alpha) \cos n\beta \right],$$

and the capacitance is

$$C = \frac{2c}{\pi} \left[\frac{Q_{-1/2}(\cosh \alpha_0)}{P_{-1/2}(\cosh \alpha_0)} + 2 \sum_{n=1}^{\infty} \frac{Q_{n-\frac{1}{2}}(\cosh \alpha_0)}{P_{n-\frac{1}{2}}(\cosh \alpha_0)} \right],$$

where $P_\nu(x)$ and $Q_\nu(x)$ are the Legendre functions of the first and second kind, and

$$c = \sqrt{l^2 - a^2}, \qquad \cosh \alpha_0 = \frac{l}{a}.$$

Hint. Introduce toroidal coordinates α, β, φ with parameter c, such that the surface of the conductor has equation $\alpha = \alpha_0$. In the course of the solution, use the integral

$$\int_0^\pi \frac{\cos n\beta \, d\beta}{\sqrt{2 \cosh \alpha_0 - 2 \cos \beta}} = Q_{n-\frac{1}{2}}(\cosh \alpha_0).$$

*494. Find the distribution of electrostatic potential on the axis of a grounded conducting torus introduced into a homogeneous electric field E_0 directed along its axis of symmetry (in the negative z-direction).

---

[17] Some of the problems can be solved more easily by using other methods (by inversion, say).

Ans.

$$u\big|_{r=0} = E_0 z - \frac{8}{\pi} E_0 \sqrt{l^2 - a^2} \sin\frac{\beta}{2} \sum_{n=1}^{\infty} \frac{nQ_{n-\frac{1}{2}}(\cosh\alpha_0)}{P_{n-\frac{1}{2}}(\cosh\alpha_0)} \sin n\beta,$$

where

$$\cosh\alpha_0 = \frac{l}{a},$$

and the dimensions l and a are the same as in Figure 138.

495. Solve the preceding problem, assuming that the external field is due to a point charge q at the center of the torus.

Ans.

$$u\big|_{r=0} = \frac{q}{z} - \frac{2q\sin\frac{1}{2}\beta}{\pi\sqrt{l^2 - a^2}} \left[\frac{Q_{-1/2}(\cosh\alpha_0)}{P_{-1/2}(\cosh\alpha_0)} + 2\sum_{n=1}^{\infty}(-1)^n \frac{Q_{n-\frac{1}{2}}(\cosh\alpha_0)}{P_{n-\frac{1}{2}}(\cosh\alpha_0)}\cos n\beta \right].$$

496. A current J flows in a ring-shaped conductor of circular cross section (see Figure 138). Find the resulting magnetic field along the z-axis, assuming that the current J is uniformly distributed over the cross section of the ring.

Ans.

$$H\big|_{r=0} = - \frac{16\sqrt{2}\left(\dfrac{l^2}{a^2} - 1\right)^{3/2}}{9\pi} \frac{J}{ca}(1 - \cos\beta)^{3/2}$$

$$\times \left\{ Q^2_{-1/2}(\cosh\alpha_0)Q^{1\prime}_{-1/2}(\cosh\alpha_0) - Q^2_{-1/2}(\cosh\alpha_0)Q^1_{-1/2}(\cosh\alpha_0) \right.$$

$$+ 2\sum_{n=1}^{\infty}[Q^2_{n-\frac{1}{2}}(\cosh\alpha_0)Q^{1\prime}_{n-\frac{1}{2}}(\cosh\alpha_0) - Q^{2\prime}_{n-\frac{1}{2}}(\cosh\alpha_0)Q^1_{n-\frac{1}{2}}(\cosh\alpha_0)]\cos n\beta \bigg\},$$

where

$$\cosh\alpha_0 = \frac{l}{a},$$

c is the velocity of light, and $Q^1_\nu(x)$, $Q^2_\nu(x)$ are associated Legendre functions of the second kind.

497. Find the distribution of a-c current along the surface of a perfect conductor shaped like a ring with circular cross section. Calculate the self-inductance L of the ring.[18]

Ans.

$$\frac{1}{L} = \frac{1}{2\pi^3\sqrt{l^2 - a^2}} \left[\frac{Q^1_{-1/2}(\cosh\alpha_0)}{P^1_{-1/2}(\cosh\alpha_0)} - 2\sum_{n=1}^{\infty} \frac{1}{4n^2 - 1} \frac{Q^1_{n-\frac{1}{2}}(\cosh\alpha_0)}{P^1_{n-\frac{1}{2}}(\cosh\alpha_0)} \right],$$

where

$$\cosh\alpha_0 = \frac{l}{a},$$

---

[18] This is the skin effect problem (see F1).

the dimensions l and a are the same as in Figure 138, and $P_\nu^1(x)$, $Q_\nu^1(x)$ are associated Legendre functions of the first and second kind. The distribution of current density along the periphery of the ring is

$$j(\beta) = -\frac{LJ}{32\pi^3} \frac{[2(\cosh\alpha_0 - \cos\beta)]^{3/2}}{\sinh\alpha_0(l^2 - a^2)} \left[\frac{1}{P_{-1/2}(\cosh\alpha_0)} + 2\sum_{n=1}^{\infty} \frac{\cos n\beta}{P_{n-1/2}^1(\cosh\alpha_0)}\right],$$

where J is the total current.

*498. Suppose a d-c current flows in a ring-shaped conductor with the dimensions shown in Figure 138, producing heat with density Q. Find the temperature distribution inside the conductor, assuming that its surface is held at temperature zero.

Ans.

$$T(\alpha, \beta) = -\frac{Q(l^2 - a^2)}{k}\left\{\frac{\sinh^2\alpha}{(2\cosh\alpha - 2\cos\beta)^2} - \sqrt{2\cosh\alpha - 2\cos\beta}\right.$$

$$\times \left[\frac{\sinh^2\alpha_0 Q''_{-1/2}(\cosh\alpha_0)}{3\pi Q_{-1/2}(\cosh\alpha_0)} Q_{-1/2}(\cosh\alpha)\right.$$

$$\left.\left.+ \frac{2\sinh^2\alpha_0}{3\pi}\sum_{n=1}^{\infty} \frac{Q''_{n-1/2}(\cosh\alpha_0)}{Q_{n-1/2}(\cosh\alpha_0)} Q_{n-1/2}(\cosh\alpha)\cos n\beta\right]\right\},$$

in terms of the Legendre function of the second kind $Q_\nu(z)$, where

$$\cosh\alpha_0 = \frac{l}{a},$$

and k is the thermal conductivity.

Hint. Subtract out the particular solution $-Qr^2/4k$ of the inhomogeneous heat conduction equation. Use the integral

$$\int_0^\pi \frac{\cos n\beta\, d\beta}{(2\cosh\alpha - 2\cos\beta)^{5/2}} = \frac{1}{3}Q''_{n-1/2}(\cosh\alpha).$$

499. Calculate the gravitational potential of a homogeneous torus of density ρ, with the dimensions shown in Figure 138, assuming that the gravitational constant equals unity.

Ans.

$$u(\alpha, \beta) = -\frac{4\rho c^2}{3}\sinh^2\alpha_0\sqrt{2\cosh\alpha - 2\cos\beta}\left\{[Q_{-1/2}(\cosh\alpha_0)Q_{-1/2}^{2\prime}(\cosh\alpha_0)\right.$$

$$- Q_{-1/2}^2(\cosh\alpha_0)Q'_{-1/2}(\cosh\alpha_0)]P_{-1/2}(\cosh\alpha)$$

$$+ 2\sum_{n=1}^{\infty}[Q_{n-1/2}(\cosh\alpha_0)Q_{n-1/2}^{2\prime}(\cosh\alpha_0)$$

$$\left.- Q_{n-1/2}^2(\cosh\alpha_0)Q'_{n-1/2}(\cosh\alpha_0)]P_{n-1/2}(\cosh\alpha)\cos n\beta\right\},$$

where $P_\nu(x)$ and $Q_\nu(x)$ are Legendre functions, $Q_\nu^2(x)$ is the associated Legendre function of the second kind, and

$$c = \sqrt{l^2 - a^2}, \qquad \cosh \alpha_0 = \frac{l}{a}.$$

500. A torus with the dimensions shown in Figure 138 is introduced into a homogeneous flow of an ideal fluid, whose direction coincides with the axis of symmetry of the torus. Solve the resulting hydrodynamical problem, and find the velocity distribution along the axis.

Ans. The stream function is

$$v = \frac{v_\infty r^2}{2} + \frac{\sinh \alpha}{\sqrt{2\cosh \alpha - 2\cos \beta}}$$

$$\times \left[\frac{c_0}{2} P_{-1/2}^1(\cosh \alpha) + \sum_{n=1}^\infty c_n P_{n-\frac{1}{2}}^1(\cosh \alpha) \cos n\beta \right],$$

where $P_\nu^1(x)$ and $Q_\nu^1(x)$ are associated Legendre functions,

$$c_n = \frac{4}{\pi P_{n-\frac{1}{2}}^1(\cosh \alpha_0)} \left\{ v_\infty (l^2 - a^2) Q_{n-\frac{1}{2}}^1(\cosh \alpha_0) \right.$$

$$\left. + \frac{2A}{(4n^2 - 1)\sinh \alpha_0} [\sinh \alpha_0 Q_{n-\frac{1}{2}}^1(\cosh \alpha_0) - 1] \right\},$$

$\cosh \alpha_0 = l/a$, and v_∞ is the velocity of the flow far from the torus. The constant A is determined from the condition

$$\int_0^\pi \frac{\partial v}{\partial \alpha} \Big|_{\alpha = \alpha_0} d\beta = 0.$$

501. Find the surface density of free charge on a thin charged conductor shaped like a spherical bowl of radius a (see Figure 139). Calculate the capacitance of the bowl (see J2, p. 250).

Ans. The charged density is

$$\sigma_i = \frac{V}{4\pi^2 a} \left[\frac{\sqrt{2\cosh \alpha - 2\cos \beta_0}}{2\cos \frac{1}{2}\beta_0} \right.$$

$$\left. - \arctan \frac{\sqrt{2\cosh \alpha - 2\cos \beta_0}}{2\cos \frac{1}{2}\beta_0} \right]$$

on the inner surface of the bowl, and

$$\sigma_o = \sigma_i + \frac{V}{4\pi a}$$

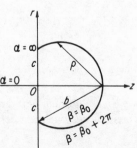

FIGURE 139

on the outer surface, where V is the potential of the bowl and

$$\sin \beta_0 = \frac{c}{a}.$$

Using the formula

$$\frac{4a^2 - b^2}{b^2 - \rho^2} = \frac{\cosh \alpha - \cos \beta_0}{2 \cos^2 \frac{1}{2}\beta_0},$$

where the distances b and ρ are shown in Figure 139, we find that

$$\sigma_i = \frac{V}{4\pi^2 a}\left[\sqrt{\frac{4a^2 - b^2}{b^2 - \rho^2}} - \arctan \sqrt{\frac{4a^2 - b^2}{b^2 - \rho^2}}\right].$$

The capacitance of the bowl is

$$C = \frac{b}{2\pi a} \sqrt{4a^2 - b^2} + \frac{2a}{\pi} \arctan \frac{b}{\sqrt{4a^2 - b^2}}.$$

Hint. To calculate the density, use the integral

$$\int_0^\infty \frac{\tau \sinh \pi\tau \cosh (\pi - \beta_0)\tau}{\cosh^2 \pi\tau} P_{-\frac{1}{2}+i\tau}(\cosh \alpha)\, d\tau = \frac{2}{\pi} \frac{\sin \frac{1}{2}\beta_0}{2 \cosh \alpha - 2 \cos \beta_0}$$

$$\times \left[1 + \frac{2 \cos \frac{1}{2}\beta_0}{\sqrt{2 \cosh \alpha - 2 \cos \beta_0}} \arctan \frac{2 \cos \frac{1}{2}\beta_0}{\sqrt{2 \cosh \alpha - 2 \cos \beta_0}}\right].$$

*502. Find the surface density of induced charge on a thin conductor shaped like a spherical bowl of radius a, due to a point charge q located at the point $r = z = 0$ (see Figure 140).

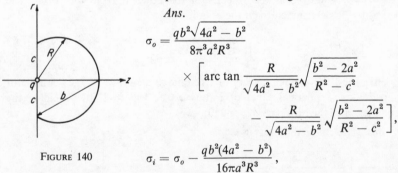

Ans.

$$\sigma_o = \frac{qb^2\sqrt{4a^2 - b^2}}{8\pi^3 a^2 R^3}$$

$$\times \left[\arctan \frac{R}{\sqrt{4a^2 - b^2}}\sqrt{\frac{b^2 - 2a^2}{R^2 - c^2}}\right.$$

$$\left. - \frac{R}{\sqrt{4a^2 - b^2}}\sqrt{\frac{b^2 - 2a^2}{R^2 - c^2}}\right],$$

FIGURE 140

$$\sigma_i = \sigma_o - \frac{qb^2(4a^2 - b^2)}{16\pi a^3 R^3},$$

where σ_o and σ_i are the charge densities on the outer and inner surfaces of the bowl.

Hint. Subtract out the potential of the point charge. To expand this potential in a Mehler-Fock integral, use the relation

$$\frac{1}{\sqrt{2 \cosh \alpha + 2 \cos \beta_0}} = \int_0^\infty \frac{\cosh \beta_0\tau}{\cosh \pi\tau} P_{-\frac{1}{2}+i\tau}(\cosh \alpha)\, d\tau.$$

503. Find the potential distribution in the space surrounding a charged conductor shaped like the "spherical zone" shown in Figure 141.

Ans.

$$u(\alpha, \beta) = V\sqrt{2\cosh\alpha - 2\cos\beta}$$

$$\times \int_0^\infty [\sinh(2\pi + \beta_0 - \beta)\tau - \cosh(\pi - \beta_0)\tau$$

$$\times \sinh(\pi - \beta)\tau] \frac{P_{-\frac{1}{2}+i\tau}(\cosh\alpha)}{\sinh(\pi + \beta_0)\tau \cosh\pi\tau} d\tau,$$

where $P_\nu(x)$ is the Legendre function of the first kind, V is the potential of the conductor and $\sin\beta_0 = c/a$.

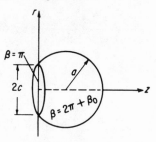

FIGURE 141

504. Use the result of the preceding problem to calculate the capacitance of a hemisphere of radius a.

Ans.

$$C = 2a\left(1 - \frac{1}{\sqrt{3}}\right).$$

505. A lens-shaped conductor at zero potential is introduced into a homogeneous electric field E_0 directed along its axis of symmetry (in the negative z-direction), as shown in Figure 142. Find the resulting potential distribution.

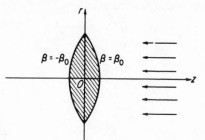

FIGURE 142

Ans.

$$u = E_0 z - 2E_0 c\sqrt{2\cosh\alpha - 2\cos\beta}$$

$$\times \int_0^\infty \frac{\tau\sinh(\pi - \beta_0)\tau}{\cosh\pi\tau}$$

$$\times \frac{\sinh\beta\tau}{\sinh\beta_0\tau} P_{-\frac{1}{2}+i\tau}(\cosh\alpha)\,d\tau.$$

506. Find the gravitational potential of a homogeneous hemisphere of density ρ and radius a.

Ans. The potential outside the hemisphere is

$$u(\alpha, \beta) = \frac{2\pi\rho a^2}{3}\sqrt{2\cosh\alpha - 2\cos\beta}$$

$$\times \int_0^\infty \left[\frac{\cosh\pi\tau - \tau^2}{\cosh(\pi\tau/2)}\sinh(\beta - \pi)\tau + (1 + 2\tau^2)\sinh\left(\frac{5\pi}{2} - \beta\right)\tau\right]$$

$$\times \frac{P_{-\frac{1}{2}+i\tau}(\cosh\alpha)}{\sinh(3\pi\tau/2)\cosh\pi\tau} d\tau$$

Hint. Inside the hemisphere, subtract out the particular solution $-2\pi\rho z^2$ of the inhomogeneous equation. Use the integral

$$\frac{3\sin^2\beta}{(2\cosh\alpha - 2\cos\beta)^{5/2}} = \cot\beta\int_0^\infty \frac{\tau\sinh(\pi - \beta)\tau}{\cosh\pi\tau}P_{-\frac{1}{2}+i\tau}(\cosh\alpha)\,d\tau$$

$$+ \int_0^\infty \frac{\tau^2\cosh(\pi - \beta)\tau}{\cosh\pi\tau}P_{-\frac{1}{2}+i\tau}(\cosh\alpha)\,d\tau.$$

7. Three-Dimensional Bipolar Coordinates

By *three-dimensional bipolar coordinates*, we mean coordinates α, β, φ related to the rectangular coordinates x, y, z by the formulas

$$x = \frac{c\sin\alpha\cos\varphi}{\cosh\beta - \cos\alpha}, \quad y = \frac{c\sin\alpha\sin\varphi}{\cosh\beta - \cos\alpha}, \quad z = \frac{c\sinh\beta}{\cosh\beta - \cos\alpha}, \quad (11)$$

where

$$0 \leqslant \alpha < \pi, \quad -\infty < \beta < \infty, \quad -\pi < \varphi \leqslant \pi,$$

and $c > 0$ is a scale factor.[19] The close resemblance between (11) and the formulas defining toroidal coordinates should be noted (see p. 233). The corresponding triply orthogonal system of surfaces consists of the spindle-shaped surfaces of revolution $\alpha = $ const, satisfying the equation

$$(r - c\cot\alpha)^2 + z^2 = \left(\frac{c}{\sin\alpha}\right)^2,$$

the spheres $\beta = $ const satisfying the equation

$$(z - c\coth\beta)^2 + r^2 = \left(\frac{c}{\sinh\beta}\right)^2,$$

and the planes $\varphi = $ const (see Figure 143). The square of the element of arc length is

$$ds^2 = \frac{c^2}{(\cosh\beta - \cos\alpha)^2}(d\alpha^2 + d\beta^2 + \sin^2\alpha\,d\varphi^2),$$

---

[19] If a point has cylindrical coordinates r, φ, z, then

$$r = \frac{c\sin\alpha}{\cosh\beta - \cos\alpha}, \quad z = \frac{c\sinh\beta}{\cosh\beta - \cos\alpha},$$

or more concisely

$$z + ir = ic\cot\frac{\alpha + i\beta}{2}.$$

and hence Laplace's equation takes the form

$$\frac{\partial}{\partial \alpha}\left(\frac{\sin \alpha}{\cosh \beta - \cos \alpha}\frac{\partial u}{\partial \alpha}\right) + \frac{\partial}{\partial \beta}\left(\frac{\sin \alpha}{\cosh \beta - \cos \alpha}\frac{\partial u}{\partial \beta}\right)$$

$$+ \frac{1}{\sin \alpha(\cosh \beta - \cos \alpha)}\frac{\partial^2 u}{\partial \varphi^2} = 0. \quad (12)$$

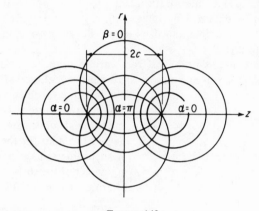

FIGURE 143

To separate variables in (12), we first introduce a new function v by making the substitution

$$u = \sqrt{2 \cosh \beta - 2 \cos \alpha}\, v,$$

as in the case of toroidal coordinates. If there is no dependence on the angle φ, it turns out that Laplace's equation (12) has particular solutions of the form

$$u = u_v = \sqrt{2 \cosh \beta - 2 \cos \alpha}\, [A_v P_v(\cos \alpha) + B_v(\cos \alpha)]$$

$$\times [C_v \cosh (v + \tfrac{1}{2})\beta + D_v \sinh (v + \tfrac{1}{2})\beta],$$

in terms of the Legendre functions of the first and second kinds, where v is a parameter and $A_v, \ldots, D_v$ are arbitrary constants (see L9, p. 232). In boundary value problems involving a region $\beta_1 < \beta < \beta_2$ bounded by two nonintersecting spheres $\beta = \beta_1$ and $\beta = \beta_2$, it is easy to see that the appropriate particular solutions are

$$u = u_n = \sqrt{2 \cosh \beta - 2 \cos \alpha}\, [M_n \cosh (n + \tfrac{1}{2})\beta$$

$$+ N_n \sinh (n + \tfrac{1}{2})\beta]P_n(\cos \alpha), \qquad n = 0, 1, 2, \ldots,$$

in terms of the Legendre polynomials $P_n(x)$, and the general solution is constructed by summing these solutions. In problems involving the region bounded by the spindle-shaped surface $\alpha = \alpha_0$, the appropriate particular solutions are obtained by choosing $\nu = -\frac{1}{2} + i\tau$ ($\tau \geqslant 0$), and are of the form

$$u = u_\tau = \sqrt{2 \cosh \beta - 2 \cos \alpha}\,[M_\tau \cos \tau\beta + N_\tau \sin \tau\beta]P_{-\frac{1}{2}+i\tau}(\pm\cos\alpha),$$

$$\tau \geqslant 0, \quad (13)$$

where the plus sign corresponds to the exterior problem ($0 \leqslant \alpha < \alpha_0$) and the minus sign to the interior problem ($\alpha_0 < \alpha \leqslant \pi$). In this case, the general solution is obtained by integrating (13) with respect to τ, and the factors M_τ and N_τ are determined by taking Fourier cosine and sine transforms with respect to β.

This section contains problems from various branches of mathematical physics which can be solved by using three-dimensional bipolar coordinates.

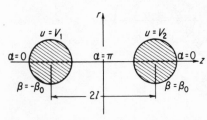

FIGURE 144

The last three problems (Probs. 512–514) involve limiting cases of bipolar and toroidal coordinates, and lead to elegant formulas for the capacitance of such objects as a pair of spheres in contact or the surface obtained by rotating a circle about a tangent line.

507. Find the electrostatic field in a spark gap consisting of two conducting spheres of radius a, with centers a distance $2l$ apart, if the spheres are at potentials V_1 and V_2 respectively (see Figure 144).

Ans. The electrostatic potential is

$$u(\alpha, \beta) = \sqrt{2 \cosh \beta - 2 \cos \alpha}\sum_{n=0}^{\infty}\left[\frac{V_2 + V_1}{2}\frac{\cosh(n+\frac{1}{2})\beta}{\cosh(n+\frac{1}{2})\beta_0}\right.$$

$$\left.+ \frac{V_2 - V_1}{2}\frac{\sinh(n+\frac{1}{2})\beta}{\sinh(n+\frac{1}{2})\beta_0}\right]e^{-(n+\frac{1}{2})\beta_0}P_n(\cos\alpha),$$

in terms of the Legendre polynomials $P_n(x)$, where

$$\cosh \beta_0 = \frac{l}{a}.$$

Hint. Use the expansion

$$\frac{1}{\sqrt{2\cosh\beta - 2\cos\alpha}} = \sum_{n=0}^{\infty}e^{-(n+\frac{1}{2})\beta}P_n(\cos\alpha).$$

*508. Find the capacitances C_{11}, C_{12} and C_{22} of a system of conductors consisting of two spheres of radii a_1 and a_2, with centers a distance $2l$ apart.[20]

---

[20] Concerning the meaning of C_{11}, C_{12} and C_{22}, see the solution, p. 370.

Assuming that the radii are equal ($a_1 = a_2 = a$), tabulate C_{12} as a function of the ratio l/a.

Ans.

$$C_{11} = c\left\{\frac{1}{2\sinh\beta_1} + \sum_{n=0}^{\infty}[e^{-(n+\frac{1}{2})\beta_1}\cosh(n+\tfrac{1}{2})(\beta_1 + \beta_2) - e^{-(n+\frac{1}{2})\beta_2}]\right.$$
$$\left.\times\frac{e^{-(n+\frac{1}{2})\beta_1}}{\sinh(n+\tfrac{1}{2})(\beta_1+\beta_2)}\right\},$$

$$C_{12} = c\sum_{n=0}^{\infty}\frac{e^{-(n+\frac{1}{2})(\beta_1+\beta_2)}}{\sinh(n+\tfrac{1}{2})(\beta_1+\beta_2)},$$

$$C_{22} = c\left\{\frac{1}{2\sinh\beta_2} + \sum_{n=0}^{\infty}[e^{-(n+\frac{1}{2})\beta_2}\cosh(n+\tfrac{1}{2})(\beta_1 + \beta_2) - e^{-(n+\frac{1}{2})\beta_1}]\right.$$
$$\left.\times\frac{e^{-(n+\frac{1}{2})\beta_2}}{\sinh(n+\tfrac{1}{2})(\beta_1+\beta_2)}\right\},$$

where β_1, β_2 and c are determined from the relations

$$\cosh\beta_1 = \frac{4l^2 + a_1^2 - a_2^2}{4la_1}, \quad \cosh\beta_2 = \frac{4l^2 - a_1^2 + a_2^2}{4la_2},$$
$$c = a_1\sinh\beta_1 = a_2\sinh\beta_2.$$

$\dfrac{l}{a}$	1.2	1.4	1.6	1.8	2.0
$\dfrac{C_{12}}{a}$	0.572	0.431	0.356	0.306	0.269

Hint. In three-dimensional bipolar coordinates α, β, φ, the surfaces of the conductors have equations $\beta = -\beta_1$ and $\beta = \beta_2$.

509. A conducting sphere of radius a is buried to a given depth in a liquid of dielectric constant ε. Find the potential distribution outside the sphere, assuming that the sphere is at potential V (see Figure 145). Calculate the capacitance of the sphere.

Ans.

$$u_1 = V\varepsilon\sqrt{2\cosh\beta - 2\cos\alpha}\sum_{n=0}^{\infty}\frac{e^{(n+\frac{1}{2})(\beta-\beta_0)}P_n(\cos\alpha)}{\sinh(n+\tfrac{1}{2})\beta_0 + \varepsilon\cosh(n+\tfrac{1}{2})\beta_0},$$
$$-\infty < \beta < 0,$$

$$u_2 = V\sqrt{2\cosh\beta - 2\cos\alpha}\sum_{n=0}^{\infty}\frac{\sinh(n+\tfrac{1}{2})\beta + \varepsilon\cosh(n+\tfrac{1}{2})\beta}{\sinh(n+\tfrac{1}{2})\beta_0 + \varepsilon\cosh(n+\tfrac{1}{2})\beta_0}$$
$$\times e^{-(n+\frac{1}{2})\beta_0}P_n(\cos\alpha), \quad 0 < \beta < \beta_0,$$

$$C = \sqrt{l^2 - a^2}$$
$$\times\left[\frac{1}{2\sinh\beta_0} + \sum_{n=0}^{\infty}\frac{\varepsilon\sinh(n+\tfrac{1}{2})\beta_0 + \cosh(n+\tfrac{1}{2})\beta_0}{\sinh(n+\tfrac{1}{2})\beta_0 + \varepsilon\cosh(n+\tfrac{1}{2})\beta_0}e^{-(2n+1)\beta_0}\right],$$

in terms of the Legendre polynomials $P_n(x)$, where

$$\cosh \beta_0 = \frac{l}{a}.$$

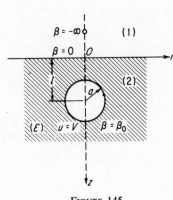

FIGURE 145

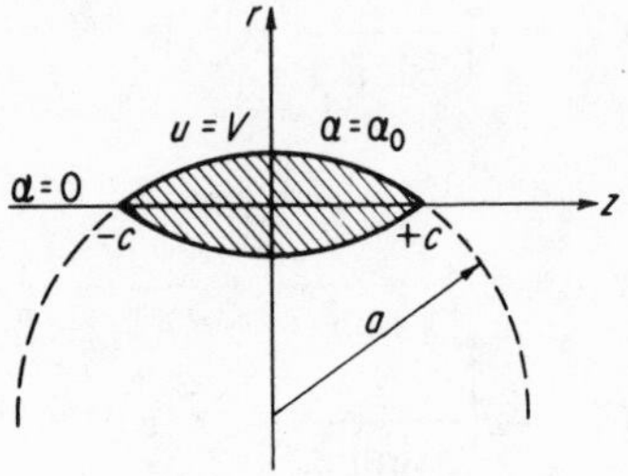

FIGURE 146

510. Find the potential distribution outside a charged spindle-shaped conductor at potential V (see Figure 146).

Ans.

$$u(\alpha, \beta) = V\sqrt{2\cosh \beta - 2\cos \alpha}$$
$$\times \int_0^\infty \frac{\cos \tau\beta}{\cosh \tau\pi} \frac{P_{-\frac12+i\tau}(-\cos \alpha_0)}{P_{-\frac12+i\tau}(\cos \alpha_0)} P_{-\frac12+i\tau}(\cos \alpha) \, d\tau,$$

in terms of the Legendre function $P_\nu(x)$, where

$$\sin \alpha_0 = \frac{c}{a}.$$

Hint. In bipolar coordinates α, β, φ, the surface of the conductor has equation $\alpha = \alpha_0$. In the course of the solution, use the integral representation

$$P_{-\frac12+i\tau}(-\cos \alpha_0) = \frac{2\cosh \pi\tau}{\pi} \int_0^\infty \frac{\cos \tau\beta \, d\beta}{\sqrt{2\cosh \beta - 2\cos \alpha_0}}.$$

511. Solve the preceding problem, assuming that the conductor is placed in a homogeneous electric field E_0 directed along the axis of rotation (in the negative z-direction).

Ans.

$$u = E_0 z - 2E_0 c\sqrt{2\cosh \beta - 2\cos \alpha} \int_0^\infty \frac{\tau}{\cosh \pi\tau}$$
$$\times \frac{P_{-\frac12+i\tau}(-\cos \alpha_0)}{P_{-\frac12+i\tau}(\cos \alpha_0)} P_{-\frac12+i\tau}(\cos \alpha) \sin \beta\tau \, d\tau.$$

***512.** Calculate the capacitance of a conductor consisting of two touching spheres of equal radius (see Figure 147).

Ans. $C = 2a \ln 2$.

Hint. Introduce *degenerate bipolar coordinates,* defined by the formula

$$z + ir = \frac{c}{\alpha + i\beta},$$

which can be obtained from the formula

$$z + ir = ic \cot \frac{\alpha + i\beta}{2}$$

(cf. footnote 19, p. 242) by replacing α, β, c by $\alpha\varepsilon$, $\beta\varepsilon$, $\frac{1}{2}c\varepsilon$ and taking the limit as $\varepsilon \to 0$. Then the surfaces of the spheres have equations $\beta = \pm\beta_0$.

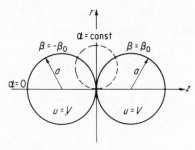

FIGURE 147

513. Calculate the capacitance of a conducting sphere of radius a lying on a plane with dielectric constant ε (see Figure 148).

Ans.

$$C = a \frac{\varepsilon + 1}{\varepsilon - 1} \ln \frac{\varepsilon + 1}{2}.$$

514. Calculate the capacitance of a conductor in the shape of the surface obtained by rotating a circle of radius a about one of its tangents (a "doughnut without a hole").

Ans.

$$C = \frac{4a}{\pi} \int_0^\infty \frac{K_0(x)}{I_0(x)} \, dx,$$

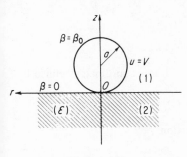

FIGURE 148

where $I_0(x)$ and $K_0(x)$ are Bessel functions of imaginary argument.

Hint. The surface of the conductor has the equation $\alpha = \alpha_0$ in degenerate bipolar coordinates (see the hint to Prob. 512).

8. Some General Problems on Separation of Variables

515. Show that a necessary and sufficient condition for being able to separate variables in Helmholtz's equation $\Delta u + k^2 u = 0$ (where Δ is the two-dimensional Laplace operator) in a system of curvilinear coordinates α, β defined by the formula

$$x + iy = f(\alpha + i\beta) \tag{14}$$

(where f is analytic) is that f be the solution of the third-order linear differential equation

$$f'''(\zeta) - \lambda f'(\zeta) = 0$$

(λ is an arbitrary constant).

516. Using the result of the preceding problem, show that apart from linear transformations (corresponding to translation and rotation of the coordinate axes or change of scale in the xy and $\alpha\beta$-planes) the only transformations of the form (14) leading to separation of variables in Helmholtz's equation are the following:

$$x + iy = e^{\alpha + i\beta} \quad \text{(polar coordinates)},$$
$$x + iy = \cosh(\alpha + i\beta) \quad \text{(elliptic coordinates)},$$
$$x + iy = (\alpha + i\beta)^2 \quad \text{(parabolic coordinates)}.$$

517. Show that Laplace's equation

$$\frac{\partial^2 u}{\partial r^2} + \frac{1}{r}\frac{\partial u}{\partial r} + \frac{1}{r^2}\frac{\partial^2 u}{\partial \varphi^2} + \frac{\partial^2 u}{\partial z^2} = 0$$

has infinitely many particular solutions of the form

$$u = r^{-1/2}A(\alpha)B(\beta)\Phi(\varphi),$$

where α, β, φ are a system of orthogonal curvilinear coordinates defined by the formula

$$z + ir = f(\alpha + i\beta),$$

and $f(\zeta)$ is a solution of the differential equation

$$f'^2(\zeta) = \sum_{k=0}^{4} \lambda_k f^k(\zeta),$$

where the λ_k are arbitrary real constants (see L2).

518. Show that all the three-dimensional coordinate systems considered in this chapter (as well as cylindrical and spherical coordinates) can be obtained as special cases of the coordinate system of the preceding problem.

Ans. Cylindrical coordinates:

$$f(\zeta) = \zeta, \quad \lambda_0 = 1, \quad \lambda_1 = \lambda_2 = \lambda_3 = \lambda_4 = 0,$$
$$\alpha = z, \quad \beta = r,$$
$$u = [AJ_\mu(\nu r) + BY_\mu(\nu r)][C\cosh \nu z + D\sinh \nu z]\frac{\cos \mu\varphi}{\sin \mu\varphi},$$

where $J_\mu(x)$ and $Y_\mu(x)$ are Bessel functions of the first and second kind.

Spherical coordinates:

$$f(\zeta) = e^{\zeta}, \quad \lambda_2 = 1, \quad \lambda_0 = \lambda_1 = \lambda_3 = \lambda_4 = 0,$$
$$\alpha = \ln r, \quad \beta = \theta,$$

$$u = [Ar^{\nu} + Br^{-\nu-1}][CP_{\nu}^{\mu}(\cos\theta) + DQ_{\nu}^{\mu}(\cos\theta)] \frac{\cos\mu\varphi}{\sin\mu\varphi},$$

where $P_{\nu}^{\mu}(x)$ and $Q_{\nu}^{\mu}(x)$ are associated Legendre functions of the first and second kind for the interval $(-1, 1)$.[21]

Prolate spheroidal coordinates:

$$f(\zeta) = c\cosh\zeta, \quad \lambda_0 = -c^2, \quad \lambda_2 = 1, \quad \lambda_1 = \lambda_3 = \lambda_4 = 0,$$

$$u = [AP_{\nu}^{\mu}(\cosh\alpha) + BQ_{\nu}^{\mu}(\cosh\alpha)][CP_{\nu}^{\mu}(\cos\beta) + DQ_{\nu}^{\mu}(\cos\beta)] \frac{\cos\mu\varphi}{\sin\mu\varphi}.$$

Oblate spheroidal coordinates:

$$f(\zeta) = c\sinh\zeta, \quad \lambda_0 = c^2, \quad \lambda_2 = 1, \quad \lambda_1 = \lambda_3 = \lambda_4 = 0,$$

$$u = [AP_{\nu}^{\mu}(i\sinh\alpha) + BQ_{\nu}^{\mu}(i\sinh\alpha)][CP_{\nu}^{\mu}(\cos\beta) + DQ_{\nu}^{\mu}(\cos\beta)] \frac{\cos\mu\varphi}{\sin\mu\varphi}.$$

Paraboloidal coordinates:

$$f(\zeta) = \frac{c\zeta^2}{2}, \quad \lambda_1 = 2c, \quad \lambda_0 = \lambda_2 = \lambda_3 = \lambda_4 = 0,$$

$$u = [AJ_{\mu}(\nu\alpha) + BY_{\mu}(\nu\alpha)][CI_{\mu}(\nu\beta) + DK_{\mu}(\nu\beta)] \frac{\cos\mu\varphi}{\sin\mu\varphi},$$

where $I_{\mu}(x)$ and $K_{\mu}(x)$ are Bessel functions of imaginary argument.

Toroidal coordinates:

$$f(\zeta) = ci\coth\frac{\zeta}{2}, \quad \lambda_0 = -\frac{c^2}{4}, \quad \lambda_2 = -\frac{1}{2}, \quad \lambda_4 = -\frac{1}{4c^2}, \quad \lambda_1 = \lambda_3 = 0,$$

$$u = \sqrt{2\cosh\alpha - 2\cos\beta}[AP_{\nu-\frac{1}{2}}^{\mu}(\cosh\alpha) + BQ_{\nu-\frac{1}{2}}^{\mu}(\cosh\alpha)]$$

$$\times [C\cos\nu\beta + D\sin\nu\beta] \frac{\cos\mu\varphi}{\sin\mu\varphi}.$$

Three-dimensional bipolar coordinates:

$$f(\zeta) = ci\cot\frac{\zeta}{2}, \quad \lambda_0 = -\frac{c^2}{4}, \quad \lambda_2 = \frac{1}{2}, \quad \lambda_4 = -\frac{1}{4c^2}, \quad \lambda_1 = \lambda_3 = 0,$$

$$u = \sqrt{2\cosh\beta - 2\cos\alpha}[AP_{\nu}^{\mu}(\cos\alpha) + BQ_{\nu}^{\mu}(\cos\alpha)]$$

$$\times [C\cosh(\nu+\tfrac{1}{2})\beta + D\sinh(\nu+\tfrac{1}{2})\beta] \frac{\cos\mu\varphi}{\sin\mu\varphi}.$$

---

[21] See e.g., L9, p. 193.

519. Prove that besides the coordinate systems listed in Prob. 518, separation of variables in Laplace's equation is also possible in coordinates defined by the formula

$$z + ir = f(\alpha + i\beta),$$

where $f(\zeta)$ is one of the Jacobian elliptic functions sn ζ, cn ζ, dn ζ.[22] Construct particular solutions of the form

$$u = r^{-1/2}A(\alpha)B(\beta)\Phi(\varphi)$$

for each of these three functions.

Ans.

1. $f(\zeta) = \text{sn } \zeta$, $\lambda_0 = 1$, $\lambda_1 = 0$, $\lambda_2 = -(1 + k^2)$, $\lambda_3 = 0$, $\lambda_4 = k^2$,

$$u = r^{-1/2}A(\alpha)B(\beta) \frac{\cos \mu\varphi}{\sin \mu\varphi},$$

where $A(\alpha)$ and $B(\beta)$ are solutions of the differential equations

$$A'' + \left[\nu + (\tfrac{1}{4} - \mu^2)\left(\frac{k'^2 \text{ sn } \alpha}{\text{cn } \alpha \text{ dn } \alpha}\right)^2\right]A = 0,$$

$$B'' - \left[\nu + (\tfrac{1}{4} - \mu^2)\left(\frac{\text{cn } i\beta \text{ dn } i\beta}{\text{sn } i\beta}\right)^2\right]B = 0,$$

and μ, ν are arbitrary parameters.

2. $f(\zeta) = \text{cn } \zeta$, $\lambda_0 = k'^2$, $\lambda_1 = 0$, $\lambda_2 = -(k'^2 - k^2)$, $\lambda_3 = 0$, $\lambda_4 = -k^2$,

$$A'' + \left[\nu + (\tfrac{1}{4} - \mu^2)\left(\frac{\text{cn } \alpha}{\text{sn } \alpha \text{ dn } \alpha}\right)^2\right]A = 0,$$

$$B'' - \left[\nu + (\tfrac{1}{4} - \mu^2)\left(\frac{\text{cn } i\beta}{\text{sn } i\beta \text{ dn } i\beta}\right)^2\right]B = 0.$$

3. $f(\zeta) = \text{dn } \zeta$, $\lambda_0 = -k'^2$, $\lambda_1 = 0$, $\lambda_2 = 1 + k'^2$, $\lambda_3 = 0$, $\lambda_4 = -1$,

$$A'' + \left[\nu + (\tfrac{1}{4} - \mu^2)\left(\frac{\text{dn } \alpha}{\text{sn } \alpha \text{ cn } \alpha}\right)^2\right]A = 0,$$

$$B'' - \left[\nu + (\tfrac{1}{4} - \mu^2)\left(\frac{\text{dn } i\beta}{\text{sn } i\beta \text{ cn } i\beta}\right)^2\right]B = 0.$$

---

[22] See L2, W2, W3, and also the paper L4, where a system of solutions of Laplace's equation suitable for solving boundary value problems for a ring of oval cross section is constructed.

520. Verify that the biharmonic equation $\Delta^2 u = 0$ (where Δ is the two-dimensional Laplace operator) has infinitely many particular solutions of the form

$$u = |f'(\alpha + i\beta)| \, A(\alpha) \frac{\cos \lambda\beta}{\sin \lambda\beta},$$

where α, β is a system of two-dimensional curvilinear coordinates defined by the formula

$$x + iy = f(\alpha + i\beta),$$

where

$$f(\zeta) = \int \frac{d\zeta}{F^2(\zeta)},$$

and $F(\zeta)$ is the solution of the differential equation

$$F''(\zeta) - \mu^2 F(\zeta) = 0$$

(λ and μ are arbitrary parameters).

521. Using the result of the preceding problem, show that the two-dimensional biharmonic equation permits separation of variables in rectangular, polar, two-dimensional bipolar and degenerate bipolar coordinates, and construct the corresponding particular solutions.

Ans. The general transformation called for here is of the form

$$f(\zeta) = \frac{a}{\mu \coth \mu\zeta + b} + d,$$

where a, b and d are arbitrary constants.

1. Rectangular coordinates:

$$f(\zeta) = \zeta, \quad \mu \to 0, \quad a = 1, \quad b = d = 0,$$
$$\alpha = x, \qquad \beta = y,$$

$$u = (A \cosh \lambda x + B \sinh \lambda x + Cx \cosh \lambda x + Dx \sin \lambda x) \frac{\cos \lambda y}{\sin \lambda y}.$$

2. Polar coordinates:

$$f(\zeta) = e^\zeta, \quad \mu = \tfrac{1}{2}, \quad a = d = 1, \quad b = -\tfrac{1}{2},$$
$$\alpha = \ln r, \qquad \beta = \varphi,$$

$$u = (Ar^\lambda + Br^{-\lambda} + Cr^{\lambda+2} - Dr^{-\lambda+2}) \frac{\cos \lambda\varphi}{\sin \lambda\varphi}.$$

3. Bipolar coordinates:

$$f(\zeta) = c \tanh \frac{\zeta}{2}, \quad \mu = \frac{1}{2}, \quad a = \frac{c}{2}, \quad b = d = 0,$$

$$u = \frac{c}{\cosh \alpha + \cos \beta} [A \cosh (\lambda + 1)\alpha + B \sinh (\lambda + 1)\alpha$$
$$+ C \cosh (\lambda - 1)\alpha + D \sinh (\lambda - 1)\alpha] \frac{\cos \lambda\beta}{\sin \lambda\beta}.$$

4. Degenerate bipolar coordinates:

$$f(\zeta) = \frac{c}{\zeta}, \quad \mu \to 0, \quad a = -\frac{c}{\mu^2}, \quad b = \frac{1}{\mu}, \quad d = \frac{c}{\mu},$$

$$u = \frac{c}{(\alpha^2 + \beta^2)^2} [A \cosh \lambda\alpha + B \sinh \lambda\alpha + C\alpha \cosh \lambda\alpha + D\alpha \sinh \lambda\alpha] \frac{\cos \lambda\beta}{\sin \lambda\beta}.$$

References

Books: Bateman (B2), Hobson (H4), Lebedev (L9), Lense (L11), Magnus and Oberhettinger (M3), Morse and Feshbach (M9), Smythe (S7), Snow (S12), Stratton et al. (see p. 358) (S18), Strutt (S19).

Papers: Bôcher (B5), Eisenhart (E1), Haentzschel (H1, H2), Lagrange (L2), Stepanov (S15), Wangerin (W2, W3).

8

INTEGRAL EQUATIONS

The use of integral equations to prove existence theorems for problems of mathematical physics, or to find approximate solutions, is a classical subject, which lies outside the scope of this book but is treated in considerable detail in the available literature. The purpose of this chapter is simply to show how integral equations can be used to find exact solutions of certain physical problems. The methods we have in mind are admittedly quite special, but very effective in the cases to which they apply, and their full possibilities do not yet seem to have been exploited. As an example of the successful application of integral equations to physical problems, we cite the work of Grinberg, summarized in his book G5, devoted to the solution of a number of interesting problems from the theory of electricity and magnetism.

This chapter consists of two sections. The first is devoted to some nonstationary problems of diffraction theory which can be reduced to the solution of familiar integral equations, e.g., Abel's equation, Volterra's equation with a difference kernel, etc. The second section, stemming from Grinberg's work, is primarily concerned with stationary problems stated in terms of electrostatics, but with obvious analogues involving magnetostatics, heat conduction or d-c current flow.

Because of their relatively greater difficulty, we omit problems whose solution requires the use of the Wiener-Hopf method, or problems which involve singular integral equations containing integrals of the Cauchy type. Concerning these topics, the reader should consult the relevant references cited at the end of the chapter (see p. 271).

I. Diffraction Theory

*522. A plane electromagnetic wave with electric field components

$$E_x = E_y = 0, \qquad E_z = f\left(t - \frac{x}{v}\right)$$

is incident on a perfectly conducting half-plane (screen) $x \geqslant 0$, $z = 0$. Denoting the components of the resulting electric field (the sum of the incident and reflected waves) by 0, 0, E and setting

$$E = f\left(t - \frac{x}{v}\right) - u,$$

show that the reflected wave u can be represented in the form

$$u = \int_{-\infty}^{\eta} \frac{\varphi(s)}{\sqrt{\xi - s}}\, ds$$

$$\left(\xi = t - \frac{x}{v}, \quad \eta = t - \frac{r}{v}, \quad r = \sqrt{x^2 + y^2}\right),$$

where the function $\varphi(s)$ satisfies Abel's integral equation

$$\int_{-\infty}^{\xi} \frac{\varphi(s)}{\sqrt{\xi - s}}\, ds = f(\xi).$$

Hint. Look for a solution of the wave equation depending only on ξ and η.

*523. Solve Prob. 522, assuming that the incident wave encounters the screen at the time $t = 0$, i.e.,

$$f(\xi) = \begin{cases} g(\xi), & \xi > 0, \\ 0, & \xi < 0. \end{cases}$$

Describe the diffraction process graphically.

Ans.

$$u = \begin{cases} \displaystyle\int_0^{\eta} \frac{\varphi(s)}{\sqrt{\xi - s}}\, ds, & \eta > 0, \\ 0, & \eta < 0, \end{cases}$$

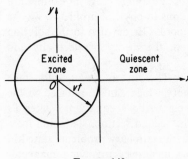

FIGURE 149

where

$$\varphi(s) = \frac{1}{\pi} \int_0^s \frac{g(\xi)}{\sqrt{s - \xi}}\, d\xi.$$

The diffraction process is illustrated in Figure 149.

524. Solve Prob. 523 for the special case where

a) $\qquad\qquad g(\xi) = 1$ (a wave with a rectilinear front);

b) $\qquad\qquad g(\xi) = \sin \omega\xi$.

Ans. In the notation of Prob. 523, the reflected wave u has the following representation in the excited zone:

a) $\qquad\qquad u = \dfrac{2}{\pi} \arcsin \sqrt{\dfrac{\eta}{\xi}}$,

b) $\qquad\qquad u = \dfrac{2}{\pi} \displaystyle\int_0^{\sqrt{\eta/(\xi-\eta)}} \dfrac{\sin \omega[\eta - (\xi - \eta)\tau^2]}{1 + \tau^2} \, d\tau$.

525. By passing to the limit $t \to \infty$ in the formulas of Prob. 524, solve the well-known Sommerfeld problem on the steady-state sinusoidal electromagnetic oscillations due to a plane wave incident on the edge of a conducting screen (see Prob. 426).

Ans.
$$u = \operatorname{Im} \{u^* e^{i\omega t}\},$$
where
$$u^* = \frac{2}{\sqrt{\pi}} e^{i\pi/4} e^{-ikx} \int_{\sqrt{k(r-x)}}^{\infty} e^{-is^2} \, ds, \quad k = \frac{\omega}{v}.$$

526. A plane electromagnetic wave with components
$$E_x = E_y = 0, \qquad E_z = f\left(t - \frac{x + a}{v}\right)$$

is incident on a perfectly conducting screen shaped like a parabolic cylinder $r = x + 2a$. Setting
$$E = f\left(t - \frac{x + a}{v}\right) - u,$$

where E is the z-component of the electric field, show that the reflected wave can be represented in the form
$$u = \int_{-\infty}^{\eta} \frac{\varphi(s)}{\sqrt{\xi - s + (2a/v)}} \, ds$$
$$\left(\xi = t - \frac{x + a}{v}, \quad \eta = t - \frac{r - a}{v}, \quad r = \sqrt{x^2 + y^2}\right),$$

where $\varphi(s)$ is the solution of the integral equation
$$\int_{-\infty}^{\xi} \frac{\varphi(s)}{\sqrt{\xi - s + (2a/v)}} \, ds = f(\xi).$$

Hint. Look for a solution of the wave equation depending only on ξ and η.

527. Solve the preceding problem of diffraction theory, assuming that the wave makes contact with the screen at the time $t = 0$ and is continuous along its front,[1] i.e.,

$$f(\xi) = \begin{cases} g(\xi), & \xi > 0, \\ 0, & \xi < 0, \end{cases} \qquad g(0) = 0.$$

Describe the diffraction process graphically.

Ans.

$$u = \begin{cases} \displaystyle\int_0^{\eta} \frac{\varphi(s)}{\sqrt{\xi - s + (2a/v)}}\, ds, & \eta > 0, \\ 0, & \eta < 0, \end{cases}$$

where

$$\varphi(s) = \frac{1}{2\pi i} \int_{\Gamma} \frac{\bar{g}}{\bar{K}}\, e^{ps}\, dp.$$

Here $\bar{g}$ and $\bar{K}$ are the Laplace transforms of the functions $g(\xi)$ and

$$K(\xi) = \left(\xi + \frac{2a}{v}\right)^{-1/2},$$

so that

$$\bar{K} = \sqrt{\pi}\, \frac{e^{2ap/v}}{\sqrt{p}}\left[1 - \Phi\left(\sqrt{\frac{2ap}{v}}\right)\right],$$

where $\Phi(x)$ is the probability integral and the path of integration Γ is a straight line parallel to the imaginary axis lying to the right of the singular points of the integrand.

The diffraction process is indicated in Figure 150. The boundary of the excited zone is the envelope of the secondary waves reflected from points of the screen, in keeping with Huygens' principle.

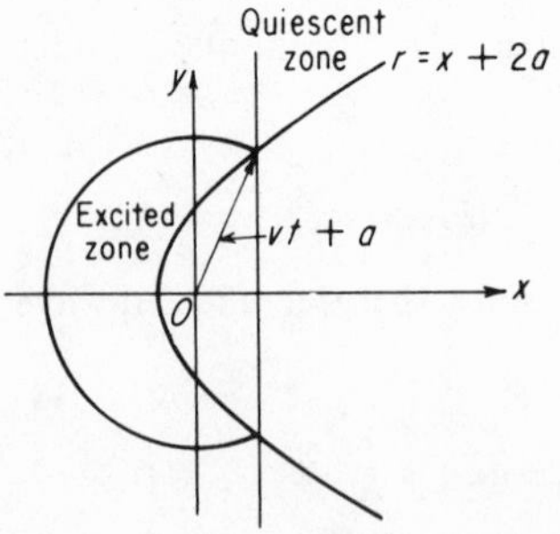

FIGURE 150

528. Suppose the incident wave in Prob. 547 has the equation

$$g(\xi) = \frac{2}{\pi} \arctan \sqrt{\frac{v\xi}{2a}}.$$

Show that the reflected wave u can then be represented in the form

$$u = \frac{2}{\pi} \arctan \sqrt{\frac{\eta}{\xi - \eta + (2a/v)}}, \qquad \eta > 0.$$

---

[1] The case of a discontinuity on the wave front can be treated by passing to the limit.

529. Consider the problem of diffraction of a plane sound wave

$$f\left(t - \frac{x+a}{v}\right).$$

by an obstacle shaped like a parabolic cylinder. Show that the reflected wave has the representation[2]

$$u = \int_{-\infty}^{\eta} \frac{\varphi(s)}{\sqrt{\xi - s + (2a/v)}}\, ds$$

$$\left(\xi = t - \frac{x+a}{v},\quad \eta = t - \frac{r-a}{v},\quad r = \sqrt{x^2 + y^2}\right),$$

where $\varphi(s)$ is the solution of the Volterra integral equation

$$\varphi(\xi) + \frac{1}{2}\sqrt{\frac{2a}{v}} \int_{-\infty}^{\xi} \frac{\varphi(s)}{[\xi - s + (2a/v)]^{3/2}}\, ds = \sqrt{\frac{2a}{v}}\, f'(\xi).$$

(see F8.)

530. Solve the preceding problem, assuming that the wave encounters the obstacle at the time $t = 0$:

$$f(\xi) = \begin{cases} g(\xi), & \xi > 0, \\ 0, & \xi < 0, \end{cases} \quad g(0) = 0.$$

Ans.

$$u = \begin{cases} \displaystyle\int_0^{\eta} \frac{\varphi(s)}{\sqrt{\xi - s + (2a/v)}}\, ds, & \eta > 0, \\ 0, & \eta < 0. \end{cases}$$

where, in the notation of Prob. 527,

$$\varphi(s) = \frac{1}{2}\sqrt{\frac{2a}{v}} \frac{1}{2\pi i} \int_{\Gamma} \frac{p\bar{g}}{1 - \frac{1}{2}\sqrt{2a/v}\, pK}\, e^{ps}\, dp.$$

*531. Consider the problem of diffraction of a plane sound wave

$$f\left(t - \frac{z+a}{v}\right)$$

by an obstacle shaped like a paraboloid of revolution $r = z + 2a$. Applying the technique of the preceding problems, show that the reflected wave has the representation

$$u = \int_{-\infty}^{\eta} \frac{\varphi(s)}{\xi - s + (2a/v)}\, ds$$

$$\left(\xi = t - \frac{z+a}{v},\quad \eta = t - \frac{r-a}{v},\quad r = \sqrt{x^2 + y^2 + z^2}\right),$$

---

[2] In problems on diffraction of acoustic waves (unlike the case of electromagnetic waves), we write the total solution in the form $f(\xi) + u$.

where $\varphi(s)$ is the solution of the Volterra integral equation

$$\varphi(\xi) + \frac{2a}{v} \int_{-\infty}^{\xi} \frac{\varphi(s)}{[\xi - s + (2a/v)]^2} \, ds = \frac{2a}{v} f'(\xi)$$

(see F8).

*532. Solve the preceding problem, assuming that the incident wave has the equation

$$f(\xi) = \begin{cases} g(\xi), & \xi > 0, \\ 0, & \xi < 0, \end{cases} \qquad g(0) = 0.$$

Ans. In the excited zone ($\eta > 0$),

$$u = \int_0^{\eta} \frac{\varphi(s)}{\xi - s + (2a/v)} \, ds,$$

where

$$\varphi(s) = \frac{a}{v} \frac{1}{2\pi i} \int_{\Gamma} \frac{p\bar{g}}{1 - (a/v)p\bar{K}} e^{ps} \, dp.$$

In the last formula, $\bar{g}$ and $\bar{K}$ are the Laplace transforms of $g(\xi)$ and the kernel

$$K(\xi) = \frac{1}{\xi + (2a/v)},$$

and the path of integration Γ is a straight line parallel to the imaginary axis lying to the right of the singular points of the integrand. Note that

$$\bar{K} = e^{-2ap/v}\mathrm{Ei}\left(-\frac{2ap}{v}\right),$$

in terms of the exponential integral $\mathrm{Ei}(x)$.

533. Consider the problem of diffraction of a plane wave by a paraboloid of revolution $r = z + 2a$ with homogeneous boundary conditions of the first kind. Show that the reflected wave has the representation

$$u = \int_{-\infty}^{\eta} \frac{\varphi(s)}{\xi - s + (2a/v)} \, ds,$$

$$\left(\xi = t - \frac{z + a}{v}, \quad \eta = t - \frac{r - a}{v}, \quad r = \sqrt{x^2 + y^2 + z^2}\right),$$

where $\varphi(s)$ is the solution of the integral equation

$$\int_{-\infty}^{\xi} \frac{\varphi(s)}{\xi - s + (2a/v)} \, ds = f(\xi).$$

534. Using the Laplace transform, solve the integral equation of Prob. 533 for the case of a wave of the form

$$f(\xi) = \begin{cases} g(\xi), & \xi > 0, \\ 0, & \xi < 0, \end{cases} \qquad g(0) = 0.$$

Ans. In the notation of Prob. 532,

$$\varphi(s) = \frac{1}{2\pi i} \int_\Gamma \frac{\bar{g}}{\bar{K}} e^{ps} \, dp.$$

2. Electrostatics

535. A conductor of arbitrary shape, bounded by a surface Σ, is introduced into a given external field $\mathbf{E}^0$ (see Figure 151). Show that the density of charge induced on the conductor satisfies the integral equation

$$\sigma(N) = \frac{E_n^0(N)}{2\pi} + \frac{1}{2\pi} \int_\Sigma \frac{\sigma(M)}{|\mathbf{r}_{MN}|^2} \cos(\mathbf{r}_{MN}, \mathbf{n}) \, dS \quad (1)$$

where M and N are two arbitrary points of the surface Σ, dS is the element of area, $\mathbf{r}_{MN}$ is the vector joining M to N, $\mathbf{n}$ is the unit exterior normal to Σ at the point N, and $E_n^0 = \mathbf{E}^0 \cdot \mathbf{n}$ is the projection of $\mathbf{E}^0$ onto $\mathbf{n}$.

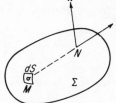

FIGURE 151

536. Show that in the special case where the surface of the conductor is an infinite plane, the solution of the integral equation (1) is given by[3]

$$\sigma(N) = \frac{E_n^0(N)}{2\pi}.$$

Use this result to find the charge density induced on a conducting plane by a point charge q placed at height h above the plane.

Ans.

$$\sigma(N) = -\frac{qh}{2\pi R^3},$$

where R is the distance from the charge to the point N of the plane.

---

[3] Naturally, this result can be found in other ways. The present method is of interest mainly because the final result is obtained practically without calculations.

537. A metallic sphere of radius a at potential V is introduced into an external electric field $\mathbf{E}^0$. Starting from the integral equation (1), show that the density of charge induced on the surface of the sphere is given by

$$\sigma(N) = \frac{E_n^0(N)}{2\pi} + \frac{V - u^0(N)}{4\pi a},$$

where u^0 is the potential of the external field. Examine the special case where the source of the field E^0 is a point charge q at distance b $(b > a)$ from the center of the sphere.

Ans.

$$\sigma(N) = \frac{V}{4\pi a} - \frac{q}{4\pi} \frac{b^2 - a^2}{aR^3},$$

where R is the distance from the charge to the given point N of the surface of the sphere.

538. Solve the preceding problem, given the total charge Q of the sphere (rather than its potential). Use the formula so obtained to solve the problem of the charge distribution on the surface of an initially uncharged insulated sphere introduced into a homogeneous external field $\mathbf{E}^0$.

Ans.

$$\sigma(N) = \frac{E_n^0(N)}{2\pi} + \frac{1}{4\pi a}\left[\frac{Q}{a} + \bar{u}^0 - u^0(N)\right],$$

where $\bar{u}^0$ is the average over the sphere of the potential of the external field:

$$\bar{u}^0 = \frac{1}{4\pi a^2} \int_\Sigma u^0(N)\, dS.$$

In the special case

$$\sigma(N) = \frac{3E}{4\pi} \cos\theta,$$

where θ is the angle between the direction of the external field $\mathbf{E}^0$ and the radius vector drawn from the center of the sphere to the point N.

539. A cylindrical conductor with cross section bounded by an arbitrary contour Γ (see Figure 152) is introduced into a given plane-parallel field $\mathbf{E}^0$. Show that the density of charge satisfies the integral equation

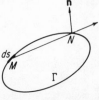

$$\sigma(N) = \frac{E_n^0(N)}{2\pi} + \frac{1}{\pi} \int_\Gamma \frac{\sigma(M)}{|\mathbf{r}_{MN}|} \cos(\mathbf{r}_{MN}, \mathbf{n})\, ds, \quad (2)$$

where M and N are two arbitrary points of the contour Γ, ds is the element of arc length, $\mathbf{r}_{MN}$ is the vector joining M to N, $\mathbf{n}$ is the unit exterior normal to Γ at the point N, and $E_n^0 = \mathbf{E}^0 \cdot \mathbf{n}$ is the projection of $\mathbf{E}^0$ onto $\mathbf{n}$.

FIGURE 152

540. Suppose a conductor shaped like an infinite circular cylinder of radius a, carrying charge Q per unit length, is introduced into an external plane-parallel field $\mathbf{E}^0$. Show that the density of induced charge on the surface of the conductor is given by

$$\sigma(N) = \frac{Q}{2\pi a} + \frac{E_n^0(N)}{2\pi}.$$

Consider the special case where

a) The external field is homogeneous;

b) The source of the external field is a line charge with charge q per unit length, placed outside the cylinder at the distance b from its axis.

Ans.

a)
$$\sigma(N) = \frac{Q}{2\pi a} + \frac{E\cos\theta}{2\pi},$$

where θ is the angle between the direction of the homogeneous field (of strength E) and the vector drawn from the center of the cylinder to the given point N on the surface of the conductor;

b)
$$\sigma(N) = \frac{Q+q}{2\pi a} - \frac{q}{2\pi}\frac{b^2 - a^2}{aR^2},$$

where R is the distance from the line charge to N.

541. Find the distribution of charge density on the inner surface of a grounded cylindrical shell of radius a, assuming that the external field is produced by line charges parallel to the axis of the cylinder passing through the points $M_k = (a_k, \varphi_k)$, $k = 1, 2, \ldots, n$.

Ans.

$$\sigma(N) = -\frac{1}{2\pi a}\sum_{k=1}^{n} q_k \frac{a^2 - a_k^2}{R_k^2},$$

where q_k is the charge per unit length of the line charge passing through the point M_k, and R_k is the distance between the points M_k and N.

*542. The electrostatic field in the region $0 < y < h$ between two grounded parallel planes is due to line sources whose free-space field is $\mathbf{E}^0$. Show that the densities $\sigma_0(x)$ and $\sigma_h(x)$ of induced charge on the planes $y = 0$ and $y = h$ satisfy the system of integral equations

$$\sigma_0(x) = \frac{1}{2\pi}E_y^0\big|_{y=0} - \frac{h}{\pi}\int_{-\infty}^{\infty}\frac{\sigma_h(\xi)}{(\xi-x)^2 + h^2}\,d\xi,$$

$$\sigma_h(x) = -\frac{1}{2\pi}E_y^0\big|_{y=h} - \frac{h}{\pi}\int_{-\infty}^{\infty}\frac{\sigma_0(\xi)}{(\xi-x)^2 + h^2}\,d\xi,$$

(3)

and then solve this system.

Ans.

$$\sigma_0(x) = \frac{1}{4\pi^2} \int_{-\infty}^{\infty} \frac{\tilde{f}_0 + e^{-|\lambda|h}\tilde{f}_h}{1 - e^{-2|\lambda|h}} e^{-i\lambda x}\, d\lambda,$$

$$\sigma_h(x) = -\frac{1}{4\pi^2} \int_{-\infty}^{\infty} \frac{\tilde{f}_h + e^{-|\lambda|h}\tilde{f}_0}{1 - e^{-2|\lambda|h}} e^{-i\lambda x}\, d\lambda,$$

where $\tilde{f}$ is the Fourier transform of $f(x)$, i.e.,[4]

$$\tilde{f} = \int_{-\infty}^{\infty} f(x)e^{i\lambda x}\, dx,$$

and

$$f_0(x) = E_y^0\big|_{y=0}, \qquad f_h(x) = E_y^0\big|_{y=h}.$$

Hint. Take the Fourier transform of each of the equations (3).

543. Solve the preceding problem for the special case where the field $\mathbf{E}^0$ is due to a line source with charge q per unit length, passing through the point $M_0 = (0, b)$.

Ans.

$$\sigma_0(x) = -\frac{q}{2h}\, \frac{\sin\dfrac{\pi b}{h}}{\cosh\dfrac{\pi x}{h} - \cos\dfrac{\pi b}{h}}, \qquad \sigma_h(x) = -\frac{q}{2h}\, \frac{\sin\dfrac{\pi b}{h}}{\cosh\dfrac{\pi x}{h} + \cos\dfrac{\pi b}{h}}.$$

Hint. To obtain the solution in closed form, use formula 15, p. 385.

544. Suppose a system of line sources, whose free-space field $\mathbf{E}^0$ has components E_r^0, E_φ^0, 0 in cylindrical coordinates, is placed inside a dihedral angle $0 < \varphi < \alpha$ with grounded conducting walls. Show that the charge densities $\sigma_0(r)$ and $\sigma_\alpha(r)$ on the walls satisfy the system of integral equations

$$\sigma_0(r) = \frac{1}{2\pi} E_\varphi^0\big|_{\varphi=0} - \frac{\sin\alpha}{\pi} \int_0^{\infty} \frac{\rho\sigma_\alpha(\rho)}{\rho^2 + r^2 - 2r\rho\cos\alpha}\, d\rho,$$

$$\sigma_\alpha(r) = -\frac{1}{2\pi} E_\varphi^0\big|_{\varphi=\alpha} - \frac{\sin\alpha}{\pi} \int_0^{\infty} \frac{\rho\sigma_0(\rho)}{\rho^2 + r^2 - 2r\rho\cos\alpha}\, d\rho,$$

and solve this system, using the Mellin transform.

Ans.

$$\sigma_0(r) = \frac{1}{4\pi^2 i} \int_{1-i\infty}^{1+i\infty} \frac{\tilde{f}_0 \sin\pi(p-1) + \tilde{f}_\alpha \sin(\pi-\alpha)(p-1)}{\sin(2\pi-\alpha)(p-1)\cdot\sin\alpha(p-1)}$$
$$\times \sin\pi(p-1)r^{-p}\, dp,$$

$$\sigma_\alpha(r) = -\frac{1}{4\pi^2 i} \int_{1-i\infty}^{1+i\infty} \frac{\tilde{f}_\alpha \sin\pi(p-1) + \tilde{f}_0 \sin(\pi-\alpha)(p-1)}{\sin(2\pi-\alpha)(p-1)\cdot\sin\alpha(p-1)}$$
$$\times \sin\pi(p-1)r^{-p}\, dp,$$

---

[4] This definition of the Fourier transform differs from the customary one by a numerical factor.

where $\bar{f}$ is the Mellin transform of $f(r)$, i.e.,

$$\bar{f} = \int_0^\infty f(r) r^{p-1} \, dr,$$

and

$$f_0(r) = E_\varphi^0 \big|_{\varphi=0}, \qquad f_\alpha(r) = E_\varphi^0 \big|_{\varphi=\alpha}.$$

Hint.

$$\int_0^\infty \frac{t^s \, dt}{t^2 - 2t \cos\alpha + 1} = \frac{\pi}{\sin\alpha} \frac{\sin(\pi - \alpha)s}{\sin \pi s}, \qquad -1 < \operatorname{Re} s < 1.$$

545. Solve the preceding problem, assuming that the field $\mathbf{E}^0$ is due to a line source with charge q per unit length, passing through the point $M_0 = (r_0, \varphi_0)$. Use the formula so obtained to find the electrostatic field due to a charged line placed at distance a from the edge of a conducting half-plane ($\alpha = 2\pi$, $r_0 = a$, $\varphi_0 = \pi$) or near a right-angular corner ($\alpha = 3\pi/2$, $r_0 = a$, $\varphi_0 = \pi$).

Ans.

$$\sigma_0(r) = -\frac{q}{\alpha r} \frac{\sin\dfrac{\pi\varphi_0}{\alpha}}{\left(\dfrac{r}{r_0}\right)^{\pi/\alpha} + \left(\dfrac{r_0}{r}\right)^{\pi/\alpha} - 2\cos\dfrac{\pi\varphi_0}{\alpha}},$$

$$\sigma_\alpha(r) = -\frac{q}{\alpha r} \frac{\sin\dfrac{\pi\varphi_0}{\alpha}}{\left(\dfrac{r}{r_0}\right)^{\pi/\alpha} + \left(\dfrac{r_0}{r}\right)^{\pi/\alpha} + 2\cos\dfrac{\pi\varphi_0}{\alpha}}.$$

546. A conductor shaped like an open surface of arbitrary form (see Figure 153) is placed in an external field $\mathbf{E}^0$. Show that the sum of the charge densities on opposite sides of the surface satisfies the integral equation of the first kind

$$\int_\Sigma \frac{\sigma(M)}{|\mathbf{r}_{MN}|} \, dS = V - u^0(N), \qquad (4)$$

where $\sigma(N) = \sigma_1(N) + \sigma_2(N)$, u_0 is the potential of the external field and V is the potential of the conductor, while the difference between the charge densities is given by the formula

$$\sigma_1(N) - \sigma_2(N) = \frac{E_n^0(N)}{2\pi} + \frac{1}{2\pi} \int_\Sigma \frac{\sigma(M)}{|\mathbf{r}_{MN}|^2} \cos(\mathbf{r}_{MN}, \mathbf{n}) \, dS.$$

$$(5)$$

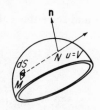

FIGURE 153

Thus, to solve the electrostatic problem completely, it is sufficient to know the solution of the integral equation (4) [see G5, Chap. 20].

547. Show that equation (5) takes the form

$$\sigma_1(N) - \sigma_2(N) = \frac{E_n^0(N)}{2\pi}$$

for a plane surface, and the form

$$\sigma_1(N) - \sigma_2(N) = \frac{E_n^0(N)}{2\pi} + \frac{V - u^0(N)}{4\pi R}$$

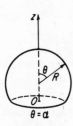

for a spherical surface of radius R, regardless of the form of the boundary curve.

548. Write the integral equation (4) for the case where the surface of the conductor is a disk of radius R or a thin spherical bowl $r = R, 0 \leqslant \theta \leqslant \alpha$ (see Figure 154), assuming that the external field has rotational symmetry with respect to the z-axis.[5]

FIGURE 154

Ans.

$$4 \int_0^R \frac{\rho \sigma(\rho)}{\rho + r} K\left(\frac{2\sqrt{\rho r}}{\rho + r}\right) d\rho = V - u^0(r), \qquad 0 \leqslant r \leqslant R$$

for the disk, and

$$2R \int_0^\alpha \frac{\sigma(\vartheta) \sin \vartheta}{\sin \tfrac{1}{2}(\vartheta + \theta)} K\left(\frac{\sqrt{\sin \vartheta \sin \theta}}{\sin \tfrac{1}{2}(\vartheta + \theta)}\right) d\vartheta = V - u^0(\theta), \qquad 0 \leqslant \theta \leqslant \alpha$$

for the bowl, where $K(k)$ is the complete elliptic integral of the first kind.

*549. Show that the integral equations of the preceding problem can be reduced to the integral equation

$$\frac{2}{\pi} \int_0^a \frac{f(y)}{x + y} K\left(\frac{2\sqrt{xy}}{x + y}\right) dy = g(x), \qquad 0 \leqslant x \leqslant a,$$

and solve this equation.

Ans.

$$f(y) = -\frac{2}{\pi} \frac{d}{dy} \int_y^a \frac{s\, ds}{\sqrt{s^2 - y^2}} \frac{d}{ds} \int_0^s \frac{g(t) t\, dt}{\sqrt{s^2 - t^2}}.$$

550. Using the results of Probs. 548 and 549, find the distribution of charge on a disk of radius R at potential V introduced into an arbitrary axially symmetric external field.

---

[5] Problems 548–555 are considered in Lebedev's paper L5.

Ans.

$$\sigma_1(r) + \sigma_2(r) = -\frac{1}{\pi^2 r}\frac{d}{dr}\int_r^R \frac{s\,ds}{\sqrt{s^2 - r^2}}\frac{d}{ds}\int_0^s [V - u^0(t)]\frac{t\,dt}{\sqrt{s^2 - t^2}},$$

$$\sigma_1(r) - \sigma_2(r) = \frac{1}{2\pi}E_z^0(r),$$

where $E_z^0(r)$ is the normal component of the external field on the surface of the disk.

551. Find the charge density on a thin conducting disk of radius R for the following cases:

a) The disk is charged to potential V, and there is no external field (freely charged disk);

b) A point charge q is placed on the axis of symmetry of the disk, at distance h from the disk.

Ans.

a) $\quad \sigma_1(r) = \sigma_2(r) = \dfrac{V}{2\pi^2}\dfrac{1}{\sqrt{R^2 - r^2}};$

b) $\quad \sigma_{1,2}(r) = -\dfrac{qh}{2\pi^2(h^2 + r^2)^{3/2}}\left[\arctan\sqrt{\dfrac{R^2 - r^2}{h^2 + r^2}} + \sqrt{\dfrac{h^2 + r^2}{R^2 - r^2}} \pm \dfrac{\pi}{2}\right],$

where σ_1 is the charge density on the side facing the charge.

552. Find the charge distribution on a thin spherical bowl $r = R$, $0 \leqslant \theta \leqslant \alpha$ at potential V placed in an arbitrary axially symmetric external field $\mathbf{E}^0$.

Ans.

$$\sigma_1(\theta) + \sigma_2(\theta) = -\frac{1}{2\pi^2 R\sin\frac{1}{2}\theta}\frac{d}{d\theta}\int_\theta^\alpha \frac{\tan\frac{1}{2}s\,ds}{\sqrt{\tan^2\frac{1}{2}s - \tan^2\frac{1}{2}\theta}}$$

$$\times \frac{d}{ds}\int_0^s \frac{[V - u^0(t)]\tan\frac{1}{2}t}{\sqrt{\tan^2\frac{1}{2}s - \tan^2\frac{1}{2}t}}\frac{dt}{\cos\frac{1}{2}t},$$

$$\sigma_1(\theta) - \sigma_2(\theta) = \frac{1}{2\pi}E_r^0(\theta) + \frac{1}{4\pi R}[V - u^0(\theta)],$$

where $u^0(\theta)$ and $E_r^0(\theta)$ are the values of the potential and the normal component of the electric field on the surface of the bowl, while σ_1 and σ_2 are the charge densities on the convex and concave sides of the surface.

553. Solve the preceding problem for the following special cases:

a) There is no external field (free charge distribution);

b) The external field is homogeneous, and the potential V is zero (a thin conducting spherical shield with a circular hole, placed in a homogeneous field $\mathbf{E}^0$).

Ans.

$$\sigma_{1,2}(\theta) = \frac{V}{4\pi^2 R}\left[\arctan\frac{\sqrt{\sin^2\frac{1}{2}\alpha - \sin^2\frac{1}{2}\theta}}{\cos\frac{1}{2}\alpha} + \frac{\cos\frac{1}{2}\alpha}{\sqrt{\sin^2\frac{1}{2}\alpha - \sin^2\frac{1}{2}\theta}} \pm \frac{\pi}{2}\right],$$

$$\sigma_{1,2}(\theta) = -\frac{3E^0\cos\theta}{4\pi^2}\left[\pm\frac{\pi}{2} + \arctan\frac{\sqrt{\sin^2\frac{1}{2}\alpha - \sin^2\frac{1}{2}\theta}}{\cos\frac{1}{2}\alpha}\right]$$
$$-\frac{E^0\cos\frac{1}{2}\alpha}{4\pi^2}\frac{3\cos\theta - 2\cos^2\frac{1}{2}\alpha}{\sqrt{\sin^2\frac{1}{2}\alpha - \sin^2\frac{1}{2}\theta}}.$$

554. Suppose a conducting plane at potential V, with a circular hole of radius R, is placed in an arbitrary axially symmetric external field (see Figure 155). Show that the problem of determining the charge distribution on the plane reduces to solving the integral equation (4), and find the distribution.

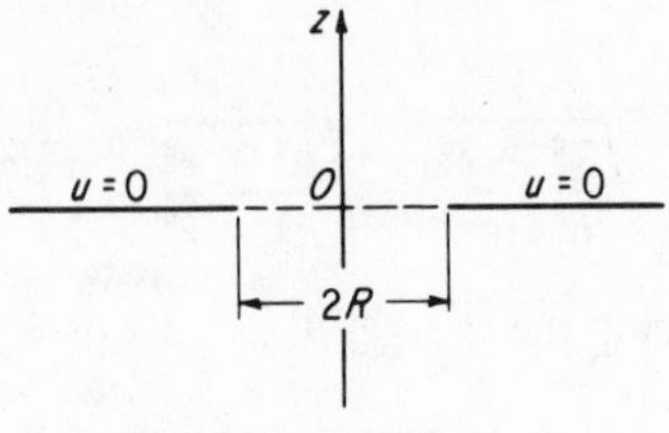

Ans.

FIGURE 155

$$\sigma_1(r) + \sigma_2(r) = \frac{1}{\pi^2}\frac{d}{dr}\int_R^r \frac{r\,ds}{\sqrt{r^2 - s^2}}$$
$$\times \frac{d}{ds}\int_s^\infty \frac{su^0(t)\,dt}{t\sqrt{t^2 - s^2}},$$

$$\sigma_1(r) - \sigma_2(r) = \frac{1}{2\pi}E_z^0(r),$$

where σ_1 is the charge density on the upper surface of the plane, while $u^0(r)$ is the potential and $E_z^0(r)$ the normal component of the external field at the point r.

555. Solve the preceding problem, assuming that the external field is due to a point charge q on the axis of symmetry of the hole at distance h from the plane.

Ans.

$$\sigma_{1,2}(r) = -\frac{qh}{2\pi^2(h^2 + r^2)^{3/2}}\left[\arctan\frac{h}{R}\sqrt{\frac{r^2 - R^2}{r^2 + h^2}} + \frac{R}{h}\sqrt{\frac{r^2 + h^2}{r^2 - R^2}} \pm \frac{\pi}{2}\right].$$

556. Show that the problem of the charge density on a grounded thin conducting half-plane, introduced into a given plane-parallel external field $\mathbf{E}^0$, reduces to the solution of the integral equation of the first kind

$$\int_0^\infty \sigma(\xi)\ln|x - \xi|\,d\xi = f(x), \qquad 0 < x < \infty, \tag{6}$$

where $\sigma(x)$ is the total charge per unit length, $\sigma(x) = \sigma_1(x) + \sigma_2(x),$[6] and $f(x) = \tfrac{1}{2}u^0(x)$, in terms of the potential $u^0(x)$ of the external field at the point x. Solve this integral equation.

Ans.

$$\sigma(x) = \frac{1}{\pi^2} \frac{d}{dx}\left[x \frac{d}{dx} \int_0^\infty \varphi(\xi) \ln \frac{\sqrt{x} + \sqrt{\xi}}{|\sqrt{x} - \sqrt{\xi}|} \, d\xi \right],$$

where

$$\varphi(x) = \frac{1}{x} [f(x) - f(0)].$$

Hint. Set $x = 0$ in (6) and subtract the result from the original equation. Then take the Mellin transform of the equation so obtained.

557. Use the result of the preceding problem to find the charge distribution on the surface of a thin conducting sheet $x > 0$, if there is a line source with charge q per unit length near the edge of the sheet.

Ans.

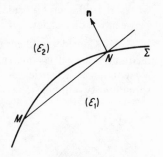

$$\sigma_{1,2} = - \frac{q}{2\pi\sqrt{r_0 x}} \frac{\sin \dfrac{\varphi_0}{2}}{1 + \dfrac{x}{r_0} \mp 2\sqrt{\dfrac{x}{r_0}} \cos \dfrac{\varphi_0}{2}},$$

where r_0, φ_0 are the polar coordinates of the point M, and the upper sign pertains to the density on the side of the sheet facing the charge.

FIGURE 156

558. Two media with dielectric constants ε_1 and ε_2 are separated by a surface Σ (see Figure 156). Consider the electrostatic field in the resulting inhomogeneous medium due to sources whose free-space field is $\mathbf{E}_0$. Show that the density of polarized charge on the surface Σ, determining the secondary field,[7] satisfies the integral equation

$$\sigma(N) = \frac{\varepsilon_1 - \varepsilon_2}{2\pi(\varepsilon_1 + \varepsilon_2)}\left[E_n^0(N) + \int_\Sigma \frac{\sigma(M)}{|\mathbf{r}_{MN}|^2} \cos(\mathbf{r}_{MN}, \mathbf{n}) \, dS \right], \qquad (7)$$

where M and N are two arbitrary points of the surface Σ, dS is the element of area, $\mathbf{r}_{MN}$ is the vector joining M to N, $\mathbf{n}$ is the unit exterior normal to

---

[6] The difference between the densities is given by the previous formula

$$\sigma_1(x) - \sigma_2(x) = \frac{1}{2\pi} E_y^0(x).$$

[7] The potential of the secondary field in each medium is given by

$$u_i(P) = \frac{1}{\varepsilon_i} \int_\Sigma \frac{\sigma(M)}{|\mathbf{r}_{MP}|} \, dS, \qquad i = 1, 2.$$

Σ at the point N, pointing from the medium with dielectric constant ε_1 to the medium with dielectric constant ε_2, and $E_n^0 = \mathbf{E}^0 \cdot \mathbf{n}$ is the projection of the external field onto $\mathbf{n}$ (see G5, Chap. 14).

559. Using the integral equation (7), find the distribution of polarized charge for the case where the surface Σ is an infinite plane (this generalizes Prob. 536).

Ans.

$$\sigma(N) = \frac{\varepsilon_1 - \varepsilon_2}{\varepsilon_1 + \varepsilon_2} \frac{E_n^0(N)}{2\pi}.$$

560. Derive the two-dimensional analogue of equation (7), corresponding to the plane-parallel electrostatic problem of an inhomogeneous medium made up of two homogeneous media with dielectric constants ε_1 and ε_2.

Ans.

$$\sigma(N) = \frac{\varepsilon_1 - \varepsilon_2}{2\pi(\varepsilon_1 + \varepsilon_2)} \left[E_n^0(N) + 2 \int_\Gamma \frac{\sigma(M)}{|\mathbf{r}_{MN}|} \cos\left(\mathbf{r}_{MN}, \mathbf{n}\right) ds \right], \tag{8}$$

where $\mathbf{n}$ is the unit normal to the contour Γ representing the interface between the dielectrics.

*561. Consider a dihedral angle whose interior $0 < \varphi < \alpha$ is filled with a medium of dielectric constant ε_1, and whose exterior $\alpha < \varphi < \pi$ is filled with a medium of dielectric constant ε_2 (see Figure 157). Show that the corresponding two-dimensional electrostatic problem reduces to solving the system of integral equations

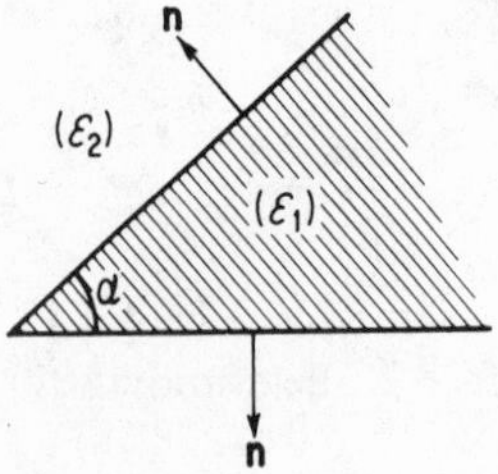

$$\sigma_0(r) = \frac{\varepsilon_1 - \varepsilon_2}{2\pi(\varepsilon_1 + \varepsilon_2)}$$

$$\times \left[-E_\varphi^0 \big|_{\varphi=0} + 2 \sin \alpha \int_0^\infty \frac{\sigma_\alpha(\rho)\rho \, d\rho}{r^2 + \rho^2 - 2r\rho \cos \alpha} \right],$$

$$\sigma_\alpha(r) = \frac{\varepsilon_1 - \varepsilon_2}{2\pi(\varepsilon_1 + \varepsilon_2)}$$

FIGURE 157

$$\times \left[E_\varphi^0 \big|_{\varphi=\alpha} + 2 \sin \alpha \int_0^\infty \frac{\sigma_0(\rho)\rho \, d\rho}{r^2 + \rho^2 - 2r\rho \cos \alpha} \right],$$

where σ_0 and σ_α are the densities of polarized charge on the faces $\varphi = 0$ and $\varphi = \alpha$. Solve this system by using the Mellin transform.

Ans.

$$\sigma_0(r) = \frac{\beta}{4\pi^2 i} \int_{1-i\infty}^{1+i\infty} \frac{\tilde{f}_\alpha \beta \sin(\pi - \alpha)(p-1) - \tilde{f}_0 \sin \pi(p-1)}{\sin^2 \pi(p-1) - \beta^2 \sin^2 (\pi - \alpha)(p-1)}$$
$$\times \, r^{-p} \sin \pi(p-1) \, dp,$$

$$\sigma_\alpha(r) = \frac{\beta}{4\pi^2 i} \int_{1-i\infty}^{1+i\infty} \frac{\tilde{f}_\alpha \sin \pi(p-1) - \tilde{f}_0 \beta \sin (\pi - \alpha)(p-1)}{\sin^2 \pi(p-1) - \beta^2 \sin^2 (\pi - \alpha)(p-1)}$$
$$\times \, r^{-p} \sin \pi(p-1) \, dp,$$

where $\bar{f}$ is the Mellin transform

$$\bar{f} = \int_0^\infty f(r) r^{p-1}\, dr,$$

and

$$f_0(r) = E_\varphi^0\big|_{\varphi=0}, \quad f_\alpha(r) = E_\varphi^0\big|_{\varphi=\alpha}, \quad \beta = \frac{\varepsilon_1 - \varepsilon_2}{\varepsilon_1 + \varepsilon_2}.$$

562. Solve the preceding problem for the special case where $\alpha = \pi/2$ and the external field $\mathbf{E}^0$ is due to a line source with charge q per unit length, located in the medium with dielectric constant ε_2 at the point $r = a$, $\varphi = \pi$.

Ans.

$$\sigma_0(r) = \frac{q}{a\varepsilon_2} \frac{\beta^2}{2\pi i} \int_{1-i\infty}^{1+i\infty} \frac{\sin^2 \tfrac{1}{2}\pi(p-1)}{\sin^2 \pi(p-1) - \beta^2 \sin^2 \tfrac{1}{2}\pi(p-1)} \left(\frac{a}{r}\right)^p dp,$$

$$\sigma_{\pi/2}(r) = \frac{q}{a\varepsilon_2} \frac{\beta}{2\pi i} \int_{1-i\infty}^{1+i\infty} \frac{\sin \tfrac{1}{2}\pi(p-1) \sin \pi(p-1)}{\sin^2 \pi(p-1) - \beta^2 \sin^2 \tfrac{1}{2}\pi(p-1)} \left(\frac{a}{r}\right)^p dp.$$

563. A slab of dielectric constant ε_1, bounded by the parallel planes $y = 0$ and $y = h$ and surrounded by a medium of dielectric constant ε_2 (see Figure 158), is introduced into an arbitrary plane-parallel field $\mathbf{E}^0$. Show that the resulting electrostatic problem reduces to solving the following system of integral equations for the polarized charge densities $\sigma_0(x)$ and $\sigma_h(x)$:

$$\sigma_0(x) = \frac{\varepsilon_1 - \varepsilon_2}{2\pi(\varepsilon_1 + \varepsilon_2)}$$
$$\times \left[-E_y^0\big|_{y=0} + 2h \int_{-\infty}^{\infty} \frac{\sigma_h(\xi)\, d\xi}{(x-\xi)^2 + h^2} \right],$$

$$\sigma_h(x) = \frac{\varepsilon_1 - \varepsilon_2}{2\pi(\varepsilon_1 + \varepsilon_2)}$$
$$\times \left[E_y^0\big|_{y=h} + 2h \int_{-\infty}^{\infty} \frac{\sigma_0(\xi)\, d\xi}{(x-\xi)^2 + h^2} \right].$$

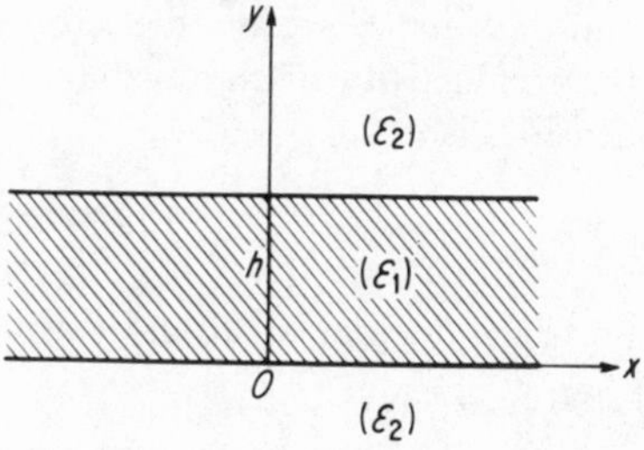

FIGURE 158

Solve this system of equations.

Ans.

$$\sigma_0(x) = \frac{\beta}{4\pi^2} \int_{-\infty}^{\infty} \frac{\beta e^{-|\lambda|h} \bar{f}_h - \bar{f}_0}{1 - \beta^2 e^{-2|\lambda|h}} e^{-i\lambda x}\, d\lambda,$$

$$\sigma_h(x) = \frac{\beta}{4\pi^2} \int_{-\infty}^{\infty} \frac{\bar{f}_h - \beta e^{-|\lambda|h} \bar{f}_0}{1 - \beta^2 e^{-2|\lambda|h}} e^{-i\lambda x}\, d\lambda,$$

where $\bar{f}$ is the Fourier transform of $f(x)$, i.e.,

$$\bar{f} = \int_{-\infty}^{\infty} f(x) e^{i\lambda x}\, dx,$$

and

$$f_0(x) = E^0_y\big|_{y=0}, \quad f_h(x) = E^0_y\big|_{y=h}, \quad \beta = \frac{\varepsilon_1 - \varepsilon_2}{\varepsilon_1 + \varepsilon_2}.$$

564. Solve the preceding problem for the special case where the external field $\mathbf{E}^0$ is due to a line source with charge q per unit length passing through the point $x = 0$, $y = h/2$.

Ans.

$$\sigma_0(x) = \sigma_h(x) = \frac{q\beta}{\pi\varepsilon_1} \int_0^\infty \frac{e^{-\lambda h/2}}{1 - \beta e^{-\lambda h}} \cos \lambda x \, d\lambda.$$

*565. A perfectly conducting half-plane $x \geqslant 0$, $y = 0$ is introduced into an external electromagnetic field with components

$$E_x = E_y = 0. \qquad E_z = E^0(x, y)e^{i\omega t}.$$

Show that the sum $j = j_1 + j_2$ of the current densities flowing in the upper and lower sides of the half-plane satisfy the integral equation

$$E(x) = \frac{\pi\omega}{c^2} \int_0^\infty H_0^{(2)}[k \,|x - \xi|]j(\xi) \, d\xi,$$

where

$$E(x) = E^0(x, 0), \qquad k = \sqrt{\frac{\varepsilon\mu\omega^2 - 4\pi\mu\omega\sigma i}{c^2}},$$

and ε, μ and σ are the dielectric constant, the magnetic permeability and the conductivity of the medium, while the difference between the current densities is

$$j_1 - j_2 = \frac{c^2 e^{-i\pi/2}}{2\pi\omega} \frac{\partial E^0(x, y)}{\partial y}\bigg|_{y=0}. \tag{9}$$

Solve the integral equation by using the transform (27), p. 196.

Ans.

$$j(x) = \frac{c^2}{2\pi\omega x} \int_0^\infty \left\{\frac{i\tau}{2} \overline{[E(x) - E(0)e^{-ikx}]} - \frac{E(0)}{\sinh \pi\tau}\right\} \frac{\tau \sinh^2 \pi\tau}{\cosh \pi\tau} e^{\pi\tau/2} H_{i\tau}^{(2)}(kx) \, d\tau,$$

where

$$\bar{f} = e^{\pi\tau/2} \int_0^\infty \frac{f(x)}{x} H_{i\tau}^{(2)}(kx) \, dx$$

[it is assumed that $f(x)$ approaches zero as $x \to 0$ in such a way that the integral converges at its lower limit].

*566. A plane electromagnetic wave with components

$$E_x = E_y = 0, \qquad E_z = E^0 e^{i(\omega t - kx)}$$

is incident on a perfectly conducting half-plane $r \geqslant 0$, $\varphi = \alpha$. Using the result of the preceding problem, find the distribution of current density on the half-plane.

Ans.

$$j_1 + j_2 = \frac{cE^0}{\pi\sqrt{\pi}}\left\{\frac{e^{-i\pi/4}\cos\frac{1}{2}\alpha}{\sqrt{2kr}}\,e^{-ikr} + \sin\alpha\,e^{i\pi/4}e^{-ikr\cos\alpha}\int_0^{\sqrt{2kr}\sin\frac{1}{2}\alpha}e^{-is^2}\,ds\right\},$$

$$j_1 - j_2 = \frac{E^0 c}{2\pi}\sin\alpha\,e^{-ikr\cos\alpha}.$$

References

Books: Grinberg (G5), Kupradze (K5), Mikhlin (M7), Morse and Feshbach (M9), Muskhelishvili (M10, M11), Noble (N1), Smirnov (S6, Vol. IV), Titchmarsh (T5), Tricomi (T9).

Papers: Fock (F3, F5), Lebedev (L5), Wiener and Hopf (W10).

Part 2

SOLUTIONS

SOLUTIONS

52. The solution of the problem reduces to the determination of the complex flow potential $w = \varphi + i\psi$, whose imaginary part is a harmonic function which equals zero on the axis of symmetry and takes the value $\psi = -v_a a$ on the walls of the channel. To determine w, we need only find a conformal mapping of the region $ABCDE$ onto the upper half-plane of the variable $\zeta = \xi + i\eta$. Suppose that in applying the Schwarz-Christoffel transformation, we make the points of the z and ζ-planes correspond in the way suggested in the hint to the problem. Then the relation between z and ζ is obtained by integrating the equation

$$\frac{dz}{d\zeta} = M(\zeta + 1)^{1/2}(\zeta + \lambda)^{-1/2}\zeta^{-1},$$

where M is a constant to be determined later. Bearing in mind that $z = ib$ if $\zeta = -1$, we find that

$$z = M\int_{-1}^{\zeta} \left(\frac{\zeta + 1}{\zeta + \lambda}\right)^{1/2}\frac{d\zeta}{\zeta} + ib,$$

where the integration is along any path joining the point $= -1$ to a given point ζ in the upper half-plane.

It follows from the condition

$$\lim_{\varepsilon \to 0}\left[z|_{\zeta=-\varepsilon} - z|_{\zeta=\varepsilon}\right] = ia$$

that $M = a\sqrt{\lambda}/\pi$, and hence it only remains to determine the value of the parameter λ. This is done by using the correspondence between the points $z = ia$ and $\zeta = -\lambda$. Since in evaluating the integral with $\zeta = -\lambda$ as its upper limit, we can integrate along the line segment joining the points $\zeta = -1$ and $\zeta = -\lambda$, on which

$$\zeta + 1 = 1 - s, \quad \zeta + \lambda = e^{i\pi}(s - \lambda), \quad \zeta = -s \quad (\lambda \leqslant s \leqslant 1),$$

275

the last requirement leads to the formula

$$1 - \frac{b}{a} = \frac{\sqrt{\lambda}}{\pi} \int_\lambda^1 \left(\frac{1-s}{s-\lambda}\right)^{1/2} \frac{ds}{s},$$

which, after carrying out the integration, implies $\lambda = (b/a)^2$ and hence $M = b/\pi$. As is easily verified, the complex potential in the ζ-plane is

$$w = -\frac{v_a a}{\pi} \ln \zeta,$$

which, together with the transformation $z = z(\zeta)$ just derived, gives a parametric solution of the problem.

To calculate the velocity along the axis of symmetry ($\xi \geqslant 0, \eta = 0$), we use the formula

$$v_x - i v_y = -\frac{dw}{dz},$$

which implies

$$\frac{v_x|_{y=0}}{v_a} = \frac{1}{\sqrt{\lambda}} \left(\frac{\xi + \lambda}{\xi + 1}\right)^{1/2},$$

where the relation between x and ξ must be established by using the transformation $z = z(\zeta)$. Choosing the path of integration to be a curve consisting of the segment $(-1, -R)$ of the real axis, an arc of a circle of radius R and the segment (R, ξ), and then taking the limit as $R \to \infty$, we find that

$$x = \frac{a\sqrt{\lambda}}{\pi} \lim_{R \to \infty} \left[\int_1^R \left(\frac{s-1}{s-\lambda}\right)^{1/2} \frac{ds}{s} - \int_\xi^R \left(\frac{s+1}{s+\lambda}\right)^{1/2} \frac{ds}{s} \right].$$

After some simple calculations, this leads to

$$\frac{x}{a} = \frac{1}{\pi} \left[\ln \frac{1 - \sqrt{\lambda}\sqrt{(\xi+1)/(\xi+\lambda)}}{1 + \sqrt{\lambda}\sqrt{(\xi+1)/(\xi+\lambda)}} + \sqrt{\lambda} \ln \frac{\sqrt{(\xi+1)/(\xi+\lambda)} + 1}{\sqrt{(\xi+1)/(\xi+\lambda)} - 1} \right].$$

The final formulas, given in the answer to the problem, are obtained by introducing the new parameter

$$t = \frac{1}{\sqrt{\lambda}} \sqrt{\frac{\xi + \lambda}{\xi + 1}}.$$

54. This problem belongs to a category which is both of considerable mathematical interest and of great importance in the applications, i.e., problems involving the formation of a jet at the boundary of an obstacle placed in a stationary plane-parallel flow of an ideal liquid. In such problems, the form of the jet is not known in advance, but must be determined from the condition that the velocity vector have a constant value on the free surface of the jet. For the case where the walls of the obstacle impeding the flow consist of line segments, an effective method of solving such problems is based on the possibility of establishing a connection between the complex potential w and the derivative dw/dz, starting from examination of the kinematic picture of the fluid motion.

Thus consider the part of the region occupied by the flow which is bounded by the axis of symmetry AB, the free boundary of the flow BC and the wall CD. The behavior of the velocity components v_x and v_y along the boundary of this region is determined by the following relations (where $v_b = Q/2b$ and $2b$ is the width of the jet at a great distance from the slot AB):

$$\text{On} \quad AB, \quad v_x = 0, \qquad -v_b \leqslant v_y \leqslant 0,$$
$$\text{On} \quad BC, \quad v_x^2 + v_y^2 = v_b^2,$$
$$\text{On} \quad CD, \quad -v_b \leqslant v_x \leqslant 0, \qquad v_y = 0.$$

Introducing the auxiliary complex variable $\zeta = dw/dz$ and taking account of the formula

$$v_x - iv_y = -\frac{dw}{dz},$$

we find on the basis of the above picture of the flow that the region $ABCD$ is mapped conformally onto the interior of the circular sector

$$|\zeta| < v_b, \qquad -\frac{\pi}{2} \leqslant \arg \zeta \leqslant 0$$

in the ζ-plane, with the boundary of the jet going into the arc of the circle Under the transformation[1]

$$\zeta = v_b(\sqrt{t} - \sqrt{t-1}),$$

this sector is mapped into the upper half-plane of the complex variable t, with the curves AB, BC and CD in the original plane going into the negative real axis, the segment $(0, 1)$ and the segment $(1, \infty)$. In the t-plane, the determination of the complex potential reduces to constructing a function analytic in the upper half-plane whose imaginary part takes the value zero on the negative real axis and the constant value $-Q/2$ on the positive real axis. It is easy to see that the solution of this problem is

$$w = \frac{Q}{2\pi} \ln t - \frac{Qi}{2},$$

---

[1] To obtain this expression, it is convenient to first transform the sector into a half-strip by using the transformation

$$s = \frac{\pi}{2} - i \ln \frac{\zeta}{v_b},$$

with the Schwarz-Christoffel transformation being applied afterwards.

which, together with the results found earlier, gives

$$\frac{dz}{dt} = \frac{1}{\zeta}\frac{dw}{dt} = \frac{Q}{2\pi v_b}\frac{\sqrt{t} + \sqrt{t-1}}{t}. \tag{1}$$

Integrating (1) and bearing in mind that the point $z = a$ must correspond to the point $t = 1$, we obtain

$$z = \frac{Q}{\pi v_b}[\sqrt{t} - 1 + \sqrt{t-1} + \arctan\sqrt{t-1}] + a, \tag{2}$$

where we choose the branch of the arc tangent which vanishes as $t \to 1$.

The functions $w = w(z)$ and $z = z(t)$ establish the required connection between the complex variable w and the variable z in parametric form. To determine the form of the boundary of the jet, we need only separate real and imaginary parts in (2), assuming that the variable t belongs to the interval $(0, 1)$. This gives the following parametric representation of the curve bounding the jet:

$$x = \frac{Q}{\pi v_b}(\sqrt{t} - 1) + a, \quad y = \frac{Q}{\pi v_b}\left[\sqrt{1-t} - \frac{1}{2}\ln\frac{1 + \sqrt{1-t}}{1 - \sqrt{1-t}}\right]$$

$$(0 \leqslant t \leqslant 1).$$

The width $2b$ of the jet at a great distance from the slot, and the corresponding value of the velocity $v_b = Q/2b$ are found from the condition that $x = b$ for $t = 0$, which implies

$$b = \frac{a}{1 + (2/\pi)}, \quad v_b = \left(1 + \frac{2}{\pi}\right)\frac{Q}{2a},$$

and immediately leads to the formulas given in the answer.

59. Guided by the hint to the problem, we construct the function $z = z(\zeta)$ mapping the interior of the rectangle onto the upper half-plane. Using the Schwarz-Christoffel transformation, we find that

$$\frac{dz}{d\zeta} = M_1(\zeta^2 - 1)^{-1/2}\left(\zeta^2 - \frac{1}{k^2}\right)^{-1/2} = M\frac{1}{\sqrt{(1 - \zeta^2)(1 - k^2\zeta^2)}} \quad (0 < k < 1),$$

which implies

$$z = M\int_0^\zeta \frac{d\zeta}{\sqrt{(1 - \zeta^2)(1 - k^2\zeta^2)}},$$

since the symmetry requires that the point $z = 0$ correspond to the point $\zeta = 0$. The values of the constants M and k are determined from the condition that the points $z = a$ and $z = a + ib$ correspond to the points $\zeta = 1$

and $\zeta = 1/k$. This leads to the formulas

$$a = M \int_0^1 \frac{d\zeta}{\sqrt{(1 - \zeta^2)(1 - k^2\zeta^2)}} = MK(k),$$

$$b = M \int_1^{1/k} \frac{d\zeta}{\sqrt{(1 - \zeta^2)(1 - k^2\zeta^2)}} = MK(k') \qquad (k' = \sqrt{1 - k^2}),$$

where $K(k)$ is the complete integral of the first kind.[2] Eliminating M, we obtain the relations

$$\frac{b}{a} = \frac{K(k')}{K(k)}, \qquad \frac{z}{a} K(k) = \int_0^\zeta \frac{d\zeta}{\sqrt{(1 - \zeta^2)(1 - k^2\zeta^2)}},$$

the first of which is an equation for determining the modulus k, while the second solves the given problem of conformal mapping.

According to the theory of elliptic functions, the inversion of the integral in the last expression is given by the formula

$$\zeta = \text{sn} \frac{Kz}{a},$$

where sn z is Jacobi's elliptic function. Under the conformal mapping, the point $z = z_q = \frac{1}{2}ib$ goes into the point $\zeta = \zeta_q = \text{sn}\,(iKb/2a) = i/\sqrt{k}$. The expression for the complex potential in the ζ-plane is

$$w = -2q \ln \frac{\zeta - \zeta_q}{\zeta - \bar{\zeta}_q} = -2q \ln \frac{\text{sn}\dfrac{Kz}{a} - \dfrac{i}{\sqrt{k}}}{\text{sn}\dfrac{Kz}{a} + \dfrac{i}{\sqrt{k}}}.$$

To calculate the distribution of charge density on the walls of the box, we use the relations

$$E_x - iE_y = -\frac{dw}{dz}, \qquad \sigma = \frac{1}{4\pi} E_n,$$

where E_n denotes the field normal to the surface of the conductor at the point where the value of the density σ is being determined. Applying these formulas, we find that the charge distribution on the wall $-a \leqslant x \leqslant a, y = 0$ is the expression given in the answer to the problem.

69. The displacement $u(x, t)$ of any point of the midline of the beam satisfies the differential equation

$$\frac{\partial^4 u}{\partial x^4} + \frac{1}{a^4} \frac{\partial^2 u}{\partial t^2} = 0$$

---

[2] To reduce the second integral to canonical form, use the substitution $\sqrt{1 - k^2\zeta^2} = k't$.

and the boundary conditions

$$u\big|_{x=0} = \frac{\partial u}{\partial x}\bigg|_{x=0} = 0, \quad \frac{\partial^2 u}{\partial x^2}\bigg|_{x=l} = 0, \quad M_0 \frac{\partial^2 u}{\partial t^2}\bigg|_{x=l} = EJ \frac{\partial^3 u}{\partial x^3}\bigg|_{x=l}.$$

To find the natural frequencies for transverse oscillations, we write

$$u(x, t) = v(x) \sin(\omega t + \varphi).$$

Then, after substituting this expression into the differential equation and the boundary conditions we find the following conditions determining the amplitude $v(x)$:

$$v^{(iv)} - \frac{\omega^2}{a^4} v = 0,$$

$$v\big|_{x=0} = v'\big|_{x=0} = 0, \quad v''\big|_{x=l} = 0, \quad v'''\big|_{x=l} = -\frac{M_0 \omega^2}{EJ}\bigg|_{x=l}.$$

The general solution of this equation is

$$v(x) = A_1 \cos \frac{\sqrt{\omega}\, x}{a} + B_1 \sin \frac{\sqrt{\omega}\, x}{a} + A_2 \cosh \frac{\sqrt{\omega}\, x}{a} + B_2 \sinh \frac{\sqrt{\omega}\, x}{a}.$$

The fact that the end $x = 0$ is clamped allows us the determine two of these constants, and leads to the expression

$$v(x) = A\left(\cos \frac{\sqrt{\omega}\, x}{a} - \cosh \frac{\sqrt{\omega}\, x}{a}\right) + B\left(\sin \frac{\sqrt{\omega}\, x}{a} - \sinh \frac{\sqrt{\omega}\, x}{a}\right).$$

Then, imposing the remaining conditions at $x = l$, we obtain the following homogeneous system of equations for the quantities A and B (for brevity, we set $\gamma = \sqrt{\omega}\, l/a$):

$$A(\cos \gamma + \cosh \gamma) + B(\sin \gamma + \sinh \gamma) = 0,$$

$$A\left[\gamma \frac{M_0}{M}(\cos \gamma - \cosh \gamma) + (\sin \gamma - \sinh \gamma)\right]$$
$$+ B\left[\gamma \frac{M_0}{M}(\sin \gamma - \sinh \gamma) - (\cos \gamma + \cosh \gamma)\right] = 0.$$

The equation determining the natural frequencies is obtained by setting the determinant of this system equal to zero. The result is

$$1 + \cos \gamma \cosh \gamma = \frac{M_0}{M} \gamma(\sin \gamma \cosh \gamma - \cos \gamma \sinh \gamma).$$

If the roots of this equation are denoted by γ_n $(n = 1, 2, \ldots)$, the natural frequencies are

$$\omega_n = \frac{a^2}{l^2} \gamma_n^2, \quad n = 1, 2, \ldots$$

83. The problem of finding the forced oscillations of the membrane under the action of a load $q \sin(\omega t + \varphi)$ distributed over a disk of radius $b < a$ can be posed as follows: Find the solution of the differential equation

$$\frac{1}{r}\frac{\partial}{\partial r}\left(r\frac{\partial u}{\partial r}\right) = \frac{1}{v^2}\frac{\partial^2 u}{\partial t^2} - \frac{q(r,t)}{T}$$

governing the oscillations of the membrane, where

$$q(r,t) = \psi(r) \sin(\omega t + \varphi),$$

$$\psi(r) = \begin{cases} q, & 0 \leqslant r \leqslant b, \\ 0, & b < r \leqslant a \end{cases} \tag{3}$$

which satisfies the boundary condition

$$u|_{r=a} = 0$$

and has the same frequency ω as the perturbing force. Writing

$$u(r,t) = w(r) \sin(\omega t + \varphi),$$

and substituting this expression into the differential equation and boundary condition, we find that

$$\frac{1}{r}\frac{d}{dr}\left(r\frac{dw}{dr}\right) + \frac{\omega^2}{v^2}w = -\frac{\psi(r)}{T}, \qquad w(a) = 0.$$

The solution of this inhomogeneous equation, obtained by variation of constants, has the form

$$w = AJ_0\left(\frac{\omega r}{v}\right) + BY_0\left(\frac{\omega r}{v}\right)$$
$$+ \frac{\pi}{2T}\int_0^r \psi(\rho)\left[J_0\left(\frac{\omega r}{v}\right)Y_0\left(\frac{\omega \rho}{v}\right) - Y_0\left(\frac{\omega r}{v}\right)J_0\left(\frac{\omega \rho}{v}\right)\right]\rho\,d\rho.$$

The constant B equals zero because of the requirement that the solution be bounded at the point $r = 0$. The constant A is determined from the condition $w(a) = 0$, which gives

$$A = \frac{\pi}{2TJ_0(\omega a/v)}\int_0^a \psi(\rho)\left[Y_0\left(\frac{\omega a}{v}\right)J_0\left(\frac{\omega \rho}{v}\right) - J_0\left(\frac{\omega a}{v}\right)Y_0\left(\frac{\omega \rho}{v}\right)\right]\rho\,d\rho.$$

After some manipulation, the desired expression for the amplitude takes the form

$$w(r) = \int_0^a \psi(\rho)G(\rho,r)\rho\,d\rho, \tag{4}$$

where

$$G(\rho, r) = \frac{\pi}{2T} \times \begin{cases} \dfrac{J_0(\omega\rho/v)}{J_0(\omega a/v)}\left[J_0\!\left(\dfrac{\omega r}{v}\right)Y_0\!\left(\dfrac{\omega a}{v}\right) - Y_0\!\left(\dfrac{\omega r}{v}\right)J_0\!\left(\dfrac{\omega a}{v}\right)\right], & \rho \leqslant r, \\[2ex] \dfrac{J_0(\omega r/v)}{J_0(\omega a/v)}\left[J_0\!\left(\dfrac{\omega \rho}{v}\right)Y_0\!\left(\dfrac{\omega a}{v}\right) - Y_0\!\left(\dfrac{\omega \rho}{v}\right)J_0\!\left(\dfrac{\omega a}{v}\right)\right], & \rho \geqslant r. \end{cases}$$

(5)

Substituting (3) and (5) into (4), and using the formulas

$$\int_0^\rho \rho J_0(\lambda\rho)\, d\rho = \frac{\rho}{\lambda} J_1(\lambda\rho), \qquad \int_0^\rho \rho Y_0(\lambda\rho)\, d\rho = \frac{\rho}{\lambda} Y_1(\lambda\rho) + \frac{2}{\pi\lambda^2}$$

and the familiar expression

$$J_0(x)Y_0'(x) - Y_0(x)J_0'(x) = \frac{2}{\pi x}$$

for the Wronskian of the Bessel functions, we finally obtain the answer on p. 48.

91. If the z-axis is parallel to the generators of the wave guide, then the only component of the electric field of the TM-wave is

$$E_z = E e^{i(\omega t - vz)}$$

(ω is the frequency of the oscillations and v is the propagation constant), whose amplitude satisfies Helmholtz's equation

$$\frac{1}{r}\frac{\partial}{\partial r}\left(r\frac{\partial E}{\partial r}\right) + \frac{1}{r^2}\frac{\partial^2 E}{\partial \varphi^2} + (k^2 - v^2)E = 0$$

($k = \omega/c = 2\pi/\lambda$, where λ is the wavelength) and the homogeneous boundary conditions

$$E\big|_{r=a} = 0, \qquad E\big|_{\varphi=0} = E_{\varphi=\alpha} = 0.$$

These equations have infinitely many nonzero solutions of the form

$$E = E_{mn} = J_{m\pi/\alpha}\!\left(\frac{\gamma_{mn}r}{a}\right)\sin\frac{m\pi\varphi}{\alpha}, \qquad m = 1, 2, \ldots,$$

where the γ_{mn} are the roots of the equation

$$J_{m\pi/\alpha}(\gamma) = 0,$$

and the value of the propagation constant corresponding to γ_{mn} is

$$v = v_{mn} = \sqrt{k^2 - \left(\frac{\gamma_{mn}}{a}\right)^2}.$$

A wave with an imaginary value of ν_{mn} falls off exponentially in the z-direction and is essentially unable to propagate in the wave guide, i.e., a wave can propagate in the guide only if ν_{mn} is real. This leads to the inequality

$$\lambda \leqslant \frac{2\pi a}{\gamma_{mn}}.$$

The maximum wavelength which can propagate in the guide is given by the formula

$$\lambda_{\max} = \frac{2\pi a}{\gamma_0},$$

where γ_0 is the smallest positive root of the equations

$$J_{m\pi/\alpha}(\gamma) = 0, \qquad m = 1, 2, \ldots$$

96. The problem reduces to integration of the equation

$$\frac{\partial^2 u}{\partial x^2} - \frac{1}{v^2}\frac{\partial^2 u}{\partial t^2} = 0,$$

with initial conditions

$$u\big|_{t=0} = f(x) = \begin{cases} \dfrac{hx}{c}, & 0 \leqslant x \leqslant c, \\[2mm] \dfrac{h(l-x)}{l-c}, & c \leqslant x \leqslant l, \end{cases} \qquad \frac{\partial u}{\partial t}\bigg|_{t=0} = 0,$$

and boundary conditions

$$u\big|_{x=0} = u\big|_{x=l} = 0.$$

Setting $u(x, t) = X(x)T(t)$ and separating variables, we arrive at the equations

$$X'' + \lambda X = 0, \qquad T'' + \lambda v^2 T = 0.$$

Solving the first of these equations with the boundary conditions $X(0) = X(l) = 0$, we find the corresponding eigenvalues and eigenfunctions

$$\lambda = \lambda_n = \frac{n^2\pi^2}{l^2}, \quad X = X_n(x) = \sin\frac{n\pi x}{l} \qquad (n = 1, 2, \ldots).$$

The solution of the second equation satisfying the conditions $T'(0) = 0$ is given by

$$T = T_n(t) = c_n \cos\frac{n\pi vt}{l}.$$

Therefore the set of particular solutions of the equation of the vibrating string satisfying all the homogeneous conditions is

$$u = u_n = c_n \sin\frac{n\pi x}{l} \cos\frac{n\pi vt}{l}, \qquad n = 1, 2, \ldots$$

According to the basic idea of the Fourier method, we now look for a solution of the given problem in the form of a series

$$u(x, t) = \sum_{n=1}^{\infty} c_n \sin \frac{n\pi x}{l} \cos \frac{n\pi vt}{l},$$

where the coefficients c_n are determined from the condition $u_t|_{=0} = f(x)$, i.e., coincide with the coefficients of the expansion of the function $f(x)$ in a Fourier series

$$f(x) = \sum_{n=1}^{\infty} c_n \sin \frac{n\pi x}{l}, \qquad 0 < x < l.$$

As is well known (see T7, p. 35),

$$c_n = \frac{2}{l} \int_0^l f(x) \sin \frac{n\pi x}{l} \, dx,$$

and hence in the present case

$$c_n = \frac{2hl^2}{n^2\pi^2 c(l-c)} \sin \frac{n\pi c}{l},$$

which leads to the answer on p. 60. It can be shown that this series represents a piecewise smooth function of the variables x and t, satisfying the equation of the vibrating string and all the initial and boundary conditions.

108. In the present case, the differential equation for longitudinal oscillations of a beam of variable cross section takes the form

$$\frac{1}{y(x)} \frac{\partial}{\partial x}\left[y(x) \frac{\partial u}{\partial x}\right] - \frac{1}{v^2} \frac{\partial^2 u}{\partial t^2} = 0,$$

where

$$y(x) = \alpha - x \tan \alpha$$

is the variable height of the cross section at x measured from the axis of symmetry of the beam. Setting $u(x, t) = X(x)T(t)$ and separating variables, we find that the factors X and T satisfy the equations

$$X'' - \frac{\tan \alpha}{y} X' + \lambda X = 0, \qquad T'' + \lambda v^2 T = 0. \tag{6}$$

The first of these equations reduces to Bessel's equation in the variable y

$$\frac{d^2 X}{dy^2} + \frac{1}{y} \frac{dX}{dy} + \left(\frac{\sqrt{\lambda}}{\tan \alpha}\right)^2 X = 0,$$

with general solution

$$X = AJ_0\left(\frac{\sqrt{\lambda} y}{\tan \alpha}\right) + BY_0\left(\frac{\sqrt{\lambda} y}{\tan \alpha}\right).$$

Using the boundary conditions

$$u\big|_{x=0} = 0, \qquad \frac{\partial u}{\partial x}\bigg|_{x=l} = 0,$$

which imply the conditions

$$X\big|_{y=a} = 0, \qquad \frac{dX}{dy}\bigg|_{y=b} = 0,$$

we obtain the eigenvalues

$$\lambda = \lambda_n = \left(\frac{\gamma_n \tan \alpha}{a}\right)^2,$$

with corresponding eigenfunctions $X(x) = X_{\gamma_n}(x)$, where

$$X_\gamma(y) = Y_0(\gamma)J_0\left(\frac{\gamma y}{a}\right) - J_0(\gamma)Y_0\left(\frac{\gamma y}{a}\right),$$

and the γ_n are consecutive positive roots of the equation $X_\gamma'(b) = 0$. Integrating the second of the equations (6) and taking account of the condition $T'(0) = 0$, we find that

$$T = c_n \cos \frac{vt\gamma_n \tan \alpha}{a}.$$

It follows that the set of particular solutions satisfying the homogeneous conditions is

$$u = u_n = c_n X_{\gamma_n}(y) \cos \frac{vt\gamma_n \tan \alpha}{a}, \qquad n = 1, 2, \ldots$$

The solution of our problem is then constructed in the form of a series

$$u(x, t) = \sum_{n=1}^{\infty} u_n,$$

where the coefficients c_n are determined from the condition

$$u\big|_{t=0} = f(x) = \sum_{n=1}^{\infty} c_n X_{\gamma_n}(y), \qquad b < y < a.$$

Using the formulas

$$\int_b^a X_{\gamma_m}(y) X_{\gamma_n}(y) y \, dy = \begin{cases} 0, & m \neq n, \\ \dfrac{b^2}{2}\left[\dfrac{a^4}{b^2\gamma_n^2} X_{\gamma_n}'^2(a) - X_{\gamma_n}^2(b)\right], & m = n, \end{cases}$$

we find that

$$c_n = \frac{2}{b^2} \frac{\int_b^a f(x) X_{\gamma_n}(y) y \, dy}{\dfrac{4a^2}{\pi^2 \gamma_n^2 b^2} - X_{\gamma_n}^2(b)},$$

where the relation

$$X'_{\gamma_n}(a) = -\frac{2}{\pi a}$$

has also been used. In this way we finally arrive at the answer on p. 64.

109. The problem reduces to integration of the system of equations

$$\frac{\partial^2 u_1}{\partial x^2} - \frac{1}{v_1^2} \frac{\partial^2 u_1}{\partial t^2} = 0, \qquad -a_1 < x < 0,$$

$$\frac{\partial^2 u_2}{\partial x^2} - \frac{1}{v_2^2} \frac{\partial^2 u_2}{\partial t^2} = 0, \qquad 0 < x < a_2,$$

with initial conditions

$$u\big|_{t=0} = f(x), \qquad \frac{\partial u}{\partial t}\bigg|_{t=0} = 0$$

and boundary conditions

$$u_1\big|_{x=-a_1} = u_2\big|_{x=a_2} = 0, \qquad u_1\big|_{x=0} = u_2\big|_{x=0},$$

$$E_1 S_1 \frac{\partial u_1}{\partial x}\bigg|_{x=0} = E_2 S_2 \frac{\partial u_2}{\partial x}\bigg|_{x=0}.$$

Separation of variables leads to the expression for the displacement

$$u(x, t) = \sum_{n=1}^{\infty} c_n X_n(x) \cos \frac{\gamma_n v_1 t}{a_1},$$

satisfying all the conditions of the problem except the first initial condition. Here

$$X_n(x) = \begin{cases} X_n^{(1)}(x) = \sin \dfrac{\gamma_n v_1 a_2}{v_2 a_1} \sin \gamma_n \left(\dfrac{x}{a_1} + 1\right), & -a_1 \leqslant x \leqslant 0, \\[3mm] X_n^{(2)}(x) = \sin \gamma_n \sin \dfrac{\gamma_n v_1 (a_2 - x)}{v_2 a_1}, & 0 \leqslant x \leqslant a_2, \end{cases} \tag{7}$$

where the γ_n are consecutive positive roots of the equation

$$S_2 \sqrt{E_2 \rho_2} \tan \gamma + S_1 \sqrt{E_1 \rho_1} \tan \frac{\gamma a_2 v_1}{a_1 v_2} = 0.$$

It can be shown that the eigenfunctions $X_n(x)$ of the problem are orthogonal on the interval $-a_1 \leqslant x \leqslant a_2$ with weight

$$r(x) = \begin{cases} S_1 \rho_1, & -a_1 \leqslant x < 0, \\ S_2 \rho_2, & 0 < x \leqslant a_2. \end{cases}$$

Therefore the initial condition $u|_{t=0} = f(x)$ implies

$$c_n = \frac{S_1 \rho_1 \int_{-a_1}^{0} f(x) X_n^{(1)}(x)\, dx + S_2 \rho_2 \int_0^{a_2} f(x) X_n^{(2)}(x)\, dx}{S_1 \rho_1 \int_{-a_1}^{0} [X_n^{(1)}(x)]^2\, dx + S_2 \rho_2 \int_0^{a_2} [X_n^{(2)}(x)]^2\, dx}. \tag{8}$$

Substituting the eigenfunctions (7) into (8), we find that the denominator becomes

$$\frac{1}{2}\left(a_1 S_1 \rho_1 \sin^2 \frac{\gamma_n v_1 a_2}{v_2 a_1} + a_2 S_2 \rho_2 \sin^2 \gamma_n\right).$$

Thus the solution finally takes the form given in the answer on p. 65.

112. To solve the problem, we have to integrate the differential equation

$$\frac{\partial^2 \theta}{\partial x^2} - \frac{1}{v^2} \frac{\partial^2 \theta}{\partial t^2} = 0$$

for torsional oscillation of the shaft, subject to the following initial and boundary conditions (J_p denotes the polar moment of inertia per unit length of the shaft)

$$\theta|_{t=0} = f(x) = \begin{cases} \dfrac{\alpha x}{a}, & 0 \leqslant x \leqslant a, \\[2mm] \dfrac{\alpha(l-x)}{l-a}, & a \leqslant x \leqslant l, \end{cases} \qquad \frac{\partial \theta}{\partial t}\bigg|_{t=0} = 0,$$

$$\theta|_{x=0} = \theta|_{x=l} = 0, \qquad \theta|_{x=a-0} = \theta|_{x=a+0},$$

$$GJ_p\left(\frac{\partial \theta}{\partial x}\bigg|_{x=a+0} - \frac{\partial \theta}{\partial x}\bigg|_{x=a-0}\right) = J_0 \frac{\partial^2 \theta}{\partial t^2}\bigg|_{x=a}.$$

Separating variables, and taking account of the fact that the ends are clamped and there is no initial velocity, we find the following particular solutions:

$$c^{(1)} \sin \sqrt{\lambda}\, x \cos \sqrt{\lambda}\, vt, \qquad 0 \leqslant x \leqslant a,$$

$$c^{(2)} \sin \sqrt{\lambda}\left(1 - \frac{x}{l}\right) \cos \sqrt{\lambda}\, vt, \qquad a \leqslant x \leqslant l.$$

Using the fact that the two sections of the shaft are joined at the point $x = a$, we obtain the eigenvalues

$$\lambda = \lambda_n = \frac{\gamma_n^2}{l^2},$$

and the corresponding eigenfunctions

$$\theta_n(x) = \begin{cases} \sin \gamma_n \left(1 - \dfrac{a}{l}\right) \sin \dfrac{\gamma_n x}{l}, & 0 \leqslant x \leqslant a, \\[3mm] \sin \dfrac{\gamma_n a}{l} \sin \gamma_n \left(1 - \dfrac{x}{l}\right), & a \leqslant x \leqslant l \end{cases}$$

$(n = 1, 2, \ldots)$, where the γ_n are consecutive positive roots of the equation

$$\sin \gamma = \frac{J_0 \gamma}{J} \sin \frac{\gamma a}{l} \sin \gamma \left(1 - \frac{a}{l}\right).$$

If the solution of the problem is written as a series

$$\theta(x, t) = \sum_{n=1}^{\infty} c_n \theta_n(x) \cos \frac{\gamma_n v t}{l},$$

then the coefficients c_n must satisfy the relation

$$f(x) = \sum_{n=1}^{\infty} c_n \theta_n(x), \qquad 0 \leqslant x \leqslant l.$$

In the present case, the functions $\psi_n(x) = \theta'_n(x)$ are orthogonal, i.e.,

$$\int_0^l \psi_m(x) \psi_n(x) \, dx$$
$$= \begin{cases} 0, & m \neq n \\[2mm] \dfrac{\gamma_n^2}{2l} \left(\left[\dfrac{a}{l} + \dfrac{\sin(2\gamma_n a/l)}{2\gamma_n}\right] \sin^2 \gamma_n \left(1 - \dfrac{a}{l}\right) \right. \\[3mm] \left. + \left[1 - \dfrac{a}{l} + \dfrac{\sin 2\gamma_n [1 - (a/l)]}{2\gamma_n}\right] \sin^2 \dfrac{\gamma_n a}{l}\right), & m = n. \end{cases}$$

It follows that

$$c_n = \frac{\int_0^l f'(x) \psi_n(x) \, dx}{\int_0^l \psi_n^2(x) \, dx},$$

which leads to the answer on p. 66.

114. The problem reduces to integrating the differential equation

$$\frac{\partial^4 u}{\partial x^4} + \frac{1}{a^4} \frac{\partial^2 u}{\partial t^2} = 0$$

for transverse oscillations of the beam, with initial conditions

$$u\big|_{t=0} = f(x), \qquad \frac{\partial u}{\partial t}\bigg|_{t=0} = g(x)$$

and boundary conditions

$$u\big|_{x=0} = \frac{\partial^2 u}{\partial x^2}\bigg|_{x=0} = u\big|_{x=l} = \frac{\partial^2 u}{\partial x^2}\bigg|_{x=l} = 0.$$

Writing $u(x, t) = X(x)T(t)$ and separating variables, we find the differential equations

$$X^{(iv)} - \lambda X = 0, \qquad T'' + a^4 \lambda T = 0$$

for the separate factors, with general solutions

$$X = A \cos \sqrt[4]{\lambda}x + B \sin \sqrt[4]{\lambda}x + C \cosh \sqrt[4]{\lambda}x + D \sinh \sqrt[4]{\lambda}x,$$
$$T = M \cos \sqrt{\lambda}a^2 t + N \sin \sqrt{\lambda}a^2 t.$$

Using the boundary conditions

$$X(0) = X''(0) = X(l) = X''(l) = 0,$$

we arrive at the eigenvalues

$$\lambda = \lambda_n = \frac{n^4 \pi^4}{l^4}$$

and eigenfunctions

$$X = X_n(x) = \sin \frac{n\pi x}{l}, \qquad n = 1, 2, \ldots$$

Determination of the constants M_n and N_n in the expansion

$$u(x, t) = \sum_{n=1}^{\infty} \left(M_n \cos \frac{n^2 \pi^2 a^2 t}{l^2} + N_n \sin \frac{n^2 \pi^2 a^2 t}{l^2} \right) \sin \frac{n\pi x}{l}$$

reduces to evaluation of the Fourier coefficients of the functions

$$f(x) = \sum_{n=1}^{\infty} M_n \sin \frac{n\pi x}{l}, \quad g(x) = \frac{\pi^2 a^2}{l^2} \sum_{n=1}^{\infty} n^2 N_n \sin \frac{n\pi x}{l} \qquad (0 < x < l).$$

118. We want the solution of the equation

$$\frac{\partial^4 u}{\partial x^4} + \frac{1}{a^4} \frac{\partial^2 u}{\partial t^2} = 0$$

satisfying the initial conditions

$$u\big|_{t=0} = f(x), \qquad \frac{\partial u}{\partial t}\bigg|_{t=0} = 0$$

and boundary conditions

$$u\big|_{x=0} = \frac{\partial u}{\partial x}\bigg|_{x=0} = \frac{\partial^2 u}{\partial x^2}\bigg|_{x=l} = \frac{\partial^3 u}{\partial x^3}\bigg|_{x=l} = 0,$$

where the initial deflection $f(x)$ is the solution of the following static problem:[3]

$$f^{(iv)}(x) = 0, \quad f(0) = f'(0) = f''(l) = 0, \quad f'''(l) = -\frac{F}{EJ}.$$

Writing $u(x, t) = X(x)T(t)$ and separating variables, we find that

$$X = A \cos \sqrt[4]{\lambda}\, x + B \sin \sqrt[4]{\lambda}\, x + C \cosh \sqrt[4]{\lambda}\, x + D \sinh \sqrt[4]{\lambda}\, x,$$

$$T = M \cos \sqrt{\lambda}\, a^2 t + N \sin \sqrt{\lambda}\, a^2 t.$$

Using the boundary conditions

$$X(0) = X'(0) = X''(l) = X'''(l) = 0,$$

we obtain the eigenvalues

$$\lambda_n = \frac{\gamma_n^4}{l^4}$$

and eigenfunctions

$$X_n(x) = (\sin \gamma_n + \sinh \gamma_n)\left(\cos \frac{\gamma_n x}{l} - \cosh \frac{\gamma_n x}{l}\right)$$

$$- (\cos \gamma_n + \cosh \gamma_n)\left(\sin \frac{\gamma_n x}{l} - \sinh \frac{\gamma_n x}{l}\right),$$

where the γ_n are consecutive positive roots of the transcendental equation $\cos \gamma \cosh \gamma + 1 = 0$.

Next we show that the functions $X_n(x)$ are orthogonal on the interval $(0, l)$. Multiplying the first of the equations

$$X_n^{(iv)} - \lambda_n X_n = 0, \qquad X_m^{(iv)} - \lambda_m X_m = 0$$

by X_m and the second by X_n, we subtract the results from each other and integrate with respect to x from 0 to l. Taking account of the boundary conditions, we obtain

$$(\lambda_n - \lambda_m)\int_0^l X_n X_m\, dx = (X_n''' X_m - X_m''' X_n + X_m'' X_n' - X_n'' X_m')\Big|_0^l = 0$$

after integration by parts. This immediately implies the required orthogonality of the functions $X_n(x)$. Using the general theory of expansions in series of orthogonal functions, we can represent the solution of the problem in the form

$$u = \sum_{n=1}^{\infty} \cos \frac{\gamma_n^2 a^2 t}{l^2}\, X_n(x) \frac{\int_0^l f(\xi) X_n(\xi)\, d\xi}{\int_0^l X_n^2(\xi)\, d\xi}. \tag{9}$$

---

[3] An explicit expression for $f(x)$ is given in Prob. 7, but will not be used in our method of solution.

The integral in the numerator is easily evaluated by replacing X_n by $X_n^{(iv)}/\lambda_n$ and integrating by parts, which gives

$$\int_0^l f(\xi) X_n(\xi) \, d\xi = \frac{1}{\lambda_n} \Bigg\{ (X_n''' f - X_n'' f' + X_n' f'' - X_n f''') \Bigg|_0^l$$

$$+ \int_0^l X_n f^{(iv)} \, d\xi \Bigg\} = \frac{X_n(l)}{\lambda_n} \frac{F}{EJ}.$$

To evaluate the integral in the denominator, we use the formula

$$\int_0^l X_n^2(\xi) \, d\xi = \frac{l}{4} [X_n^2(l) + X_n''^2(l) - 2X_n'(l) X_n'''(l)],$$

(see T2, p. 336), which in the present case takes the form

$$\int_0^l X_n^2(\xi) \, d\xi = \frac{l}{4} X_n^2(l).$$

Substituting these integrals into (9), we find that

$$u(x, t) = \frac{4Fl^3}{EJ} \sum_{n=1}^{\infty} \frac{X_n(x)}{\gamma_n^4 X_n(l)} \cos \frac{\gamma_n^2 a^2 t}{l^2}.$$

The form of the solution given in the answer on p. 68 is obtained after making the substitution

$$X_n(l) = 2(\cos \gamma_n \sinh \gamma_n - \sin \gamma_n \cosh \gamma_n).$$

120. The problem reduces to integration of the equation

$$\frac{1}{r} \frac{\partial}{\partial r} \left(r \frac{\partial u}{\partial r} \right) - \frac{1}{v^2} \frac{\partial^2 u}{\partial t^2} = 0$$

for a vibrating membrane, with initial conditions

$$u\big|_{t=0} = 0, \quad \frac{\partial u}{\partial t}\bigg|_{t=0} = f(r) = \begin{cases} \dfrac{P}{\pi \varepsilon^2 \rho}, & 0 \leqslant r < \varepsilon, \\ 0, & \varepsilon < r \leqslant a, \end{cases}$$

and boundary condition

$$u\big|_{r=a} = 0.$$

Writing $u(r, t) = R(r)T(t)$ and separating variables, we arrive at the equations

$$\frac{1}{r} \frac{d}{dr} \left(r \frac{dR}{dr} \right) + \lambda R = 0, \qquad T'' + \lambda v^2 T = 0.$$

The permissible values of the parameter λ are obtained from the requirement that the first of these equations have solutions which are bounded in the region $0 \leqslant r < a$ and satisfy the boundary condition $R(a) = 0$. This leads to the eigenvalues and eigenfunctions

$$\lambda = \lambda_n = \frac{\gamma_n^2}{a^2}, \quad R = R_n(r) = J_0\left(\frac{\gamma_n r}{a}\right) \qquad (n = 1, 2, \ldots),$$

where the γ_n are consecutive positive roots of the equation $J_0(\gamma) = 0$.

The solution of the equation for a vibrating membrane satisfying all the homogeneous conditions is

$$u(r, t) = \sum_{n=1}^{\infty} c_n \sin \frac{\gamma_n vt}{a} J_0\left(\frac{\gamma_n r}{a}\right).$$

The constants c_n are determined from the condition

$$\frac{\partial u}{\partial t}\bigg|_{t=0} = f(r),$$

which, after substitution of the series for $u(r, t)$, takes the form

$$f(r) = \frac{v}{a} \sum_{n=1}^{\infty} c_n \gamma_n J_0\left(\frac{\gamma_n r}{a}\right), \qquad 0 \leqslant r < a.$$

According to the well-known formula for the coefficients of expansions in Fourier-Bessel series (see T7, p. 221), we have

$$\frac{v\gamma_n}{a} c_n = \frac{2}{a^2 J_1^2(\gamma_n)} \int_0^a f(r) J_0\left(\frac{\gamma_n r}{a}\right) r \, dr = \frac{2P J_1(\gamma_n \varepsilon/a)}{\pi \varepsilon \rho a \gamma_n J_1^2(\gamma_n)},$$

which implies the answer on p. 69.

124. We want the solution of the equation

$$\frac{1}{r}\frac{\partial}{\partial r}\left\{r\frac{\partial}{\partial r}\left[\frac{1}{r}\frac{\partial}{\partial r}\left(r\frac{\partial u}{\partial r}\right)\right]\right\} + \frac{1}{b^4}\frac{\partial^2 u}{\partial t^2} = 0$$

for transverse oscillations of a plate which satisfies the initial conditions

$$u\big|_{t=0} = f(r), \qquad \frac{\partial u}{\partial t}\bigg|_{t=0} = g(r),$$

and boundary conditions

$$u\big|_{r=a} = 0, \qquad \frac{\partial u}{\partial r}\bigg|_{r=a} = 0.$$

Separating variables, we obtain

$$u(r, t) = R(r)T(t),$$

$$\frac{1}{r}\frac{d}{dr}\left\{r\frac{d}{dr}\left[\frac{1}{r}\frac{d}{dr}\left(r\frac{dR}{dr}\right)\right]\right\} - \lambda R = 0,$$

$$T'' + b^4 \lambda T = 0.$$

The functions $R(r)$ remaining finite at the center of the plate are of the form

$$R(r) = AJ_0(\sqrt[4]{\lambda}\, r) + BI_0(\sqrt[4]{\lambda}\, r).$$

It follows from the boundary conditions

$$R(a) = R'(a) = 0$$

that

$$\lambda = \lambda_n = \frac{\gamma_n^4}{a^4}, \qquad R = R_{\gamma_n}(r),$$

where

$$R_\gamma(r) = I_0(\gamma)J_0\left(\frac{\gamma r}{a}\right) - J_0(\gamma)I_0\left(\frac{\gamma r}{a}\right),$$

and the γ_n are consecutive positive roots of the equation $R_\gamma'(a) = 0$.

The eigenfunctions $R_{\gamma_n}(r)$ are orthogonal on the interval $(0, a)$ with weight r, since

$$\frac{1}{a}(\lambda_m - \lambda_n)\int_0^a rR_nR_m\,dr$$

$$= \left\{R_n\frac{d}{dr}\left[\frac{1}{r}\frac{d}{dr}\left(r\frac{dR_m}{dr}\right)\right] - R_m\frac{d}{dr}\left[\frac{1}{r}\frac{d}{dr}\left(r\frac{dR_n}{dr}\right)\right] + R_m'R_n'' - R_n'R_m''\right\}\bigg|_{r=a} = 0,$$

where we introduce the abbreviation $R_{\gamma_k} \equiv R_k$ and use the boundary conditions for the function $R_n(r)$. The solution of the problem is given by the formula

$$u(r, t) = \sum_{n=1}^\infty \frac{R_n(r)}{\int_0^a R_n^2(\rho)\rho\,d\rho}\left[\cos\frac{\gamma_n^2b^2t}{a^2}\int_0^a f(\rho)R_n(\rho)\rho\,d\rho\right.$$

$$\left. + \frac{a^2}{b^2\gamma_n^2}\sin\frac{\gamma_n^2b^2t}{a^2}\int_0^a g(\rho)R_n(\rho)\rho\,d\rho\right].$$

The value of the integral

$$\int_0^a R_n^2\rho\,d\rho$$

can be found from the relation

$$\int_0^a R_n^2\rho\,d\rho = \frac{a^6}{4}\left\{R_n''^2 - R_n'\frac{d}{dr}\left[\frac{1}{r}\frac{d}{dr}\left(r\frac{dR_n}{dr}\right)\right] - \frac{R_n'R_n''}{\gamma_n} - R_n'R_n'''\right\}\bigg|_{r=a},$$

which takes the form

$$\int_0^a R_n^2(\rho)\rho\,d\rho = \frac{a^6}{4}R_n''^2(a) = a^2I_0^2(\gamma_n)J_0^2(\gamma_n)$$

after some simple calculations.

132. The problem reduces to integration of the equation

$$\frac{1}{r}\frac{\partial}{\partial r}\left(r\frac{\partial u}{\partial r}\right) + \frac{1}{\gamma^2}\frac{\partial^2 u}{\partial\varphi^2} = -\frac{q}{T},$$

with zero boundary conditions

$$u|_{r=a} = u|_{\varphi=0} = u|_{\varphi=\pi} = 0.$$

If we write $u = u_1 + u_2$, where

$$u_1 = -\frac{qr^2}{2T} \sin^2 \varphi,$$

the function u_2 must satisfy Laplace's equation

$$\frac{1}{r} \frac{\partial}{\partial r}\left(r \frac{\partial u_2}{\partial r}\right) + \frac{1}{r^2} \frac{\partial^2 u_2}{\partial \varphi^2} = 0,$$

with the boundary conditions

$$u_2|_{\varphi=0} = u_2|_{\varphi=\pi} = 0, \qquad u_2|_{r=a} = \frac{qa^2}{2T} \sin^2 \varphi.$$

The substitution $u_2(r, \lambda) = R(r)\Phi(\varphi)$ leads to the equations

$$\frac{1}{r} \frac{d}{dr}\left(r \frac{dR}{dr}\right) - \frac{\lambda}{r^2} R = 0, \qquad \frac{d^2\Phi}{d\varphi^2} + \lambda\Phi = 0,$$

with general solutions

$$R = Ar^{\sqrt{\lambda}} + Br^{-\sqrt{\lambda}}, \qquad \Phi = C \cos \sqrt{\lambda}\, \varphi + D \sin \sqrt{\lambda}\, \varphi.$$

By satisfying the homogeneous boundary conditions

$$\Phi(0) = \Phi(\pi) = 0,$$

we obtain the eigenvalues $\lambda_n = n^2$ and the eigenfunctions

$$\Phi_n(\varphi) = \sin n\varphi, \qquad n = 1, 2, \ldots$$

Because of the finiteness of the solution for $r = 0$, the constant B must be set equal to zero.

Thus the function u_2 can be represented as the sum of the series

$$u_2 = \sum_{n=1}^{\infty} A_n \left(\frac{r}{a}\right)^n \sin n\varphi. \tag{10}$$

It follows from the boundary condition for $r = a$ that

$$\frac{qa^2}{2T} \sin^2 \varphi = \sum_{n=1}^{\infty} A_n \sin n\varphi, \qquad 0 < \varphi < \pi,$$

and hence, by the theory of Fourier series,

$$A_n = \frac{qa^2}{\pi T} \int_0^\pi \sin^2 \varphi \sin n\varphi \, d\varphi = \begin{cases} 0, & n = 2k, \\ \dfrac{4qa^2}{\pi T} \dfrac{1}{(2k+1)[4 - (2k+1)^2]}, & n = 2k + 1. \end{cases}$$

Substituting the values of the coefficients A_n into (10), we obtain

$$u(r, \varphi) = -\frac{qr^2 \sin^2 \varphi}{2T} + \frac{4qa^2}{\pi T} \sum_{n=0}^{\infty} \left(\frac{r}{a}\right)^{2n+1} \frac{\sin(2n+1)\varphi}{(2n+1)[4-(2n+1)^2]}.$$

This result can be written in closed form by using the expansions

$$\sum_{n=0}^{\infty} \frac{\rho^{2n+1}}{2n+1} \sin(2n+1)\varphi = \frac{1}{2} \arctan \frac{2\rho \sin \varphi}{1-\rho^2},$$

$$\sum_{n=0}^{\infty} \frac{\rho^{2n+1}}{2n+1} \cos(2n+1)\varphi = \frac{1}{4} \ln \frac{1 + 2\rho \cos \varphi + \rho^2}{1 - 2\rho \cos \varphi + \rho^2},$$

where $|\rho| \leqslant 1$, $|\varphi| < \pi$. After some manipulation, we obtain

$$\sum_{n=1}^{\infty} \frac{\rho^{2n+1} \sin(2n+1)\varphi}{(2n+1)[4-(2n+1)^2]} = \frac{1}{8}\left[1 - \frac{1}{2}\left(\rho^2 + \frac{1}{\rho^2}\right)\cos 2\varphi\right] \arctan \frac{2\rho \sin \varphi}{1-\rho^2}$$

$$+ \frac{1}{32}\left(\frac{1}{\rho^2} - \rho^2\right) \sin 2\varphi \ln \frac{1 + 2\rho \cos \varphi + \rho^2}{1 - 2\rho \cos \varphi + \rho^2} - \frac{1}{8}\left(\frac{1}{\rho} - \rho\right) \sin \varphi,$$

which immediately implies the form of the solution given in the answer on p. 74.

136. We want the solution of the equilibrium equation

$$\Delta^2 u = \frac{q}{D}$$

for a semicircular plate which satisfies the boundary conditions

$$u\big|_{\varphi=0} = \frac{\partial^2 u}{\partial \varphi^2}\bigg|_{\varphi=0} = u\big|_{\varphi=\pi} = \frac{\partial^2 u}{\partial \varphi^2}\bigg|_{\varphi=\pi} = 0, \qquad u\big|_{r=a} = \frac{\partial u}{\partial r}\bigg|_{r=a} = 0.$$

Setting

$$u = \frac{qa^4}{24D}\left(\frac{r^4}{a^4}\sin^4 \varphi - v\right),$$

we find that the function $v(r, \varphi)$ satisfies the homogeneous biharmonic equation $\Delta^2 v = 0$ with the boundary conditions

$$v\big|_{\varphi=0} = \frac{\partial^2 v}{\partial \varphi^2}\bigg|_{\varphi=0} = v\big|_{\varphi=\pi} = \frac{\partial^2 v}{\partial \varphi^2}\bigg|_{\varphi=\pi} = 0, \qquad v\big|_{r=a} = \sin^4 \varphi,$$

$$\frac{\partial v}{\partial r}\bigg|_{r=a} = \frac{4}{a}\sin^4 \varphi.$$

We can separate variables in the biharmonic equation by looking for particular solutions of the form

$$v = v_\mu(r, \varphi) = (A \cos \mu\varphi + B \sin \mu\varphi)R(r).$$

It follows from the boundary conditions for $\varphi = 0$ and $\varphi = \pi$ that

$$A = 0, \quad \mu = n \quad (n = 1, 2, \ldots).$$

Moreover,

$$R(r) = C_n r^n + D_n r^{n+2},$$

since the deflection at the center of the plate is bounded. Therefore, summing particular solutions, we find that

$$v(r, \varphi) = \sum_{n=1}^{\infty} \left[M_n \left(\frac{r}{a} \right)^n + N_n \left(\frac{r}{a} \right)^{n+2} \right] \sin n\varphi. \tag{11}$$

The values of the constants M_n and N_n are determined from the boundary conditions on the arc $r = a$:

$$\sin^4 \varphi = \sum_{n=1}^{\infty} (M_n + N_n) \sin n\varphi,$$

$$\frac{4}{a} \sin^4 \varphi = \sum_{n=1}^{\infty} \left[\frac{nM_n}{a} + \frac{(n+2)N_n}{a} \right] \sin n\varphi, \quad 0 < \varphi < \pi.$$

This gives the system

$$M_n + N_n = \frac{2}{\pi} \int_0^\pi \sin^4 \varphi \sin n\varphi \, d\varphi,$$

$$nM_n + (n+2)N_n = \frac{8}{\pi} \int_0^\pi \sin^4 \varphi \sin n\varphi \, d\varphi.$$

After some simple calculations, we find that

$$J_n = \int_0^\pi \sin^4 \varphi \sin n\varphi \, d\varphi$$

$$= \frac{1 - (-1)^n}{8} \left[\frac{1}{2} \left(\frac{1}{n+4} + \frac{1}{n-4} \right) - 2 \left(\frac{1}{n+2} + \frac{1}{n-2} \right) + \frac{3}{n} \right],$$

which implies

$$M_n = \frac{n-2}{\pi} J_n, \quad N_n = \frac{4-n}{\pi} J_n.$$

Using the expansion

$$\sum_{n=0}^{\infty} \frac{(-1)^n}{2n+1} \rho^{2n+1} = \frac{1}{2} \arctan \frac{2\rho}{1 - \rho^2}, \quad |\rho| < 1,$$

we can sum the series (11) for $\varphi = \pi/2$, thereby expressing the deflection of the axis of symmetry of the plate in closed form in terms of elementary functions.

140. To reduce the problem to a special case of the Neumann problem, we subtract out the velocity potential of the source, by setting $u = u_0 + u_1$,

where

$$u_0 = \frac{m}{4\pi\rho} + \text{const}$$

(ρ is the distance from the source to an arbitrary point of the flow). Then the function u_1 must be a solution of the equation

$$\frac{\partial}{\partial r}\left(r^2 \frac{\partial u_1}{\partial r}\right) + \frac{1}{\sin\theta}\frac{\partial}{\partial\theta}\left(\sin\theta\frac{\partial u_1}{\partial\theta}\right) = 0,$$

which is regular outside the sphere, and satisfies the boundary condition

$$\left.\frac{\partial u_1}{\partial r}\right|_{r=a} = -\left.\frac{\partial u_0}{\partial r}\right|_{r=a}$$

and the condition $u_1|_{r\to\infty} \to 0$ at infinity. Setting $u_1(r, \theta) = R(r)\Theta(\theta)$ and separating variables, we arrive at the equations

$$(r^2 R')' - \lambda R = 0, \qquad \frac{1}{\sin\theta}(\sin\theta \cdot \Theta')' + \lambda\Theta = 0.$$

This equation has finite solutions for $\theta = 0$ and $\theta = \pi$ if and only if

$$\lambda = \lambda_n = n(n+1), \qquad n = 0, 1, 2, \ldots,$$

which determines the eigenvalues of the problem. The corresponding eigenfunctions are

$$\Theta_n(\theta) = P_n(\cos\theta),$$

where $P_n(x)$ is the Legendre polynomial of degree n. Similarly, for $R_n(r)$ we obtain

$$R_n(r) = A_n r^n + B_n r^{-n-1},$$

where $A_n = 0$, because of the condition at infinity.

Thus we find that

$$u_1(r, \theta) = \sum_{n=0}^{\infty} B_n r^{-n-1} P_n(\cos\theta), \tag{12}$$

and to determine the constants B_n, we need only satisfy the boundary condition

$$\frac{m}{4\pi}\left.\frac{\partial(1/\rho)}{\partial r}\right|_{r=a} = \sum_{n=0}^{\infty}(n+1)B_n a^{-n-2} P_n(\cos\theta), \qquad 0 \leqslant \theta \leqslant \pi.$$

In the present case, we can calculate the coefficients B_n in (12) by differentiating the expansion of the generating function

$$\frac{1}{\rho} = \frac{1}{\sqrt{r^2 - 2br\cos\theta + b^2}} = \frac{1}{b}\sum_{n=0}^{\infty}\left(\frac{r}{b}\right)^n P_n(\cos\theta),$$

thereby obtaining

$$B_n = \frac{m}{4\pi}\frac{n}{n+1}\frac{a^{2n+1}}{b^{n+1}}. \tag{13}$$

Substitution of (13) into (12) gives

$$u = \frac{m}{4\pi}\left[\frac{1}{\rho} + \frac{a}{br}\sum_{n=1}^{\infty}\frac{n}{n+1}\left(\frac{a^2}{br}\right)^n P_n(\cos\theta)\right]. \tag{14}$$

This series can be summed by integrating the expansion

$$\frac{r}{\bar{\rho}} = \frac{r}{\sqrt{r^2 - 2\bar{b}r\cos\theta + \bar{b}^2}} = \sum_{n=0}^{\infty}\left(\frac{\bar{b}}{r}\right)^n P_n(\cos\theta), \qquad \bar{b} = \frac{a^2}{b} < a$$

with respect to the parameter $\bar{b}$ from 0 to $\bar{b}$, which leads to the relation

$$\sum_{n=0}^{\infty}\frac{\bar{b}^{n+1}}{n+1}\frac{P_n(\cos\theta)}{r^{n+1}} = \int_0^{\bar{b}}\frac{d\bar{b}}{\sqrt{r^2 - 2\bar{b}r\cos\theta + \bar{b}^2}} = \ln\frac{r(1+\cos\theta)}{\bar{\rho} + r\cos\theta - \bar{b}}. \tag{15}$$

Writing (14) in the form

$$u = \frac{m}{4\pi}\left[\frac{1}{\rho} + \frac{a}{br}\sum_{n=1}^{\infty}\left(\frac{\bar{b}}{r}\right)^n P_n(\cos\theta) - \frac{1}{a}\sum_{n=1}^{\infty}\frac{\bar{b}^{n+1}}{n+1}\frac{P_n(\cos\theta)}{r^{n+1}}\right]$$

and using (15), we arrive at the expression given in the answer on p. 77.

145. The problem reduces to solving the system of equations

$$\frac{\partial^2 T_1}{\partial x^2} = \frac{c_1\rho_1}{k_1}\frac{\partial T_1}{\partial t}, \qquad 0 < x < a_1,$$

$$\frac{\partial^2 T_2}{\partial x^2} = \frac{c_2\rho_2}{k_2}\frac{\partial T_2}{\partial t}, \qquad a_1 < x < a_1 + a_2,$$

with initial condition

$$T\big|_{t=0} = T_0$$

and boundary conditions

$$T_1\big|_{x=0} = T_2\big|_{x=a_1+a_2} = 0, \qquad T_1\big|_{x=a_1} = T_2\big|_{x=a_1},$$

$$k_1\frac{\partial T_1}{\partial x}\bigg|_{x=a_1} = k_2\frac{\partial T_2}{\partial x}\bigg|_{x=a_1}.$$

Application of the Fourier method leads to the expression

$$T(x, t) = \sum_{n=1}^{\infty} C_n e^{-\gamma n^2 t/b_1 a_1^2} X_n(x),$$

for the required temperature distribution, satisfying the homogeneous boundary conditions for arbitrary values of the coefficients C_n, where

$$X_n(x) = \begin{cases} X_n^{(1)}(x) = \sin\dfrac{a_2\sqrt{b_2}\,\gamma_n}{a_1\sqrt{b_1}}\sin\dfrac{\gamma_n x}{a_1}, & 0 \leqslant x \leqslant a_1, \\[4mm] X_n^{(2)}(x) = \sin\gamma_n\sin\dfrac{\sqrt{b_2}\,\gamma_n(a_1 + a_2 - x)}{\sqrt{b_1}\,a_1}, & a_1 \leqslant x \leqslant a_1 + a_2, \end{cases} \tag{16}$$

and the γ_n are consecutive positive roots of the transcendental equation

$$\sqrt{b_2}\, k_2 \tan \gamma + \sqrt{b_1}\, k_1 \tan \frac{a_2\sqrt{b_2}\,\gamma}{a_1\sqrt{b_1}} = 0 \quad . \left(b_i = \frac{c_i \rho_i}{k_i}, \quad i = 1, 2 \right).$$

By the usual procedure, it can be shown that the eigenfunctions $X_n(x)$ are orthogonal on the interval $0 \leqslant x \leqslant a_1 + a_2$ with weight

$$r(x) = \begin{cases} c_1 \rho_1, & 0 \leqslant x < a_1, \\ c_2 \rho_2, & a_1 < x \leqslant a_1 + a_2. \end{cases}$$

Therefore the coefficients C_n can be calculated by using the formula

$$C_n = \frac{c_1 \rho_1 \int_0^{a_1} f(\xi) X_n^{(1)}(\xi)\, d\xi + c_2 \rho_2 \int_{a_1}^{a_1+a_2} f(\xi) X_n^{(2)}(\xi)\, d\xi}{c_1 \rho_1 \int_0^{a_1} [X_n^{(1)}(\xi)]^2\, d\xi + c_2 \rho_2 \int_{a_1}^{a_1+a_2} [X_n^{(2)}(\xi)]^2\, d\xi}. \tag{17}$$

Substituting from (16), we find that the denominator of (17) equals

$$\frac{1}{2}\left[a_1 c_1 \rho_1 \sin^2 \frac{a_2\sqrt{b_2}\,\gamma}{a_1\sqrt{b_1}} + a_2 c_2 \rho_2 \sin^2 \gamma_n \right].$$

Then setting $f(\xi) = T_0$ in (17) and making some simple calculations, we arrive at the answer on p. 79.

148. We want the solution of the equation

$$\frac{1}{r}\frac{\partial}{\partial r}\left(r \frac{\partial T}{\partial r} \right) = \frac{\partial T}{\partial \tau}$$

satisfying the initial condition

$$T|_{\tau=0} = f(r)$$

and the boundary condition

$$\left.\frac{\partial T}{\partial r}\right|_{r=a} = 0.$$

Setting $T(r, \tau) = R(r)\Theta(\tau)$ and separating variables, we obtain

$$\frac{1}{r}(rR')' + \lambda R = 0, \qquad \Theta' + \lambda \Theta = 0.$$

Then

$$R = J_0(\sqrt{\lambda}\, r)$$

is the solution of the first equation which is finite on the axis of the cylinder. From the boundary condition $R'(a) = 0$, we find the eigenvalues

$$\lambda = \lambda_n = \frac{\gamma_n^2}{a^2}$$

and corresponding eigenfunctions

$$R = R_n = J_0\left(\frac{\gamma_n r}{a}\right),$$

where $\gamma_0 = 0, \gamma_1, \ldots, \gamma_n, \ldots$ are consecutive nonnegative roots of the equation $J_1(\gamma) = 0$. The general solution of the second equation is

$$\Theta_n = c_n e^{-\gamma_n^2 \tau/a^2},$$

and the expression

$$T(r, \tau) = \sum_{n=0}^{\infty} c_n e^{-\gamma_n^2 \tau/a^2} J_0\left(\frac{\gamma_n r}{a}\right),$$

obtained by summation of particular solutions, satisfies all the conditions of the problem, except the initial condition. Since the eigenfunctions are orthogonal with weight r on the interval $(0, a)$, it follows from the initial condition that the coefficients c_n are given by

$$c_n = \frac{\int_0^a f(\rho) J_0\left(\frac{\gamma_n \rho}{a}\right) \rho \, d\rho}{\int_0^a J_0^2\left(\frac{\gamma_n \rho}{a}\right) \rho \, d\rho} = \frac{2}{a^2 J_0^2(\gamma_n)} \int_0^a f(\rho) J_0\left(\frac{\gamma_n \rho}{a}\right) \rho \, d\rho.$$

153. The problem reduces to integration of the differential equation

$$\frac{1}{r^2} \frac{\partial}{\partial r}\left(r^2 \frac{\partial T}{\partial r}\right) = \frac{\partial T}{\partial \tau},$$

with initial condition

$$T|_{\tau=0} = f(r)$$

and boundary condition

$$T|_{r=a} = 0.$$

Setting $T(r, \tau) = R(r)\Theta(\tau)$ and separating variables, we obtain the equations

$$\frac{1}{r^2}(r^2 R')' + \lambda R = 0, \qquad \Theta' + \lambda\Theta = 0,$$

whose general solutions are

$$R = A \frac{\sin\sqrt{\lambda}\,r}{r} + B \frac{\cos\sqrt{\lambda}\,r}{r}, \qquad \Theta = Ce^{-\lambda\tau}.$$

From the condition that R be finite at the center of the sphere, we find that $B = 0$, while the boundary condition $R(a) = 0$ leads to the eigenvalues

$$\lambda = \lambda_n = \frac{n^2\pi^2}{a^2}$$

and corresponding eigenfunctions

$$R = R_n = \frac{1}{r} \sin \frac{n\pi r}{a}.$$

Summing particular solutions, we obtain

$$T(r, \tau) = \frac{1}{r} \sum_{n=1}^{\infty} c_n e^{-n^2\pi^2\tau/a^2} \sin \frac{n\pi r}{a}.$$

The coefficients c_n must be determined from the initial condition

$$T\big|_{\tau=0} = f(r) = \frac{1}{r} \sum_{n=1}^{\infty} c_n \sin \frac{n\pi r}{r}, \qquad 0 < r < a,$$

which, by the theory of Fourier series, implies

$$c_n = \frac{2}{a} \int_0^a f(\rho) \sin \frac{n\pi\rho}{a} \, \rho \, d\rho.$$

This leads at once to the answer on p. 81.

169. This temperature distribution problem leads to integration of Laplace's equation

$$\frac{1}{r}\frac{\partial}{\partial r}\left(r\frac{\partial T}{\partial r}\right) + \frac{\partial^2 T}{\partial z^2} = 0,$$

with boundary conditions

$$T\big|_{r=a} = T_0, \qquad \left(\frac{\partial T}{\partial z} \pm hT\right)\bigg|_{z=\pm l/2} = 0.$$

Writing $T(r, z) = R(r)Z(z)$ and separating variables, we obtain the ordinary differential equations

$$\frac{1}{r}(rR')' - \lambda R = 0, \qquad Z'' + \lambda Z = 0,$$

with general solutions

$$R = AI_0(\sqrt{\lambda}\,r) + BK_0(\sqrt{\lambda}\,r), \qquad Z = C\cos\sqrt{\lambda}\,z + D\sin\sqrt{\lambda}\,z.$$

The constant B equals zero because of the requirements that the temperature be finite on the axis of the cylinder. The boundary conditions

$$(Z' \pm hZ)\big|_{z=\pm l/2} = 0$$

lead to the eigenvalues

$$\lambda = \lambda_n = \frac{4\gamma_n^2}{l^2}$$

and eigenfunctions

$$Z = Z_n = \cos \frac{2\gamma_n z}{l},$$

where the γ_n are consecutive positive roots of the equation

$$\tan \gamma = \frac{hl}{2\gamma}.$$

The expression

$$T(r, z) = \sum_{n=1}^{\infty} c_n I_0\left(\frac{2\gamma_n r}{l}\right) \cos \frac{2\gamma_n z}{l} \qquad (18)$$

satisfies Laplace's equation and the boundary conditions on the ends of the cylinder. To determine the constants c_n, we use the boundary condition on the lateral surface

$$T\big|_{r=a} = T_0 = \sum_{n=1}^{\infty} c_n I_0\left(\frac{2\gamma_n a}{l}\right) \cos \frac{2\gamma_n z}{l}, \qquad -\frac{l}{2} < z < \frac{l}{2}.$$

Because of the orthogonality of the eigenfunctions, this gives

$$c_n I_0\left(\frac{2\gamma_n a}{l}\right) = \frac{T_0 \int_0^{l/2} \cos \frac{2\gamma_n z}{l} \, dz}{\int_0^{l/2} \cos^2 \frac{2\gamma_n z}{l} \, dz} = \frac{2T_0 \sin \gamma_n}{\gamma_n + \sin \gamma_n \cos \gamma_n}. \qquad (19)$$

Substituting (19) into (18), we obtain the answer on p. 87.

175. We have to integrate the system of differential equations

$$\frac{1}{r} \frac{\partial}{\partial r}\left(r \frac{\partial T_1}{\partial r}\right) + \frac{\partial^2 T_1}{\partial z^2} = 0, \qquad -h_1 < z < 0,$$

$$\frac{1}{r} \frac{\partial}{\partial r}\left(r \frac{\partial T_2}{\partial r}\right) + \frac{\partial^2 T_2}{\partial z^2} = 0, \qquad 0 < z < h_2,$$

with boundary conditions

$$T\big|_{r=a} = T_0, \qquad T_1\big|_{z=-h_1} = T_2\big|_{z=h_2} = 0,$$

$$T_1\big|_{z=0} = T_2\big|_{z=0} = 0, \qquad k_1 \frac{\partial T_1}{dz}\bigg|_{z=0} = k_2 \frac{\partial T_2}{\partial z}\bigg|_{z=0}$$

Application of the Fourier method leads to the expression

$$T(r, z) = \sum_{n=1}^{\infty} c_n Z_n(z) I_0\left(\frac{\gamma_n r}{h_1}\right) \qquad (20)$$

for the required temperature distribution, satisfying all the conditions of the problem except the boundary condition on the lateral surface, where

$$
Z_n(z) = \begin{cases} Z_n^{(1)}(z) = \sin \dfrac{\gamma_n h_2}{h_1} \sin \dfrac{\gamma_n(z + h_1)}{h_1}, & -h_1 \leqslant z \leqslant 0, \\[3mm] Z_n^{(2)}(z) = \sin \gamma_n \sin \dfrac{\gamma_n(h_2 - z)}{h_1}, & 0 \leqslant z \leqslant h_2, \end{cases}
$$

and the γ_n are consecutive positive roots of the equation

$$
\tan \gamma + \frac{k_1}{k_2} \tan \frac{\gamma h_2}{h_1} = 0.
$$

The eigenfunctions $Z_n(z)$ are orthogonal on the interval $(-h_1, h_2)$ with weight

$$
r(z) = \begin{cases} k_1, & -h_1 \leqslant z < 0, \\ k_2, & 0 < z \leqslant h_2. \end{cases}
$$

To see this, we multiply the equations

$$
Z_n'' + \frac{\gamma_n^2}{a^2} Z_n = 0, \qquad Z_m'' + \frac{\gamma_m^2}{a^2} Z_m = 0
$$

by $r(z)Z_m(z)$ and $r(z)Z_n(z)$, respectively, subtract the results from each other, and then integrate with respect to z from $-h_1$ to h_2. This gives

$$
\frac{\gamma_n^2 - \gamma_m^2}{a^2} \int_{-h_1}^{h} Z_m Z_n r \, dz = \int_{-h_1}^{h_2} r[Z_m'' Z_n - Z_n'' Z_m] \, dz
$$
$$
= k_1(Z_m^{(1)\prime} Z_n^{(1)} - Z_n^{(1)\prime} Z_m^{(1)}) \Big|_{-h_1}^{0} + k_2(Z_m^{(2)\prime} Z_n^{(2)} - Z_n^{(2)\prime} Z_m^{(2)}) \Big|_{0}^{h_2} = 0,
$$

where we have used the boundary conditions

$$
Z_p^{(1)}(-h_1) = Z_p^{(2)}(h_2) = 0, \quad Z_p^{(1)}(0) = Z_p^{(2)}(0), \quad k_1 Z_p^{(1)\prime}(0) = k_2 Z_p^{(2)\prime}(0).
$$

The orthogonality of the functions $Z_n(z)$, together with the condition $T|_{r=a} = T_0$, implies

$$
I_0\!\left(\frac{\gamma_n a}{h_1}\right) c_n = T_0 \frac{k_1 \int_{-h_1}^{0} Z_n^{(1)}(\zeta) \, d\zeta + k_2 \int_{0}^{h_2} Z_n^{(2)}(\zeta) \, d\zeta}{k_1 \int_{-h_1}^{0} [Z_n^{(1)}(\zeta)]^2 \, d\zeta + k_2 \int_{0}^{h_2} [Z_n^{(2)}(\zeta)]^2 \, d\zeta}. \tag{21}
$$

Evaluating the integrals in (21), we obtain

$$
c_n = \frac{2T_0 \tan \gamma_n \left(\cos \gamma_n - \cos \dfrac{\gamma_n h_2}{h_1}\right)}{\gamma_n I_0\!\left(\dfrac{\gamma_n a}{h_1}\right) \left(\dfrac{k_1}{k_2} \sin^2 \dfrac{\gamma_n h_2}{h_1} + \dfrac{h_2}{h_1} \sin^2 \gamma_n\right)}.
$$

Substitution of these coefficients into (20) gives the answer on p. 89.

176. The problem reduces to integration of Laplace's equation

$$\frac{\partial}{\partial r}\left(r^2 \frac{\partial T}{\partial r}\right) + \frac{1}{\sin \theta} \frac{\partial}{\partial \theta}\left(\sin \theta \frac{\partial T}{\partial \theta}\right) = 0,$$

with boundary condition

$$T|_{r=a} = f(\theta) = \begin{cases} T_0, & 0 < \theta < \alpha, \\ 0, & \alpha < \theta \leqslant \pi. \end{cases}$$

The required harmonic function is constructed as a series

$$T(r, \theta) = \sum_{n=0}^{\infty} c_n \left(\frac{r}{a}\right)^n P_n(\cos \theta), \tag{22}$$

where $P_n(x)$ is the Legendre polynomial of degree n. Because of the boundary condition, the coefficients c_n must coincide with the expansion coefficients of the function $f(\theta)$ with respect to the Legendre polynomials, i.e.,

$$f(\theta) = \sum_{n=1}^{\infty} c_n P_n(\cos \theta), \qquad 0 \leqslant \theta \leqslant \pi,$$

which implies

$$c_n = \frac{2n+1}{2} \int_0^\pi f(\theta) P_n(\cos \theta) \sin \theta \, d\theta = \frac{2n+1}{2} T_0 \int_{\cos \alpha}^1 P_n(x) \, dx.$$

For $n = 0$ we immediately find

$$\int_{\cos \alpha}^1 P_0(x) \, dx = 1 - \cos \alpha.$$

For arbitrary n, we use the recurrence formula

$$(2n + 1)P_n(x) = P'_{n+1}(x) - P'_{n-1}(x),$$

obtaining

$$\int_{\cos \alpha}^1 P_n(x) \, dx = \frac{1}{2n+1} [P_{n-1}(\cos \alpha) - P_{n+1}(\cos \alpha)], \qquad n = 1, 2, \ldots.$$

Substituting the values of c_n obtained in this way into (22), we find the answer on p. 90.

190. To solve the problem, we find it convenient to assume that the current J is uniformly distributed with density $J/2\varepsilon h$ over a small section $|y| < \varepsilon$ of the sheet (where h is the thickness of the sheet), afterwards taking the limit as $\varepsilon \to 0$. Then the problem reduces to integration of the two-dimensional Laplace equation

$$\frac{\partial^2 u}{\partial x^2} + \frac{\partial^2 u}{\partial y^2} = 0,$$

with boundary conditions

$$\frac{\partial u}{\partial y}\bigg|_{y=\pm b} = 0, \quad \frac{\partial u}{\partial x}\bigg|_{x=\pm a} = f(y) = \begin{cases} -\dfrac{J}{2\varepsilon\sigma h}, & |y| < \varepsilon, \\ 0, & |y| > \varepsilon. \end{cases}$$

Application of the Fourier method leads to an expansion of the form

$$u = c_0 x + \sum_{n=1}^{\infty} c_n \sinh \frac{n\pi x}{b} \cos \frac{n\pi y}{b} + \text{const,}$$

whose coefficients are calculated from the formulas

$$c_0 = \frac{1}{b} \int_0^b f(y)\, dy, \quad c_n = \frac{2}{n\pi \cosh(n\pi a/b)} \int_0^b f(y) \cos \frac{n\pi y}{b}\, dy$$

$$(n = 1, 2, \ldots).$$

Substituting for $f(y)$, evaluating the integrals and taking the limit as $\varepsilon \to 0$, we find the answer on p. 95.

192. The potential of the surface current must satisfy Laplace's equation

$$\frac{1}{a^2} \frac{\partial^2 u}{\partial \varphi^2} + \frac{\partial^2 u}{\partial z^2} = 0 \quad (-\pi < \varphi \leqslant \pi, \quad -l < z < l)$$

(cf. Prob. 21). To formulate boundary conditions for the problem, we first assume that the current J is distributed with constant density over the section $|\varphi| < \varepsilon$, $z = \pm l$, so that

$$\frac{\partial u}{\partial z}\bigg|_{z=\pm l} = f(\varphi) = \begin{cases} \dfrac{J}{2a\varepsilon h\sigma}, & |\varphi| < \varepsilon, \\ 0, & |\varphi| > \varepsilon. \end{cases}$$

Writing

$$u_n = \Phi_n(z)Z_n(z), \quad n = 0, 1, 2, \ldots,$$

and separating variables, we obtain

$$\Phi_n(\varphi) = A_n \cos n\varphi + B_n \sin n\varphi,$$

$$Z_n(z) = C_n \cosh \frac{nz}{a} + D_n \sinh \frac{nz}{a}, \quad Z_0 = C_0 + D_0 z,$$

where we use the required periodicity of the solution in the angular variable φ. Because of the symmetry with respect to the plane $\varphi = 0$, the coefficients B_n vanish, and hence the solution of the problem can be written as a series

$$u(\varphi, z) = \sum_{n=1}^{\infty} \left(M_n \cosh \frac{nz}{a} + N_n \sinh \frac{nz}{a} \right) \cos n\varphi + N_0 z + \text{const.}$$

The boundary conditions give

$$M_n = 0, \quad f(\varphi) = \sum_{n=1}^{\infty} \frac{n}{a} N_n \cosh \frac{nl}{a} \cos n\varphi + N_0,$$

i.e.,

$$N_0 = \frac{1}{\pi} \int_0^\pi f(\varphi) \, d\varphi = \frac{J}{2\pi a h \sigma},$$

$$\frac{n N_n}{a} \cosh \frac{nl}{a} = \frac{2}{\pi} \int_0^\pi f(\varphi) \cos n\varphi \, d\varphi = \frac{2}{\pi} \frac{J}{2 a \varepsilon h \sigma} \frac{\sin n\varepsilon}{n}.$$

Taking the limit as $\varepsilon \to 0$, we obtain

$$N_n = \frac{J}{\pi n h \sigma \cosh (nl/a)},$$

which leads to the answer on p. 95.

196. The problem reduces to integration of the system of differential equations

$$\frac{1}{r} \frac{\partial}{\partial r} \left(r \frac{\partial u_i}{\partial r} \right) + \frac{1}{r^2} \frac{\partial^2 u_i}{\partial \varphi^2} = 0, \quad \begin{pmatrix} b < r < \infty, & i = 1, \\ a < r < b, & i = 2, \\ 0 < r < a, & i = 3, \end{pmatrix} \quad -\pi < \varphi \leqslant \pi$$

for the potential of the magnetic field, with boundary conditions

$$u_1|_{r\to\infty} \to H_0 x + \text{const},$$

$$\frac{\partial u_1}{\partial \varphi}\bigg|_{r=b} = \frac{\partial u_2}{\partial \varphi}\bigg|_{r=b}, \qquad \frac{\partial u_1}{\partial r}\bigg|_{r=b} = \mu \frac{\partial u_2}{\partial r}\bigg|_{r=b},$$

$$\frac{\partial u_2}{\partial \varphi}\bigg|_{r=a} = \frac{\partial u_3}{\partial \varphi}\bigg|_{r=a}, \qquad \mu \frac{\partial u_2}{\partial r}\bigg|_{r=a} = \frac{\partial u_3}{\partial r}\bigg|_{r=a}.$$

Bearing in mind that the required potentials are even periodic functions of the variable φ, we represent the functions u_i as series

$$u_1 = H_0 r \cos \varphi + \sum_{n=1}^{\infty} A_n r^{-n} \cos n\varphi,$$

$$u_2 = C_0 \ln r + \sum_{n=1}^{\infty} (B_n r^n + C_n r^{-n}) \cos n\varphi,$$

$$u_3 = \sum_{n=1}^{\infty} D_n r^n \cos n\varphi$$

(where arbitrary additive constants have been omitted). Because of the boundary conditions, the constants C_0 and C_n ($n \geqslant 2$) vanish. To determine the remaining constants, we use the system of equations

$$A_1 - b^2 B_1 - C_1 = -H_0 b^2,$$

$$A_1 + \mu b^2 B_1 - \mu C_1 = H_0 b^2,$$

$$a^2 B_1 + C_1 - a^2 D_1 = 0,$$

$$\mu a^2 B_1 - \mu C_1 - a^2 D_1 = 0.$$

It follows that

$$A_1 = -\frac{H_0 b^2 (\mu^2 - 1)(b^2 - a^2)}{\Delta}, \qquad C_1 = \frac{2H_0 a^2 b^2 (\mu - 1)}{\Delta},$$

$$B_1 = \frac{2H_0 b^2 (\mu + 1)}{\Delta}, \qquad D_1 = \frac{4H_0 b^2 \mu}{\Delta},$$

where

$$\Delta = b^2 (\mu + 1)^2 - a^2 (\mu - 1)^2.$$

Substituting these coefficients into the series for the u_i, we obtain the answer on p. 97.

201. In this problem, it is convenient to characterize the magnetic field by a vector potential which in each of the media (air, magnetic material, air) has a single component

$$A_\varphi^{(i)} = A^{(i)}(r, \theta), \qquad i = 1, 2, 3.$$

Setting

$$A^{(1)} = A_0 + A_1, \quad A^{(2)} = A_2, \quad A^{(3)} = A_3,$$

where

$$A_0(r, \theta) = \frac{2Jr_0}{c} \int_0^\pi \frac{\cos \varphi \, d\varphi}{\sqrt{r^2 - 2rr_0 \sin \theta \cos \varphi + r_0^2}} \qquad (23)$$

is the vector potential of the loop, we reduce the problem to determination of the functions A_i satisfying the differential equations[4]

$$\frac{\partial}{\partial r}\left(r^2 \frac{\partial A_1}{\partial r}\right) + \frac{1}{\sin \theta} \frac{\partial}{\partial \theta}\left(\sin \theta \frac{\partial A_1}{\partial \theta}\right) - \frac{A_1}{\sin^2 \theta} = 0, \qquad 0 < r < a,$$

$$\frac{\partial}{\partial r}\left(r^2 \frac{\partial A_2}{\partial r}\right) + \frac{1}{\sin \theta} \frac{\partial}{\partial \theta}\left(\sin \theta \frac{\partial A_2}{\partial \theta}\right) - \frac{A_2}{\sin^2 \theta} = 0, \qquad a < r < b,$$

$$\frac{\partial}{\partial r}\left(r^2 \frac{\partial A_3}{\partial r}\right) + \frac{1}{\sin \theta} \frac{\partial}{\partial \theta}\left(\sin \theta \frac{\partial A_3}{\partial \theta}\right) - \frac{A_3}{\sin^2 \theta} = 0, \qquad b < r < \infty$$

---

[4] Note that

$$(\Delta \mathbf{A})_\varphi = \Delta A_\varphi - \frac{A_\varphi}{r^2 \sin^2 \theta}.$$

and the boundary conditions

$$A^{(1)}\big|_{r=a} = A^{(2)}\big|_{r=a}, \qquad \frac{\partial}{\partial r}(rA^{(1)})\big|_{r=a} = \frac{1}{\mu}\frac{\partial}{\partial r}(rA^{(2)})\big|_{r=a},$$

$$A^{(2)}\big|_{r=b} = A^{(3)}\big|_{r=b}, \quad A^{(3)}\big|_{r\to\infty} \to 0, \quad \frac{1}{\mu}\frac{\partial}{\partial r}(rA^{(2)})\big|_{r=b} = \frac{\partial}{\partial r}(rA^{(3)})\big|_{r=b}.$$

Looking for solutions of these equations of the form $A = R(r)\Theta(\theta)$, we obtain the equations

$$(r^2 R')' - \lambda R = 0, \qquad \frac{1}{\sin\theta}(\sin\theta \cdot \Theta')' + \left(\lambda - \frac{1}{\sin^2\theta}\right)\Theta = 0. \quad (24)$$

The permissible values of the parameter λ are determined from the condition that the second of the equations (24) have solutions which are regular in the closed interval $0 \leqslant \theta \leqslant \pi$. This requirement leads to the eigenvalues and corresponding eigenfunctions

$$\lambda = \lambda_n = n(n+1), \quad \Theta = \Theta_n = P_n^1(\cos\theta) \qquad (n = 0, 1, 2, \ldots),$$

where the $P_n^1(x)$ are associated Legendre functions of the first kind. The general solution of the first of the equations (24) is

$$R = R_n = Mr^n + Nr^{-n-1}.$$

Taking account of the behavior of the functions A_i near $r = 0$ and $r = \infty$, we find that they can be represented as series of the form

$$A_1(r, \theta) = \sum_{n=0}^{\infty} A_n r^n P_n^1(\cos\theta),$$

$$A_2(r, \theta) = \sum_{n=0}^{\infty} (B_n r^n + C_n r^{-n-1}) P_n^1(\cos\theta),$$

$$A_3(r, \theta) = \sum_{n=0}^{\infty} D_n r^{-n-1} P_n^1(\cos\theta).$$

The vector potential of the source can also be represented as an expansion in terms of Legendre functions, by starting from the formula

$$\frac{1}{\sqrt{r^2 - 2rr_0\sin\theta\cos\varphi + r_0^2}} = \frac{1}{r} + \sum_{n=1}^{\infty} P_n(\sin\theta\cos\varphi)\left(\frac{r_0}{r}\right)^n, \qquad r > r_0. \quad (25)$$

Using the addition formula for spherical harmonics

$$P_n(\sin\theta\cos\varphi) = P_n(0)P_n(\cos\theta) + 2\sum_{m=1}^{n} \frac{\Gamma(n-m+1)}{\Gamma(n+m+1)}$$
$$\times P_n^m(0)P_n^m(\cos\theta)\cos m\varphi,$$

and substituting (25) into the integral (23) for A_0, we find after some simple calculations that

$$A_0\big|_{r>r_0} = \frac{2\pi J}{c} \sum_{n=0}^{\infty} \frac{1}{n(n+1)} P_n^1(0) P_n^1(\cos\theta)\left(\frac{r_0}{r}\right)^{n+1}.$$

Then the boundary conditions lead to the following system of equations for determining the coefficients A_n, B_n, C_n and D_n:

$$a^n A_n - a^n B_n - a^{-n-1} C_n = -\frac{2\pi J}{c}\frac{P_n^1(0)}{n(n+1)}\left(\frac{r_0}{a}\right)^{n+1},$$

$$(n+1)a^n A_n - \frac{n+1}{\mu}a^n B_n + \frac{n}{\mu}a^{-n-1}C_n = \frac{2\pi J}{c}\frac{P_n^1(0)}{n+1}\left(\frac{r_0}{a}\right)^{n+1},$$

$$b^n B_n + b^{-n-1}C_n - b^{-n-1}D_n = 0,$$

$$\frac{n+1}{\mu}b^n B_n - \frac{n}{\mu}b^{-n-1}C_n + nb^{-n-1}D_n = 0.$$

Solving this system we obtain

$$D_n = \frac{2\pi J\mu}{cn(n+1)}$$

$$\times \frac{(2n+1)^2 P_n^1(0) r_0^{n+1}}{[n(\mu+1)+1][n(\mu+1)+\mu] - (a/b)^{2n+1}n(n+1)(\mu-1)^2},$$

which leads to the solution in the region outside the shield given on p. 99, if we bear in mind that $P_{2k}^1(0) = 0$.

206. The magnetic field in the spherical resonator has only a φ-component with complex amplitude $H_\varphi = H(r,\theta)$. Writing $H = H_0 + H_1$, where[5]

$$H_0 = \frac{P\sin\theta}{cr^2}(1 + ikr)e^{-ikr}$$

is the magnetic field of the source, we find that H_1 satisfies the equation

$$\Delta H_1 + \left(k^2 - \frac{1}{r^2\sin^2\theta}\right)H_1 = 0.$$

Next we introduce a new unknown function $u = u(r,\theta)$ such that

$$H_1 = \frac{\partial u}{\partial\theta}.$$

---

[5] This expression can be obtained from the relations

$$H_0 = (\operatorname{curl} \mathbf{A}^{(0)})_\varphi, \quad A_r^{(0)} = \frac{P}{c}\frac{e^{-ikr}}{r}\cos\theta, \quad A_\theta^{(0)} = -\frac{P}{c}\frac{e^{-ikr}}{r}\sin\theta.$$

Then u is the solution of Helmholtz's equation

$$\frac{1}{r^2}\frac{\partial}{\partial r}\left(r^2\frac{\partial u}{\partial r}\right) + \frac{1}{r^2\sin\theta}\frac{\partial}{\partial\theta}\left(\sin\theta\frac{\partial u}{\partial\theta}\right) + k^2 u = 0$$

which is regular inside the sphere. Since the tangential component of the electric field

$$E_\theta = \frac{i}{kr}\frac{\partial}{\partial r}(rH)$$

must vanish on the surface of the sphere, it follows that

$$\frac{\partial}{\partial r}\left(r\frac{\partial u}{\partial\theta}\right)\Bigg|_{r=a} = -\frac{\partial}{\partial r}(rH_\theta)\Bigg|_{r=a}.$$

Using the Fourier method to solve the differential equation for u, we find that

$$u(r,\theta) = \frac{1}{\sqrt{r}}\sum_{n=0}^{\infty} c_n J_{n+\frac{1}{2}}(kr)P_n(\cos\theta), \tag{26}$$

in terms of the Legendre polynomials $P_n(x)$ and the Bessel functions of half-integral order $J_{n+\frac{1}{2}}(x)$. Using the boundary condition and the familiar relation

$$J_{3/2}(z) = \sqrt{\frac{2}{\pi z}}\left(\frac{\sin z}{z} - \cos z\right),$$

we find that the coefficients c_n equal

$$c_n\big|_{n\neq 1} = 0, \qquad c_1 = \sqrt{\frac{\pi k}{2}}\frac{Pk}{c}e^{-ika}\frac{1 + ika - k^2 a^2}{(1 - k^2 a^2)\sin ka - ka\cos ka}.$$

Substituting these values of c_n into (26) and differentiating with respect to θ, we arrive at the expression for $H_\varphi = H(r,\theta)$ given on p. 101.

210. The problem reduces to solving the equation

$$\frac{\partial^2 u}{\partial x^2} - \frac{1}{v^2}\frac{\partial^2 u}{\partial t^2} = -\frac{q(x,t)}{T} \tag{27}$$

of the vibrating string, with zero initial conditions

$$u\big|_{t=0} = \frac{\partial u}{\partial t}\Bigg|_{t=0} = 0$$

and homogeneous boundary conditions

$$u\big|_{x=0} = u\big|_{x=l} = 0$$

of the first kind. We look for a solution in the form of an expansion

$$u(x, t) = \sum_{n=1}^{\infty} \frac{\bar{u}_n}{\int_0^l X_n^2(x)\, dx} X_n(x),$$

with respect to the eigenfunctions $X_n(x)$ of the corresponding homogeneous problem, where the weight r equals 1 and

$$\bar{u}_n = \int_0^l u X_n(x)\, dx.$$

The functions $X_n(x)$ are the nontrivial solutions of the equation

$$X'' + \lambda X = 0$$

satisfying the homogeneous boundary conditions

$$X(0) = X(l) = 0.$$

Such solutions exist for

$$\lambda = \lambda_n = \frac{n^2 \pi^2}{l^2} \qquad n = 1, 2 \ldots,$$

and are of the form

$$X = X_n(x) = \sin \frac{n\pi x}{l}.$$

To determine the coefficients $\bar{u}_n$, we multiply (27) by $X_n(x)$ and integrate with respect to x from 0 to l.[6] Integrating by parts twice and taking account of the boundary conditions, we obtain

$$\bar{u}_n'' + \left(\frac{n\pi v}{l}\right)^2 \bar{u}_n = \frac{v^2}{T} \int_0^l q(x, t) \sin \frac{n\pi x}{l}\, dx.$$

The solution of this equation can be found by variation of constants:

$$\bar{u}_n = A_n \cos \frac{n\pi v t}{l} + B_n \sin \frac{n\pi v t}{l} + \frac{vl}{n\pi T} \int_0^t \sin \frac{n\pi v(t - \tau)}{l}\, d\tau \int_0^l q(\xi, \tau) \sin \frac{n\pi \xi}{l}\, d\xi.$$

To calculate the constants A_n and B_n, we use the initial conditions for the function $\bar{u}_n$, which are obtained by multiplying the original initial conditions by $X_n(x)$ and integrating with respect to x from 0 to l. The result is

$$\bar{u}_n(0) = \bar{u}_n'(0) = 0,$$

which implies $A_n = B_n = 0$. In this way, we arrive at the answer on p. 108.

---

[6] In the interest of using a unified approach, we follow the general scheme on p. 105. For problems of the type under consideration, this method is entirely equivalent to that described on p. 104.

217. We have to solve the inhomogeneous equation

$$\frac{\partial^4 u}{\partial x^4} + \frac{1}{a^4} \frac{\partial^2 u}{\partial t^2} = \frac{q \sin \omega t}{EJ} \tag{28}$$

for transverse oscillations of the beam, with zero initial conditions

$$u\big|_{t=0} = \frac{\partial u}{\partial t}\bigg|_{t=0} = 0$$

and homogeneous boundary conditions

$$u\big|_{x=\pm l} = \frac{\partial u}{\partial x}\bigg|_{x=\pm l} = 0.$$

A feature of this problem is that it involves an expansion in terms of eigenfunctions of a fourth-order differential operator. Following the usual method, we represent the solution as an expansion

$$u(x, t) = \sum_{n=1}^{\infty} \frac{\bar{u}_n}{\int_{-l}^{l} r X_n^2(x)\, dx} X_n(x)$$

with respect to the eigenfunctions $X_n(x)$ of the homogeneous problem, where

$$\bar{u}_n = \int_{-l}^{l} r u X_n(x)\, dx.$$

In the present case, the weight $r = 1$, and the functions $X_n(x)$ are the solutions of the equation

$$X^{(iv)} - \lambda X = 0$$

satisfying the boundary conditions

$$X(\pm l) = X'(\pm l) = 0.$$

Simple calculations show that[7]

$$\lambda = \lambda_n = \frac{\gamma_n^4}{l^4}, \qquad n = 1, 2, \ldots,$$

$$X_n(x) = \cosh \gamma_n \cos \frac{\gamma_n x}{l} - \cos \gamma_n \cosh \frac{\gamma_n x}{l},$$

where the γ_n are consecutive positive roots of the equation

$$\tan \gamma + \tanh \gamma = 0.$$

To determine the functions $\bar{u}_n$, we multiply (28) by $X_n(x)$ and integrate with respect to x from $-l$ to l. Integrating by parts four times and taking account of the boundary conditions we obtain

$$\bar{u}_n'' + \left(\frac{\gamma_n a}{l}\right)^4 \bar{u}_n = \frac{qa^4}{EJ} \sin \omega t \int_{-l}^{+l} X_n(x)\, dx.$$

---

[7] Concerning the orthogonality of the functions $X_n(x)$, see the solution of Prob. 118.

The solution of this equation satisfying the zero initial conditions

$$\bar{u}_n(0) = \bar{u}'_n(0) = 0$$

obtained by multiplying the original initial conditions by $X_n(x)$ and integrating with respect to x from $-l$ to l, is given by

$$\bar{u}_n = \frac{qa^2l^2}{EJ\gamma_n^2} \frac{\omega \sin \dfrac{a^2\gamma_n^2 t}{l^2} - \dfrac{\gamma_n^2 a^2}{l^2} \sin \omega t}{\omega^2 - \left(\dfrac{a^2\gamma_n^2}{l^2}\right)^2} \int_{-l}^{l} X_n(x)\, dx.$$

The final form of the solution, as given in the answer on p. 110, is found by taking account of the easily verified formulas

$$\int_{-l}^{l} X_n(x)\, dx = \frac{4l \sin \gamma_n \cosh \gamma_n}{\gamma_n},$$

$$\int_{-l}^{l} X_n^2(x)\, dx = \frac{l}{2} X_n''^2(l) = 2l \cosh^2 \gamma_n \cos^2 \gamma_n.$$

222. The problem reduces to integrating the equation

$$\frac{\partial^4 u}{\partial x^4} + \frac{1}{a^4} \frac{\partial^2 u}{\partial t^2} = 0 \qquad (29)$$

for the oscillating beam, with zero initial conditions

$$u\big|_{t=0} = \frac{\partial u}{\partial t}\bigg|_{t=0} = 0$$

and inhomogeneous boundary conditions

$$u\big|_{x=0} = \frac{\partial u}{\partial x}\bigg|_{x=0} = \frac{\partial^2 u}{\partial x^2}\bigg|_{x=l} = 0, \qquad \frac{\partial^3 u}{\partial x^3}\bigg|_{x=l} = -\frac{P(t)}{EJ}.$$

Applying Grinberg's method, we look for a solution in the form of an expansion

$$u(x,\, t) = \sum_{n=1}^{\infty} \frac{\bar{u}_n}{\int_0^l r X_n^2(x)\, dx} X_n(x)$$

with respect to the eigenfunctions $X_n(x)$ of the homogeneous problem, where

$$\bar{u}_n = \int_0^l r u X_n(x)\, dx.$$

Explicit expressions for the functions $X_n(x)$ are obtained by solving the equation

$$X_n^{(iv)} - \lambda X = 0,$$

with homogeneous boundary conditions

$$X(0) = X'(0) = X''(l) = X'''(l) = 0.$$

This gives

$$\lambda = \lambda_n = \frac{\gamma_n^4}{l^4}, \qquad n = 1, 2, \dots,$$

$$X_n(x) = (\sin \gamma_n + \sinh \gamma_n)\left(\cos \frac{\gamma_n x}{l} - \cosh \frac{\gamma_n x}{l}\right)$$

$$- (\cos \gamma_n + \cosh \gamma_n)\left(\sin \frac{\gamma_n x}{l} - \sinh \frac{\gamma_n x}{l}\right),$$

where the γ_n are consecutive positive roots of the equation $\cos \gamma \cosh \gamma + 1 = 0$. The functions $X_n(x)$ are orthogonal on the interval $(0, l)$ with weight $r = 1$ (see the solution to Prob. 118), and the integral of $X_n^2(x)$ is

$$\int_0^l X_n^2(x)\, dx = \frac{l}{4} X_n^2(l) = l(\sinh \gamma_n + \sin \gamma_n)^2.$$

To determine the coefficients $\bar{u}_n$, we multiply (29) by $X_n(x)$ and integrate with respect to x from 0 to l. After a bit of manipulation, we arrive at the equation

$$\bar{u}_n'' + \left(\frac{\gamma_n a}{l}\right)^4 \bar{u}_n = \frac{a^4}{EJ} P(t) X_n(l).$$

The solution of this equation satisfying the initial conditions

$$\bar{u}_n(0) = \bar{u}_n'(0) = 0,$$

obtained by multiplying the original initial conditions by $X_n(x)$ and integrating with respect to x from 0 to l, is given by

$$\bar{u}_n = \frac{a^2 l^2}{EJ\gamma_n^2} X_n(l) \int_0^t P(\tau) \sin \frac{\gamma_n^2 a^2(t - \tau)}{l^2}\, d\tau,$$

and immediately leads to the answer on p. 112.

223. Clearly we can express the dependence of the external load on the coordinate x and the time t in the form

$$q(x, t) = \begin{cases} \dfrac{A \sin \omega t}{2\varepsilon} & \text{for } vt - \varepsilon < x < vt + \varepsilon \\ 0 & \text{otherwise,} \end{cases}$$

where $\varepsilon > 0$ is arbitrarily small. Let

$$u(x, t) = \frac{2}{l} \sum_{n=1}^{\infty} \bar{u}_n \sin \frac{n\pi x}{l},$$

where

$$\bar{u}_n = \int_0^l u \sin \frac{n\pi x}{l} \, dx.$$

Multiplying the equation for the oscillations by $\sin (n\pi x/l)$, integrating with respect to x from 0 to l, and taking account of the boundary conditions, we obtain

$$\bar{u}_n'' + \left(\frac{n\pi a}{l}\right)^4 \bar{u}_n = \frac{a^4}{EJ} \int_0^l q(x, t) \sin \frac{n\pi x}{l} \, dx,$$

or

$$\bar{u}_n'' + \left(\frac{n\pi a}{l}\right)^4 \bar{u}_n = \frac{Aa^4}{EJ} \sin \omega t \sin \frac{n\pi v t}{l}$$

$$= \frac{Aa^4}{2EJ}\left[\cos\left(\omega - \frac{n\pi v}{l}\right)t - \cos\left(\omega + \frac{n\pi v}{l}\right)t\right]$$

after passing to the limit $\varepsilon \to 0$. The general solution of this equation is

$$\bar{u}_n = A_n \cos \frac{a^2 n^2 \pi^2 t}{l^2} + B_n \sin \frac{a^2 n^2 \pi^2 t}{l^2}$$

$$+ \frac{Aa^4}{2EJ}\left[H_1 \cos\left(\omega - \frac{n\pi v}{l}\right)t - H_2 \cos\left(\omega + \frac{n\pi v}{l}\right)t\right], \quad (30)$$

where we introduce the abbreviation

$$H_{1,2} = \frac{1}{\left(\dfrac{n\pi a}{l}\right)^4 - \left(\omega \mp \dfrac{n\pi v}{l}\right)^2}.$$

Using the initial conditions

$$u\big|_{t=0} = \frac{\partial u}{\partial t}\bigg|_{t=0} = 0,$$

we find that

$$\bar{u}_n\big|_{t=0} = \bar{u}_n'\big|_{t=0} = 0,$$

and hence

$$B_n = 0, \qquad A_n = -\frac{Aa^4}{2EJ}(H_1 - H_2).$$

Substituting these values of the coefficients into (30), and letting M denote the mass of the beam, we obtain the answer on p. 112.

225. To solve the problem, we assume that the external load is distributed over the membrane with density

$$q(r, t) = \begin{cases} \dfrac{p \sin \omega t}{2\varepsilon} & \text{for } b - \varepsilon < r < b + \varepsilon, \\ 0 & \text{otherwise,} \end{cases}$$

where $\varepsilon > 0$ is arbitrarily small. The deflection $u(r, t)$ of an arbitrary point of the membrane is then the solution of the inhomogeneous equation

$$\frac{1}{r}\frac{\partial}{\partial r}\left(r\frac{\partial u}{\partial r}\right) - \frac{1}{v^2}\frac{\partial^2 u}{\partial t^2} = -\frac{q(r, t)}{T},$$

with homogeneous initial and boundary conditions

$$u\big|_{t=0} = \frac{\partial u}{\partial t}\bigg|_{t=0} = 0, \qquad u\big|_{r=a} = 0.$$

The desired solution is constructed as an expansion

$$u = \sum_{n=1}^{\infty} \frac{\bar{u}_n}{\int_0^a rR_n^2(r)\,dr} R_n(r)$$

with respect to the eigenfunctions $R_n(r)$ of the homogeneous problem. The latter are the solutions of the equation

$$(rR')' + \lambda rR = 0$$

which are unbounded in the closed interval $[0, a]$ and satisfy the boundary condition $R(a) = 0$. As usual, $\bar{u}_n$ denotes the integral

$$\bar{u}_n = \int_0^a ruR_n(r)\,dr.$$

It is easily verified that the eigenvalues and eigenfunctions are given by

$$\lambda = \lambda_n = \frac{\gamma_n^2}{a^2}, \qquad n = 1, 2, \ldots,$$

$$R = R_n(r) = J_0\left(\frac{\gamma_n r}{a}\right),$$

where $J_0(x)$ is the Bessel function and the γ_n are consecutive positive roots of the equation $J_0(\gamma) = 0$. Applying the usual method for determining the coefficients $\bar{u}_n$, we obtain

$$\bar{u}_n'' + \left(\frac{\gamma_n v}{a}\right)^2 \bar{u}_n = \frac{pv^2 \sin \omega t}{2\varepsilon T}\int_{b-\varepsilon}^{b+\varepsilon} J_0\left(\frac{\gamma_n r}{a}\right) r\,dr,$$

or

$$\bar{u}_n'' + \left(\frac{\gamma_n v}{a}\right)^2 \bar{u}_n = \frac{pbv^2}{T} J_0\left(\frac{\gamma_n b}{a}\right) \sin \omega t$$

after passing to the limit $\varepsilon \to 0$. Integrating this equation with zero initial conditions

$$\bar{u}_n(0) = \bar{u}_n'(0) = 0,$$

we find that

$$\bar{u}_n = \frac{2abpv}{\omega T} \frac{\sin \dfrac{\gamma_n vt}{a} - \dfrac{\gamma_n v}{\omega a} \sin \omega t}{1 - \left(\dfrac{\gamma_n v}{\omega a}\right)^2} \frac{J_0\left(\dfrac{\gamma_n b}{a}\right)}{\gamma_n},$$

which leads to the answer on p. 113.

227. To solve the problem, we regard the concentrated load as the limiting case of a load distributed over a disk of small radius ε. Then, to determine the transverse oscillations of the plate due to this load, we integrate the equation

$$\frac{1}{r}\frac{\partial}{\partial r}\left\{r\frac{\partial}{\partial r}\left[\frac{1}{r}\frac{\partial}{\partial r}\left(r\frac{\partial u}{\partial r}\right)\right]\right\} + \frac{1}{b^4}\frac{\partial^2 u}{\partial t^2} = \frac{q(r,t)}{D},$$

where

$$q(r,t) = \begin{cases} \dfrac{A\sin\omega t}{\pi\varepsilon^2}, & 0 \leqslant r < \varepsilon, \\ 0, & \varepsilon < r \leqslant a, \end{cases}$$

subject to zero initial conditions and the boundary conditions

$$u|_{r=a} = \frac{\partial u}{\partial r}\bigg|_{r=a} = 0.$$

Let

$$u(r,t) = \sum_{n=1}^{\infty} \frac{\bar{u}_n}{\displaystyle\int_0^a r R_n^2(r)\,dr} R_n(r),$$

where the $R_n(r)$ are the eigenfunctions of the homogeneous problem and

$$\bar{u}_n = \int_0^a ru R_n(r)\,dr.$$

The functions $R_n(r)$ are the solutions of the differential equation

$$\frac{1}{r}\frac{d}{dr}\left\{r\frac{d}{dr}\left[\frac{1}{r}\frac{d}{dr}\left(r\frac{dR}{dr}\right)\right]\right\} - \lambda R = 0$$

which are bounded for $r = 0$ and satisfy the conditions

$$R(a) = R'(a) = 0.$$

Therefore

$$R = R_{\gamma_n}, \qquad R_\gamma(r) = I_0(\gamma)J_0\left(\frac{\gamma r}{a}\right) - J_0(\gamma)I_0\left(\frac{\gamma r}{a}\right),$$

where the γ_n are consecutive positive roots of the equation $R'_\gamma(a) = 0$. The corresponding eigenvalues are

$$\lambda = \lambda_n = \frac{\gamma_n^4}{a^4}, \qquad n = 1, 2, \ldots,$$

and the functions $R_{\gamma_n}(r)$ are orthogonal on the interval $(0, a)$ with weight r (see the solution to Prob. 124).

Multiplying the original differential equation by $rR_{\gamma_n}(r)$, integrating with respect to r from 0 to a, and then integrating by parts four times, we obtain

$$\bar{u}_n'' + \left(\frac{\gamma_n b}{a}\right)^4 \bar{u}_n = \frac{b^4}{D} \frac{A \sin \omega t}{\pi \varepsilon^2} \int_0^\varepsilon r R_{\gamma_n}(r) \, dr \,,$$

or

$$\frac{Ab^4}{2\pi D} R_{\gamma_n}(0) \sin \omega t.$$

after taking the limit as $\varepsilon \to 0$. The solution of this equation satisfying the boundary conditions

$$\bar{u}_n(0) = \bar{u}_n'(0) = 0$$

is

$$\bar{u}_n = \frac{Aa^2b^2}{2\pi \omega D \gamma_n^2} R_{\gamma_n}(0) \frac{\sin \dfrac{\gamma_n^2 b^2 t}{a^2} - \dfrac{\gamma_n^2 b^2}{\omega a^2} \sin \omega t}{1 - \left(\dfrac{\gamma_n^2 b^2}{\omega a^2}\right)^2} \cdot$$

Substituting these values of $\bar{u}_n$ into the series for $u(r, t)$ and using the formula

$$\int_0^a r R_{\gamma_n}^2(r) \, dr = a^2 I_0^2(\gamma_n) J_0^2(\gamma_n),$$

we finally arrive at the form of the solution given in the answer on p. 114.

230. To solve the problem, we replace the line load p by a load uniformly distributed over the strip $-\varepsilon < x < \varepsilon$, $-b \leqslant y \leqslant b$ of width 2ε, i.e., we reduce the problem to integration of Poisson's equation

$$\frac{\partial^2 u}{\partial x^2} + \frac{\partial^2 u}{\partial y^2} = -\frac{q(x, y)}{T} = \begin{cases} -\dfrac{p}{2\varepsilon T}, & |x| < \varepsilon, \\ 0, & |x| > \varepsilon, \end{cases} \tag{31}$$

with homogeneous boundary conditions of the first kind:

$$u\big|_{x=\pm a} = u\big|_{y=\pm b} = 0.$$

Two forms of the solution can be found. To obtain the first, we represent the displacement as a series with respect to the eigenfunctions of the corresponding homogeneous problem which depend on the variable x:

$$u = \frac{2}{a} \sum_{n=0}^\infty \bar{u}_n \cos \frac{(2n + 1)\pi x}{2a}, \qquad \bar{u}_n = \int_0^a u \cos \frac{(2n + 1)\pi x}{2a} \, dx.$$

Multiplying (31) by $\cos[(2n+1)\pi x/2a]$, integrating with respect to x from 0 to a, and then integrating by parts twice, we obtain[8]

$$\bar{u}_n'' - \left[\frac{(2n+1)\pi}{2a}\right]^2 \bar{u}_n = -\frac{p}{2\varepsilon T}\int_0^\varepsilon \cos\frac{(2n+1)\pi x}{2a}\,dx,$$

or

$$\bar{u}_n'' - \left[\frac{(2n+1)\pi}{2a}\right]^2 \bar{u}_n = -\frac{p}{2T}$$

after taking the limit as $\varepsilon \to 0$. The solution of this equation satisfying the conditions $\bar{u}_n|_{y=\pm b} = 0$ is

$$\bar{u}_n = \frac{2pa^2}{\pi^2 T(2n+1)^2}\left[1 - \frac{\cosh\dfrac{(2n+1)\pi y}{2a}}{\cosh\dfrac{(2n+1)\pi b}{2a}}\right],$$

which immediately implies formula (12), p. 115.

The second form of the solution is obtained by expanding u in a series with respect to eigenfunctions which depend on the variable y:

$$u = \frac{2}{b}\sum_{n=0}^\infty \bar{u}_n \cos\frac{(2n+1)\pi y}{2b}, \qquad \bar{u}_n = \int_0^b u\cos\frac{(2n+1)\pi y}{2b}\,dy.$$

This time the coefficients $\bar{u}_n$ are functions of the variable x (rather than of y), and are determined by the equation

$$\bar{u}_n'' - \left[\frac{(2n+1)\pi}{2b}\right]^2\bar{u}_n = f(x) = \begin{cases} \dfrac{pb(-1)^{n+1}}{\pi\varepsilon T(2n+1)}, & |x| < \varepsilon, \\ 0, & |x| > \varepsilon. \end{cases}$$

The solution of this equation satisfying the conditions $\bar{u}_n|_{x=\pm a} = 0$ is

$$\bar{u}_n = \frac{b}{\pi(2n+1)}\left[\int_0^x f(\xi)\sinh\frac{(2n+1)\pi(x-\xi)}{2b}\,d\xi \right.$$
$$\left. - \frac{\cosh\dfrac{(2n+1)\pi x}{2b}}{\cosh\dfrac{(2n+1)\pi a}{2b}}\int_0^a f(\xi)\sinh\frac{(2n+1)\pi(a-\xi)}{2b}\,d\xi\right].$$

Substituting for $f(\xi)$ and taking the limit as $\varepsilon \to 0$, we obtain

$$\bar{u}_n = \frac{2pb^2(-1)^n}{\pi^2 T(2n+1)^2}\frac{\sinh\dfrac{(2n+1)\pi(a-|x|)}{2b}}{\cosh\dfrac{(2n+1)\pi a}{2b}},$$

----

[8] We also take account of the boundary condition $u|_{x=a} = 0$ and the relation $(\partial u/\partial x)|_{x=0} = 0$ implied by the symmetry of the problem.

which implies formula (13), p. 115. Which form of the solution to use in making calculations depends on which series converges more rapidly (this depends primarily on the ratio a/b of the sides of the rectangle).

240. The problem reduces to solving the biharmonic equation

$$\frac{\partial^4 u}{\partial x^4} + 2\frac{\partial^4 u}{\partial x^2\, \partial y^2} + \frac{\partial^4 u}{\partial y^4} = 0,$$

with boundary conditions

$$u\big|_{x=0} = u\big|_{x=a} = 0, \qquad \frac{\partial^2 u}{\partial x^2}\bigg|_{x=0} = \frac{\partial^2 u}{\partial y^2}\bigg|_{x=a} = -\frac{m}{D},$$

$$u\big|_{y=\pm b/2} = \frac{\partial u}{\partial y}\bigg|_{y=\pm b/2} = 0.$$

It is easy to construct a function

$$u^* = \frac{mx(a-x)}{2D}$$

satisfying both the differential equation and the boundary conditions at $x = 0$ and $x = a$. If we set

$$u = u^* + v,$$

then the new unknown function v must be a solution of the homogeneous biharmonic equation satisfying homogeneous boundary conditions in x:

$$v\big|_{x=0} = v\big|_{x=a} = \frac{\partial^2 v}{\partial x^2}\bigg|_{x=0} = \frac{\partial^2 v}{\partial x^2}\bigg|_{x=a} = 0.$$

This enables us to use the Fourier method, where the boundary conditions in the variables y take the form

$$v\big|_{y=\pm b/2} = u^*(x), \qquad \frac{\partial v}{\partial y}\bigg|_{y=\pm b/2} = 0.$$

Taking account of the boundary conditions in the variable x, we look for particular solutions of the biharmonic equation $\Delta^2 v = 0$ of the form

$$v = v_n(y) \sin \frac{n\pi x}{a}, \qquad n = 1, 2, \ldots$$

The amplitude v_n must then be a solution of the differential equation

$$v_n^{(iv)} - \left(\frac{n\pi}{a}\right)^2 v_n'' + \left(\frac{n\pi}{a}\right)^4 v_n = 0$$

which is even in y, and hence

$$v_n = A_n \cosh \frac{n\pi y}{a} + B_n y \sinh \frac{n\pi y}{a}.$$

Writing v as a series

$$v = \sum_{n=1}^{\infty} \left(A_n \cosh \frac{n\pi y}{a} + B_n y \sinh \frac{n\pi y}{a} \right) \sin \frac{n\pi x}{a},$$

we determine the coefficients A_n and B_n from the remaining conditions

$$v\big|_{y=b/2} = u^*(x), \qquad \frac{\partial v}{\partial y}\bigg|_{y=b/2} = 0.$$

This requires expanding the known function $u^*(x)$ in a Fourier series with respect to $\sin(n\pi x/a)$. In this way, we eventually arrive at the answer on p. 119.

241. Suppose the line load p is replaced by a load uniformly distributed over the sector $-\varepsilon < \varphi < \varepsilon$, $0 < r < a$ with central angle 2ε, where $\varepsilon > 0$ is arbitrarily small. Then the problem reduces to solving the inhomogeneous biharmonic equation

$$\left[\frac{1}{r} \frac{\partial}{\partial r} \left(r \frac{\partial}{\partial r} \right) + \frac{1}{r^2} \frac{\partial^2}{\partial \varphi^2} \right] \left[\frac{1}{r} \frac{\partial}{\partial r} \left(r \frac{\partial u}{\partial r} \right) + \frac{1}{r^2} \frac{\partial^2 u}{\partial \varphi^2} \right] = \begin{cases} \dfrac{p}{\varepsilon a D}, & |\varphi| < \varepsilon, \\[2mm] 0, & |\varphi| > \varepsilon, \end{cases}$$

with homogeneous boundary conditions

$$u\big|_{r=a} = \frac{\partial u}{\partial r}\bigg|_{r=a} = 0.$$

With our way of measuring angles, u is an even function of φ and hence can be written as a cosine series

$$u(r, \varphi) = \frac{1}{\pi} \bar{u}_0 + \frac{2}{\pi} \sum_{n=1}^{\infty} \bar{u}_n \cos n\varphi,$$

where

$$\bar{u}_n = \int_0^{\pi} u \cos n\varphi \, d\varphi.$$

To find $\bar{u}_n$, we multiply the equation for u by $\cos n\varphi$ and integrate with respect to φ from 0 to π. Then, integrating by parts four times, we find that

$$\left(\frac{d^2}{dr^2} + \frac{1}{r} \frac{d}{dr} - \frac{n^2}{r^2} \right)^2 \bar{u}_n = \frac{p \sin n\varepsilon}{\varepsilon a D n},$$

where the right-hand side can be replaced by p/aD after taking the limit as $\varepsilon \to 0$. We are interested in the solution of this equation which is regular for $r = 0$, i.e.,

$$\bar{u}_n = A_n r^n + B_n r^{n+2} + \bar{u}_n^*,$$

where

$$\bar{u}_n^* = \frac{p}{aD} \frac{r^4}{(4 - n^2)(16 - n^2)},$$

except for the cases $n = 2$ and $n = 4$:

$$\bar{u}_2^* = \frac{p}{aD}\frac{r^4 \ln r}{48}, \qquad \bar{u}_4^* = -\frac{p}{aD}\frac{r^4 \ln r}{96}.$$

The constants A_n and B_n are determined from the conditions

$$\bar{u}_n(a) = \bar{u}_n'(a) = 0.$$

242. The problem reduces to integration of the heat conduction equation

$$\frac{\partial^2 T}{\partial x^2} = \frac{\partial T}{\partial \tau}, \tag{32}$$

with the zero initial condition

$$T|_{\tau=0} = 0$$

and inhomogeneous boundary conditions

$$-k\frac{\partial T}{\partial x}\bigg|_{x=0} = q, \qquad T|_{x=a} = T_0.$$

It is easy to see that the linear function

$$T^* = T_0 + \frac{q}{k}(a - x)$$

is a solution of (32) satisfying both inhomogeneous boundary conditions in the variable x. Therefore, writing

$$T = T^* - u,$$

we find that u satisfies the differential equation

$$\frac{\partial^2 u}{\partial x^2} = \frac{\partial u}{\partial \tau}$$

with initial condition

$$u|_{\tau=0} = T^*(x) = T_0 + \frac{q}{k}(a - x)$$

and homogeneous boundary conditions

$$\frac{\partial u}{\partial x}\bigg|_{x=0} = 0, \qquad u|_{x=a} = 0.$$

Application of the Fourier method gives

$$u = \sum_{n=0}^{\infty} c_n e^{-(2n+1)^2\pi^2\tau/(2a)^2} \cos\frac{(2n + 1)\pi x}{2a},$$

where the c_n are the coefficients of the Fourier expansion of $T^*(x)$ with respect to the functions $\cos [(2n + 1)\pi x/2a]$.

247. We have to solve the inhomogeneous equation

$$\frac{1}{r}\frac{\partial}{\partial r}\left(r\frac{\partial T}{\partial r}\right) = \frac{\partial T}{\partial \tau} - \frac{Q}{k}, \tag{33}$$

with the zero initial condition

$$T|_{\tau=0} = 0$$

and homogeneous boundary condition of the third kind:

$$\left(\frac{\partial T}{\partial r} + hT\right)\bigg|_{r=a} = 0.$$

Suppose the solution is of the form

$$T = \sum_{n=1}^{\infty} \frac{\bar{T}_n}{\int_0^a rR_n^2(r)\,dr} R_n(r),$$

where

$$\bar{T}_n = \int_0^a TR_n(r)r\,dr,$$

in terms of the eigenfunctions $R_n(r)$ of the homogeneous problem. The latter must satisfy the equation

$$\frac{1}{r}(rR')' + \lambda R = 0, \tag{34}$$

the boundary condition

$$R'(a) + hR(a) = 0$$

and the requirement that $R(0)$ be bounded. Solutions of the required type exist if

$$\lambda = \lambda_n = \frac{\gamma_n^2}{a^2}, \qquad n = 1, 2, \ldots,$$

where the γ_n are consecutive positive roots of the equation

$$\gamma J_1(\gamma) = ahJ_0(\gamma).$$

The corresponding solutions of (34) are

$$R = R_n(r) = J_0\left(\frac{\gamma_n r}{a}\right).$$

These functions are orthogonal with weight r on the interval $(0, a)$, and moreover

$$\int_0^a rR_n^2(r)\,dr = \frac{a^2}{2}[J_0^2(\gamma_n) + J_1^2(\gamma_n)] = \frac{a^2}{2}J_0^2(\gamma_n)\left[1 + \left(\frac{ha}{\gamma_n}\right)^2\right].$$

To find the functions $\bar{T}_n$, we multiply (33) by $R_n(r)$ and integrate from 0 to a. Then, integrating by parts twice and taking account of the boundary conditions, we find that

$$\bar{T}'_n + \left(\frac{\gamma_n}{a}\right)^2 \bar{T}_n = \frac{Qa^2}{k} \frac{J_1(\gamma_n)}{\gamma_n}. \tag{35}$$

The solution of (35) satisfying the condition $\bar{T}_n = 0$ is

$$\bar{T}_n = \frac{Qa^4}{k} \frac{J_1(\gamma_n)}{\gamma_n^3} [1 - e^{-\gamma_n^2 \tau / a^2}].$$

Therefore the desired temperature distribution can be represented as a series

$$T(r, \tau) = \frac{2Qa^2}{k} \sum_{n=1}^{\infty} \frac{J_1(\gamma_n) J_0(\gamma_n r/a)}{\gamma_n^3 [J_0^2(\gamma_n) + J_1^2(\gamma_n)]} [1 - e^{-\gamma_n^2 \tau / a^2}].$$

This form of the solution is suitable only for small values of τ, i.e., during the initial stages of the heating. For large values of τ, it is convenient to subtract out the terms of the series which are independent of time, by using the formula

$$\frac{1}{8}\left(1 - \frac{r^2}{a^2}\right) + \frac{1}{4ah} = \sum_{n=1}^{\infty} \frac{J_1(\gamma_n)}{\gamma_n^3} \frac{J_0(\gamma_n r/a)}{J_0^2(\gamma_n) + J_1^2(\gamma_n)}.$$

Then $T(r, \tau)$ takes the form given in the answer on p. 121.

261. The problem reduces to finding the solution of the equation

$$\frac{\partial^2 T}{\partial x^2} + \frac{\partial^2 T}{\partial y^2} = -\frac{Q}{k}$$

which satisfies the boundary conditions of the second kind[9]

$$\frac{\partial T}{\partial x}\bigg|_{x=\pm a} = \frac{\partial T}{\partial y}\bigg|_{y=0} = 0, \qquad \frac{\partial T}{\partial y}\bigg|_{y=b} = f(x) = \begin{cases} -\dfrac{Qab}{kc}, & |x| < c, \\ 0, & |x| > c. \end{cases}$$

The solution can be obtained in two different forms, either as a series with respect to the eigenfunctions $X_n(x)$ satisfying homogeneous boundary conditions in the variable x, or as a series with respect to the eigenfunctions $Y_n(y)$ satisfying homogeneous boundary conditions at the end points of the interval $0 \leqslant y \leqslant b$. The first form of the solution is

$$T(x, y) = \frac{1}{a}\bar{T}_0 + \frac{2}{a}\sum_{n=1}^{\infty} \bar{T}_n \cos \frac{n\pi x}{a}, \tag{36}$$

---

[9] The density q of the heat current through the section $|x| < c$, $y = b$ can be expressed in terms of the density Q of heat produced inside the bar by using the condition $qc = Qab$ for solvability of the problem.

where

$$\bar{T}_n = \int_0^a T \cos \frac{n\pi x}{a}\, dx.$$

To determine $\bar{T}_n$, we multiply the original inhomogeneous equation by $\cos (n\pi x/a)$ and integrate with respect to x from 0 to a. This gives

$$\bar{T}_n'' - \left(\frac{n\pi}{a}\right)^2 \bar{T}_n = \begin{cases} -\dfrac{Qa}{k}, & n = 0, \\ 0, & n \geqslant 1, \end{cases}$$

which implies

$$\bar{T}_0 = -\frac{Qay^2}{2k} + A_0 + B_0 y,$$

$$\bar{T}_n = A_n \cosh \frac{n\pi y}{a} + B_n \sinh \frac{n\pi y}{a}, \qquad n \geqslant 1.$$

Using the boundary conditions in y, we find that

$$\frac{d\bar{T}_n}{dy}\bigg|_{y=0} = 0, \qquad \frac{d\bar{T}_n}{dy}\bigg|_{y=b} = \int_0^a f(x) \cos \frac{n\pi x}{a}\, dx = -\frac{Qa^2 b}{n\pi kc} \sin \frac{n\pi c}{a},$$

which leads to the following values of the constants:[10]

$$A_n = -\frac{Qa^3 b}{n^2 \pi^2 kc} \frac{\sin (n\pi c/a)}{\sinh (n\pi b/a)}, \qquad B_n = 0.$$

Substituting A_n and B_n into (36), we obtain formula (14), p. 126.

To obtain the other form of the solution, we set

$$T(x, y) = \frac{1}{b}\bar{T}_0 + \frac{2}{b}\sum_{n=1}^{\infty} \bar{T}_n \cos \frac{n\pi y}{b},$$

where

$$\bar{T}_n = \int_0^b T \cos \frac{n\pi y}{b}\, dy.$$

Then, by the same procedure as before, we obtain the differential equation

$$\bar{T}_n'' - \left(\frac{n\pi}{b}\right)^2 \bar{T}_n = (-1)^{n+1} f(x) \tag{37}$$

determining the coefficients $\bar{T}_n$. The solution of (37) satisfying the conditions[11]

$$\frac{d\bar{T}_n}{dx}\bigg|_{x=0} = \frac{d\bar{T}_n}{dx}\bigg|_{x=a} = 0$$

---

[10] The constant A_0 remains indeterminate.

[11] The desired solution $T(x, y)$ is an even function of x, and hence, from now on, we need only consider the region $0 \leqslant x \leqslant a$.

can be found by variation of constants, and turns out to be

$$T_n = \frac{Qab^3}{c} \frac{(-1)^{n+1}}{n^2\pi^2 k} \left[\frac{\cosh\dfrac{n\pi x}{b}}{\sinh\dfrac{n\pi a}{b}} \left(\sinh\frac{n\pi a}{b} - \sinh\frac{n\pi(a-c)}{b} \right) \right.$$

$$\left. + \begin{cases} 1 - \cosh\dfrac{n\pi x}{b}, & x < c, \\ \cosh\dfrac{n\pi(x-c)}{b} - \cosh\dfrac{n\pi x}{b}, & x > c, \end{cases} \right], \qquad n = 1, 2, \ldots$$

Similar calculations for the case $n = 0$ lead to the following expression:

$$T_0 = \text{const} + \begin{cases} \dfrac{Qb}{2kc}(a-c)x^2, & x < c, \\ \dfrac{Qb}{k}\left(ax - \dfrac{x^2}{2}\right), & x > c. \end{cases}$$

After some manipulation, we find that

$$T(x, y) = -\frac{Qx^2}{2k} + \begin{cases} \dfrac{Qax^2}{2kc}, & |x| < c, \\ \dfrac{Qa\,|x|}{k}, & |x| > c, \end{cases} + \text{const}$$

$$+ \frac{2Qab^2}{\pi^2 kc} \sum_{n=1}^{\infty} \frac{(-1)^n \cos\dfrac{n\pi y}{b}}{n^2 \sinh\dfrac{n\pi a}{b}}$$

$$\times \begin{cases} \sinh\dfrac{n\pi(a-c)}{b}\cosh\dfrac{n\pi x}{b} - \sinh\dfrac{n\pi a}{b}, & |x| < c, \\ -\sinh\dfrac{n\pi c}{b}\cosh\dfrac{n\pi(a-|x|)}{b}, & |x| > c. \end{cases}$$

The form of the solution given in the answer on p. 126 is obtained if we improve the convergence by using the formula

$$\sum_{n=1}^{\infty} \frac{(-1)^{n+1}}{n^2}\cos nx = \frac{\pi^2}{12} - \frac{x^2}{4}, \qquad -\pi < x < \pi$$

to carry out partial summation of the series.

269. To solve the problem, we assume that heat is produced with uniform density $Q/\pi\varepsilon^2$ inside a cylinder of arbitrarily small radius ε. Then the problem reduces to integrating Poisson's equation

$$\frac{1}{r}\frac{\partial}{\partial r}\left(r\frac{\partial T}{\partial r}\right) + \frac{\partial^2 T}{\partial z^2} = f(r) = \begin{cases} -\dfrac{Q}{\pi k\varepsilon^2}, & 0 \leqslant r < \varepsilon \\ 0, & \varepsilon < r \leqslant a, \end{cases}$$

with boundary conditions

$$T|_{r=a} = 0, \qquad \left(\frac{\partial T}{\partial z} \pm hT\right)\Bigg|_{z=\pm l} = 0.$$

Expanding the solution in a series of eigenfunctions of the corresponding homogeneous problem depending on the variable r, we find that

$$T(r, z) = \frac{2}{a^2}\sum_{n=1}^{\infty}\frac{\bar{T}_n}{J_1^2(\gamma_n)}J_0\left(\frac{\gamma_n r}{a}\right), \qquad \bar{T}_n = \int_0^a TJ_0\left(\frac{\gamma_n r}{a}\right)r\,dr,$$

where the γ_n are consecutive positive roots of the equation $J_0(\gamma) = 0$. Multiplying the original equation by $rJ_0(\gamma_n r/a)$ and integrating with respect to r from 0 to a, we obtain

$$\bar{T}_n'' - \left(\frac{\gamma_n}{a}\right)^2\bar{T}_n = -\frac{Qa}{\pi\varepsilon k}\frac{J_1(\gamma_n\varepsilon/a)}{\gamma_n},$$

or

$$\bar{T}_n'' - \left(\frac{\gamma_n}{a}\right)^2\bar{T}_n = -\frac{Q}{2\pi k}$$

after taking the limit as $\varepsilon \to 0$. The solution of this equation satisfying the boundary conditions

$$\left(\frac{d\bar{T}_n}{dz} \pm h\bar{T}_n\right)\Bigg|_{z=\pm l} = 0$$

is

$$\bar{T}_n = \frac{Qa^2}{2\pi k\gamma_n^2}\left[1 - \frac{ah\cosh(\gamma_n z/a)}{\gamma_n\sinh(\gamma_n l/a) + ah\cosh(\gamma_n l/a)}\right],$$

which leads to the answer given on p. 130.

272. This problem of electrostatics reduces to finding a solution of Laplace's equation

$$\frac{\partial^2 u}{\partial x^2} + \frac{\partial^2 u}{\partial y^2} = 0$$

satisfying the following inhomogeneous boundary conditions of the first kind:

$$u|_{x=0} = V, \qquad u|_{y=\pm b} = f(x) = \begin{cases} V, & 0 \leqslant x < a, \\ 0, & a < x < \infty. \end{cases}$$

Following Grinberg's method, we look for a solution in the form of an expansion with respect to the eigenfunctions of the corresponding homogeneous problem,[12] i.e.,

$$u(x, y) = \frac{2}{b} \sum_{n=0}^{\infty} \bar{u}_n \cos \frac{(2n + 1)\pi y}{2b},$$

where

$$\bar{u}_n = \int_0^b u \cos \frac{(2n + 1)\pi y}{2b} \, dy.$$

To determine the unknown quantities $\bar{u}_n$, we multiply Laplace's equation by $\cos [(2n + 1)\pi y/2b]$ and integrate with respect to y from 0 to b. Taking account of the boundary conditions, we obtain

$$\bar{u}_n'' - \left[\frac{(2n + 1)\pi}{2b}\right]^2 \bar{u}_n = (-1)^{n+1} \frac{(2n + 1)\pi}{2b} f(x). \tag{38}$$

We want the solution of (38) which is bounded at infinity and satisfies the condition

$$\bar{u}_n\big|_{x=0} = \int_0^b V \cos \frac{(2n + 1)\pi y}{2b} \, dy = \frac{2bV(-1)^n}{(2n + 1)\pi}.$$

It is easy to see that this solution can be written in the form

$$\bar{u}_n = \begin{cases} \bar{u}_n^{(1)} = B_n \sinh \dfrac{(2n + 1)\pi x}{2b} + \dfrac{2bV(-1)^n}{(2n + 1)\pi}, & x < a, \\[2mm] \bar{u}_n^{(2)} = C_n e^{-(2n+1)\pi x/2b}, & x > a, \end{cases}$$

where the values of the constants B_n and C_n are determined from the "contact conditions"

$$\bar{u}_n\big|_{x=a-0} = \bar{u}_n\big|_{x=a+0}, \qquad \bar{u}_n'\big|_{x=a-0} = \bar{u}_n'\big|_{x=a+0},$$

which imply

$$B_n = \frac{2bV(-1)^{n+1}}{(2n + 1)\pi} e^{-(2n+1)\pi a/2b}, \qquad C_n = \frac{2bV(-1)^n}{(2n + 1)\pi} \cosh \frac{(2n + 1)\pi a}{2b}.$$

Substitution of these values of the coefficients into $\bar{u}_n$ leads to the following series solution of the problem:

$$u\big|_{x<a} = \frac{4V}{\pi} \sum_{n=0}^{\infty} \frac{(-1)^n}{2n + 1} \left[1 - e^{-(2n+1)\pi b/2a} \sinh \frac{(2n + 1)\pi x}{2a}\right] \cos \frac{(2n + 1)\pi y}{2b},$$

$$u\big|_{x>a} = \frac{4V}{\pi} \sum_{n=0}^{\infty} \frac{(-1)^n}{2n + 1} \cosh \frac{(2n + 1)\pi a}{2b} e^{-(2n+1)\pi x/2b} \cos \frac{(2n + 1)\pi y}{2b}.$$

---

[12] Choosing the other form of the solution leads to an expansion in a Fourier sine integral over the integral $(0, \infty)$.

To obtain the final form of the solution, we improve the convergence by using the formula

$$\sum_{n=0}^{\infty} \frac{(-1)^n}{2n + 1} \cos (2n + 1)x = \frac{\pi}{4}, \qquad |x| < \frac{\pi}{2}$$

to sum the slowly convergent part of the first series. It would be noted that the solution can also be written in closed form.

277. To solve the problem, we first assume that the charge q is uniformly distributed with density ρ over an arbitrarily small cylinder $0 \leqslant r < \delta$, $c - \frac{1}{2}\varepsilon < z < c + \frac{1}{2}\varepsilon$, i.e., we reduce the problem to integration of Poisson's equation

$$\frac{1}{r} \frac{\partial}{\partial r}\left(r \frac{\partial u}{\partial r}\right) + \frac{\partial^2 u}{\partial z^2} = -4\pi\rho(r, z),$$

where

$$\rho(r, z) = \begin{cases} \dfrac{q}{\pi\delta^2\varepsilon} & \text{for} \quad 0 \leqslant r < \delta, \quad c - \tfrac{1}{2}\varepsilon < z < c + \tfrac{1}{2}\varepsilon, \\ 0 & \text{otherwise,} \end{cases}$$

subject to the boundary conditions

$$u\big|_{r=a} = u\big|_{z=0} = u\big|_{z=l} = 0.$$

One of the two possible forms of the solution is an expansion with respect to the functions $J_0(\gamma_n r/a)$, which are the eigenfunctions of the corresponding homogeneous problem, i.e.,

$$u(r, z) = \frac{2}{a^2} \sum_{n=1}^{\infty} \frac{\bar{u}_n}{J_1^2(\gamma_n)} J_0\left(\frac{\gamma_n r}{a}\right), \qquad \bar{u}_n = \int_0^a u J_0\left(\frac{\gamma_n r}{a}\right) r \, dr,$$

where the γ_n are consecutive positive roots of the equation $J_0(\gamma) = 0$. Multiplying the original equation by $r J_0(\gamma_n r/a)$ and integrating with respect to r from 0 to a, we arrive at the equation

$$\bar{u}_n'' - \left(\frac{\gamma_n}{a}\right)^2 \bar{u}_n = -4\pi \int_0^a \rho(t, z) J_0\left(\frac{\gamma_n t}{a}\right) t \, dt, \tag{39}$$

which is to be solved with zero boundary conditions

$$\bar{u}_n\big|_{z=0} = \bar{u}_n\big|_{z=l} = 0.$$

The general solution of (39) satisfying the first of these conditions is

$$\bar{u}_n = A_n \frac{\sinh (\gamma_n z/a)}{\sinh (\gamma_n l/a)} - \frac{4\pi a}{\gamma_n} \int_0^z \sinh \frac{\gamma_n(z - \zeta)}{a} \, d\zeta \int_0^a \rho(t, \zeta) J_0\left(\frac{\gamma_n t}{a}\right) t \, dt.$$

Using the other boundary condition to calculate A_n, and then passing to the limit δ, $\varepsilon \to 0$, we obtain

$$A_n = \frac{2aq}{\gamma_n} \sinh \frac{\gamma_n(l-c)}{a}.$$

Thus the coefficients $\bar{u}_n$ are equal to

$$\bar{u}_n = \frac{2aq}{\gamma_n \sinh \dfrac{\gamma_n l}{a}} \times \begin{cases} \sinh \dfrac{\gamma_n(l-c)}{a} \sinh \dfrac{\gamma_n z}{a}, & 0 \leqslant z \leqslant c, \\[2mm] \sinh \dfrac{\gamma_n c}{a} \sinh \dfrac{\gamma_n(l-c)}{a}, & c \leqslant z \leqslant l, \end{cases}$$

which immediately leads to the answer on p. 134.

The other form of the solution can be obtained by expanding $u(r, z)$ in a series with respect to the eigenfunctions in the variable z, i.e.,

$$u(r, z) = \frac{2}{l} \sum_{n=1}^{\infty} \bar{u}_n \sin \frac{n\pi z}{l}, \qquad \bar{u}_n = \int_0^l u \sin \frac{n\pi z}{l} \, dz.$$

282. Since the potential distribution must be an odd function of the coordinate z, the problem reduces to solving Laplace's equation

$$\frac{1}{r} \frac{\partial}{\partial r} \left(r \frac{\partial u}{\partial r} \right) + \frac{\partial^2 u}{\partial z^2} = 0,$$

with boundary conditions

$$u\big|_{z=0} = 0, \quad u\big|_{z=l} = V, \quad u\big|_{r=a} = f(z),$$

where

$$f(z) = \begin{cases} V, & \delta < z < l, \\[2mm] V \sin \dfrac{\pi z}{2\delta}, & 0 < z < \delta. \end{cases}$$

To obtain homogeneous boundary conditions in the variable z, we set

$$u = V \frac{z}{l} - v.$$

Then the function $v(r, z)$ will be the solution of Laplace's equation satisfying the homogeneous conditions

$$v\big|_{z=0} = v\big|_{z=l} = 0$$

and the following boundary condition on the lateral surface

$$v\big|_{r=a} = \varphi(z) = \begin{cases} V\left(\dfrac{z}{l} - 1 \right), & \delta < z < l, \\[2mm] V\left(\dfrac{z}{l} - \sin \dfrac{\pi z}{2\delta} \right), & 0 < z < \delta. \end{cases}$$

To find the function v, we can now use the Fourier method, which, after separation of variables and determination of eigenvalues and eigenfunctions, leads to the expansion

$$v(r, z) = \sum_{n=1}^{\infty} c_n I_0\left(\frac{n\pi r}{l}\right) \sin \frac{n\pi z}{l},$$

where the coefficients c_n are found from the condition

$$v|_{r=a} = \varphi(z).$$

After determining the potential $u(r, z)$, the electric field on the axis of the lens can be calculated from the formula

$$E_z|_{r=0} = -\left.\frac{\partial u}{\partial z}\right|_{r=0}.$$

289. Suppose the current is distributed with uniform density over the arbitrarily small area

$$\theta_0 - \frac{\delta}{2} < \theta < \theta_0 + \frac{\delta}{2}, \qquad |\varphi| < \frac{\varepsilon}{2}.$$

Then the problem reduces to integration of the equation

$$\frac{1}{a^2 \sin \theta} \frac{\partial}{\partial \theta}\left(\sin \theta \frac{\partial u}{\partial \theta}\right) + \frac{1}{a^2 \sin \theta} \frac{\partial^2 u}{\partial \varphi^2} = f(\theta, \varphi),$$

$$0 \leqslant \theta \leqslant \frac{\pi}{2}, \quad -\pi \leqslant \varphi < \pi \quad (40)$$

(see Prob. 21, p. 14), where

$$f(\theta, \varphi) = \begin{cases} -\dfrac{J}{\sigma h a^2 \, \delta\varepsilon \sin \theta} & \text{for} \quad \theta_0 - \dfrac{\delta}{2} < \theta < \theta_0 + \dfrac{\delta}{2}, \quad |\varphi| < \dfrac{\varepsilon}{2}, \\ 0 & \text{otherwise,} \end{cases}$$

subject to the boundary condition

$$u|_{\theta=\pi/2} = 0.$$

If we introduce a new variable by writing

$$\psi = \tan \frac{\theta}{2}, \qquad f(\theta, \varphi) = F(\psi, \varphi),$$

then (40) takes the simpler form

$$\frac{1}{\psi} \frac{\partial}{\partial \psi}\left(\psi \frac{\partial u}{\partial \psi}\right) + \frac{1}{\psi^2} \frac{\partial^2 u}{\partial \varphi^2} = \frac{4a^2}{(1 + \psi^2)^2} F(\psi, \varphi). \qquad 0 < \psi < 1,$$

whose solution can be constructed as a Fourier series

$$u = \frac{1}{\pi} \bar{u}_0 + \frac{2}{\pi} \sum_{n=1}^{\infty} \bar{u}_n \cos n\varphi,$$

where

$$\bar{u}_n = \int_0^{\pi} u \cos n\varphi \, d\varphi.$$

To determine the coefficients $\bar{u}_n$, we follow the usual approach, obtaining the differential equation

$$\frac{1}{\psi} \frac{d}{d\psi} \left(\psi \frac{d\bar{u}_n}{d\psi} \right) - \frac{n^2}{\psi^2} \bar{u}_n = \frac{4a^2}{(1 + \psi^2)^2} \int_0^{\pi} F(\psi, \eta) \cos n\eta \, d\eta$$

whose general solution is

$$\bar{u}_n = A_n \psi^n + B_n \psi^{-n}$$
$$+ \frac{2a^2}{n} \int_0^{\psi} \frac{\xi}{(1 + \xi^2)^2} \left[\left(\frac{\psi}{\xi} \right)^n - \left(\frac{\psi}{\xi} \right)^{-n} \right] d\xi \int_0^{\pi} F(\xi, \eta) \cos n\eta \, d\eta.$$

The constants A_n and B_n are determined from the boundary condition

$$\bar{u}_n|_{\psi=1} = 0$$

and the condition that $\bar{u}_n$ be bounded for $\psi = 0$. Passing to the limit δ, $\varepsilon \to 0$, we find that[13]

$$A_n = - \frac{J}{4\sigma h n} (\psi_0^n - \psi_0^{-n}), \qquad \psi_0 = \tan \frac{\theta_0}{2},$$

$$B_n = 0, \qquad n = 1, 2, \ldots,$$

which implies

$$\bar{u}_n = - \frac{J}{4\sigma h n} \times \begin{cases} (\psi_0^n - \psi_0^{-n})\psi^n, & 0 \leqslant \psi \leqslant \psi_0, \\ (\psi^n - \psi^{-n})\psi_0^n, & \psi_0 \leqslant \psi \leqslant 1, \end{cases} \qquad (n = 1, 2, \ldots).$$

For $n = 0$ we have

$$\bar{u}_0 = \frac{J}{2\sigma h} \times \begin{cases} \ln \psi_0, & 0 \leqslant \psi \leqslant \psi_0, \\ \ln \psi, & \psi_0 \leqslant \psi \leqslant 1. \end{cases}$$

Therefore the desired solution has the following series representation:

$$u|_{0 \leqslant \psi \leqslant \psi_0} = \frac{J}{2\pi\sigma h} \left[\ln \psi_0 + \sum_{n=1}^{\infty} (\psi_0^{-n} - \psi_0^n)\psi^n \frac{\cos n\varphi}{n} \right],$$

$$u|_{\psi_0 \leqslant \psi \leqslant 1} = \frac{J}{2\pi\sigma h} \left[\ln \psi + \sum_{n=1}^{\infty} (\psi^{-n} - \psi^n)\psi_0^n \frac{\cos n\varphi}{n} \right].$$

---

[13] The coefficients B_n vanish for arbitrary values of δ and ε.

Using the formula

$$\sum_{n=1}^{\infty} \frac{\rho^n \cos nx}{n} = -\frac{1}{2} \ln (1 - 2\rho \cos x + \rho^2) \qquad (|\rho| < 1, 0 < x < \pi)$$

to sum the series, we arrive at the answer given on p. 138.

296. In this problem it is convenient to characterize the electromagnetic field by the vector potential $\mathbf{A}e^{i\omega t}$, whose complex amplitude has components $A_x = A_y = 0$, $A_z = A(r, z)$. Suppose the current in the dipole is replaced by a current distributed over the volume of an arbitrarily small cylinder

$$0 \leqslant r < \delta, \qquad -\frac{\varepsilon}{2} < z < \frac{\varepsilon}{2}.$$

Then $A(r, z)$ is determined by the differential equation

$$\frac{1}{r}\frac{\partial}{\partial r}\left(r\frac{\partial A}{\partial r}\right) + \frac{\partial^2 A}{\partial z^2} + k^2 A = -\frac{4\pi j}{c},$$

where

$$j = \begin{cases} \dfrac{J}{\pi\delta^2} & \text{for} \quad 0 \leqslant r < \delta, \quad -\dfrac{\varepsilon}{2} < z < \dfrac{\varepsilon}{2}, \\ 0 & \text{otherwise.} \end{cases}$$

The tangential component of the electric field must vanish on the surface of the resonator, and hence

$$\frac{\partial A}{\partial z}\bigg|_{z=0} = \frac{\partial A}{\partial z}\bigg|_{z=l} = 0, \qquad \frac{\partial}{\partial r}\left(r\frac{\partial A}{\partial r}\right)\bigg|_{r=a} = 0.$$

We look for a solution of the problem in the form of a Fourier cosine series

$$A(r, z) = \frac{1}{l}\bar{A}_0 + \frac{2}{l}\sum_{n=1}^{\infty}\bar{A}_n \cos\frac{n\pi z}{l},$$

where

$$\bar{A}_n = \int_0^l A \cos\frac{n\pi z}{l}\, dz.$$

The usual argument implies

$$\frac{1}{r}\frac{d}{dr}\left(r\frac{d\bar{A}_n}{dr}\right) - \left(\frac{n^2\pi^2}{l^2} - k^2\right)\bar{A}_n = -\frac{4\pi}{c}\int_0^l j(r, \zeta)\cos\frac{n\pi\zeta}{l}\, d\zeta,$$

$$\frac{d}{dr}\left(r\frac{d\bar{A}_n}{dr}\right)\bigg|_{r=a} = 0,$$

where the last condition is equivalent to

$$\bar{A}_n\big|_{r=a} = 0,$$

because of the differential equation for $\bar{A}_n$. Using the method of variation of constants, we find that

$$\bar{A}_n = \frac{4\pi}{c} \frac{1}{I_0(\alpha_n a)} \int_0^a G_n(r, \rho)\rho \, d\rho \int_0^l j(\rho, \zeta) \cos \frac{n\pi\zeta}{l} \, d\zeta,$$

$$\alpha_n = \sqrt{\frac{n^2\pi^2}{l^2} - k^2},$$

$$G_n(r, \rho) = \begin{cases} I_0(\alpha_n\rho)[I_0(\alpha_n a)K_0(\alpha_n r) - K_0(\alpha_n a)I_0(\alpha_n r)], & \rho \leqslant r, \\ I_0(\alpha_n r)[I_0(\alpha_n a)K_0(\alpha_n\rho) - K_0(\alpha_n a)I_0(\alpha_n\rho)], & \rho \geqslant r. \end{cases}$$

Then, passing to the limit δ, $\varepsilon \to 0$ and bearing in mind that $\lim_{\varepsilon \to 0} J\varepsilon = 0$, we arrive at the expression

$$\bar{A}_n = \frac{\pi P}{c} \frac{1}{I_0(\alpha_n a)} [I_0(\alpha_n a)K_0(\alpha_n r) - K_0(\alpha_n a)I_0(\alpha_n r)],$$

which immediately implies the answer on p. 141.

303. We want the solution of Laplace's equation

$$\frac{\partial^2 T}{\partial x^2} + \frac{\partial^2 T}{\partial y^2} = 0 \qquad (0 \leqslant x < \infty, 0 \leqslant y < \infty)$$

satisfying the boundary conditions

$$T|_{y=0} = 0, \qquad \frac{\partial T}{\partial x}\bigg|_{x=0} = f(y) = \begin{cases} -\dfrac{q}{k}, & 0 < y < b, \\ 0, & b < y < \infty. \end{cases}$$

Application of the Fourier method leads to the particular solutions

$$T = T_\lambda = B_\lambda e^{-\lambda x} \sin \lambda y, \qquad \lambda \geqslant 0,$$

which are bounded in the quadrant $0 \leqslant x < \infty$, $0 \leqslant y < \infty$ and vanish for $y = 0$. Integrating with respect to the parameter λ, we obtain

$$T(x, y) = \int_0^\infty B_\lambda e^{-\lambda x} \sin \lambda y \, d\lambda,$$

where the coefficient B_λ is determined from the boundary condition

$$\frac{\partial T}{\partial x}\bigg|_{x=0} = f(y) = \int_0^\infty B_\lambda \lambda \sin \lambda y \, d\lambda, \qquad 0 \leqslant y < \infty.$$

Because of the theorem on expansion in a Fourier sine integral, we have

$$B_\lambda = -\frac{2}{\pi\lambda} \int_0^\infty f(y) \sin \lambda y \, dy = \frac{2q}{\pi k} \frac{1 - \cos \lambda b}{\lambda^2},$$

which is the same as the expression for $T(x, y)$ given on p. 150.

313. To avoid the difficulties associated with the fact that the logarithmic potential does not go to zero at infinity, we look for the components of the electric field in the two media:

$$E_{x1}, E_{y1}, E_{x2}, E_{y2}.$$

Setting

$$E_{x1} = E_x^{(0)} + E_x^{(1)}, \quad E_{y1} = E_y^{(0)} + E_y^{(1)}, \quad E_{x2} \equiv E_x^{(2)}, \quad E_{y2} \equiv E_y^{(2)},$$

where $E^{(0)}$ is the field due to the charged wire in an unbounded medium of dielectric constant ε_1, with components

$$E_x^{(0)} = \frac{qx}{\varepsilon_1[x^2 + (y-a)^2]}, \qquad E_y^{(0)} = \frac{q(y-a)}{\varepsilon_1[x^2 + (y-a)^2]},$$

we obtain the system of differential equations

$$\frac{\partial E_x^{(i)}}{\partial x} + \frac{\partial E_y^{(i)}}{\partial y} = 0, \qquad \frac{\partial E_x^{(i)}}{\partial y} - \frac{\partial E_y^{(i)}}{\partial x} = 0, \qquad (41)$$

which, together with the boundary conditions

$$[E_x^{(0)} + E_x^{(1)}]_{y=0} = E_x^{(2)}|_{y=0}, \quad \varepsilon_1[E_y^{(0)} + E_y^{(1)}]_{y=0} = \varepsilon_2 E_y^{(2)}|_{y=0},$$
$$E_x^{(1)}, E_y^{(1)}|_{y\to+\infty} \to 0, \quad E_x^{(2)}, E_y^{(2)}|_{y\to-\infty} \to 0, \quad E_x^{(i)}, E_y^{(i)}|_{x\to\pm\infty} \to 0, \qquad (42)$$

determine the functions $E_x^{(i)}, E_y^{(i)}$ $(i = 1, 2)$. A convenient way of solving (41) is to use the method of integral transforms, by taking the sine transform of $E_x^{(i)}$ and the cosine transform of $E_y^{(i)}$ $(i = 1, 2)$.[14] Thus we multiply the first of the equations (41) and the second of each pair of boundary conditions (42) by $\cos \lambda x$, and the second of the equations (41) and the first of each pair of boundary conditions by $\sin \lambda x$. Then, integrating from 0 to ∞, we find that

$$\frac{d\bar{E}_y^{(i)}}{dy} + \lambda \bar{E}_x^{(i)} = 0, \qquad \frac{d\bar{E}_x^{(i)}}{dy} + \lambda \bar{E}_y^{(i)} = 0,$$

$$[\bar{E}_x^{(2)} - \bar{E}_x^{(1)}]_{y=0} = \frac{\pi q}{2\varepsilon_1} e^{-\lambda a}, \qquad [\varepsilon_2 \bar{E}_y^{(2)} - \varepsilon_1 \bar{E}_y^{(1)}]_{y=0} = -\frac{\pi q}{2} e^{-\lambda a}, \qquad (43)$$

$$\bar{E}_x^{(1)}, \bar{E}_y^{(1)}|_{y\to+\infty} \to 0, \qquad \bar{E}_x^{(2)}, \bar{E}_y^{(2)}|_{y\to-\infty} \to 0,$$

where

$$\bar{E}_x^{(i)} = \int_0^\infty E_x^{(i)} \sin \lambda x \, dx, \qquad \bar{E}_y^{(i)} = \int_0^\infty E_y^{(i)} \cos \lambda x \, dx.$$

The solution of the system (43) is

$$\bar{E}_x^{(1)} = \bar{E}_y^{(1)} = \frac{\pi q}{2\varepsilon_1} \frac{\varepsilon_1 - \varepsilon_2}{\varepsilon_1 + \varepsilon_2} e^{-\lambda(a+y)},$$

$$\bar{E}_x^{(2)} = -\bar{E}_y^{(2)} = \frac{\pi q}{\varepsilon_1 + \varepsilon_2} e^{-\lambda(a-y)}.$$

---

[14] Note that the cosine transform of $E_y^{(i)}$ and the sine transform of $E_x^{(i)}$ vanish, because of the symmetry of the problem.

Using the inversion formulas

$$E_x^{(i)} = \frac{2}{\pi} \int_0^\infty \bar{E}_x^{(i)} \sin \lambda x \, d\lambda, \qquad E_y^{(i)} = \frac{2}{\pi} \int_0^\infty \bar{E}_y^{(i)} \cos \lambda x \, d\lambda,$$

and making a few simple calculations, we arrive at the expressions for the components of the electric field given in the answer on p. 154.

321. The electric field has only a z-component, whose complex amplitude we denote by $E(x, y)$. If we regard the current as distributed over an arbitrarily small rectangle $a - \delta < x < a + \delta$, $|y| < \varepsilon$, then the solution of the problem reduces to integration of the inhomogeneous Helmholtz equation

$$\Delta E + k^2 E = \frac{4\pi i \omega}{c^2} j(x, y), \tag{44}$$

where

$$j(x, y) = \begin{cases} \dfrac{J}{4\delta\varepsilon} & \text{for} \quad a - \delta < x < a + \delta, \qquad |y| < \varepsilon, \\ 0 & \text{otherwise}, \end{cases}$$

with boundary conditions

$$E\big|_{x=0} = E_{y=\pm b} = 0, \qquad E\big|_{x\to\infty} \to 0.$$

To solve the problem, we first make a Fourier sine transform, carrying (44) into the ordinary differential equation

$$\bar{E}'' - (\lambda^2 - k^2)\bar{E} = \frac{4\pi i \omega}{c^2} \int_0^\infty j(\xi, y) \sin \lambda \xi \, d\xi \tag{45}$$

for the quantity

$$\bar{E} = \int_0^\infty E \sin \lambda x \, dx.$$

The solution of (45) satisfying the boundary conditions

$$\bar{E}\big|_{y=\pm b} = 0$$

can be obtained by variation of constants. Then, taking the limit as $\delta, \varepsilon \to 0$, we find after some simple calculations that

$$\bar{E} = -\frac{2\pi i k J}{c\sqrt{\lambda^2 - k^2}} \frac{\sinh \sqrt{\lambda^2 - k^2}\,(b - |y|)}{\cosh \sqrt{\lambda^2 - k^2}\,b} \sin \lambda a.$$

This immediately leads to the answer on p. 157, if we use the inversion formula

$$E = \frac{2}{\pi} \int_0^\infty \bar{E} \sin \lambda x \, d\lambda.$$

324. We want the stresses σ_x, τ_{xy} and σ_y satisfying the system of equations

$$\frac{\partial \sigma_x}{\partial x} + \frac{\partial \tau_{xy}}{\partial y} = 0, \qquad \frac{\partial \tau_{xy}}{\partial x} + \frac{\partial \sigma_y}{\partial y} = 0, \qquad \text{(equilibrium equations)}$$

$$\frac{\partial^2 \sigma_x}{\partial y^2} - 2\frac{\partial^2 \tau_{xy}}{\partial x\,\partial y} + \frac{\partial^2 \sigma_y}{\partial x^2} = 0 \qquad \text{(compatibility equation)}$$

and the boundary conditions

$$\sigma_y|_{y=0} = f(x), \qquad \tau_{xy}|_{y=0} = g(x).$$

Introducing Fourier transforms

$$\bar{F} = \int_{-\infty}^{\infty} F e^{i\lambda x}\, dx$$

of the unknown functions, we multiply each of the equations (46) by $e^{i\lambda x}$ and integrate with respect to x from $-\infty$ to ∞, taking account of the behavior of the stresses as $x \to \pm\infty$.[15] This gives the system of ordinary differential equations

$$-i\lambda\bar{\sigma}_x + \bar{\tau}'_{xy} = 0, \qquad -i\lambda\bar{\tau}_{xy} + \bar{\sigma}'_y = 0,$$
$$\bar{\sigma}''_x + 2i\lambda\bar{\tau}'_{xy} - \lambda^2\bar{\sigma}_y = 0,$$

which must be solved with the boundary conditions

$$\bar{\sigma}_y|_{y=0} = \bar{f}, \qquad \bar{\tau}_{xy}|_{y=0} = \bar{g}$$

and the conditions at infinity.

$$\bar{\sigma}_x, \bar{\tau}_{xy}, \bar{\sigma}_y \to 0 \quad \text{as} \quad y \to \infty.$$

The solution of the system (47) satisfying all the conditions of the problem is

$$\bar{\tau}_{xy} = (A + B\lambda y)e^{-|\lambda|y},$$

$$\bar{\sigma}_x = \frac{1}{i}\left[B(1 - |\lambda|\,y) - A\frac{|\lambda|}{\lambda}\right]e^{-|\lambda|y},$$

$$\bar{\sigma}_y = \frac{1}{i}\left[\frac{|\lambda|}{\lambda}A + B(1 + |\lambda|\,y)\right]e^{-|\lambda|y},$$

where the constants A and B have the form

$$A = \bar{g}, \qquad B = i\bar{f} - \frac{|\lambda|}{\lambda}\bar{g}.$$

---

[15] We assume that the stresses and their first derivatives approach zero at infinity. It should be noted that the problem cannot be solved in this way for the Airy stress function, since the latter cannot be expanded as a Fourier integral.

The final form of the solution given on p. 158, involving various integrals, is found by using the inversion formula

$$F = \frac{1}{2\pi} \int_{-\infty}^{\infty} \bar{F} e^{-i\lambda x} \, d\lambda$$

to go back from the quantities $\bar{\tau}_{xy}$, $\bar{\sigma}_x$, $\bar{\sigma}_y$ to the stresses τ_{xy}, σ_x, σ_y themselves.

328. Replacing the concentrated force P by a load uniformly distributed over the arbitrarily small rectangle

$$-\frac{\delta}{2} < x < \frac{\delta}{2}, \qquad b - \frac{\varepsilon}{2} < y < b + \frac{\varepsilon}{2},$$

we reduce the problem to integration of the inhomogeneous biharmonic equation

$$\frac{\partial^4 u}{\partial x^4} + 2 \frac{\partial^4 u}{\partial x^2 \, \partial y^2} + \frac{\partial^4 u}{\partial y^4} = \frac{q(x, y)}{D}, \tag{48}$$

where

$$q(x, y) = \begin{cases} \dfrac{P}{\delta\varepsilon} & \text{for} \quad -\dfrac{\delta}{2} < x < \dfrac{\delta}{2}, \quad b - \dfrac{\varepsilon}{2} < y < b + \dfrac{\varepsilon}{2}, \\ 0 & \text{otherwise}, \end{cases}$$

subject to the boundary conditions

$$u|_{y=0} = \frac{\partial u}{\partial y}\bigg|_{y=0} = 0.$$

Taking the Fourier transform of (48), where

$$\bar{u} = \int_0^\infty u \cos \lambda x \, dx,$$

we obtain the following equation for $\bar{u}$:[16]

$$\bar{u}^{(iv)} - 2\lambda^2 \bar{u}'' + \lambda^4 \bar{u} = \frac{1}{D} \int_0^\infty q(\xi, y) \cos \lambda \xi \, d\xi. \tag{49}$$

The general solution of (49) can be obtained by variation of constants, and has the form

$$\bar{u}(y) = (A + B\lambda y)e^{-\lambda y} + (C + E\lambda y)e^{\lambda y}$$

$$+ \frac{1}{2D\lambda^3} \int_0^y K_\lambda(y, \eta) \, d\eta \int_0^\infty q(\xi, \eta) \cos \lambda \xi \, d\xi,$$

where

$$K_\lambda(y, \eta) = \lambda(y - \eta) \cosh \lambda(y - \eta) - \sinh \lambda(y - \eta).$$

---

[16] It is assumed that u and its first three derivatives with respect to x go to zero as $x \to \infty$.

At this stage, it is convenient to simplify the calculations by taking the limit as δ, $\varepsilon \to 0$. The result is

$$\bar{u}\big|_{0\leqslant y\leqslant b} = (A + B\lambda y)e^{-\lambda y} + (C + E\lambda y)e^{\lambda y},$$

$$\bar{u}\big|_{b\leqslant y<\infty} = (A + B\lambda y)e^{-\lambda y} + (C + E\lambda y)e^{\lambda y} + \frac{P}{4D\lambda^3} K_\lambda(y, b).$$

The constants C and E are determined from the condition

$$\bar{u}\big|_{y\to\infty} \to 0,$$

which gives

$$C = \frac{P}{8D\lambda^3}(1 + \lambda b)e^{-\lambda b}, \qquad E = -\frac{P}{8D\lambda^3} e^{-\lambda b}.$$

The other two constants are found from the boundary conditions

$$\bar{u}\big|_{y=0} = \bar{u}'\big|_{y=0} = 0,$$

which implies

$$A = -C, \qquad B = -2C - E.$$

The value of the deflection $u(x, y)$ is obtained by using the inversion formula. To find the bending moment and the shear force

$$-M = D\frac{\partial^2 u}{\partial y^2}\bigg|_{y=0}, \qquad -N = D\frac{\partial^3 u}{\partial y^3}\bigg|_{y=0}$$

on the clamped edge, we differentiate the expression

$$\bar{u}\big|_{y\leqslant b} = \frac{Pe^{-\lambda b}}{4D\lambda^3}[(1 + \lambda b + \lambda^2 by)\sinh\lambda y - \lambda y(1 + \lambda b)\cosh\lambda y],$$

obtaining

$$\frac{\partial^2\bar{u}}{\partial y^2}\bigg|_{y=0} = \frac{Pbe^{-\lambda b}}{2D}, \qquad \frac{\partial^3\bar{u}}{\partial y^3}\bigg|_{y=0} = -\frac{(1 + \lambda b)Pe^{-\lambda b}}{2D}.$$

The values of M and N are then found by substituting the corresponding values of $\bar{M}$ and $\bar{N}$ into the appropriate inversion formulas.

334. The problem reduces to integration of the heat conduction equation

$$\frac{1}{r}\frac{\partial}{\partial r}\left(r\frac{\partial T}{\partial r}\right) = \frac{\partial T}{\partial\tau}, \qquad 0 \leqslant r < \infty,$$

with the initial condition

$$T\big|_{\tau=0} = f(r).$$

Writing $T = R(r)\Theta(\tau)$ and separating variables, we obtain

$$\frac{1}{r}(rR')' + \lambda R = 0, \qquad \Theta' + \lambda\Theta = 0.$$

Integrating these equations, and taking account of the boundedness of T as $r \to 0$, we find that

$$T = T_\lambda = c_\lambda e^{-\lambda \tau} J_0(\sqrt{\lambda}\, r).$$

It follows from the boundedness of T as $r \to \infty$ that the parameter λ can only take positive values $\lambda = \mu^2$. This leads to the following set of particular solutions depending continuously on μ:

$$T = T_\mu = c_\mu e^{-\mu^2 \tau} J_0(\mu r), \qquad 0 \leqslant \mu < \infty.$$

The general solution is then constructed as an integral of the form

$$T(r, \tau) = \int_0^\infty c_\mu e^{-\mu^2 \tau} J_0(\mu r)\, d\mu. \tag{50}$$

The coefficients c_μ are determined from the initial condition, which gives

$$c_\mu = \mu \int_0^\infty f(r) J_0(\mu r) r\, dr, \tag{51}$$

if we take account of Hankel's integral theorem. Substituting (51) into (50), reversing the order of integration and then integrating with respect to μ, we find the form of the solution given in the answer on p. 162.

335. We want the solution of the equation

$$\frac{1}{r} \frac{\partial}{\partial r}\left(r \frac{\partial T}{\partial r} \right) = \frac{\partial T}{\partial \tau}, \qquad a < r < \infty \tag{52}$$

satisfying the initial condition $T|_{\tau=0} = 0$ and the boundary conditions[17]

$$T|_{r=a} = T_0, \qquad T|_{r \to \infty} \to 0.$$

Writing

$$\varphi_\lambda(r) = J_0(\lambda a) Y_0(\lambda r) - Y_0(\lambda a) J_0(\lambda r),$$

we carry out a "Weber transform" by multiplying (52) by $r\varphi_\lambda(r)$ and integrating from a to ∞. Taking account of the behavior of the various functions as $r \to \infty$ and the relations

$$\varphi_\lambda(a) = 0, \qquad \varphi_\lambda'(a) = \frac{2}{\pi a},$$

we find that

$$\frac{d\bar{T}}{d\tau} - \lambda^2 \bar{T} = \frac{2T_0}{\pi}, \tag{53}$$

where

$$\bar{T} = \int_a^\infty T r \varphi_\lambda(r)\, dr.$$

---

[17] It is assumed that $\sqrt{r}\, T$ and $\sqrt{r}(\partial T/\partial r)$ approach zero as $r \to \infty$, and that the integral

$$\int_a^\infty \sqrt{r}\, |T|\, dr$$

converges.

The solution of (53) satisfying the condition $\bar{T}|_{\tau=0}$ is

$$\bar{T} = \frac{2T_0}{\pi\lambda^2}(1 - e^{-\lambda^2\tau}).$$

To determine the solution $T(r, \tau)$ from its Weber transform, we use the inversion formula

$$T(r, \tau) = \int_0^\infty \frac{\bar{T}\,\varphi_\lambda(r)\lambda\,d\lambda}{J_0^2(\lambda a) + Y_0^2(\lambda a)}$$

[cf. formula (15), p. 161].

351. As is well known (see T4, p. 343), in the case of axially symmetric problems of elasticity theory, the stresses can be expressed in terms of a solution $u(r, z)$ of the biharmonic equation (it is assumed that there are no body forces). To subtract out the singularity at the point of application of the force, we write

$$u = u_0 + u_1,$$

where

$$u_0 = \frac{P}{8\pi(1 - \nu)}\sqrt{r^2 + (z - a)^2}$$

is the stress function corresponding to a concentrated force P applied to an infinite elastic body, and u_1 is a biharmonic function regular in the region $z > 0$. Since the unknown stress σ_z is related to the function u by the formula

$$\sigma_z = \frac{\partial}{\partial z}\left[(2 - \nu)\,\Delta u - \frac{\partial^2 u}{\partial z^2}\right],$$

to solve the problem we need only find the quantities Δu_1 and $\partial^2 u_1/\partial z^2$. The first quantity is harmonic in the region $z > 0$ and can be written as an integral

$$\Delta u_1 = \int_0^\infty A_\lambda e^{-\lambda z} J_0(\lambda r)\lambda\,d\lambda, \tag{54}$$

while the second quantity is biharmonic in the region $z > 0$ and can be written in the form

$$\frac{\partial^2 u_1}{\partial z^2} = \int_0^\infty (B_\lambda + C_\lambda z)e^{-\lambda z} J_0(\lambda r)\lambda\,d\lambda \tag{55}$$

(note that the integrand is biharmonic). Comparing the result of differentiating (54) twice with respect to z with the result of applying the operator

$$\Delta = \frac{1}{r}\frac{\partial}{\partial r}\left(r\frac{\partial}{\partial r}\right) + \frac{\partial^2}{\partial z^2}$$

to (55), we find that $C_\lambda = -\frac{1}{2}\lambda A_\lambda$. To determine the remaining constants, we have to use the boundary conditions

$$\sigma_z\big|_{z=0} = \tau_{rz}\big|_{z=0} = 0,$$

which can be written as conditions on the function u_1:[18]

$$\frac{\partial}{\partial z}\left[(2-\nu)\,\Delta u_1 - \frac{\partial^2 u_1}{\partial z^2}\right]\bigg|_{z=0} = -\frac{\partial}{\partial z}\left[(2-\nu)\,\Delta u_0 - \frac{\partial^2 u_0}{\partial z^2}\right]\bigg|_{z=0},$$

$$\left[(1-\nu)\,\Delta u_1 - \frac{\partial^2 u_1}{\partial z^2}\right]\bigg|_{z=0} = -\left[(1-\nu)\,\Delta u_0 - \frac{\partial^2 u_0}{\partial z^2}\right]\bigg|_{z=0}.$$

Performing the differentiations on the right, expanding the results in Hankel integrals and substituting from (54) and (55), we obtain a system of linear equations determining the constants A_λ and B_λ. The formula given in the answer on p. 168 is obtained after evaluating certain integrals of a familiar type.

355. The problem reduces to integration of the one-dimensional heat conduction equation

$$\frac{\partial^2 T}{\partial x^2} = \frac{\partial T}{\partial \tau},$$

with the initial condition

$$T\big|_{\tau=0} = 0$$

the boundary condition

$$-k\frac{\partial T}{\partial x}\bigg|_{x=0} = q(\tau)$$

and the condition at infinity

$$T\big|_{x\to\infty} \to 0.$$

Introducing the Laplace transform

$$\bar{T} = \int_0^\infty T e^{-p\tau}\,d\tau,$$

we multiply the differential equation and boundary conditions by $e^{-p\tau}$ and integrate with respect to τ from 0 to ∞. If we take account of the initial condition, this gives

$$\bar{T}'' - p\bar{T} = 0, \quad -k\bar{T}'\big|_{x=0} = \bar{q}, \quad \bar{T}\big|_{x\to\infty} \to 0,$$

which implies

$$\bar{T} = \frac{\bar{q}}{k\sqrt{p}}\,e^{-\sqrt{p}\,x}, \qquad \mathrm{Re}\,\sqrt{p} > 0.$$

---

[18] The second of these equations follows from the formula

$$\tau_{rz} = \frac{\partial}{\partial r}\left[(1-\nu)\Delta u - \frac{\partial^2 u}{\partial z^2}\right].$$

The problem is now solved by using the Fourier-Mellin inversion theorem

$$T = \frac{1}{2\pi i k} \int_\Gamma e^{p\tau - \bar{p}x} \bar{q} \frac{dp}{\sqrt{p}},$$

where Γ is a straight line parallel to the imaginary axis lying to the right of all the singular points of the integrand. In Case a, where $q = q_0 = \text{const}$, we have

$$T = \frac{q_0}{k} \frac{1}{2\pi i} \int_\Gamma e^{p\tau - \sqrt{px}} \frac{dp}{p\sqrt{p}}. \quad (56)$$

As the next step, we calculate the derivative

$$\frac{\partial T}{\partial x} = -\frac{q_0}{k} \frac{1}{2\pi i} \int_\Gamma e^{p\tau - \sqrt{px}} \frac{dp}{p}.$$

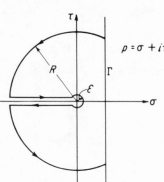

FIGURE 159

Applying Cauchy's integral theorem to the contour shown in Figure 159, and then taking the limit as $\varepsilon \to 0$, $R \to \infty$, we obtain[19]

$$\frac{\partial T}{\partial x} = -\frac{q_0}{k} \left[1 - \frac{2}{\pi} \int_0^\infty e^{-r\tau} \sin \sqrt{r}\, x \frac{dr}{r} \right]$$

$$= -\frac{q_0}{k} \left[1 - \Phi\left(\frac{x}{2\sqrt{\tau}}\right) \right],$$

where $\Phi(x)$ is the probability integral. It follows that

$$T = -\frac{q_0}{k} \int_\infty^x \left[1 - \Phi\left(\frac{\xi}{2\sqrt{\tau}}\right) \right] d\xi,$$

and the final form of the solution given in the answer on p. 171 is obtained from this formula by integrating by parts.

In Case b,

$$\bar{q} = \frac{q_0\omega}{\omega^2 + p^2},$$

$$T = \frac{q_0\omega}{k} \frac{1}{2\pi i} \int_\Gamma e^{p\tau - \sqrt{px}} \frac{dp}{\sqrt{p}\,(p^2 + \omega^2)}.$$

The temperature of the surface of the body can be found by using the convolution theorem, which gives

$$T\big|_{x=0} = \frac{q_0}{k} \frac{1}{2\pi i} \int_\Gamma \frac{\omega}{\omega^2 + p^2} e^{p\tau} \frac{dp}{\sqrt{p}} = \frac{q_0}{k\sqrt{\pi}} \int_0^\tau \sin \omega(\tau - s) \frac{ds}{\sqrt{s}}$$

and leads at once to the answer on p. 171.

---

[19] Direct application of the method of contour integration to the integral (56) itself is impossible, since the corresponding integral along the circle of radius ε becomes infinite as $\varepsilon \to 0$.

357. Using the Laplace transform, we write the solution in the form of a contour integral

$$T = T_0 \left[1 - \frac{h}{2\pi i} \int_\Gamma \frac{e^{p\tau - \sqrt{p}x}}{p(h + \sqrt{p})} \, dp \right],$$

where Γ is a straight line parallel to and on the right of the imaginary axis, and $\sqrt{p}$ denotes the branch of the square root whose real part is positive.[20] A simple way of calculating the integral

$$J = \frac{1}{2\pi i} \int_\Gamma \frac{e^{p\tau - \sqrt{p}x}}{p(h + \sqrt{p})} \, dp$$

is to make the substitution

$$\frac{1}{\sqrt{p} + h} = \int_0^\infty e^{-s(\sqrt{p} + h)} \, ds$$

and then reverse the order of integration. Together with the result obtained in the solution of Prob. 355, this gives

$$J = \int_0^\infty e^{-hs} \, ds \, \frac{1}{2\pi i} \int_\Gamma e^{p\tau - \sqrt{p}(x+s)} \frac{dp}{p} = \int_0^\infty e^{-hs} \left[1 - \Phi\left(\frac{x+s}{2\sqrt{\tau}} \right) \right] \, ds.$$

Integrating by parts, we find that

$$J = \frac{1}{h} \left[1 - \Phi\left(\frac{x}{2\sqrt{\tau}} \right) \right] - \frac{1}{h} e^{hx + h^2\tau} \left[1 - \Phi\left(h\sqrt{\tau} + \frac{x}{2\sqrt{\tau}} \right) \right],$$

which leads at once to the answer on p. 172.

371. The problem reduces to finding a solution of the equation

$$\frac{1}{r} \frac{\partial}{\partial r} \left(r \frac{\partial T}{\partial r} \right) = \frac{\partial T}{\partial \tau}$$

satisfying the initial conditions

$$T\big|_{\tau=0} = \begin{cases} 0, & r < a, \\ T_0, & r = a, \end{cases}$$

and the boundary condition

$$\left(\alpha \frac{\partial T}{\partial \tau} + \frac{1}{a} \frac{\partial T}{\partial r} \right)\bigg|_{r=a, \tau>0} = 0.$$

Taking Laplace transforms and using the initial condition, we obtain the equation

$$\frac{1}{r} \frac{d}{dr} \left(r \frac{d\bar{T}}{dr} \right) - p\bar{T} = 0$$

---

[20] For this branch, $\sqrt{p} \neq -h$, and hence $p = 0$ is the only singular point of the integrand.

and the condition

$$\left(\alpha(p\bar{T} - T_0) + \frac{1}{a}\bar{T}'\right)\Bigg|_{r=a} = 0,$$

which together imply

$$\bar{T} = \frac{\alpha a T_0 I_0(\sqrt{p}\, r)}{\sqrt{p}[\alpha a \sqrt{p}\, I_0(\sqrt{p}\, a) + I_1(\sqrt{p}\, a)]}.$$

The solution of the problem is given by the inversion formula

$$T = \frac{\alpha a T_0}{2\pi i} \int_\Gamma \frac{I_0(\sqrt{p}\, r) e^{p\tau}\, dp}{\sqrt{p}[\alpha a \sqrt{p}\, I_0(\sqrt{p}\, a) + I_1(\sqrt{p}\, a)]}.$$

The contour integral can be evaluated by residues, since the integrand is single-valued. The singular points of the integrand consist of poles at the points $p = 0$ and $p = p_n = -\gamma_n^2/a^2$, where the γ_n are consecutive positive roots of the equation

$$J_1(\gamma) + \alpha J_0(\gamma) = 0.$$

Calculating the residues at these points, we immediately find the answer on p. 177.[21] As in other problems with boundary conditions involving time derivatives, the solution of this problem is greatly simplified by the use of Laplace transforms.

375. In the first region $0 \leqslant r < \infty$, $0 < z < \infty$, the concentration $C_1(r, z, t)$ satisfies the equation

$$\frac{1}{r}\frac{\partial}{\partial r}\left(r\frac{\partial C_1}{\partial r}\right) + \frac{\partial^2 C_1}{\partial z^2} = \frac{1}{D}\frac{\partial C_1}{\partial t}, \tag{57}$$

the initial condition

$$C_1\big|_{t=0},$$

the boundary condition

$$\frac{\partial C_1}{\partial z}\bigg|_{z=0} = \begin{cases} f(t), & 0 \leqslant r < a, \\ 0, & a < r < \infty \end{cases} \tag{58}$$

[where $f(t)$ is a function to be determined later], and the conditions at infinity

$$C_1\big|_{r \to \infty} \to 0, \qquad C_1\big|_{z \to \infty} \to 0.$$

In the second region (the tube), the concentration $C_2(z, t)$ satisfies the one-dimensional equation

$$\frac{\partial^2 C_2}{\partial z^2} = \frac{1}{D}\frac{\partial C_2}{\partial t}, \tag{59}$$

---

[21] It is easy to see that the integral along the large circle of radius R completing the contour of integration goes to zero as $R \to \infty$.

the initial condition

$$C_2|_{t=0} = C_0,$$

and the boundary conditions

$$\frac{\partial C_2}{\partial z}\bigg|_{z=0} = f(t), \qquad \frac{\partial C_2}{\partial z}\bigg|_{z=-l} = 0. \qquad (60)$$

Taking first the Laplace transform and then the Hankel transform of (57) and (58), and using the initial condition and the condition $C_1|_{r\to\infty} \to 0$, we obtain

$$\frac{d^2\bar{\bar{C}}_1}{dz^2} - \left(\frac{p}{D} + \lambda^2\right)\bar{\bar{C}}_1 = 0,$$

$$\frac{d\bar{\bar{C}}_1}{dz}\bigg|_{z=0} = \frac{\bar{f}}{\lambda}\, aJ_1(\lambda a), \qquad \bar{\bar{C}}_1\big|_{z\to\infty} \to 0,$$

where a single overbar denotes the Laplace transform and a double overbar the Laplace transform followed by the Hankel transform. Integrating the equation for $\bar{\bar{C}}_1$, we obtain

$$\bar{\bar{C}}_1 = -\frac{\bar{f}aJ_1(\lambda a)}{\lambda\sqrt{\lambda^2 + \dfrac{p}{D}}}\, e^{-\sqrt{\lambda^2+(p/D)}\,z},$$

which implies

$$\bar{C}_1\big|_{r=z=0} = \int_0^\infty \bar{\bar{C}}_1 \lambda\, d\lambda = \frac{\bar{f}}{\sqrt{p/D}}\,[e^{-a\sqrt{p/D}} - 1]$$

after inverting the Hankel transform. Similarly, taking the Laplace transform of (59) and (60), we find that

$$\bar{C}_2 = \frac{C_0}{p} + \frac{\bar{f}\cosh\sqrt{p/D}(z+l)}{\sqrt{p/D}\,\sinh\sqrt{p/D}\,l}.$$

In the present approximation, we can find the unknown quantity f by using the relation

$$\bar{C}_1\big|_{r=z=0} = \bar{C}_2\big|_{z=0},$$

which implies

$$f = \frac{C_0}{\sqrt{pD}(e^{-a\sqrt{p/D}} - 1 - \coth\sqrt{p/D}\,l)}.$$

The amount of substance M in the tube can now be calculated from the formula

$$M = \int_{-l}^0 C_2(z, t)\, dz = M_0 + \frac{C_0}{2\pi i}\int_\Gamma \frac{e^{pt}\, dp}{p\sqrt{p/D}[e^{-a\sqrt{p/D}} - 1 - \coth\sqrt{p/D}\,l]},$$

where $M_0 = C_0 l$ is the initial amount of substance inside the tube. Integrating along the contour shown in Figure 159, we obtain the answer on p. 178.

386. We want the solution of the system

$$L\frac{\partial I_1}{\partial t} + \frac{\partial u_1}{\partial x} = 0, \quad C\frac{\partial u_1}{\partial t} + \frac{\partial I_1}{\partial x} = 0 \quad (0 < x < l),$$

$$L\frac{\partial I_2}{\partial t} + \frac{\partial u_2}{\partial x} = 0, \quad C\frac{\partial u_2}{\partial t} + \frac{\partial I_2}{\partial x} = 0 \quad (l < x < \infty)$$

satisfying zero initial conditions and the boundary conditions

$$u_1\big|_{x=0} = Ee^{-\alpha t}, \qquad u_2\big|_{x\to\infty} \to 0,$$

$$u_1\big|_{x=l} = u_2\big|_{x=l}, \qquad I_1\big|_{x=l} = I_2\big|_{x=l} + \frac{u_2}{R_0}\bigg|_{x=l}.$$

Eliminating the variable t by taking Laplace transforms, we obtain

$$Lp\bar{I}_1 + \frac{d\bar{u}_1}{dx} = 0, \qquad Cp\bar{u}_1 + \frac{d\bar{I}_1}{dx} = 0 \quad (0 < x < l),$$

$$Lp\bar{I}_2 + \frac{d\bar{u}_2}{dx} = 0, \qquad Cp\bar{u}_2 + \frac{d\bar{I}_2}{dx} = 0 \quad (l < x < \infty),$$

$$\bar{u}_1\big|_{x=0} = \frac{E_0}{p+\alpha}, \qquad \bar{u}_2\big|_{x\to\infty} \to 0,$$

$$\bar{u}_1\big|_{x=l} = \bar{u}_2\big|_{x=l}, \qquad \bar{I}_1\big|_{x=l} = \bar{I}_2\big|_{x=l} + \frac{\bar{u}_2}{R_0}\bigg|_{x=l}.$$

These equations can be solved for $\bar{u}_1$, $\bar{u}_2$, $\bar{I}_1$ and $\bar{I}_2$. In particular, for $\bar{u}_2$ we obtain the expression

$$\bar{u}_2 = \frac{E_0}{p+\alpha} \frac{e^{-p(x-l)/v}}{\cosh pT + [1 + (Z/R_0)]\sinh pT},$$

where $v = 1/\sqrt{LC}$ is the propagation velocity, $T = l/v$ is the time it takes a wave to traverse the part of the line going from $x = 0$ to $x = l$, and $Z = \sqrt{L/C}$ is the wave resistance. Then the Fourier-Mellin inversion formula leads to the following representation of u_2 as a contour integral:

$$u_2(x, t) = \frac{E_0}{2\pi i}\int_\Gamma \frac{e^{[t-(x/v)+T]}}{\cosh pT + [1 + (Z/R_0)]\sinh pT}\frac{dp}{p+\alpha}, \quad x \geqslant l.$$

The most interesting form of the solution can be obtained by using the expansion

$$\frac{1}{\cosh pT + [1 + (Z/R_0)]\sinh pT} = \frac{2R_0 e^{-pT}}{2R_0 + Z}\sum_{n=0}^{\infty}\left(\frac{Z}{2R_0 + Z}\right)^n e^{-2npT}$$

and then integrating term by term. This gives

$$u_2(x, t) = \frac{2R_0E_0}{2R_0 + Z} \sum_{n=0}^{\infty} \left(\frac{Z}{2R_0 + Z}\right)^n \frac{1}{2\pi i} \int_{\Gamma} e^{p[t-(x/v)-2nT]} \frac{dp}{p + \alpha} .$$

According to the formula

$$\frac{1}{2\pi i} \int_{\Gamma} e^{p\tau} \frac{dp}{p + \alpha} = \begin{cases} 0, & \tau < 0, \\ e^{-\alpha\tau}, & \tau > 0, \end{cases}$$

all the terms of this series vanish for fixed x and t, starting from some value of n. In particular, we have

$$u_2\big|_{0 < t < x/v} = 0,$$

$$u_2\big|_{x/v < t < 2T+(x/v)} = \frac{2R_0E_0}{2R_0 + Z} e^{-\alpha[t-(x/v)]},$$

$$u_2\big|_{2T+(x/v) < t < 4T+(x/v)} = \frac{2R_0E_0}{2R_0 + Z}\left[e^{-\alpha[t-(x/v)]} + \frac{Z}{2R_0 + Z} e^{-\alpha[t-(x/v)-2T]}\right],$$

and so on. The general result given in the answer on p. 183 can easily be obtained by induction, with the help of the formula for summing a finite geometric series. The jumps in the voltage can be interpreted as the arrival at the point x of successive refracted waves.

402. The problem involves integration of the equation

$$\Delta^2 u + \frac{\partial^2 u}{\partial \tau^2} = 0 \qquad (\tau = \sqrt{D/\rho}\, t, \quad 0 \leqslant r < \infty),$$

subject to the conditions

$$\frac{\partial u}{\partial r}\bigg|_{r=0} = 0, \quad \left(r \frac{\partial}{\partial r}(\Delta u)\right)\bigg|_{r=0} = f(\tau) = \begin{cases} \dfrac{P\sqrt{D}}{2\pi\sqrt{\rho}\,\varepsilon}, & 0 \leqslant \tau < \varepsilon, \\ 0, & \tau > \varepsilon, \end{cases} \quad (\varepsilon \to 0)$$

at the point of application of the force, and to the condition at infinity

$$u\big|_{r \to \infty} \to 0.$$

Going over to the Laplace transform $\bar{u}$, we find that $\bar{u}$ satisfies the differential equation

$$\Delta^2 \bar{u} + p^2 \bar{u} = 0, \tag{61}$$

the boundary conditions

$$\frac{d\bar{u}}{dr}\bigg|_{r=0} = 0, \quad \left(r \frac{d}{dr}(\Delta \bar{u})\right)\bigg|_{r=0} = \frac{P}{2\pi D}, \quad \bar{u}\big|_{r \to \infty} \to 0$$

and the condition at infinity

$$\bar{u}\big|_{r\to\infty} \to 0.$$

The solution of (61) vanishing at infinity is

$$\bar{u} = AK_0(r\sqrt{p}\, e^{i\pi/4}) + BK_0(r\sqrt{p}\, e^{-i\pi/4}),$$

where $\sqrt{p}$ denotes the principal branch of the square root ($|\arg p| < \pi$). Taking account of the behavior of the Macdonald function near the point $r = 0$ and using the boundary conditions, we find that

$$B = -A = \frac{P}{4\pi i p\sqrt{D\rho}}.$$

Application of the convolution theorem gives

$$u(r, \tau) = \frac{P}{4\pi i\sqrt{D\rho}} \int_0^\tau F(\tau)\, d\tau,$$

where

$$F(\tau) = \frac{1}{2\pi i} \int_\Gamma [K_0(r\sqrt{p}\, e^{-i\pi/4}) - K_0(r\sqrt{p}\, e^{i\pi/4})]e^{p\tau}dp.$$

Transforming this expression by integrating along the contour shown in Figure 159, we eventually obtain[22]

$$\begin{aligned}
F(\tau) &= \frac{1}{2\pi i} \int_0^\infty [K_0(r\sqrt{\rho}\, e^{3i\pi/4}) - K_0(r\sqrt{\rho}\, e^{i\pi/4}) \\
&\quad - K_0(r\sqrt{\rho}e^{-i\pi/4}) + K_0(r\sqrt{\rho}\, e^{-3i\pi/4})]e^{-\rho\tau}\, d\rho \\
&= \frac{1}{2} \int_0^\infty e^{-\rho\tau}[I_0(r\sqrt{\rho}\, e^{i\pi/4}) - I_0(r\sqrt{\rho}\, e^{-i\pi/4})]\, d\rho.
\end{aligned}$$

Using the formula

$$\int_0^\infty e^{-a^2x^2} J_0(bx)x\, dx = \frac{e^{-b^2/4a^2}}{2a^2},$$

to evaluate the integral, we find that

$$F(\tau) = -\frac{1}{i\tau} \sin\frac{r^2}{4\tau},$$

which immediately leads to the answer on p. 188, after integrating with respect to τ.

---

[22] Note that the function $K_0(z)$ is analytic in the z-plane cut along the line $(-\infty, 0)$, and use the formula

$$K_0(e^{i\pi}y) = K_0(y) - i\pi I_0(y).$$

406. Taking first the Laplace transform of the original differential equation and boundary conditions, and then the Fourier transform with respect to the variable y, we find that

$$\frac{d^2\bar{\bar{u}}}{dx^2} - \left(\frac{p^2}{v^2} + bp + \lambda^2\right)\bar{\bar{u}} = 0,$$

$$\bar{\bar{u}}\Big|_{x=0} = \frac{1}{p}F, \qquad \bar{\bar{u}}\Big|_{x\to\infty} \to 0,$$

where

$$\bar{\bar{u}} = \int_{-\infty}^{\infty} e^{i\lambda y}\, dy \int_0^{\infty} u e^{-pt}\, dt, \qquad F = \int_{-\infty}^{\infty} f(\eta) e^{i\lambda\eta}\, d\eta.$$

It follows that

$$\bar{\bar{u}} = \frac{1}{p} F e^{-\sqrt{(p/v)^2 + bp + \lambda^2}\, x},$$

where the radical denotes the branch of the square root which has positive real part. Using the inversion formula

$$u = \frac{1}{2\pi i}\int_\Gamma e^{pt}\, dp\, \frac{1}{2\pi}\int_{-\infty}^{\infty}\bar{\bar{u}}\, e^{-i\lambda y}\, d\lambda$$

and reversing the order of integration, after substituting for $\bar{\bar{u}}$ and F, we obtain

$$u = \frac{1}{2\pi}\int_{-\infty}^{\infty} f(\eta)\, d\eta\, \frac{1}{2\pi i}\int_\Gamma e^{pt}\frac{dp}{p}\int_{-\infty}^{\infty} e^{i\lambda(\eta - y)} e^{-\sqrt{(p/v)^2 + bp + \lambda^2}\, x}\, d\lambda.$$

The inner integral can be evaluated by using the formula[23]

$$\int_0^{\infty} e^{-\alpha\sqrt{\lambda^2 + z^2}}\cos\beta\lambda\, d\lambda = \frac{\alpha z}{\sqrt{\alpha^2 + \beta^2}} K_1(\sqrt{\alpha^2 + \beta^2}\, z) \tag{62}$$

involving Macdonald's function $K_1(x)$. Then the solution can be represented as the following double integral:

$$u = \frac{x}{\pi}\int_{-\infty}^{\infty}\frac{f(\eta)\, d\eta}{\sqrt{x^2 + (y - \eta)^2}}$$

$$\times \frac{1}{2\pi i}\int_\Gamma \sqrt{\frac{p^2}{v^2} + bp}\, K_1\left(\sqrt{\frac{p^2}{v^2} + bp}\,\sqrt{x^2 + (y - \eta)^2}\right) e^{pt}\frac{dp}{p}. \tag{63}$$

---

[23] To deduce (62), substitute $\mu = 1$, $\nu = -\tfrac{1}{2}$ into formula (5.15.6) on p. 134 of L9, recalling that

$$J_{-1/2}(z) = \sqrt{\frac{2}{\pi z}}\cos z, \qquad K_{1/2}(z) = \sqrt{\frac{\pi}{2z}}\, e^{-z}.$$

We now use Cauchy's theorem to evaluate the contour integral

$$J(t, \mu) = \frac{1}{2\pi i} \int_\Gamma K_0(\mu \sqrt{p} \sqrt{p + c}) e^{pt}\, dp,$$

bypassing the branch points $p = -c$ and $p = 0$ both on the upper and lower branches of the cut. The result is

$$J(t, \mu) = \begin{cases} \dfrac{e^{-ct/2}}{\sqrt{t^2 - \mu^2}} \cosh \dfrac{c\sqrt{t^2 - \mu^2}}{2}, & t > \mu, \\ 0, & t < \mu, \end{cases} \tag{64}$$

where in the course of the calculations, we use the formula

$$\int_0^{\pi/2} J_0(z_1 \sin x) \cosh(z_2 \cos x) \sin x\, dx = \frac{\sinh \sqrt{z_2^2 - z_1^2}}{\sqrt{z_2^2 - z_1^2}}, \qquad z_2 > z_1$$

[easily deduced from formula (4.455) of R2, p. 240 by setting $p = 0$, $q = -\frac{1}{2}$. It follows from (63) and (64) that

$$u\big|_{t > x/v} = -\frac{x}{\pi v} \int_{y - \sqrt{v^2 t^2 - x^2}}^{y + \sqrt{v^2 t^2 - x^2}} \frac{f(\eta)\, d\eta}{\sqrt{x^2 + (y - \eta)^2}} \frac{\partial}{\partial \mu} \int_\mu^t J(\tau, \mu)\, d\tau,$$

$$u\big|_{t < x/v} = 0,$$

where

$$\mu = \frac{1}{v} \sqrt{x^2 + (y - \eta)^2}, \qquad c = v^2 b.$$

The answer on p. 189 is easily obtained by evaluating the inner integral.

407. This problem can easily be solved by using the Mellin transform. Suppose the function T being sought is such that

$$T = O(1), \qquad r \frac{\partial T}{\partial r} = O(1) \qquad \text{as} \quad r \to 0,$$

$$T = O(r^{-s}), \qquad r \frac{\partial T}{\partial r} = O(r^{-s}) \qquad \text{as} \quad r \to \infty, \tag{65}$$

where s is some positive number.[24] Multiplying Laplace's equation $\Delta T = 0$ by r^{p+1}, where p is a complex number such that $0 < \operatorname{Re} p < s$, and integrating the result from 0 to ∞, we obtain

$$\left(r^{p+1} \frac{\partial T}{\partial r} - p r^p T \right)\bigg|_0^\infty + p^2 \bar{T} + \frac{d^2 \bar{T}}{d\varphi^2} = 0, \tag{66}$$

---

[24] The existence of a solution with these properties can be anticipated from physical considerations. After the solution has been obtained, we can easily verify that it actually satisfies all the conditions of the problem.

where

$$\bar{T} = \int_0^\infty T r^{p-1} \, dr \qquad (67)$$

is the Mellin transform of the function $\bar{T}$. It follows from (65) that the term

$$\left(r^{p+1} \frac{\partial T}{\partial r} - p r^p T \right) \Big|_0^\infty$$

vanishes if $0 < \operatorname{Re} p < s$,[25] thereby reducing (66) to

$$\frac{d^2 \bar{T}}{d\varphi^2} + p^2 \bar{T} = 0,$$

together with the boundary conditions

$$\bar{T}\big|_{\varphi=0} = 0, \qquad \bar{T}\big|_{\varphi=\alpha} = T_0 \frac{a^p}{p}$$

(implied by those obeyed by the function T). Thus we see at once that

$$\bar{T} = T_0 \frac{a^p}{p} \frac{\sin p}{\sin p\alpha},$$

which, in particular, implies that $s = \pi/\alpha$.

The temperature distribution T is now determined by using the inversion formula

$$T = \frac{T_0}{2\pi i} \int_{\sigma - i\infty}^{\sigma + i\infty} \frac{\sin p\varphi}{\sin p\alpha} \left(\frac{a}{r} \right)^p \frac{dp}{p},$$

where $0 < \sigma < \pi/\alpha$. The line integral can be evaluated by using residue theory, after completing the contour of integration by the arc of a circle of sufficiently large radius, lying to the left of the line $\operatorname{Re} p = \sigma$ if $r < a$ and to the right of this line if $r > a$. After some simple calculations, we obtain

$$\frac{T}{T_0} = \begin{cases} \dfrac{\varphi}{\alpha} + \dfrac{1}{\pi} \displaystyle\sum_{n=1}^\infty \frac{(-1)^n}{n} \left(\frac{r}{a} \right)^{n\pi/\alpha} \sin \frac{n\pi\varphi}{\alpha}, & 0 < r < a, \\[3mm] \dfrac{1}{\pi} \displaystyle\sum_{n=1}^\infty \frac{(-1)^{n+1}}{n} \left(\frac{a}{r} \right)^{n\pi/\alpha} \sin \frac{n\pi\varphi}{\alpha}, & a < r < \infty. \end{cases}$$

Using the formula

$$\sum_{n=1}^\infty \frac{(-1)^{n+1}}{n} \rho^n \sin nx = \arctan \frac{\rho \sin x}{1 + \rho \cos x}, \qquad \rho^2 < 1,$$

to sum the series, we arrive at the single analytical expression for the function $T(r, \varphi)$ given in the answer on p. 190.

---

[25] Note that it also follows from (65) that the integral is analytic in the strip $0 < \operatorname{Re} p < s$, being uniformly convergent in every closed subset of the strip.

415. If we replace the concentrated load by a uniformly distributed load with density

$$q(r, \varphi) = \begin{cases} \dfrac{P}{r_0 \, \delta \varepsilon} & \text{for } r_0 - \dfrac{\delta}{2} < r < r_0 + \dfrac{\delta}{2}, \quad \varphi_0 - \dfrac{\varepsilon}{2} < \varphi < \varphi_0 + \dfrac{\varepsilon}{2}, \\ 0 & \text{otherwise,} \end{cases}$$

where δ and ε are arbitrarily small positive numbers, then the problem reduces to solving the inhomogeneous biharmonic equation

$$\Delta^2 u = \frac{q(r, \varphi)}{D} \qquad (0 < r < \infty, \quad 0 < \varphi < \alpha), \tag{68}$$

subject to the boundary conditions

$$u\big|_{\varphi=0} = \frac{\partial u}{\partial \varphi}\bigg|_{\varphi=0} = u\big|_{\varphi=\alpha} = \frac{\partial u}{\partial \varphi}\bigg|_{\varphi=\alpha} = 0.$$

Multiplying (68) by r^{p+2} (where p is a suitably chosen complex number), and integrating from 0 to ∞, we find that [26]

$$\left\{ r^{p+2} \frac{\partial}{\partial r} \Delta u - (p+1) r^{p+1} \Delta u + (p+1)^2 r^p \frac{\partial u}{\partial r} - (p-1)(p+1)^2 r^{p-1} u \right\}\bigg|_0^\infty$$

$$+ (p-1)^2 (p+1)^2 \bar{u} + [(p-1)^2 + (p+1)^2] \frac{d^2 \bar{u}}{d\varphi^2} + \frac{d^4 \bar{u}}{d\varphi^4}$$

$$= \frac{1}{D} \int_0^\infty q(r, \varphi) r^{p+2} \, dr, \tag{69}$$

where

$$\bar{u} = \int_0^\infty u r^{p-2} \, dr. \tag{70}$$

Suppose the function u is such that the quantities $r^{-1}u$, $\partial u/\partial r$, $r \Delta u$ and $r^2(\partial \Delta u/\partial r)$ are all $O(r^{s_1})$ as $r \to 0$ and all $O(r^{-s_2})$ as $r \to \infty$, where $s_1 > 0$, $s_2 > 0$. Then the integrated term $|\{\ldots\}|_0^\infty$ in (69) vanishes if $-s_1 < \operatorname{Re} p < s_2$, thereby reducing (69) to the ordinary differential equation [27]

$$\frac{d^4 \bar{u}}{d\varphi^4} + [(p-1)^2 + (p+1)^2] \frac{d^2 \bar{u}}{d\varphi^2} + (p-1)^2 (p+1)^2 \bar{u}$$

$$= \frac{1}{D} \int_0^\infty q(r, \varphi) r^{p+2} \, dr.$$

---

[26] In problems of elasticity theory involving integration of the biharmonic equation, it is best to use a modification of the Mellin transform, in which the exponent p is replaced by $p - 1$.

[27] By the same token, the integral (70) is analytic in the strip $-s_1 < \operatorname{Re} p < s_2$, being uniformly convergent in every closed subset of the strip.

Using the method of variation of constants, we find that

$$\bar{u} = A \cos (p - 1)\varphi + B \sin (p - 1)\varphi + C \cos (p + 1)\varphi + E \sin (p + 1)\varphi$$
$$+ \frac{1}{4Dp} \int_0^\varphi \left[\frac{\sin (p - 1)(\varphi - t)}{p - 1} - \frac{\sin (p + 1)(\varphi - t)}{p + 1} \right] dt \int_0^\infty q(\rho, t)\rho^{p+1} \, d\rho,$$

where the boundary conditions

$$\bar{u}\Big|_{\varphi=0} = \frac{d\bar{u}}{d\varphi}\Big|_{\varphi=0} = \bar{u}\Big|_{\varphi=\alpha} = \frac{d\bar{u}}{d\varphi}\Big|_{\varphi=\alpha} = 0$$

serve to determine the coefficients A, B, C and E. Passing to the limit δ, $\varepsilon \to 0$ and solving for these coefficients, we find that $\bar{u}$ is a meromorphic function with poles at the points where the expression $p^2 \sin^2 \alpha - \sin^2 p\alpha$ vanishes, and moreover that the number $s_1 = s_2$ is the smallest root of the equation

$$p^2 \sin^2 \alpha - \sin^2 p\alpha = 0.$$

The bending moment M and the shear stress N along the edge $\varphi = 0$ can be determined from the relations

$$\overline{Mr^2}\Big|_{\varphi=0} = -D \frac{d^2\bar{u}}{d\varphi^2}\Big|_{\varphi=0}, \qquad \overline{Nr^3}\Big|_{\varphi=0} = -D \frac{d^3\bar{u}}{d\varphi^3}\Big|_{\varphi=0}.$$

Using the inversion formula for the Mellin transform, and choosing the imaginary axis as the path of integration, we find, after a certain amount of calculation, that M and N are the same as in the answer on p. 193.

418. Following the Fourier method, we look for particular solutions of Laplace's equation of the form

$$T = R(r)\Phi(\varphi) \sin \frac{n\pi z}{l}.$$

Separating variables and integrating the resulting equations, we find that

$$T = [AI_{\sqrt{-\lambda}}(n\pi r/\lambda) + BK_{\sqrt{-\lambda}}(n\pi r/\lambda)] [C \cosh \sqrt{\lambda}\, \varphi + D \sinh \sqrt{\lambda}\, \varphi] \sin \frac{n\pi z}{l},$$

where $I_\nu(x)$ and $K_\nu(x)$ are cylinder functions of imaginary argument. Because of the behavior of $I_\nu(x)$ and $K_\nu(x)$ as $r \to 0$ and $r \to \infty$, the boundedness of the solutions T requires that $A = 0$ and $\lambda > 0$. Thus the particular solutions needed to solve the boundary value problem, which has a continuous spectrum $(0, \infty)$, are of the form

$$T = T_\tau = (M_\tau \cosh \tau\varphi + N_\tau \sinh \tau\varphi) K_{i\tau}\left(\frac{n\pi r}{l}\right) \sin \frac{n\pi z}{l}, \qquad 0 < \tau < \infty.$$

The general solution is constructed by integrating with respect to the parameter τ. Noting that $T|_{\varphi=0} = 0$, we have

$$T = \sin \frac{n\pi z}{l} \int_0^\infty N_\tau \sinh \tau\varphi \, K_{i\tau}\left(\frac{n\pi r}{l}\right) d\tau.$$

The coefficient N_τ is determined from the boundary condition

$$T|_{\varphi=\alpha} = f(r) \sin \frac{n\pi z}{l},$$

which gives

$$f(r) = \int_0^\infty N_\tau \sinh \tau\alpha \, K_{i\tau}\left(\frac{n\pi r}{l}\right) d\tau, \qquad 0 < r < \infty. \tag{71}$$

For a certain class of functions $f(r)$, we can invert (71), obtaining[28]

$$N_\tau \sinh \tau\alpha = \frac{2}{\pi^2} \tau \sinh \pi\tau \int_0^\infty f(r) \frac{K_{i\tau}(n\pi r/l)}{r} \, dr.$$

The conditions for using this formula are usually satisfied, except that $f(r)$ may not go to zero sufficiently rapidly as $r \to 0$. If $f(0) \neq 0$, then in most cases encountered in practice we can use the formula[29]

$$N_\tau \sinh \tau\alpha = \frac{2}{\pi} f(0) + \frac{2}{\pi^2} \tau \sinh \pi\tau \int_0^\infty [f(r) - f(0)e^{-n\pi r/l}] \frac{K_{i\tau}(n\pi r/l)}{r} \, dr \tag{72}$$

to determine N_τ (see L9, pp. 150–153). Assuming that the conditions imposed on $f(r)$ are sufficient to guarantee the applicability of (72), we arrive at the result given in the answer on p. 196.

422. We subtract out the source potential, by writing

$$u = \frac{q}{R} - v,$$

where R is the distance from the charge q to an arbitrary point of space. Then the problem reduces to integration of Laplace's equation

$$\frac{1}{r} \frac{\partial}{\partial r}\left(r \frac{\partial v}{\partial r}\right) + \frac{1}{r^2} \frac{\partial^2 v}{\partial \varphi^2} + \frac{\partial^2 v}{\partial z^2} = 0$$

$$(0 < r < \infty, \quad 0 < \varphi < 2\pi, -\infty < z < \infty),$$

with boundary conditions

$$v|_{\varphi=0} = \frac{q}{R}\bigg|_{\varphi=0}, \qquad v|_{\varphi=2\pi} = \frac{q}{R}\bigg|_{\varphi=2\pi}.$$

---

[28] This follows from formula (24), p. 195.
[29] Implied by formulas (24) and (26), p. 195.

Expanding the function v in a Fourier cosine integral, i.e., setting

$$v = \frac{2}{\pi} \int_0^\infty \bar{v} \cos \sigma z \, d\sigma,$$

we find that $\bar{v}$ satisfies the equation

$$\frac{1}{r} \frac{\partial}{\partial r} \left(r \frac{\partial \bar{v}}{\partial r} \right) + \frac{1}{r^2} \frac{\partial^2 \bar{v}}{\partial \varphi^2} - \sigma^2 \bar{v} = 0 \qquad (73)$$

and the boundary conditions

$$\bar{v}\big|_{\varphi=0} = \left(\overline{\frac{q}{R}} \right)\Big|_{\varphi=0} = \int_0^\infty \frac{q}{R}\Big|_{\varphi=0} \cos \sigma z \, dz,$$

$$\bar{v}\big|_{\varphi=2\pi} = \int_0^\infty \frac{q}{R}\Big|_{\varphi=2\pi} \cos \sigma z \, dz,$$

which, after evaluating the integrals take the form

$$\bar{v}\big|_{\varphi=0} = \bar{v}\big|_{\varphi=2\pi} = q K_0(\sigma\sqrt{r^2 + r_0^2 - 2rr_0 \cos \varphi_0}),$$

in terms of Macdonald's function $K_0(x)$. Using the Fourier method to integrate (73), we represent $\bar{v}$ as an integral

$$\bar{v} = q \int_0^\infty M_{\sigma,\tau} \frac{\cosh (\pi - \varphi)\tau}{\cosh \pi\tau} K_{i\tau}(\sigma r) \, d\tau$$

(see the solution to Prob. 418), where the coefficient $M_{\sigma,\tau}$ is determined by the condition

$$K_0(\sigma\sqrt{r^2 + r_0^2 - 2rr_0 \cos \varphi_0}) = \int_0^\infty M_{\sigma,\tau} K_{i\tau}(\sigma r) \, d\tau, \qquad 0 < r < \infty. \quad (74)$$

To avoid the difficulties associated with direct application of the inversion theorem, we write the left-hand side of (74) in the form

$$K_0(\sigma\sqrt{r^2 + r_0^2 - 2rr_0 \cos \varphi_0})$$
$$= [K_0(\sigma\sqrt{r^2 + r_0^2 - 2rr_0 \cos \varphi_0}) - K_0(\sigma r_0)]$$
$$+ K_0(\sigma r_0)[1 - e^{-\sigma r}] + K_0(\sigma r_0)e^{-\sigma r}$$

and use the formula

$$e^{-\sigma r} = \frac{2}{\pi} \int_0^\infty K_{i\tau}(\sigma r) \, d\tau.$$

Then the inversion formula implies

$$M_{\sigma,\tau} = \frac{2}{\pi} K_0(\sigma r_0) + \frac{2}{\pi^2} K_0(\sigma r_0)\tau \sinh \pi\tau \int_0^\infty \frac{1 - e^{-\sigma r}}{r} K_{i\tau}(\sigma r) \, dr$$

$$+ \frac{2}{\pi^2} \tau \sinh \pi\tau \int_0^\infty [K_0(\sigma\sqrt{r^2 + r_0^2 - 2rr_0 \cos \varphi_0}) - K_0(\sigma r_0)]K_{i\tau}(\sigma r) \frac{dr}{r}.$$

$$(75)$$

The integrals appearing in (75) can be calculated by using the formulas

$$\int_0^\infty [K_0(\sqrt{x^2 + y^2 - 2xy \cos \gamma}) - K_0(y)]K_{i\tau}(x) \frac{dx}{x}$$

$$= \frac{\pi}{\tau}\left[K_{i\tau}(y) \frac{\cosh (\pi - \gamma)\tau}{\sinh \pi\tau} - \frac{1}{2} \frac{K_0(y)}{\sinh \frac{1}{2}\pi\tau}\right],$$

$$\int_0^\infty (1 - e^{-x})K_{i\tau}(x) \frac{dx}{x} = \frac{\pi}{2\tau} \frac{\tanh \frac{1}{4}\pi\tau}{\cosh \frac{1}{2}\pi\tau},$$

which lead to the result

$$M_{\sigma,\tau} = \frac{2}{\pi} \cosh (\pi - \varphi_0)\tau \, K_{i\tau}(\sigma r_0).$$

Thus the solution of the problem takes the form

$$v = \frac{4q}{\pi^2} \int_0^\infty \cos \sigma z \, d\sigma \int_0^\infty \frac{\cosh (\pi - \varphi)\tau}{\cosh \pi\tau} \cosh (\pi - \varphi_0)\tau \, K_{i\tau}(\sigma r_0)K_{i\tau}(\sigma r) \, d\tau.$$

$$(76)$$

Substituting the known integral representation

$$K_{i\tau}(x)K_{i\tau}(y) = \int_0^\infty K_0(\sqrt{x^2 + y^2 + 2xy \cosh s}) \cos \tau s \, ds$$

into (76), reversing the order of integration and evaluating inner integrals,[30] we find that

$$v = \frac{q}{2\pi} \frac{1}{\sqrt{2rr_0}} \int_\lambda^\infty \left[\frac{1}{\cosh \frac{1}{2}s - \cos \frac{1}{2}(\varphi + \varphi_0)} + \frac{1}{\cosh \frac{1}{2}s + \cos \frac{1}{2}(\varphi - \varphi_0)}\right]$$

$$\times \frac{\sinh \frac{1}{2}s \, ds}{\sqrt{\cosh s - \cosh \lambda}}, \quad (77)$$

where

$$\cosh \lambda = \frac{z^2 + r^2 + r_0^2}{2rr_0}.$$

To obtain the final form of the solution, given in the answer on p. 198, we evaluate the integral in (77) by making the substitution

$$\cosh \frac{s}{2} = \cosh \frac{\lambda}{2} \cosh t.$$

---

[30] The formulas

$$\int_0^\infty \frac{\cos \tau s \, ds}{\sqrt{\cosh \lambda + \cosh s}} = \frac{1}{\sinh \pi\tau} \int_\lambda^\infty \frac{\sin \tau s \, ds}{\sqrt{\cosh s - \cosh \lambda}},$$

$$\int_0^\infty \frac{\cosh \psi\tau}{\sinh 2\pi\tau} \sin \tau s \, d\tau = \frac{1}{4} \frac{\sinh \frac{1}{2}s}{\cosh \frac{1}{2}s + \cos \frac{1}{2}\psi}, \quad 0 < \psi < 2\pi$$

are used in the course of the calculation.

434. Introducing elliptic coordinates α and β, where

$$x = c \cosh \alpha \cos \beta, \qquad y = c \sinh \alpha \sin \beta$$

and c is the eccentricity of the given ellipse, we assume that the charge q is uniformly distributed over the curvilinear rectangle

$$0 < \alpha < \delta, \qquad \beta^* - \frac{\varepsilon}{2} < |\beta| < \beta^* + \frac{\varepsilon}{2}.$$

The problem then reduces to integration of Poisson's equation

$$\frac{\partial^2 u}{\partial \alpha^2} + \frac{\partial^2 u}{\partial \beta^2} = -4\pi\rho h^2, \tag{78}$$

where

$$\rho = \rho(\alpha, \beta) = \begin{cases} \dfrac{q}{2h^2\delta\varepsilon} & \text{for} \quad 0 < \alpha < \delta, \ \beta^* - \dfrac{\varepsilon}{2} < |\beta| < \beta^* + \dfrac{\varepsilon}{2}, \\ 0 & \text{otherwise} \end{cases}$$

is the charge density inside the elliptic cylinder, and

$$h = c\sqrt{\cosh^2 \alpha - \cos^2 \beta}$$

is the metric coefficient. Since u must be even in the variable β, we look for a solution of the form

$$u = \frac{1}{\pi}\,\bar{u}_0 + \frac{2}{\pi}\sum_{n=1}^{\infty}\bar{u}_n \cos n\beta,$$

where

$$\bar{u}_n = \int_0^{\pi} u \cos n\beta \, d\beta.$$

Multiplying (78) by $\cos n\beta$ and integrating from 0 to π, we obtain the equation

$$\bar{u}_n'' - n^2\bar{u}_n = -4\pi\int_0^{\pi}\rho h^2 \cos n\beta \, d\beta,$$

whose solution is easily found by variation of constants:

$$\bar{u}_n = A_n \cosh n\alpha + B_n \sinh n\alpha - \frac{4\pi}{n}\int_0^{\pi}\cos n\beta \, d\beta\int_0^{\alpha}\rho h^2 \sinh n(\alpha - \xi) \, d\xi.$$

The condition that the components of the electric field be bounded at the foci of the ellipse implies $B_n = 0$. The value of the second constant A_n is determined from the boundary condition

$$\bar{u}_n\big|_{\alpha=\alpha_0} = 0,$$

where α_0 is the value of the coordinate α on the surface of the cylinder.

Taking the limit as δ, $\varepsilon \to 0$, we find that

$$\bar{u}_n = \frac{2\pi q}{n} \cos n\beta* \frac{\sinh n(\alpha_0 - \alpha)}{\cosh n\alpha_0}, \qquad n = 0, 1, 2, \ldots,$$

which immediately implies the answer on p. 205.

437. Since the regular solutions of the two-dimensional Laplace equation are of the form

$$u = u_n = A_n \cosh n\alpha \cos n\beta + B_n \sinh n\alpha \sin n\beta, \qquad n = 0, 1, 2, \ldots$$

inside the ellipse $\alpha = \alpha_0$,[31] and of the form

$$u = u_n = e^{-n\alpha}(C_n \cos n\beta + D_n \sin n\beta), \qquad n = 0, 1, 2, \ldots$$

outside the ellipse, we look for a magnetic potential of the form

$$u^{(1)} = H_0(x \cos \gamma + y \sin \gamma) + \sum_{n=1}^{\infty} e^{-n\alpha}(C_n \cos n\beta + D_n \sin n\beta)$$

in the air, and

$$u^{(2)} = \sum_{n=1}^{\infty} (A_n \cosh n\alpha \cos n\beta + B_n \sinh n\alpha \sin n\beta)$$

in the magnetic medium (arbitrary additive constants are omitted). The values of the coefficients $A_n, \ldots, D_n$ are determined from the condition that the tangential component of the magnetic field and the normal component of the magnetic induction be continuous on the boundary surface, i.e.,

$$\left.\frac{\partial u^{(1)}}{\partial \beta}\right|_{\alpha=\alpha_0} = \left.\frac{\partial u^{(2)}}{\partial \beta}\right|_{\alpha=\alpha_0}, \qquad \left.\frac{\partial u^{(1)}}{\partial \alpha}\right|_{\alpha=\alpha_0} = \mu \left.\frac{\partial u^{(2)}}{\partial \alpha}\right|_{\alpha=\alpha_0}.$$

This gives

$$C_1 = (1 - \mu)H_0 c \frac{e^{\alpha_0} \cosh \alpha_0 \sinh \alpha_0 \cos \gamma}{\cosh \alpha_0 + \mu \sinh \alpha_0},$$

$$D_1 = (1 - \mu)H_0 c \frac{e^{\alpha_0} \cosh \alpha_0 \sinh \alpha_0 \sin \gamma}{\sinh \alpha_0 + \mu \cosh \alpha_0},$$

$$A_1 = \frac{H_0 c e^{\alpha_0} \cos \gamma}{\cosh \alpha_0 + \mu \sinh \alpha_0}, \qquad B_1 = \frac{H_0 c e^{\alpha_0} \sin \gamma}{\sinh \alpha_0 + \mu \cosh \alpha_0},$$

where all the other coefficients vanish. The final expressions for $u^{(1)}$ and $u^{(2)}$ given on p. 206 are obtained by using the relations

$$\cosh \alpha_0 = \frac{a}{c}, \qquad \sinh \alpha_0 = \frac{b}{c}.$$

---

[31] The other combinations of products of hyperbolic and trigonometric functions lead to infinite values of grad u at the foci of the ellipse.

443. To make the problem homogeneous, we write the torsion function as a sum

$$u = -y^2 + v.$$

Then v is a solution of Laplace's equation regular inside the cut ellipse (i.e., in the region $|\alpha| < \alpha_0$, $0 < \beta < \pi$) and satisfying the boundary conditions

$$v|_{\beta=0} = v_\beta|_{=\pi} = 0, \qquad v|_{\alpha=\pm\alpha_0} = (c \sinh \alpha_0 \sin \beta)^2.$$

Applying the Fourier method and using the evenness of v in the variable α, we construct a solution of the form

$$v = \sum_{n=1}^{\infty} A_n \frac{\cosh n\alpha}{\cosh n\alpha_0} \sin n\beta.$$

The constants A_n are determined from the boundary condition for $\alpha = \alpha_0$, which gives

$$A_n = \begin{cases} \dfrac{8c^2}{\pi} \dfrac{\sinh^2 \alpha_0}{n(4 - n^2)}, & n = 1, 3, 5, \ldots \\ 0 & n = 2, 4, 6, \ldots \end{cases}$$

Thus the torsion function is given by the series

$$u = -y^2 + \frac{8b^2}{\pi} \sum_{n=0}^{\infty} \frac{\cosh (2n + 1)\alpha}{\cosh (2n + 1)\alpha_0} \frac{\sin (2n + 1)\beta}{(2n + 3)(1 - 4n^2)}, \tag{79}$$

while the torsional rigidity can be calculated from the formula

$$C = 4G \int_0^{\alpha_0} \int_0^{\pi} uh^2 \, d\alpha \, d\beta, \tag{80}$$

where

$$h = c\sqrt{\cosh^2 \alpha - \cos^2 \beta}$$

is the metric coefficient. Substituting (79) into (80) and evaluating the double integral, we arrive at the expression given in the answer on p. 209.

449. Choosing a system of parabolic coordinates α, β such that the surface of the cylinder has equation $\beta = \beta_0$, and regarding the charge q as uniformly distributed over the small area bounded by the curves $|\alpha| = \delta$, $\beta = \varepsilon$, we reduce the problem to integration of Poisson's equation

$$\frac{\partial^2 u}{\partial \alpha^2} + \frac{\partial^2 u}{\partial \beta^2} = -4\pi\rho h^2 \qquad (-\infty < \alpha < \infty, 0 < \beta < \beta_0),$$

where

$$\rho = \begin{cases} \dfrac{q}{2h^2 \delta\varepsilon} & \text{for} \quad |\alpha| < \delta, \beta < \varepsilon, \\ 0 & \text{otherwise,} \end{cases}$$

and

$$h = c\sqrt{\alpha^2 + \beta^2}$$

is the metric coefficient. The problem is solved by taking the Fourier cosine transform. Writing

$$\bar{u} = \int_0^\infty u \cos \lambda\alpha \, d\alpha,$$

we multiply Poisson's equation by $\cos \lambda\alpha$ and integrate from 0 to ∞. This gives the equation

$$\bar{u}'' - \lambda^2 \bar{u} = -4\pi \int_0^\infty \rho h^2 \cos \lambda\alpha \, d\alpha,$$

whose solution is easily found by variation of constants:

$$\bar{u} = A \cosh \lambda\beta + B \sinh \lambda\beta - \frac{4\pi}{\lambda} \int_0^\infty \cos \lambda\alpha \, d\alpha \int_0^\beta \rho h^2 \sinh \lambda(\beta - \eta) \, d\eta.$$

The requirement that grad u be bounded at the focus of the parabola implies $B = 0$. The value of the constant A is determined from the boundary condition

$$\bar{u}|_{\beta=\beta_0} = 0,$$

which, in the limit δ, $\varepsilon \to 0$, gives

$$A = \frac{2\pi q}{\lambda} \tanh \lambda\beta_0.$$

The corresponding value of $\bar{u}$ is

$$\bar{u} = \frac{2\pi q}{\lambda} \frac{\sinh \lambda(\beta_0 - \beta)}{\cosh \lambda\beta_0},$$

and the final answer (see p. 211) is obtained by using the inversion formula

$$u = \frac{2}{\pi} \int_0^\infty \bar{u} \cos \lambda\alpha \, d\lambda.$$

457. If we introduce bipolar coordinates α, β as shown in Figure 122, p. 215, and represent the torsion function u as a sum

$$u = -y^2 + v,$$

the problem reduces to determining the function v which is harmonic in the domain $\alpha_0 < \alpha < \infty$, $0 < \beta < 2\pi$ and satisfies the conditions

$$v|_{\beta=0} = v|_{\beta=2\pi} = 0, \qquad v|_{\alpha=\alpha_0} = y^2|_{\alpha=\alpha_0} = \frac{c^2 \sin^2 \beta}{(\cosh \alpha_0 + \cos \beta)^2},$$

$$v|_{\alpha \to \infty} \to 0.$$

The solution is constructed as a series

$$v = \sum_{n=1}^\infty C_n \, e^{-n(\alpha-\alpha_0)/2} \sin \frac{n\beta}{2},$$

whose coefficients, according to the theory of Fourier series, are given by

$$C_n = \frac{c^2}{\pi} \int_0^{2\pi} \frac{\sin^2 \beta \sin \frac{1}{2} n\beta}{(\cosh \alpha_0 + \cos \beta)^2} \, d\beta.$$

To calculate the torsional rigidity, we use the relation

$$C = G \left[-2 \int_S y^2 \, ds + \int_0^{2\pi} \left(y^2 \frac{\partial v}{\partial \alpha} - v \frac{\partial y^2}{\partial \alpha} \right) \Big|_{\alpha = \alpha_0} d\beta \right],$$

implied by the formula given in the hint to the problem (see p. 215) after setting $\psi = -y^2$.

471. Setting

$$u = \frac{q}{R} + u_1,$$

where R is the distance from the source to an arbitrary point of space, we reduce the problem to integration of Laplace's equation

$$\Delta u_1 = 0,$$

with the boundary condition

$$u_1 \big|_{\alpha=0} = -\frac{q}{R} \Big|_{\alpha=0} = -\frac{q}{c\sqrt{\sinh^2 \alpha_0 + \sin^2 \beta}}, \qquad \sinh \alpha_0 = \frac{d}{a}$$

and the condition at infinity

$$u_1 \big|_{\alpha \to \infty} \to 0.$$

In keeping with the discussion on p. 222, we look for a secondary potential in the form of a series

$$u_1 = \sum_{n=0}^{\infty} A_n Q_n (i \sinh \alpha) P_n (\cos \beta),$$

where the coefficients A_n are determined from the boundary condition. Using the theorem on expansion of an arbitrary function in a series of Legendre polynomials, we find that

$$A_n = -\frac{2n+1}{2Q_n(0)} \frac{q}{c} \int_{-1}^1 \frac{P_n(x) \, dx}{\sqrt{\cosh^2 \alpha_0 - x^2}} \qquad (81)$$

for even n, while $A_n = 0$ for odd n. To evaluate (81) for even n, we use the integral

$$J_n = \int_{-1}^1 \frac{P_{2n}(x) \, dx}{\sqrt{b^2 - x^2}}, \qquad b > 1,$$

which can be evaluated by expanding $(b^2 - x^2)^{-1/2}$ in a power series and then integrating term by term. Using well-known formulas, we find that

$$J_n = \frac{1}{b}\int_{-1}^{1}\frac{P_{2n}(x)\,dx}{\sqrt{1 - (x/b)^2}} = \frac{1}{b}\sum_{m=0}^{\infty}\frac{\Gamma(m + .\frac{1}{2})}{\Gamma(\frac{1}{2})\Gamma(m + 1)}\frac{1}{b^{2m}}\int_{-1}^{1}x^{2m}P_{2n}(x)\,dx$$

$$= \frac{2^{2n+1}}{b}\sum_{m=n}^{\infty}\frac{\Gamma(m + \frac{1}{2})\Gamma(2m + 1)\Gamma(m + n + 1)}{\Gamma(\frac{1}{2})\Gamma(m + 1)\Gamma(m - n + 1)\Gamma(2m + 2n + 2)}\frac{1}{b^{2m}}$$

$$= \frac{\Gamma^2(n + \frac{1}{2})}{\sqrt{\pi}\, b^{2n+1}\Gamma(2n + \frac{3}{2})}\sum_{k=0}^{\infty}\frac{(n + \frac{1}{2})_k(n + \frac{1}{2})_k}{k!\,(2n + \frac{3}{2})_k}\left(\frac{1}{b^2}\right)^k,$$

where

$$(\lambda)_k = \frac{\Gamma(\lambda + k)}{\Gamma(\lambda)}.$$

The result can be expressed in terms of the hypergeometric function

$$F(\alpha, \beta; \gamma; z) = \sum_{k=0}^{\infty}\frac{(\alpha)_k(\beta)_k}{k!\,(\gamma)_k}z^k,$$

i.e.,

$$J_n = \frac{\Gamma^2(n + \frac{1}{2})}{\sqrt{\pi}\, b^{2n+1}\Gamma(2n + \frac{3}{2})}F\left(n + \frac{1}{2}, n + \frac{1}{2}; 2n + \frac{3}{2}; \frac{1}{b^2}\right).$$

Using the familiar formula

$$F(\alpha, \beta; \gamma; z) = (1 - z)^{-\alpha}F\left(\alpha, \gamma - \beta; \gamma; \frac{z}{z - 1}\right),$$

we find that

$$J_n = \frac{1}{\sqrt{\pi}}\frac{\Gamma^2(n + \frac{1}{2})}{\Gamma(2n + \frac{3}{2})}\frac{1}{\sinh^{2n+1}\alpha_0}F\left(n + \frac{1}{2}, n + 1; 2n + \frac{3}{2}; -\frac{1}{\sinh^2\alpha_0}\right),$$

or

$$J_n = 2iP_{2n}(0)Q_{2n}(i\sinh\alpha_0),$$

because of the definition of the Legendre function of the second kind. Thus the required values of the coefficients A_n are

$$A_{2n} = \frac{2q}{\pi c}(4n + 1)Q_{2n}(i\sinh\alpha_0),$$

which leads to the potential distribution

$$u = \frac{q}{R} + \frac{2}{\pi}\frac{q}{c}\sum_{n=0}^{\infty}(4n + 1)Q_{2n}(i\sinh\alpha_0)Q_{2n}(i\sinh\alpha)P_{2n}(\cos\beta),$$

if we note that

$$\frac{P_{2n}(0)}{Q_{2n}(0)} = \frac{2i}{\pi}.$$

The distribution of charge on the surface of the disk is now found by differentiation, according to the formula

$$\sigma = -\frac{1}{4\pi}\left(\frac{1}{c\sqrt{\sinh^2\alpha + \cos^2\beta}}\frac{\partial u}{\partial\alpha}\right)\bigg|_{\alpha=0}.$$

481. The problem reduces to solving the system of equations

$$\Delta u^{(1)} = -4\pi\rho \quad (0 \leqslant \alpha < \alpha_0), \qquad \Delta u^{(2)} = 0 \quad (\alpha_0 < \alpha < \infty)$$

for the gravitational potentials $u^{(1)}$ and $u^{(2)}$, with boundary conditions

$$u^{(1)}\big|_{\alpha=\alpha_0} = u^{(2)}\big|_{\alpha=\alpha_0}, \qquad \frac{\partial u^{(1)}}{\partial\alpha}\bigg|_{\alpha=\alpha_0} = \frac{\partial u^{(2)}}{\partial\alpha}\bigg|_{\alpha=\alpha_0}.$$

Setting

$$u^{(1)} = u_0 + u_1, \qquad u^{(2)} = u_2,$$

where

$$u_0 = -\pi\rho r^2 = -\pi\rho c^2 \sinh^2\alpha \sin^2\beta,$$

and noting that u_1 is harmonic inside the spheroid ($\alpha < \alpha_0$), while u_2 is harmonic outside the spheroid ($\alpha > \alpha_0$), we have

$$u_1 = \sum_{n=0}^{\infty} A_n P_n(\cosh\alpha)P_n(\cos\beta), \qquad u_2 = \sum_{n=0}^{\infty} B_n Q_n(\cosh\alpha)P_n(\cos\beta).$$

Using the boundary conditions, we obtain the formulas

$$-\tfrac{2}{3}\pi\rho c^2 \sinh^2\alpha_0[1 - P_2(\cos\beta)] + \sum_{n=0}^{\infty} A_n P_n(\cosh\alpha_0)P_n(\cos\beta)$$
$$= \sum_{n=0}^{\infty} B_n Q_n(\cosh\alpha_0)P_n(\cos\beta),$$

$$-\tfrac{4}{3}\pi\rho c^2 \cosh\alpha_0[1 - P_2(\cos\beta)] + \sum_{n=0}^{\infty} A_n P_n'(\cosh\alpha_0)P_n(\cos\beta)$$
$$= \sum_{n=0}^{\infty} B_n Q_n'(\cosh\alpha_0)P_n(\cos\beta)$$

determining the coefficients,[32] which imply that

$$A_n = B_n = 0, \qquad n = 1, 3, 4, 5, \ldots$$

Thus A_0, B_0, A_2, B_2 satisfy the system of equations

$$A_0 P_0(\cosh\alpha_0) - B_0 Q_0(\cosh\alpha_0) = \tfrac{2}{3}\pi\rho c^2 \sinh^2\alpha_0,$$
$$A_0 P_0'(\cosh\alpha_0) - B_0 Q_0'(\cosh\alpha_0) = \tfrac{4}{3}\pi\rho c^2 \cosh\alpha_0,$$
$$A_2 P_2(\cosh\alpha_0) - B_2 Q_2(\cosh\alpha_0) = -\tfrac{2}{3}\pi\rho c^2 \sinh^2\alpha_0,$$
$$A_2 P_2'(\cosh\alpha_0) - B_2 Q_2'(\cosh\alpha_0) = -\tfrac{4}{3}\pi\rho c^2 \cosh\alpha_0,$$

---

[32] Note that

$$\sin^2\beta = \tfrac{2}{3}[1 - P_2(\cos\beta)].$$

whose solution is[33]

$$A_0 = \tfrac{2}{3}\pi\rho c^2 \sinh^2 \alpha_0 \left(1 + 2 \cosh \alpha_0 \ln \coth \frac{\alpha_0}{2}\right),$$

$$A_2 = \tfrac{4}{3}\pi\rho c^2 \sinh^2 \alpha_0 \left(1 - \cosh \alpha_0 \ln \coth \frac{\alpha_0}{2}\right),$$

$$B_0 = \tfrac{4}{3}\pi\rho c^2 \cosh \alpha_0 \sinh^2 \alpha_0, \qquad B_2 = -\tfrac{4}{3}\pi\rho c^2 \cosh \alpha_0 \sinh^2 \alpha_0.$$

Substituting for B_0 and B_2 in the formula for u_2, we find that the gravitational potential outside the spheroid is

$$u_2 = \pi\rho \, \frac{ab^2}{c} \left\{ [2(\sin^2 \beta - \sinh^2 \alpha) + 3 \sinh^2 \alpha \sin^2 \beta] \ln \coth \frac{\alpha}{2} \right.$$
$$\left. + \cosh \alpha (3 \cos^2 \beta - 1) \right\}.$$

To obtain an asymptotic representation of the gravitational potential for small eccentricity c, we introduce spherical coordinates R and θ, and use the formulas

$$z = \frac{c}{2}(e^\alpha + e^{-\alpha}) \cos \beta = R \cos \theta, \qquad r = \frac{c}{2}(e^\alpha - e^{-\alpha}) \sin \beta = R \sin \theta.$$

Solving for α and β, we find that

$$\cosh \alpha = \frac{1}{2}\frac{R}{c} \left\{ \sqrt{1 + \frac{2c}{R}\cos \theta + \frac{c^2}{R^2}} + \sqrt{1 - \frac{2c}{R}\cos \theta + \frac{c^2}{R^2}} \right\},$$

$$\cos \beta = \frac{1}{2}\frac{R}{c} \left\{ \sqrt{1 + \frac{2c}{R}\cos \theta + \frac{c^2}{R^2}} - \sqrt{1 - \frac{2c}{R}\cos \theta + \frac{c^2}{R^2}} \right\}.$$

It follows that as $c \to 0$,

$$\cosh \alpha = \frac{R}{c}\left[1 + \frac{c^2}{2R^2}\sin^2 \theta\right] + O\!\left(\frac{c^2}{R^2}\right),$$

$$\cos \beta = \cos \theta + O\!\left(\frac{c^2}{R^2}\right).$$

---

[33] Here we use the expression

$$P_n(z)Q_n'(z) - Q_n(z)P_n'(z) = \frac{1}{1 - z^2}$$

for the Wronskian of the Legendre functions, as well as the formulas

$$P_0(z) = 1, \qquad P_2(z) = \tfrac{1}{2}(3z^2 - 1),$$

$$Q_0(z) = \frac{1}{2}\ln \frac{z+1}{z-1}, \qquad Q_2(z) = \frac{1}{2}\left[\frac{3z^2-1}{2}\ln \frac{z+1}{z-1} - 3z\right].$$

Using these formulas and the exact solution found previously, we obtain

$$u_2 = M\left[\frac{1}{R} + \frac{c^2}{5R^3} P_2(\cos \theta) + O\left(\frac{c^4}{R^4}\right)\right],$$

where M is the mass of the spheroid (cf. the answer on p. 228).

483. If we write

$$u = u_0 + u_1,$$

where $u_0 = q/R$ is the source potential (R is the distance from the charge to an arbitrary point of space), and introduce spheroidal coordinates α, β, such that the hyperbola has the equation $\beta = \beta_0$, then the problem reduces to finding the function u_1 which is harmonic in the region $0 \leqslant \beta < \beta_0$ and satisfies the boundary condition

$$u_1|_{\beta=\beta_0} = -\frac{q}{R}\bigg|_{\beta=\beta_0} = -\frac{q}{c(\cosh \alpha - \cos \beta_0)}.$$

In prolate spheroidal coordinates, Laplace's equation takes the form

$$\frac{1}{\sinh \alpha} \frac{\partial}{\partial \alpha}\left(\sinh \alpha \frac{\partial u_1}{\partial \alpha}\right) + \frac{1}{\sin \beta} \frac{\partial}{\partial \beta}\left(\sin \beta \frac{\partial u_1}{\partial \beta}\right) = 0$$

if we assume that u_1 is independent of φ. Setting

$$u_1 = A(\alpha)B(\beta),$$

we obtain the equations

$$\frac{1}{\sinh \alpha} (\sinh \alpha \cdot A')' + \lambda A = 0, \qquad \frac{1}{\sin \beta} (\sin \beta \cdot B')' - \lambda B = 0$$

for the separate factors. Therefore

$$u_1 = [MP_\nu(\cosh \alpha) + NQ_\nu(\cosh \alpha)][CP_\nu(\cos \beta) + DQ_\nu(\cos \beta)], \quad (82)$$

where $P_\nu(z)$ and $Q_\nu(z)$ are Legendre functions of the first and second kind, and ν is an auxiliary parameter related to λ by the formula

$$\lambda = -\nu(\nu + 1).$$

Taking account of the behavior of the Legendre functions near the points $z = 1$ and $z = \infty$, we see that in order for the solutions (82) to represent bounded real functions in the region $0 \leqslant \alpha < \infty, 0 \leqslant \beta < \beta_0$, the parameter ν must be chosen equal to $-\frac{1}{2} + i\tau$ ($\tau > 0$), and the constants N and D

must be set equal to zero.[34] Thus we arrive at the particular solutions

$$u_1 = u_{1,\tau} = C_\tau P_{-\frac{1}{2}+i\tau}(\cosh \alpha)P_{-\frac{1}{2}+i\tau}(\cos \beta), \qquad \tau \geqslant 0.$$

To construct the general solution, we integrate over the parameter τ, obtaining

$$u_1 = \int_0^\infty C_\tau P_{-\frac{1}{2}+i\tau}(\cosh \alpha)P_{-\frac{1}{2}+i\tau}(\cos \beta)\, d\tau,$$

where the coefficients C_τ must satisfy the boundary condition

$$u_1\big|_{\beta=\beta_0} = -\frac{q}{c(\cosh \alpha - \cos \beta_0)}$$

$$= \int_0^\infty C_\tau P_{-\frac{1}{2}+i\tau}(\cos \beta_0)P_{-\frac{1}{2}+i\tau}(\cosh \alpha)\, d\tau, \qquad \alpha > 0.$$

Using the inversion formula implied by the Mehler-Fock theorem, we find that

$$C_\tau = -\frac{q\tau \tanh \pi\tau}{cP_{-\frac{1}{2}+i\tau}(\cos \beta_0)} \int_0^\infty \frac{\sinh \alpha}{\cosh \alpha - \cos \beta_0} P_{-\frac{1}{2}+i\tau}(\cosh \alpha)\, d\alpha$$

$$= -\frac{q}{c} \frac{\tau \tanh \pi\tau}{P_{-\frac{1}{2}+i\tau}(\cos \beta_0)} \int_1^\infty \frac{P_{-\frac{1}{2}+i\tau}(\xi)}{\xi - \cos \beta_0}\, d\xi.$$

Evaluating the integral, we arrive at the formula for the electrostatic potential given in the answer on p. 229.[35]

---

[34] In particular, we use the formula

$$P_\nu(\cosh \alpha) = \frac{\Gamma(\nu + 1)}{\sqrt{\pi}\,\Gamma(\nu + \frac{3}{2})} e^{-(\nu+1)\alpha} \tan \pi\nu F(\nu + 1, \tfrac{1}{2}; \nu + \tfrac{3}{2}; e^{-2\alpha})$$

$$+ \frac{\Gamma(\frac{1}{2} + \nu)}{\sqrt{\pi}\,\Gamma(1 + \nu)} e^{\nu\alpha} F(-\nu, \tfrac{1}{2}; \tfrac{1}{2} - \nu; e^{-2\alpha}),$$

which shows that a bounded solution in the interval $(0, \infty)$ exists only if $-1 < \mathrm{Re}\,\nu < 0$. The extra requirement that u_1 be real compels us to set $\nu = -\frac{1}{2} + i\tau$.

[35] To prove the formula

$$J = \int_1^\infty \frac{P_{-\frac{1}{2}+i\tau}(\xi)}{\xi - \cos \beta_0}\, d\xi = \frac{\pi}{\cosh \pi\tau} P_{-\frac{1}{2}+i\tau}(-\cos \beta_0)$$

(see L7), use the integral representation

$$P_{-\frac{1}{2}+i\tau}(\cosh \alpha) = \frac{2}{\pi} \int_0^\alpha \frac{\cos \tau s\, ds}{\sqrt{2\cosh \alpha - 2\cosh s}},$$

and then reverse the order of integration with respect to α and s. After evaluating the inner integral, this gives

$$J = 2\int_0^\infty \frac{\cos \tau s\, ds}{\sqrt{2\cosh s - 2\cos \beta_0}} = \frac{\pi}{\cosh \pi\tau} P_{-\frac{1}{2}+i\tau}(-\cos \beta_0),$$

where we have used another integral representation of $P_{-\frac{1}{2}+i\tau}(x)$.

494. Bearing in mind that the potential u can be represented in the form $u = E_0 z + u_1$, where u_1 is harmonic outside the torus and goes to zero at infinity, we look for a solution of the form

$$u = E_0 z + \sqrt{2 \cosh \alpha - 2 \cos \beta} \sum_{n=1}^{\infty} A_n \frac{P_{n-1/2}(\cosh \alpha)}{P_{n-1/2}(\cosh \alpha_0)} \sin n\beta.$$

The coefficients A_n are found from the boundary condition and coincide with the coefficients of the function

$$-2E_0 c \sin \beta (2 \cosh \alpha_0 - 2 \cos \beta)^{-3/2}$$

when expanded in a Fourier sine series in the interval $(0, \pi)$. Thus we find that

$$A_n = -\frac{4E_0 c}{\pi} \int_0^{\pi} \frac{\sin \beta \sin n\beta \, d\beta}{(2 \cosh \alpha_0 - 2 \cos \beta)^{3/2}}.$$

Integrating by parts and using the formula given in the hint to Prob. 493, we arrive at the answer given on p. 237.

498. Setting

$$T = -\frac{Qr^2}{4k} + u, \qquad r^2 = x^2 + y^2,$$

we reduce the problem to finding the function u. This function is harmonic outside the torus $(0 < \alpha < \alpha_0)$ and goes to zero at infinity (i.e., as $\alpha \to 0$, $\beta \to 0$). We look for a solution of the form

$$u = \sqrt{2 \cosh \alpha - 2 \cos \beta} \sum_{n=0}^{\infty} A_n \frac{Q_{n-1/2}(\cosh \alpha)}{Q_{n-1/2}(\cosh \alpha_0)} \cos n\beta,$$

where the coefficients A_n are determined from the boundary condition

$$u\big|_{\alpha=\alpha_0} = \frac{Q}{4k} r^2 \big|_{\alpha=\alpha_0}.$$

It follows from the theory of Fourier series that

$$A = \frac{Qc^2 \sinh^2 \alpha_0}{k\pi} \int_0^{\pi} \frac{d\beta}{(2 \cosh \alpha_0 - 2 \cos \beta)^{5/2}},$$

$$A_n = \frac{2Qc^2 \sinh^2 \alpha_0}{k\pi} \int_0^{\pi} \frac{\cos n\beta \, d\beta}{(2 \cosh \alpha_0 - 2 \cos \beta)^{5/2}}, \qquad n = 1, 2, \ldots$$

Evaluating these integrals, we eventually arrive at the answer on p. 238.

502. If we subtract out the singularity at the point $r = z = 0$ by setting

$$u = \frac{q}{\sqrt{r^2 + z^2}} + u_1,$$

the potential u_1 of the secondary field is harmonic in the region $0 < \alpha < \infty$, $\beta_0 < \beta < 2\pi + \beta_0$ outside the conductor, and vanishes as $\alpha \to 0$, $\beta \to 2\pi$.

The function u_1 can be represented in the form of an integral

$$u_1 = -\frac{q}{c}\sqrt{2\cosh\alpha - 2\cos\beta}\int_0^\infty M_\tau \frac{\cosh(\pi + \beta_0 - \beta)}{\cosh\pi\tau}P_{-\frac{1}{2}+i\tau}(\cosh\alpha)\,d\tau.$$

It follows from the boundary condition

$$u|_{\beta=\beta_0} = u|_{\beta=2\pi+\beta_0} = 0$$

that M_τ coincides with the coefficients of the expansion of the function

$$(2\cosh\alpha + 2\cos\beta_0)^{-1/2}$$

in a Mehler-Fock integral with respect to the functions $P_{-\frac{1}{2}+i\tau}(\cosh\alpha)$, i.e.,

$$(2\cosh\alpha + 2\cos\beta_0)^{-1/2} = \int_0^\infty M_\tau P_{-\frac{1}{2}+i\tau}(\cosh\alpha)\,d\tau, \qquad \alpha \geqslant 0.$$

In the present case, we cannot determine M_τ directly by using the inversion formula implied by the Mehler-Fock theorem, since the function being expanded does not belong to the class for which the theorem holds (see L9, p. 228). However, it can be shown without recourse to the Mehler-Fock theorem (ibid., p. 229) that

$$(2\cosh\alpha + 2\cos\beta_0)^{-1/2} = \int_0^\infty \frac{\cosh\beta_0\tau}{\cosh\pi\tau}P_{-\frac{1}{2}+i\tau}(\cosh\alpha)\,d\tau,$$

and hence

$$M_\tau = \frac{\cosh\beta_0\tau}{\cosh\pi\tau}.$$

Therefore the solution of the problem is

$$u_1 = -\frac{q}{c}\sqrt{2\cosh\alpha - 2\cos\beta}$$

$$\times \int_0^\infty \frac{\cosh\beta_0\tau\cosh(\pi + \beta_0 - \beta)\tau}{\cosh^2\pi\tau}P_{-\frac{1}{2}+i\tau}(\cosh\alpha)\,d\tau.$$

The charge density on the inner and outer surfaces of the spherical bowl are given by

$$\sigma_i = -\frac{\cosh\alpha - \cos\beta_0}{4\pi c}\frac{\partial u}{\partial\beta}\bigg|_{\beta=\beta_0}, \qquad \sigma_o = \frac{\cosh\alpha - \cos\beta_0}{4\pi c}\frac{\partial u}{\partial\beta}\bigg|_{\beta=2\pi+\beta_0}.$$

Performing the differentiation with respect to β and evaluating the resulting integrals by replacing the Legendre function by its integral representation

$$P_{-\frac{1}{2}+i\tau}(\cosh\alpha) = \frac{2}{\pi}\cosh\pi\tau\int_0^\infty\frac{\cos\tau\psi\,d\psi}{\sqrt{2\cosh\psi + 2\cosh\alpha}},$$

we eventually arrive at the closed-form expressions for σ_o and σ_i given in the answer on p. 240.

508. To calculate the capacitances, we must first solve the electrostatic problem, assuming that the spheres have arbitrary given potentials V_1 and V_2. Introducing a system of bipolar coordinates α, β, φ in which the spheres under consideration have equations $\beta = -\beta_1$ and $\beta = \beta_2$, we reduce the problem to determining a function u which is harmonic in the region $-\beta_1 < \beta < \beta_2$ and goes to zero as $\alpha \to 0$, $\beta \to 0$. The desired solution can be constructed in the form of a series

$$u = \sqrt{2 \cosh \beta - 2 \cos \alpha} \sum_{n=0}^{\infty} [A_n \cosh (n + \tfrac{1}{2})\beta + B_n \sinh (n + \tfrac{1}{2})\beta] P_n(\cos \alpha),$$

where the coefficients A_n and B_n are found from the boundary conditions

$$u\big|_{\beta=-\beta_1} = V_1, \qquad u\big|_{\beta=\beta_2} = V_2,$$

which immediately lead to a system of linear equations for A_n and B_n if we use the familiar expansion

$$\frac{1}{\sqrt{2 \cosh \beta - 2 \cos \alpha}} = \sum_{n=0}^{\infty} e^{-(n+\frac{1}{2})\beta} P_n(\cos \alpha), \qquad \beta > 0.$$

After determining u, the charges on each conductor can be calculated from the formulas

$$Q_1 = -\frac{1}{2} \int_0^{\pi} \left(r \frac{\partial u}{\partial \beta} \right)\bigg|_{\beta=-\beta_1} d\alpha, \qquad Q_2 = \frac{1}{2} \int_0^{\pi} \left(r \frac{\partial u}{\partial \beta} \right)\bigg|_{\beta=\beta_2} d\alpha.$$

To find the capacitances C_{11}, C_{12} and C_{22}, we use the relations

$$C_{11} = Q_1\big|_{V_1=V_2=1}, \quad C_{12} = Q_1\big|_{V_1=0, V_2=-1} = Q_2\big|_{V_1=-1, V_2=0}, \quad C_{22} = Q_2\big|_{V_1=V_2=1}.$$

512. In the new coordinate system, the problem reduces to solving the equation

$$\frac{\partial^2 u}{\partial \alpha^2} + \frac{\partial^2 u}{\partial \beta^2} - \frac{\alpha^2 - \beta^2}{\alpha(\alpha^2 + \beta^2)} \frac{\partial u}{\partial \alpha} - \frac{2\beta}{\alpha^2 + \beta^2} \frac{\partial u}{\partial \beta} = 0,$$

with the boundary conditions

$$u\big|_{\beta=\pm\beta_0} = V.$$

Variables can be separated by setting

$$u = \sqrt{\alpha^2 + \beta^2} \, A(\alpha)B(\beta).$$

Integrating the resulting equations for $A(\alpha)$ and $B(\beta)$, and noting that u must be even in β and bounded at $\alpha = 0$, we arrive at the following particular solutions:

$$u = u_\mu = M_\mu \sqrt{\alpha^2 + \beta^2} \, J_0(\mu\alpha) \cosh \mu\beta, \qquad \mu > 0.$$

The solution is then constructed in the form

$$u = V\sqrt{\alpha^2 + \beta^2} \int_0^\infty N_\mu \frac{\cosh \mu\beta}{\cosh \mu\beta_0} J_0(\mu\alpha).$$

Using the well-known formula

$$\frac{1}{\sqrt{\alpha^2 + \beta^2}} = \int_0^\infty e^{-\mu\beta} J_0(\mu\alpha)\, d\mu, \qquad \beta > 0,$$

and taking account of the boundary condition $u|_{\beta=\beta_0} = V$, we find that $N_\mu = e^{-\mu\beta_0}$, and hence

$$u = V\sqrt{\alpha^2 + \beta^2} \int_0^\infty e^{-\mu\beta_0} \frac{\cosh \mu\beta}{\cosh \mu\beta_0} J_0(\mu\alpha)\, d\mu. \tag{83}$$

To calculate the total charge Q on the conductor, we start from the relation

$$Q = \int_0^\infty \left(r \frac{\partial u}{\partial \beta} \right)\Big|_{\beta=\beta_0} d\alpha. \tag{84}$$

After substituting (83) into (84) and reversing the order of integration, we obtain the expression[36]

$$Q = cV \int_0^\infty \frac{e^{-\mu\beta_0}}{\cosh \mu\beta_0}\, d\mu = 2Va \ln 2.$$

The capacitance C is now determined from the formula $Q = VC$.

522–523. If we set

$$E = f\left(t - \frac{x}{v} \right) - u,$$

the problem reduces to finding a solution of the wave equation

$$\frac{\partial^2 u}{\partial x^2} + \frac{\partial^2 u}{\partial y^2} = \frac{1}{v^2} \frac{\partial^2 u}{\partial t^2}$$

satisfying zero initial conditions, the boundary condition

$$u|_S = f\left(t - \frac{x}{v} \right)\Big|_S$$

and the condition that u vanish at infinity. Introducing new variables

$$\xi = t - \frac{x}{v}, \quad \eta = t - \frac{r}{v} \quad (r = \sqrt{x^2 + y^2}),$$

---

[36] Note that the parameter c is related to the radius a by the formula $c = 2a\beta_0$.

and looking for a solution which is a function only of ξ and η, we obtain the equation

$$\frac{\partial^2 u}{\partial \xi\, \partial \eta} + \frac{1}{2(\xi - \eta)} \frac{\partial u}{\partial \eta} = 0,$$

whose solution is

$$u = \int_{-\infty}^{\eta} \frac{\varphi(s)}{\sqrt{\xi - s}}\, ds + \psi(\xi),$$

where φ and ψ are arbitrary functions. Moreover $\psi(\xi) = 0$, since

$$u|_{r \to \infty} = u|_{\eta \to -\infty} = 0,$$

and therefore

$$u = \int_{-\infty}^{\eta} \frac{\varphi(s)}{\sqrt{\xi - s}}\, ds. \tag{85}$$

On the screen $\eta = \xi$, $u = f(\xi)$, and hence $\varphi(s)$ must satisfy Abel's integral equation

$$\int_{-\infty}^{\xi} \frac{\varphi(s)}{\sqrt{\xi - s}}\, ds = f(\xi),$$

with solution[37]

$$\varphi(s) = \frac{1}{\pi} \frac{d}{ds} \int_{-\infty}^{s} \frac{f(\xi)}{\sqrt{s - \xi}}\, d\xi.$$

If $s < 0$, then the integrand vanishes identically by hypothesis, which implies $\varphi(s) = 0$.

It follows from (85) that

$$u|_{\eta < 0} \equiv 0,$$

i.e., the excited zone is bounded by the circle $\eta = 0$ (see Figure 149, p. 254). Outside the excited zone,

$$E = f\left(t - \frac{x}{v}\right),$$

and in particular, $E \equiv 0$ for $x > vt$. For $\eta > 0$, i.e., in the excited zone,

$$u = \int_{0}^{\eta} \frac{\varphi(s)}{\sqrt{\xi - s}}\, ds,$$

where

$$\varphi(s) = \frac{1}{\pi} \frac{d}{ds} \int_{0}^{s} \frac{g(\xi)}{\sqrt{s - \xi}}\, d\xi.$$

---

[37] See e.g., S6, Vol. II, p. 220.

531–532. In the present problem, considerations like those given in the solution of Probs. 522–523 show that the reflected wave can be represented in the form

$$u = \int_{-\infty}^{\eta} \frac{\varphi(s)}{\xi - s + (2a/v)}\, ds, \tag{86}$$

where $\varphi(s)$ satisfies the Volterra integral equation

$$\varphi(\xi) + \frac{2a}{v} \int_{-\infty}^{\xi} \frac{\varphi(s)}{[\xi - s + (2a/v)]^2}\, ds = \frac{2a}{v} f'(\xi), \tag{87}$$

and the variables ξ and η are defined by

$$\xi = t - \frac{z + a}{v}, \quad \eta = t - \frac{r - a}{v} \quad (r = \sqrt{x^2 + y^2 + z^2}).$$

Assuming that

$$f(\xi) = \begin{cases} g(\xi), & \xi > 0, \\ 0, & \xi < 0, \end{cases} \quad g(0) = 0,$$

we find that $\varphi(s) = 0$ if $s < 0$, and hence $u = 0$ if $\eta < 0$. Thus the boundary of the excited zone is determined by the equation $\eta = 0$. The value of u inside the excited zone ($\eta > 0$) is given by (86) and (87) with $f(\xi)$ replaced by $g(\xi)$ and the intervals of integration $(-\infty, \xi)$, $(-\infty, \eta)$ replaced by $(0, \xi)$, $(0, \eta)$.

The integral equation

$$\varphi(\xi) + \frac{2a}{v} \int_0^{\xi} \frac{\varphi(s)}{[\xi - s + (2a/v)]^2}\, ds = \frac{2a}{v} g'(\xi), \quad \xi > 0 \tag{88}$$

belongs to the class which can be solved readily by the use of the Laplace transform.[38] Writing

$$\bar{f} = \int_0^{\infty} f e^{-p\xi}\, d\xi,$$

multiplying (88) by $e^{-p\xi}$ and integrating with respect to ξ from 0 to ∞, we find that

$$\bar{\varphi} = \frac{\dfrac{a}{v} p \bar{g}}{1 - \dfrac{a}{v} p \bar{R}},$$

where

$$\bar{R} = \int_0^{\infty} \frac{e^{-p\xi}}{\xi + (2a/v)}\, d\xi = -e^{2ap/v}\, \mathrm{Ei}\left(-\frac{2ap}{v}\right),$$

---

[38] In the applications, it is sometimes more convenient to construct the resolvent of the given equation, without recourse to the Laplace transform (see e.g., F8).

and Ei (z) is the exponential integral. The final answer can be obtained by using the inversion formula

$$\varphi(\xi) = \frac{1}{2\pi i} \int_\Gamma \overline{\varphi} e^{p\xi} \, dp,$$

where the line Γ lies to the right of the singular points of the function $\overline{\varphi}$.

542. To find integral equations for the charge densities, first let $M = x$ be a fixed point in the plane $y = 0$. Then

$$\cos(\mathbf{r}_{MN}, \mathbf{n}) = 0$$

if the variable point $N = \xi$ also belongs to the plane $y = 0$, while

$$\cos(\mathbf{r}_{MN}, \mathbf{n}) = -\frac{h}{|\mathbf{r}_{MN}|}, \qquad |\mathbf{r}_{MN}| = \sqrt{(\xi - x)^2 + h^2}$$

if $N = \xi$ belongs to the plane $y = h$. Therefore the integral equation (2) on p. 260 takes the form

$$\sigma_0(x) = \frac{1}{2\pi} E_y^0 \big|_{y=0} - \frac{h}{\pi} \int_{-\infty}^\infty \frac{\sigma_h(\xi)}{(\xi - x)^2 + h^2} \, d\xi.$$

In just the same way, choosing $M = x$ in the plane $y = h$, we obtain the integral equation

$$\sigma_h(x) = -\frac{1}{2\pi} E_y^0 \big|_{y=h} - \frac{h}{\pi} \int_{-\infty}^\infty \frac{\sigma_0(\xi)}{(\xi - x)^2 + h^2} \, d\xi.$$

This system of integral equations can be solved by using Fourier transforms. Multiplying each equation by $e^{i\lambda x}$ and integrating with respect to x from $-\infty$ to ∞, we obtain

$$\overline{\sigma}_0 = \frac{\overline{f}_0}{2\pi} - \frac{h}{\pi} \int_{-\infty}^\infty e^{i\lambda x} \, dx \int_{-\infty}^\infty \frac{\sigma_h(\xi)}{(\xi - x)^2 + h^2} \, d\xi,$$

$$\overline{\sigma}_h = -\frac{\overline{f}_h}{2\pi} - \frac{h}{\pi} \int_{-\infty}^\infty e^{i\lambda x} \, dx \int_{-\infty}^\infty \frac{\sigma_0(\xi)}{(\xi - x)^2 + h^2} \, d\xi,$$

where

$$\overline{f} = \int_{-\infty}^\infty f(x) e^{i\lambda x} \, dx$$

and

$$f_0(x) = E_y^0 \big|_{y=0}, \qquad f_h(x) = E_y^0 \big|_{y=h}$$

(for brevity). Reversing the order of integration and using the well-known formula

$$\int_0^\infty \frac{\cos \lambda \eta}{\eta^2 + h^2} \, d\eta = \frac{\pi}{2h} e^{-|\lambda| h},$$

we obtain the following system of linear algebraic equations for $\bar{\sigma}_0$ and $\bar{\sigma}_h$:

$$\bar{\sigma}_0 + e^{-|\lambda|h}\bar{\sigma}_h = \frac{\bar{f}_0}{2\pi}, \qquad e^{-|\lambda|h}\bar{\sigma} + \bar{\sigma}_h = -\frac{\bar{f}_h}{2\pi}.$$

It follows that

$$\bar{\sigma}_0 = \frac{1}{2\pi}\frac{\bar{f}_0 + e^{-|\lambda|h}\bar{f}_h}{1 - e^{-2|\lambda|h}}, \qquad \bar{\sigma}_h = -\frac{1}{2\pi}\frac{\bar{f}_h + e^{-|\lambda|h}\bar{f}_0}{1 - e^{-2|\lambda|h}}.$$

The answer on p. 262 is now an immediate consequence of the inversion formula

$$f(x) = \frac{1}{2\pi}\int_{-\infty}^{\infty} \bar{f}e^{-i\lambda x}\, d\lambda.$$

549. The integral equation

$$\frac{2}{\pi}\int_0^a \frac{f(x)}{x+y}\, K\left(\frac{2\sqrt{xy}}{x+y}\right) dx = g(y), \qquad 0 \leqslant y \leqslant a \tag{89}$$

can be solved as follows: Writing (89) in the form

$$\frac{2}{\pi}\int_0^y \frac{f(x)}{y}\frac{1}{1+(x/y)} K\left[\frac{2\sqrt{x/y}}{1+(x/y)}\right] dx$$

$$+ \frac{2}{\pi}\int_y^a \frac{f(x)}{x}\frac{1}{1+(y/x)} K\left[\frac{2\sqrt{y/x}}{1+(y/x)}\right] dx = g(y),$$

we make a Landen transformation

$$\frac{1}{1+k} K\left(\frac{2\sqrt{k}}{1+k}\right) = K(k),$$

obtaining

$$\frac{2}{\pi}\int_0^y \frac{f(x)}{y} K\left(\frac{x}{y}\right) dx + \frac{2}{\pi}\int_y^a \frac{f(x)}{x} K\left(\frac{y}{x}\right) dx = g(y). \tag{90}$$

Because of the formulas

$$\frac{1}{y} K\left(\frac{x}{y}\right) = \int_0^x \frac{ds}{\sqrt{(x^2 - s^2)(y^2 - s^2)}},$$

$$\frac{1}{x} K\left(\frac{y}{x}\right) = \int_0^y \frac{ds}{\sqrt{(x^2 - s^2)(y^2 - s^2)}},$$

(90) becomes

$$\frac{2}{\pi}\int_0^y f(x)\, dx \int_0^x \frac{ds}{\sqrt{(x^2 - s^2)(y^2 - s^2)}}$$

$$+ \frac{2}{\pi}\int_y^a f(x)\, dx \int_0^y \frac{ds}{\sqrt{(x^2 - s^2)(y^2 - s^2)}} = g(y).$$

The expression on the left can be represented as a double integral over the trapezoid bounded by the lines $s = 0$, $s = x$, $s = y$ and $x = a$. Changing the order of integration in this integral, we obtain

$$\frac{2}{\pi} \int_0^y \frac{ds}{\sqrt{y^2 - s^2}} \int_s^a \frac{f(x)}{\sqrt{x^2 - s^2}} \, dx = g(y). \tag{91}$$

If we write

$$\int_s^a \frac{f(x)}{\sqrt{x^2 - s^2}} \, dx = \Psi(s), \qquad 0 \leqslant s \leqslant a,$$

(91) goes into Schlömilch's integral equation

$$\frac{2}{\pi} \int_0^y \frac{\Psi(s)}{\sqrt{y^2 - s^2}} \, ds = g(y), \qquad 0 \leqslant y \leqslant a,$$

with solution

$$\Psi(s) = \frac{d}{ds} \int_0^s \frac{g(t)t}{\sqrt{s^2 - t^2}} \, dt$$

(see W8, p. 229). To deduce $f(x)$ from a knowledge of $\Psi(s)$, we use the formula

$$f(x) = -\frac{2}{\pi} \frac{d}{dx} \int_x^a \frac{\Psi(s)s}{\sqrt{s^2 - x^2}} \, ds$$

(see B1). Substituting for $\Psi(s)$, we arrive at the answer on p. 264.

561. To find integral equations for the virtual charge densities on the planes $\varphi = 0$ and $\varphi = \alpha$, we note that if $M = r$ is a fixed point in the first of these planes and if $N = \rho$ is an arbitrary point of the interface between the two dielectrics, then

$$\cos{(\mathbf{r}_{MN}, \mathbf{n})} = \begin{cases} \dfrac{\rho \sin \alpha}{\sqrt{r^2 + \rho^2 - 2r\rho \cos \alpha}} & \text{if } N \text{ belongs to the plane } \varphi = \alpha, \\ 0 & \text{if } N \text{ belongs to the plane } \varphi = 0. \end{cases}$$

Applying formula (7), p. 267, we find that

$$\sigma_0(r) = -\frac{\beta}{2\pi} E_\varphi^0 \big|_{\varphi=0} + \frac{\beta}{\pi} \sin \alpha \int_0^\infty \frac{\sigma_\alpha(\rho)\rho}{r^2 + \rho^2 - 2r\rho \cos \alpha} \, d\rho,$$

where

$$\beta = \frac{\varepsilon_1 - \varepsilon_2}{\varepsilon_1 + \varepsilon_2}.$$

By a similar argument, if we choose the fixed point M in the plane $\varphi = \alpha$, then

$$\sigma_\alpha(r) = \frac{\beta}{2\pi} E_\varphi^0 \big|_{\varphi=\alpha} + \frac{\beta}{\pi} \sin \alpha \int_0^\infty \frac{\sigma_0(\rho)\rho}{r^2 + \rho^2 - 2r\rho \cos \alpha} \, d\rho.$$

This system of integral equations can be solved by using Mellin transforms. Multiplying each equation by r^{p-1} and integrating with respect to r from 0 to ∞, we eventually obtain the system of linear algebraic equations

$$\bar{\sigma}_0 - \beta \frac{\sin{(\pi - \alpha)(p-1)}}{\sin{\pi(p-1)}} \bar{\sigma}_\alpha = -\frac{\beta}{2\pi} \bar{f}_0,$$

$$-\beta \frac{\sin{(\pi - \alpha)(p-1)}}{\sin{\pi(p-1)}} \bar{\sigma}_0 + \bar{\sigma}_\alpha = \frac{\beta}{2\pi} \bar{f}_\alpha \tag{92}$$

for $\bar{\sigma}_0$ and $\bar{\sigma}_\alpha$, where

$$\bar{f} = \int_0^\infty f(r) r^{p-1}\, dr$$

and

$$f_0(r) = E_\varphi^0 \big|_{\varphi=0}, \qquad f_\alpha(r) = E_\varphi^0 \big|_{\varphi=\alpha}$$

(for brevity). To guarantee the convergence of the integrals appearing in (92), we choose p to be a complex number of the form $p = 1 + i\tau (-\infty < \tau < \infty)$.[39] Solving the system (92) for $\bar{\sigma}_0$ and $\bar{\sigma}_\alpha$, we find the values of the charge densities by using the inversion formulas

$$\sigma_0 = \frac{1}{2\pi i} \int_{1-i\infty}^{1+i\infty} \bar{\sigma}_0 r^{-p}\, dp, \qquad \sigma_\alpha = \frac{1}{2\pi i} \int_{1-i\infty}^{1+i\infty} \bar{\sigma}_\alpha r^{-p}\, dp.$$

565. The requirement that the tangential component of the electric field be zero on the surface of the conductor leads to the integral equation of the first kind

$$\int_0^\infty H_0^{(2)}(k\,|x - \xi|) j(\xi)\, d\xi = \frac{c^2}{\pi\omega} E(x), \qquad 0 \leqslant x < \infty, \tag{93}$$

where $j(\xi)$ is the total density of current flowing on both sides of the half-plane, and $E(x)$ is the tangential component of the external field at the point x.[40] This integral equation can be solved by using the integral transform (27), p. 196. To reduce (93) to a form suitable for application of this method, we multiply the equation

$$\int_0^\infty H_0^{(2)}(k\xi) j(\xi)\, d\xi = \frac{c^2}{\pi\omega} E(0)$$

---

[39] Each of the densities σ_0 and σ_α is $O(r^{-s_1})$ as $r \to 0$ and $O(r^{-s_2})$ as $r \to \infty$, where $s_1 < 1$ and $s_2 > 1$. The functions f_0 and f_α are assumed to be $O(1)$ as $r \to 0$ and $O(r^{-s})$ as $r \to \infty$, where $s > 1$.

[40] Here we have

$$j(\xi) = j_z\big|_{y=+0} + j_z\big|_{y=-0} = j_1 + j_2, \qquad E(x) = E_z^0\big|_{y=0}.$$

The difference between the current densities is given by formula (9), p. 270, implied by the conservation law for the circulation of the magnetic field around a closed contour.

by e^{-ikx} and subtract it from (93). The result is

$$\int_0^\infty [H_0^{(2)}(k\,|x-\xi|) - e^{-ikx}H_0^{(2)}(k\xi)]j(\xi)\,d\xi = \frac{c^2}{\pi\omega}[E(x) - E(0)e^{-ikx}]. \quad (94)$$

Multiplying both sides of (94) by

$$e^{\pi\tau/2}\frac{H_{i\tau}^{(2)}(kx)}{x}$$

and integrating with respect to x from 0 to ∞, we obtain

$$\int_0^\infty j(\xi)\,d\xi \int_0^\infty \frac{H_0^{(2)}(k\,|x-\xi|) - e^{-ikx}H_0^{(2)}(k\xi)}{x}\,e^{\pi\tau/2}H_{i\tau}^{(2)}(kx)\,dx$$

$$= \frac{c^2 e^{\pi\tau/2}}{\pi\omega}\int_0^\infty \frac{E(x) - e^{-ikx}E(0)}{x}H_{i\tau}^{(2)}(kx)\,dx, \quad (95)$$

assuming that it is legitimate to reverse the order of integration. The inner integral in the left-hand side can be evaluated after making the preliminary transformation

$$\int_0^\infty \frac{H_0^{(2)}(k\,|x-\xi|) - e^{-ikx}H_0^{(2)}(k\xi)}{x}\,e^{\pi\tau/2}H_{i\tau}^{(2)}(kx)\,dx$$

$$= \int_0^\infty \frac{H_0^{(2)}(k\,|x-\xi|) - H_0^{(2)}(k\xi)}{x}\,e^{\pi\tau/2}H_{i\tau}^{(2)}(kx)\,dx$$

$$+ H^{(2)}(k\xi)\int_0^\infty \frac{1 - e^{-ikx}}{x}\,e^{\pi\tau/2}H_{i\tau}^{(2)}(dx) = I_1 + I_2.$$

It then follows from the addition theorem for Hankel functions, [see L9, formula (5.12.11), p. 126] that

$$I_1 = H_0^{(2)}(k\xi)\left\{\int_0^\xi \frac{J_0(kx) - 1}{x}\,e^{\pi\tau/2}H_{i\tau}^{(2)}(kx)\,dx - \int_\xi^\infty \frac{e^{\pi\tau/2}H_{i\tau}^{(2)}(kx)}{x}\,dx\right\}$$

$$+ J_0(k\xi)\int_\xi^\infty \frac{H_0^{(2)}(kx)}{x}\,e^{\pi\tau/2}H_{i\tau}^{(2)}(kx)\,dx$$

$$+ 2\sum_{m=1}^\infty \left\{H_m^{(2)}(k\xi)\int_0^\xi \frac{J_m(kx)}{x}\,e^{\pi\tau/2}H_{i\tau}^{(2)}(kx)\,dx\right.$$

$$\left. + J_m(k\xi)\int_\xi^\infty \frac{H_m^{(2)}(kx)}{x}\,e^{\pi\tau/2}H_{i\tau}^{(2)}(kx)\,dx\right\}.$$

The integrals on the right can be evaluated in closed form, eventually leading to the expressions

$$I_1 = \frac{1}{i\tau \sinh \frac{1}{2}\pi\tau} H_0^{(2)}(k\xi) - \frac{2}{\pi i\tau^2} e^{\pi\tau/2} H_{i\tau}^{(2)}(k\xi)\left[1 + 2\sum_{m=1}^{\infty} \frac{\tau^2}{m^2 + \tau^2}\right]$$

$$= \frac{1}{i\tau \sinh \frac{1}{2}\pi\tau} H_0^{(2)}(k\xi) - \frac{2}{i\tau} e^{\pi\tau/2} \coth \pi\tau \, H_{i\tau}^{(2)}(k\xi),$$

$$I_2 = H_0^{(2)}(k\xi)\left[\frac{2}{i\tau \sinh \pi\tau} - \frac{1}{i\tau \sinh \frac{1}{2}\pi\tau}\right].$$

Thus we finally have

$$\int_0^\infty \frac{H_0^{(2)}(k\,|x - \xi|) - e^{-ikx}H_0^{(2)}(k\xi)}{x} \, e^{\pi\tau/2} H_{i\tau}^{(2)}(kx) \, dx$$

$$= \frac{2}{i\tau \sinh \pi\tau} [H_0^{(2)}(k\xi) - e^{\pi\tau/2} \coth \pi\tau \, H_{i\tau}^{(2)}(k\xi)].$$

If we now introduce the integral transform

$$\bar{\varphi}(\tau) = \int_0^\infty \varphi(x) e^{\pi\tau/2} \frac{H_{i\tau}^{(2)}(kx)}{x}, \qquad 0 \leqslant \tau < \infty,$$

(95) takes the form

$$\frac{2}{i\tau \sinh \pi\tau}\left[\frac{c^2}{\pi\omega} E(0) - \overline{\xi j(\xi)} \cosh \pi\tau\right] = \frac{c^2}{\pi\omega} \overline{[E(x) - e^{-ikx}E(0)]},$$

which implies

$$\overline{\xi j(\xi)} = \frac{c^2}{\pi\omega}\left\{\frac{E(0)}{\cosh \pi\tau} - \frac{i\tau \tanh \pi\tau}{2} \overline{[E(x) - e^{-ikx}E(0)]}\right\}.$$

The final form of the solution given in the answer on p. 270 is obtained by using the formula[41]

$$\varphi(x) = -\frac{1}{2}\int_0^\infty \bar{\varphi}(\tau) e^{\pi\tau/2}\tau \sinh \pi\tau \, H_{i\tau}^{(2)}(kx) \, d\tau, \qquad 0 < x < \infty.$$

566. The problem reduces to solving the integral equation

$$\int_0^\infty H_0^{(2)}(k\,|r - \rho|)j(\rho) \, d\rho = \frac{c^2 E^0}{\pi\omega} e^{-ikr\cos\alpha}, \qquad 0 \leqslant r < \infty.$$

We assume that k is of the form $k = |k|\,e^{-i\gamma}(0 < \gamma < \pi/2)$ as in Prob. 565,

---

[41] The applicability of this formula is guaranteed by the requirement that k be a complex number of the form $k = |k|\,e^{-i\gamma}$ $(0 < \gamma < \pi/2)$, and that the external field be due to line sources located in the finite part of the xy-plane.

and also that the angle between the half-plane and the direction of propagation of the incident wave is less than γ.[42] Then the problem can be solved by using the general formulas obtained in Prob. 565, with

$$x = r, \quad \xi = \rho, \quad E(x) = E^0 e^{-ikr} \cos \alpha.$$

Bearing in mind that

$$\overline{[E(r) - e^{-ikr}E(0)]} = E^0 \int_0^\infty (e^{-ikr \cos \alpha} - e^{-ikr})e^{\pi\tau/2} \frac{H_{i\tau}^{(2)}(kr)}{r} \, dr$$

$$= \frac{2iE^0}{\tau \sinh \pi\tau} (\cosh \tau\alpha - 1),$$

we have

$$\rho \, j(\rho) = -\frac{c^2 E^0}{2\pi\omega} \int_0^\infty \tau \tanh \pi\tau \, e^{\pi\tau/2} \cosh \tau\alpha \, H_{i\tau}^{(2)}(k\rho) \, d\tau.$$

The last integral can be written in the form

$$\rho \, j(\rho) = \frac{ic^2 E^0}{2\pi\omega} \frac{d}{d\alpha} [\psi(\pi + \alpha) - \psi(\pi - \alpha)],$$

where

$$\psi(\beta) = -\frac{1}{2i} \int_0^\infty \frac{\cosh \beta\tau}{\cosh \pi\tau} e^{\pi\tau/2} H_{i\tau}^{(2)}(k\rho) \, d\tau, \qquad |\beta| < \pi + \gamma.$$

In the paper K3, it is shown that

$$\psi(\beta) = e^{ik\rho \cos \beta} \left[-\frac{1}{2} + \frac{e^{i\pi/4}}{\sqrt{\pi}} \int_0^{\sqrt{2k\rho} \cos \frac{1}{2}\beta} e^{-is^2} \, ds \right].$$

Using this formula and performing the differentiation with respect to α, we obtain the expression for $j = j_1 + j_2$ given in the answer on p.271.

----

[42] These restrictions are needed to guarantee the convergence of the integrals and to justify using the inversion formula, but can be dropped in the final results. In particular, the expressions for the current densities j_1 and j_2 found here are also valid for real k, in which case they coincide with the corresponding formulas for the Sommerfeld problem (see Prob. 427).

MATHEMATICAL APPENDIX

I. Special Functions Appearing in the Text

Certain basic functions

The gamma function

$$\Gamma(z) = \int_0^\infty e^{-t} t^{z-1}\, dt, \qquad \text{Re } z > 0.$$

The probability integral

$$\Phi(z) = \frac{2}{\sqrt{\pi}} \int_0^z e^{-t^2}\, dt.$$

The Fresnel integrals

$$C(z) = \int_0^z \cos \frac{\pi t^2}{2}\, dt, \qquad S(z) = \int_0^z \sin \frac{\pi t^2}{2}\, dt.$$

The exponential integral

$$\text{Ei}\,(z) = \int_{-\infty}^z \frac{e^t}{t}\, dt, \qquad 0 < \arg z < 2\pi.$$

The sine integral

$$\text{Si}\,(z) = \int_0^z \frac{\sin t}{t}\, dt.$$

The cosine integral

$$\text{Ci}\,(z) = \int_\infty^z \frac{\cos t}{t}\, dt, \qquad |\arg z| < \pi.$$

Orthogonal polynomials

The Legendre polynomials

$$P_n(x) = \frac{1}{2^n n!} \frac{d^n}{dx^n} (x^2 - 1)^n, \qquad n = 0, 1, 2, \dots$$

The Hermite polynomials

$$H_n(x) = (-1)^n e^{x^2} \frac{d^n}{dx^n} e^{-x^2}, \qquad n = 0, 1, 2, \dots$$

The Laguerre polynomials

$$L_n^\alpha(x) = e^x \frac{x^{-\alpha}}{n!} \frac{d^n}{dx^n} (e^{-x} x^{n+\alpha}), \qquad n = 0, 1, 2, \dots,$$

$$L_n^0(x) = L_n(x).$$

Cylinder functions

The Bessel function of the first kind

$$J_\nu(z) = \sum_{k=0}^\infty \frac{(-1)^k (z/2)^{\nu+2k}}{\Gamma(k+1)\Gamma(k+\nu+1)}, \qquad |\arg z| < \pi.$$

The Bessel function of the second kind

$$Y_\nu(z) = \frac{J_\nu(z)\cos \nu\pi - J_{-\nu}(z)}{\sin \nu\pi}, \qquad |\arg z| < \pi.$$

The Bessel function of the second kind of integral order ($n = 0, 1, 2, \dots$)

$$Y_n(z) = \lim_{\nu \to n} Y_\nu(z) = \frac{2}{\pi} J_n(z) \ln \frac{z}{2} - \frac{1}{\pi} \sum_{k=0}^{n-1} \frac{(n-k-1)!}{k!} \left(\frac{z}{2}\right)^{2k-n}$$

$$- \frac{1}{\pi} \sum_{k=0}^\infty \frac{(-1)^k (z/2)^{n+2k}}{k!\,(n+k)!} [\psi(k+1) + \psi(k+n+1)], \qquad |\arg z| < \pi,$$

where $\psi(z)$ is the logarithmic derivative of the gamma function (the first sum is omitted if $n = 0$).

The first and second Hankel functions

$$H_\nu^{(1)}(z) = J_\nu(z) + i Y_\nu(z), \quad H_\nu^{(2)}(z) = J_\nu(z) - i Y_\nu(z), \qquad |\arg z| < \pi.$$

The Bessel function of imaginary argument

$$I_\nu(z) = \sum_{k=0}^\infty \frac{(z/2)^{\nu+2k}}{\Gamma(k+1)\Gamma(k+\nu+1)}, \qquad |\arg z| < \pi.$$

The Macdonald function

$$K_\nu(z) = \frac{\pi}{2} \frac{I_{-\nu}(z) - I_\nu(z)}{\sin \nu\pi}, \qquad |\arg z| < \pi.$$

The Macdonald function of integral order ($n = 0, 1, 2, \ldots$)

$$K_n(z) = \lim_{\nu \to n} K_\nu(z) = (-1)^{n+1} I_n(z) \ln \frac{z}{2}$$
$$+ \frac{1}{2} \sum_{k=0}^{n-1} \frac{(-1)^k (n-k-1)!}{k!} \left(\frac{z}{2}\right)^{2k-n}$$
$$+ \frac{1}{2} (-1)^n \sum_{k=0}^{\infty} \frac{(z/2)^{n+2k}}{k! \, (k+n)!} [\psi(k+1) + \psi(k+n+1)],$$
$$|\arg z| < \pi,$$

where $\psi(z)$ is the logarithmic derivative of the gamma function (the first sum is omitted if $n = 0$).

Spherical harmonics

The Legendre functions of the first and second kinds

$$P_\nu(z) = F\left(-\nu, \nu+1; 1; \frac{1-z}{2}\right), \quad |z-1| < 2,$$

$$Q_\nu(z) = \frac{\sqrt{\pi}\, \Gamma(1+\nu)}{\Gamma\left(\nu + \frac{3}{2}\right)(2z)^{\nu+1}} \, F\left(\frac{\nu}{2}+1, \frac{\nu}{2}+\frac{1}{2}; \nu+\frac{3}{2}; \frac{1}{z^2}\right),$$

$$|z| > 1, \quad |\arg z| < \pi,$$

where

$$F(\alpha, \beta; \gamma; z) = \sum_{k=0}^{\infty} \frac{(\alpha)_k (\beta)_k}{(\gamma)_k k!} z^k, \quad |z| < 1,$$

$$(\lambda)_0 = 1, \quad (\lambda)_k = \frac{\Gamma(\lambda+k)}{\Gamma(\lambda)} = \lambda(\lambda+1)\cdots(\lambda+k-1)$$

is the hypergeometric series.[1]

For real x in the interval $(-1, 1)$, the Legendre function of the second kind is defined by the formula

$$Q_\nu(x) = \tfrac{1}{2}[Q_\nu(x+i0) + Q_\nu(x-i0)].$$

Analytic expressions for the spherical harmonics appearing in this book can be found in H4, L9 and M2.

The associated Legendre functions

$$P_\nu^m(z) = (z^2 - 1)^{m/2} \frac{d^m P_\nu(z)}{dz^m}, \quad Q_\nu^m(z) = (z^2 - 1)^{m/2} \frac{d^m Q_\nu(z)}{dz^m},$$

$$|\arg(z-1)| < \pi \quad (m = 1, 2, \ldots).$$

---

[1] The functions $P_\nu(z)$ and $Q_\nu(z)$ are defined outside the indicated regions by using analytic continuation (see e.g., L9, Sec. 7.3).

The associated Legendre functions for the interval $(-1, 1)$

$$P_\nu^m(x) = (-1)^m (1 - x^2)^{m/2} \frac{d^m P_\nu(x)}{dx^m},$$

$$Q_\nu^m(x) = (-1)^m (1 - x^2)^{m/2} \frac{d^m Q_\nu(x)}{dx^m}.$$

Elliptic integrals and functions

The elliptic integrals of the first and second kinds

$$F(\varphi, k) = \int_0^\varphi \frac{d\varphi}{\sqrt{1 - k^2 \sin^2 \varphi}}, \qquad E(\varphi, k) = \int_0^\varphi \sqrt{1 - k^2 \sin^2 \varphi} \, d\varphi.$$

The *complete* elliptic integrals of the first and second kinds

$$K(k) = F\left(\frac{\pi}{2}, k\right), \qquad E(k) = E\left(\frac{\pi}{2}, k\right).$$

The Jacobian elliptic functions

$$\operatorname{sn} z = \sin \varphi, \quad \operatorname{cn} z = \cos \varphi, \quad \operatorname{dn} z = \sqrt{1 - k^2 \sin^2 \varphi},$$

where φ is the inversion of the elliptic integral of the first kind, i.e.,

$$z = F(\varphi, k) = \int_0^\varphi \frac{d\varphi}{\sqrt{1 - k^2 \sin^2 \varphi}}.$$

Further information on special functions can be found in such books as Erdélyi et al. (E2), Gray and Mathews (G2), Hobson (H4), Jackson (J1), Lebedev (L9), Lense (L11, L12), MacRobert (M2), Magnus and Oberhettinger (M3), McLachlan (M5), Ryshik and Gradstein (R2), Smirnov (S6, Vol. III, Pt. 2), Snow (S12), Watson (W4), and Whittaker and Watson (W8).

2. Some Expansions in Series of Orthogonal Functions

1. $\displaystyle\sum_{n=1}^\infty \frac{\sin (n\pi x/a)}{n} = \frac{\pi}{2}\left(1 - \frac{x}{a}\right), \qquad 0 < x \leqslant a.$

2. $\displaystyle\sum_{n=1}^\infty \frac{\cos (n\pi x/a)}{n} = -\ln \left(2 \sin \frac{\pi x}{2a}\right), \qquad 0 < x \leqslant a.$

3. $\displaystyle\sum_{n=1}^\infty (-1)^{n-1} \frac{\sin (n\pi x/a)}{n} = \frac{\pi x}{2a}, \qquad 0 \leqslant x < a.$

4. $\displaystyle\sum_{n=1}^\infty (-1)^{n-1} \frac{\cos (n\pi x/a)}{n} = \ln \left(2 \cos \frac{\pi x}{2a}\right), \qquad 0 \leqslant x < a.$

5. $\displaystyle\sum_{n=0}^{\infty} \frac{\sin\left[(2n+1)\pi x/2a\right]}{2n+1} = \frac{\pi}{4}$, $0 < x \leqslant a$.

6. $\displaystyle\sum_{n=0}^{\infty} \frac{\cos\left[(2n+1)\pi x/2a\right]}{2n+1} = \frac{1}{2}\ln\cot\frac{\pi x}{2a}$, $0 < x < a$.

7. $\displaystyle\sum_{n=0}^{\infty}(-1)^n \frac{\sin\left[(2n+1)\pi x/2a\right]}{2n+1} = \frac{1}{2}\ln\tan\frac{\pi x}{2a}$, $0 < x < a$.

8. $\displaystyle\sum_{n=0}^{\infty}(-1)^n \frac{\cos\left[(2n+1)\pi x/2a\right]}{2n+1} = \frac{\pi}{4}$, $0 \leqslant x < a$.

9. $\displaystyle\sum_{n=1}^{\infty} \frac{\cos\left(n\pi x/a\right)}{n^2} = \pi^2\left(\frac{1}{6} - \frac{x}{2a} + \frac{x^2}{4a^2}\right)$, $0 \leqslant x \leqslant a$.

10. $\displaystyle\sum_{n=1}^{\infty}(-1)^{n-1} \frac{\cos\left(n\pi x/a\right)}{n^2} = \pi^2\left(\frac{1}{12} - \frac{x^2}{4a^2}\right)$, $0 \leqslant x \leqslant a$.

11. $\displaystyle\sum_{n=0}^{\infty} \frac{\cos\left[(2n+1)\pi x/2a\right]}{(2n+1)^2} = \frac{\pi^2}{8}\left(1 - \frac{x}{a}\right)$, $0 \leqslant x \leqslant a$.

12. $\displaystyle\sum_{n=0}^{\infty}(-1)^n \frac{\sin\left[(2n+1)\pi x/2a\right]}{(2n+1)^2} = \frac{\pi^2 x}{8a}$, $0 \leqslant x \leqslant a$.

13. $\displaystyle\sum_{n=1}^{\infty} \frac{\sin\left(n\pi x/a\right)}{n^3} = \pi^3\left(\frac{x}{6a} - \frac{x^2}{4a^2} + \frac{x^3}{12a^3}\right)$, $0 \leqslant x \leqslant a$.

14. $\displaystyle\sum_{n=1}^{\infty}(-1)^{n-1} \frac{\sin\left(n\pi x/a\right)}{n^3} = \frac{\pi^3 x}{12a}\left(1 - \frac{x^2}{a^2}\right)$, $0 \leqslant x \leqslant a$.

15. $\displaystyle\sum_{n=0}^{\infty} \frac{\sin\left[(2n+1)\pi x/2a\right]}{(2n+1)^3} = \frac{\pi^3 x}{16a}\left(1 - \frac{x}{2a}\right)$, $0 \leqslant x \leqslant a$.

16. $\displaystyle\sum_{n=0}^{\infty}(-1)^n \frac{\cos\left[(2n+1)\pi x/2a\right]}{(2n+1)^3} = \frac{\pi^3}{32}\left(1 - \frac{x^2}{a^2}\right)$, $0 \leqslant x \leqslant a$.

17. $\displaystyle\sum_{n=1}^{\infty} \frac{J_0(\gamma_n r/a)}{\gamma_n J_1(\gamma_n)} = \frac{1}{2}$, $0 \leqslant r < a$,

where the γ_n are the positive roots of the equation $J_0(\gamma) = 0$.

18. $\displaystyle\sum_{n=1}^{\infty} \frac{J_0(\gamma_n r/a)}{\gamma_n^3 J_1(\gamma_n)} = \frac{1}{8}\left(1 - \frac{r^2}{a^2}\right)$, $0 \leqslant r \leqslant a$,

where the γ_n are the positive roots of the equation $J_0(\gamma) = 0$.

19. $\displaystyle\sum_{n=1}^{\infty} \frac{J_0(\gamma_n r/a)}{\gamma_n^2 J_0(\gamma_n)} = \frac{1}{4}\left(\frac{r^2}{a^2} - \frac{1}{2}\right)$, $0 \leqslant r \leqslant a$,

where the γ_n are the positive roots of the equation $J_1(\gamma) = 0$.

20. $x^{2m} = \dfrac{P_0(x)}{2m + 1}$

$$+ \sum_{n=1}^{m} (4n + 1) \frac{2m(2m - 2) \cdots (2m - 2n + 2)}{(2m + 1)(2m + 3) \cdots (2m + 2n + 1)} P_{2n}(x),$$

$$m = 1, 2, \ldots$$

21. $x^{2m+1} = \dfrac{3P_1(x)}{2m + 3}$

$$+ \sum_{n=1}^{m} (4n + 3) \frac{2m(2m - 2) \cdots (2m - 2n + 2)}{(2m + 3)(2m + 5) \cdots (2m + 2n + 3)} P_{2n+1}(x),$$

$$m = 1, 2, \ldots$$

22. $\dfrac{1}{\sqrt{1 - 2tx + x^2}} = \sum_{n=0}^{\infty} t^n P_n(x), \qquad -1 \leqslant x \leqslant 1, \quad |t| < 1.$

For various other expansions in orthogonal functions (and series of a different kind), we refer to the handbooks by Jolley (J5) and Ryshik and Gradstein (R2).

3. Some Definite Integrals Frequently Encountered in the Applications

1. $\displaystyle\int_0^\infty \frac{x^{\nu-1}}{1 + x}\, dx = \frac{\pi}{\sin \pi \nu}, \qquad 0 < \mathrm{Re}\, \nu < 1.$

2. $\displaystyle\int_0^\infty \frac{x^\nu\, dx}{1 - 2x \cos \varphi + x^2} = \frac{\pi}{\sin \pi \nu} \frac{\sin (\pi - \varphi)\nu}{\sin \varphi}, \qquad -1 < \mathrm{Re}\, \nu < 1,$

$$0 < \varphi < 2\pi.$$

3. $\displaystyle\int_0^\infty \frac{\sin ax}{x}\, dx = \frac{\pi}{2}, \qquad a > 0.$

4. $\displaystyle\int_0^\infty \frac{\cos ax - \cos bx}{x}\, dx = \ln \frac{b}{a}, \qquad a > 0, \quad b > 0.$

5. $\displaystyle\int_0^\infty \frac{\cos bx}{a^2 + x^2}\, dx = \frac{\pi}{2a} e^{-ab}, \qquad a > 0, \quad b > 0.$

6. $\displaystyle\int_0^\infty e^{-ax} x^{\nu-1}\, dx = \frac{\Gamma(\nu)}{a^\nu}, \qquad a > 0, \quad \mathrm{Re}\, \nu > 0.$

7. $\int_0^\infty e^{-ax} \cos bx \, dx = \dfrac{a}{a^2 + b^2}$, $a > 0$.

8. $\int_0^\infty e^{-ax} \sin bx \, dx = \dfrac{b}{a^2 + b^2}$, $a > 0$.

9. $\int_0^\infty e^{-a^2 x^2} \, dx = \dfrac{\sqrt{\pi}}{2a}$, $a > 0$.

10. $\int_0^\infty e^{-a^2 x^2} \cos bx \, dx = \dfrac{\sqrt{\pi}}{2a} e^{-b^2/4a^2}$, $a > 0$.

11. $\int_0^\infty e^{-a^2 x^2 - (b^2/x^2)} \, dx = \dfrac{\sqrt{\pi}}{2a} e^{-2ab}$, $a > 0$, $b > 0$.

12. $\int_0^\infty \sin x^2 \, dx = \int_0^\infty \cos x^2 \, dx = \dfrac{1}{2}\sqrt{\dfrac{\pi}{2}}$.

13. $\int_0^\infty \dfrac{\sinh px}{\cosh qx} \sin rx \, dx = \dfrac{\pi}{q} \dfrac{\sinh \dfrac{\pi r}{2a} \sin \dfrac{\pi p}{2a}}{\cosh \dfrac{\pi r}{q} + \cos \dfrac{\pi p}{q}}$, $0 \leqslant p < q$.

14. $\int_0^\infty \dfrac{\cosh px}{\sinh qx} \sin rx \, dx = \dfrac{\pi}{2q} \dfrac{\sinh \dfrac{\pi r}{q}}{\cosh \dfrac{\pi r}{q} + \cos \dfrac{\pi p}{q}}$, $0 \leqslant p < q$.

15. $\int_0^\infty \dfrac{\sinh px}{\sinh qx} \cos rx \, dx = \dfrac{\pi}{2q} \dfrac{\sin \dfrac{\pi p}{q}}{\cosh \dfrac{\pi r}{q} + \cos \dfrac{\pi p}{q}}$, $0 \leqslant p < q$.

16. $\int_0^\infty \dfrac{\cosh px}{\cosh qx} \cos rx \, dx = \dfrac{\pi}{q} \dfrac{\cosh \dfrac{\pi r}{2q} \cos \dfrac{\pi p}{2q}}{\cosh \dfrac{\pi r}{q} + \cos \dfrac{\pi p}{q}}$, $0 \leqslant p < q$.

17. $\int_0^\infty \dfrac{\cosh px}{(\cosh x)^q} \, dx = 2^{q-2} \dfrac{\Gamma\left(\dfrac{q+p}{2}\right)\Gamma\left(\dfrac{q-p}{2}\right)}{\Gamma(q)}$, $0 \leqslant p < q$.

18. $\displaystyle\int_0^\infty e^{-ax} J_0(bx)\, dx = \frac{1}{\sqrt{a^2 + b^2}}, \qquad a > 0.$

19. $\displaystyle\int_0^\infty e^{-a^2 x^2} J_\nu(bx) x^{\nu+1}\, dx = \frac{b^\nu}{(2a^2)^{\nu+1}} e^{-b^2/4a^2}, \qquad b > 0, \quad \mathrm{Re}\,\nu > -1.$

20. $\displaystyle\int_0^\infty \frac{x^{\nu+1} J_\nu(bx)}{(x^2 + a^2)^{\mu+1}}\, dx = \frac{a^{\nu-\mu} b^\mu}{2^\mu \Gamma(\mu + 1)} K_{\nu-\mu}(ab),$

$$a > 0, \quad b > 0, \quad -1 < \mathrm{Re}\,\nu < 2\,\mathrm{Re}\,\mu + \tfrac{3}{2}.$$

21. $\displaystyle\int_0^\infty \frac{K_\mu(a\sqrt{x^2 + y^2})}{(x^2 + y^2)^{\mu/2}} J_\nu(bx) x^{\nu+1}\, dx$

$$= \frac{b^\nu}{a^\mu} \left(\frac{\sqrt{a^2 + b^2}}{y}\right)^{\mu-\nu-1} K_{\mu-\nu-1}(y\sqrt{a^2 + b^2}),$$

$$a > 0, \quad b > 0, \quad y > 0, \quad \mathrm{Re}\,\nu > -1.$$

Among the handbooks on definite integrals, we cite those by Dwight (D2) and Ryshik and Gradstein (R2), as well as the celebrated compendium of Bierens de Haan (B4).

4. Expansion of Some Differential Operators in Orthogonal Curvilinear Coordinates

General formulas

Let (q_1, q_2, q_3) be a system of orthogonal curvilinear coordinates related to Cartesian coordinates (x, y, z) by the formula

$$x = x(q_1, q_2, q_3), \quad y = y(q_1, q_2, q_3), \quad z = z(q_1, q_2, q_3).$$

Suppose the square of the element of arc length in the given system is

$$ds^2 = h_1^2\, dq_1^2 + h_2^2\, dq_2^2 + h_3^2\, dq_3^2,$$

where the h_i are the metric coefficients

$$h_i = \sqrt{\left(\frac{\partial x}{\partial q_i}\right)^2 + \left(\frac{\partial y}{\partial q_i}\right)^2 + \left(\frac{\partial z}{\partial q_i}\right)^2}, \qquad i = 1, 2, 3.$$

Then the differential operators $(\mathrm{grad}\,u)_{q_i}$, $\mathrm{div}\,\mathbf{A}$, Δu, $(\mathrm{curl}\,\mathbf{A})_{q_i}$ [where u and $\mathbf{A}$ are given functions of the coordinates, and the index q_i denotes the corresponding vector component] take the form

$$(\text{grad } u)_{q_i} = \frac{1}{h_i} \frac{\partial u}{\partial q_i},$$

$$\text{div } \mathbf{A} = \frac{1}{h_1 h_1 h_3} \left\{ \frac{\partial}{\partial q_1} (h_2 h_3 A_{q_1}) + \frac{\partial}{\partial q_2} (h_3 h_1 A_{q_2}) + \frac{\partial}{\partial q_3} (h_1 h_2 A_{q_3}) \right\},$$

$$\Delta u = \frac{1}{h_1 h_2 h_3} \left\{ \frac{\partial}{\partial q_1} \left(\frac{h_2 h_3}{h_1} \frac{\partial u}{\partial q_1} \right) + \frac{\partial}{\partial q_2} \left(\frac{h_3 h_1}{h_2} \frac{\partial u}{\partial q_2} \right) + \frac{\partial}{\partial q_3} \left(\frac{h_1 h_2}{h_3} \frac{\partial u}{\partial q_3} \right) \right\},$$

$$(\text{curl } \mathbf{A})_{q_1} = \frac{1}{h_2 h_3} \left\{ \frac{\partial}{\partial q_2} (h_3 A_{q_3}) - \frac{\partial}{\partial q_3} (h_2 A_{q_2}) \right\},$$

$$(\text{curl } \mathbf{A})_{q_2} = \frac{1}{h_3 h_1} \left\{ \frac{\partial}{\partial q_3} (h_1 A_{q_1}) - \frac{\partial}{\partial q_1} (h_3 A_{q_3}) \right\},$$

$$(\text{curl } \mathbf{A})_{q_3} = \frac{1}{h_1 h_2} \left\{ \frac{\partial}{\partial q_1} (h_2 A_{q_2}) - \frac{\partial}{\partial q_2} (h_1 A_{q_1}) \right\}.$$

Cylindrical coordinates

$$x = r \cos \varphi, \quad y = r \sin \varphi, \quad z = z,$$

$$(0 \leqslant r < \infty, \quad -\pi < \varphi \leqslant \pi, \quad -\infty < z < \infty),$$

$$ds^2 = dr^2 + r^2 \, d\varphi^2 + dz^2, \qquad h_r = 1, \quad h_\varphi = r, \quad h_z = 1,$$

$$(\text{grad } u)_r = \frac{\partial u}{\partial r}, \quad (\text{grad } u)_\varphi = \frac{1}{r} \frac{\partial u}{\partial \varphi}, \quad (\text{grad } u)_z = \frac{\partial u}{\partial z},$$

$$\text{div } \mathbf{A} = \frac{1}{r} \frac{\partial}{\partial r} (r A_r) + \frac{1}{r} \frac{\partial A_\varphi}{\partial \varphi} + \frac{\partial A_z}{\partial z},$$

$$\Delta u = \frac{1}{r} \frac{\partial}{\partial r} \left(r \frac{\partial u}{\partial r} \right) + \frac{1}{r^2} \frac{\partial^2 u}{\partial \varphi^2} + \frac{\partial^2 u}{\partial z^2},$$

$$(\text{curl } \mathbf{A})_r = \frac{1}{r} \frac{\partial A_z}{\partial \varphi} - \frac{\partial A_\varphi}{\partial z},$$

$$(\text{curl } \mathbf{A})_\varphi = \frac{\partial A_r}{\partial z} - \frac{\partial A_z}{\partial r},$$

$$(\text{curl } \mathbf{A})_z = \frac{1}{r} \frac{\partial}{\partial r} (r A_\varphi) - \frac{1}{r} \frac{\partial A_r}{\partial \varphi}.$$

The coordinates of the vector $\Delta \mathbf{A} = \text{grad div } \mathbf{A} - \text{curl curl } \mathbf{A}$ are:

$$(\Delta \mathbf{A})_r = \Delta A_r - \frac{1}{r^2} A_r - \frac{2}{r^2} \frac{\partial A_\varphi}{\partial \varphi},$$

$$(\Delta \mathbf{A})_\varphi = \Delta A_\varphi - \frac{1}{r^2} A_\varphi + \frac{2}{r^2} \frac{\partial A_r}{\partial \varphi},$$

$$(\Delta \mathbf{A})_z = \Delta A_z.$$

Spherical coordinates

$$x = r \sin\theta \cos\varphi, \quad y = r \sin\theta \sin\varphi, \quad z = r \cos\theta,$$

$$(0 \leqslant r < \infty, \quad 0 \leqslant \theta \leqslant \pi, \quad -\pi < \varphi \leqslant \pi),$$

$$ds^2 = dr^2 + r^2\, d\theta^2 + r^2 \sin^2\theta\, d\varphi^2, \quad h_r = 1, \quad h_\theta = r, \quad h_\varphi = r \sin\theta,$$

$$(\text{grad } u)_r = \frac{\partial u}{\partial r}, \quad (\text{grad } u)_\theta = \frac{1}{r} \frac{\partial u}{\partial \theta}, \quad (\text{grad } u)_\varphi = \frac{1}{r \sin\theta} \frac{\partial u}{\partial \varphi},$$

$$\text{div } \mathbf{A} = \frac{1}{r^2} \frac{\partial}{\partial r} (r^2 A_r) + \frac{1}{r \sin\theta} \frac{\partial}{\partial \theta} (A_\theta \sin\theta) + \frac{1}{r \sin\theta} \frac{\partial A_\varphi}{\partial \varphi},$$

$$\Delta u = \frac{1}{r^2} \frac{\partial}{\partial r}\left(r^2 \frac{\partial u}{\partial r}\right) + \frac{1}{r^2 \sin\theta} \frac{\partial}{\partial \theta}\left(\sin\theta \frac{\partial u}{\partial \theta}\right) + \frac{1}{r^2 \sin^2\theta} \frac{\partial^2 u}{\partial \varphi^2},$$

$$(\text{curl } \mathbf{A})_r = \frac{1}{r \sin\theta} \frac{\partial}{\partial \theta} (A_\varphi \sin\theta) - \frac{1}{r \sin\theta} \frac{\partial A_\theta}{\partial \varphi},$$

$$(\text{curl } \mathbf{A})_\theta = \frac{1}{r \sin\theta} \frac{\partial A_r}{\partial \varphi} - \frac{1}{r} \frac{\partial}{\partial r} (r A_\varphi),$$

$$(\text{curl } \mathbf{A})_\varphi = \frac{1}{r} \frac{\partial}{\partial r} (r A_\theta) - \frac{1}{r} \frac{\partial A_r}{\partial \theta},$$

$$(\Delta \mathbf{A})_r = \Delta A_r - \frac{2}{r^2} A_r - \frac{2}{r^2 \sin\theta} \frac{\partial}{\partial \theta} (A_\theta \sin\theta) - \frac{2}{r^2 \sin\theta} \frac{\partial A_\varphi}{\partial \varphi},$$

$$(\Delta \mathbf{A})_\theta = \Delta A_\theta - \frac{1}{r^2 \sin^2\theta} A_\theta + \frac{2}{r^2} \frac{\partial A_r}{\partial \theta} - \frac{2 \cos\theta}{r^2 \sin^2\theta} \frac{\partial A_\varphi}{\partial \varphi},$$

$$(\Delta \mathbf{A})_\varphi = \Delta A_\varphi - \frac{1}{r^2 \sin^2\theta} A_\varphi + \frac{2}{r^2 \sin\theta} \frac{\partial A_r}{\partial \varphi} + \frac{2 \cos\theta}{r^2 \sin^2\theta} \frac{\partial A_\theta}{\partial \varphi}.$$

Expressions for the above differential operators in other special orthogonal curvilinear coordinate systems can be found in Chapter 7 of this book, and in the handbook by Magnus and Oberhettinger (M3).

VARIATIONAL AND RELATED METHODS[1]

Many, and perhaps most, mathematical problems encountered in science and engineering are difficult or impossible to solve by analytical methods. It is also found that explicitly obtained exact solutions are often too cumbersome for interpretation and numerical evaluation. Therefore, in these instances, it is either necessary or convenient to employ approximate methods which yield accurate numerical estimates of the solution. The recent development of high speed electronic digital computers has made practical the successful application of many of these methods to complex problems.

This supplement contains a collection of typical problems that illustrate a special class of approximate methods. They are related either directly or indirectly to the variational formulation of physical problems. Almost all of the examples are concerned with boundary value problems for ordinary or partial differential equations. However, with suitable and sometimes trivial modifications, the methods presented can often be applied to other situations, e.g., eigenvalue problems or problems involving integral equations or integro-differential equations.

The selection of problems was, in large measure, influenced by the amount of computational work necessary to obtain a solution. Hence, by necessity, they are essentially "simple." However, the methods employed can usually be applied directly to more complicated problems, the only additional difficulty being that the calculations are more involved.

The supplement is independent of the main body of the book in the sense

---

[1] This supplement was written by Edward L. Reiss, Courant Institute of Mathematical Sciences, New York University.

that all equation and problem numbers refer only to those in the supplement. Literature references, indicated in brackets, are to items in the references section on p. 412.

I. Variational Methods

I.1. FORMULATION OF VARIATIONAL PROBLEMS[2]

Physical problems can frequently be formulated mathematically as minimum problems, as well as in terms of differential or integral equations. The solution is then a function, selected from a certain class called *admissible functions*, which minimizes a specified functional[3] with respect to all admissible functions. For example, *Hamilton's principle* is an alternative to Newton's equations of motion as a formulation of the laws of mechanics. It states that if $\mathbf{u}(\mathbf{x}, t)$ is a vector describing the motion of a mechanical system,[4] then between any two times t_0 and t_1, the actual (stable) motion is an admissible vector which coincides with the actual motion at $t = t_0$ and $t = t_1$ and makes the functional

$$\int_{t_0}^{t_1} (T - U)\, dt$$

a minimum. Here T and U are the kinetic and potential energy functionals of the system. The admissibility conditions usually take the form of boundary conditions and of continuity requirements on $\mathbf{u}$ and its derivatives. If the mechanical system is in equilibrium, so that $T \equiv 0$ and $\mathbf{u}$ is independent of t, then Hamilton's principle becomes the *principle of minimum potential energy*: the actual (stable) displacement of the system is an admissible vector that minimizes the potential energy functional.

It is usually not difficult to show that the admissible function (or vector) that minimizes the functional is the solution of a system of differential equations (or sometimes integro-differential equations, or integral equations) called the *Euler equations* for the functional. Thus in mechanics we obtain

---

[2] For a fuller discussion of the calculus of variations and its applications, see [3, 4, 8].

[3] Here we use the general term *functional* to denote any mapping of a set of functions (e.g., admissible functions) into real numbers. Thus, for example,

$$F = \int_0^1 f(x)\, dx$$

is a functional, where $f(x)$ is any piecewise continuous function on the unit interval. In our applications, the domain of the functional is the set of admissible functions.

[4] We use bold face to indicate a vector, and $\mathbf{x} = (x_1, x_2, \ldots, x_p)$ is the vector of p independent variables. The function $\mathbf{u}$ is a vector-valued function of $p + 1$ variables.

the equations of motion as the Euler equations for Hamilton's principle and the equilibrium equations as the Euler equations for the principle of minimum potential energy.

To illustrate these remarks, consider the functional

$$I[u] = \int_a^b [p(x)u'^2 + q(x)u^2 + 2f(x)u] \, dx \tag{1}$$

for the scalar function $u(x)$ of the single variable x. Here a prime is used to denote differentiation. The admissibility conditions are the following: $u(x)$, $u'(x)$ and $u''(x)$ are continuous functions in the closed interval $[a, b]$ which satisfy the boundary conditions

$$u(a) = u_0, \qquad u(b) = u_1, \tag{2}$$

where u_0 and u_1 are prescribed numbers. The prescribed functions $p(x)$, $p'(x)$, $q(x)$ and $f(x)$ are continuous in $[a, b]$. We shall now show that if the admissible function $u(x)$ minimizes I, i.e., $I[u] \leqslant I[v]$ for all admissible functions v, then u is a solution of the Euler equation

$$Lu \equiv (pu')' - qu = f, \qquad a \leqslant x \leqslant b. \tag{3}$$

To see this, we introduce $\bar{u}(x)$, the *variation* of u, namely a function defined in the interval $[a, b]$, which has a continuous second derivative and satisfies the homogeneous boundary conditions (2) [i.e., vanishes at the end points a and b], but is otherwise arbitrary. Consider the admissible function $v = u + \varepsilon\bar{u}$, where the real number ε is a parameter. Then

$$J(\varepsilon) = I[v] = I[u + \varepsilon\bar{u}]$$

is a quadratic function of ε, given by

$$J(\varepsilon) = I[u] + \varepsilon I_1[u, \bar{u}] + \frac{\varepsilon^2}{2} I_2[\bar{u}], \tag{4}$$

where

$$I_1 = \frac{dJ(0)}{d\varepsilon}, \qquad I_2 = \frac{d^2J(0)}{d\varepsilon^2}.$$

It is easy to show, by using integration by parts and the conditions $\bar{u}(a) = \bar{u}(b) = 0$, that

$$I_1 = 2\int_a^b (pu'\bar{u}' + qu\bar{u} + f\bar{u}) \, dx = -2\int_a^b (Lu - f)\bar{u} \, dx. \tag{5}$$

Since $I[v]$ is minimized when $\varepsilon = 0$ and hence $J(\varepsilon)$ is a minimum at $\varepsilon = 0$, it follows from (4) that $I_1[u, \bar{u}] = 0$ for all variations $\bar{u}(x)$. Thus we conclude from (5) that u satisfies the Euler equation (3).

1.1.1. Determine the Euler equation corresponding to the functional (1) if $u(a) = u_0$ and no conditions are specified at $x = b$. Determine the boundary conditions that u must satisfy at $x = b$ in order to minimize I (they are called *natural boundary conditions*.)

Ans. $Lu = f, \qquad u'(b) = 0.$

Hint. Let $\bar{u}(a) = 0$ and $\bar{u}(b)$ be arbitrary.[5]

1.1.2. Let $F(x, u, u')$ be a specified twice continuously differentiable function of its arguments x, u and u'. Determine the Euler equation of the functional

$$I[u] = \int_a^b F(x, u, u') \, dx,$$

assuming as admissibility conditions that u, u' and u'' are continuous in $[a, b]$ and satisfy (2).

Ans.

$$\left(\frac{\partial F}{\partial u'}\right)' - \frac{\partial F}{\partial u} = 0.$$

1.1.3. Let $F(x, u, u', \ldots, u^{(n)})$ be a specified twice continuously differentiable function of its arguments, where $u^{(n)} \equiv d^n u/dx^n$. Determine the Euler equation of the functional

$$I[u] = \int_a^b F(x, u, u', \ldots, u^{(n)}) \, dx,$$

assuming as admissibility conditions that u, $u', \ldots, u^{(n)}$ are continuous in $[a, b]$ and have prescribed values at $x = a, b$.

Ans.

$$\frac{\partial F}{\partial u} - \left(\frac{\partial F}{\partial u'}\right)' + \left(\frac{\partial F}{\partial u''}\right)'' - \cdots + (-1)^n \left(\frac{\partial F}{\partial u^{(n)}}\right)^{(n)} = 0.$$

1.1.4. Determine the Euler equation of the functional

$$I[u] = \iint_D [u_x^2 + u_y^2 + 2f(x, y)u] \, dx \, dy$$

for the functions $u(x, y)$ defined on the domain D in the xy-plane bounded by the contour C (the subscripts denote the corresponding partial derivatives, e.g., $u_x \equiv \partial u/\partial x$). The admissibility conditions are that u and its first and second partial derivatives be continuous and that u satisfy the boundary condition

$$u = \varphi(s) \text{ on } C, \tag{6}$$

where s is arc length along C and φ is a specified function on C. The function f is prescribed and continuous on C.

Ans. $\Delta u \equiv u_{xx} + u_{yy} = f(x, y).$

---

[5] For a discussion of more general boundary conditions, see [1], pp. 203–207.

1.1.5. In Prob. 1.1.4, alter the admissibility conditions so that (6) is satisfied on a subarc C_1 of C. On $C_2 \equiv C - C_1$, there are no specified boundary conditions. Determine the Euler equation and the natural boundary condition that must be satisfied on C_2 in order that u minimize I.

Ans.　　　　　$\Delta u = f(x, y), \qquad u_n = 0$ on $C_2,$

where the subscript n denotes differentiation with respect to the unit outward normal $\mathbf{n}$ to D.

1.1.6. Determine the Euler equation of the functional

$$I = \iint_D [a(x, y)u_x^2 + b(x, y)u_y^2 + c(x, y)u^2 + 2f(x, y)u] \, dx \, dy,$$

using the admissibility conditions of Prob. 1.1.4. Here a, b, c and f are prescribed continuous functions on D, and a and b have continuous first partial derivatives.

Ans.

$$(au_x)_x + (bu_y)_y - cu = f.$$

1.1.7. Determine the Euler equation and natural boundary condition for the functional

$$I = \iint_D [a(x, y)(u_x^2 + u_y^2) + c(x, y)u^2 + 2f(x, y)u] \, dx \, dy$$

$$+ \int_C a[x(s), y(s)][A(s)u^2 - 2\varphi(s)u] \, ds,$$

where C is the contour bounding D, the functions a, c and f are prescribed and continuous on D, and a has continuous first partial derivatives. The prescribed functions $A(s)$ and $\varphi(s)$ are continuous on C.

Ans.

$$(au_x)_x + (au_y)_y - cu = f, \qquad u_n + Au = \varphi.$$

1.1.8. Determine the Euler equation of the functional

$$I[u] = \iint_D [(\Delta u)^2 - 2f(x, y)u] \, dx \, dy,$$

where f is a prescribed continuous function on D. The admissible functions $u(x, y)$ have continuous partial derivatives up to and including the fourth order, and satisfy the boundary conditions

$$u = \varphi(s), \qquad u_n = \psi(s) \qquad \text{on } C. \tag{7}$$

Ans.

$$\Delta^2 u \equiv u_{xxxx} + 2u_{xxyy} + u_{yyyy} = f(x, y).$$

Hint. Let $\bar{u} = \bar{u}_n = 0$ for x, y on C.

1.1.9. Determine the Euler equation of the functional

$$I[u] = \iint\limits_{D} [(\Delta u)^2 - 2(1 - \nu)(u_{xx}u_{yy} - u_{xy}^2) - 2fu] \, dx \, dy,$$

where the admissible functions have the same continuity properties as in Prob. 1.1.8 and the condition (7) is satisfied on the subarc C_1 of C. The remaining part of the boundary $C_2 \equiv C - C_1$ is "free," i.e., no conditions are specified on C_2. Determine the natural boundary conditions on C_2. When $C_2 = 0$, compare the results with those obtained in the previous problem. The constant ν is a specified number in the range $0 \leqslant \nu < \frac{1}{2}$.

Ans.

$$\Delta^2 u = f, \quad \nu\,\Delta u + (1 - \nu)(u_{xx}n_1^2 + 2u_{xy}n_1n_2 + u_{yy}n_2^2) = 0,$$

$$(\Delta u)_n + (1 - \nu)[(u_{yy} - u_{xx})n_1n_2 + u_{xy}(n_1^2 - n_2^2)]_s,$$

where n_1 and n_2 are the x and y-components of the outward unit normal to D, and the subscript s denotes differentiation with respect to arc length s along C.

Hint. On C_2, $\bar{u}$ and $\bar{u}_n$ are arbitrary.

1.1.10. Determine admissibility conditions and a functional whose Euler equation and natural boundary condition yield the following boundary value problem for the region D in the xu-plane with contour C:

$$\Delta^2 u = f(x, y) \qquad \text{for } x, y \text{ in } D,$$

$$u = 0, \quad \nu\,\Delta u + (1 - \nu)(u_{xx}n_1^2 + 2u_{xy}n_1n_2 + u_{yy}n_2^2) = 0 \qquad \text{for } x, y \text{ on } C.$$

Ans. The functional is given in the preceding problem. The admissible functions have the same continuity properties as in Prob. 1.1.8, and in addition, $u = 0$ on C.

1.2. THE RITZ METHOD [14]

The minimum property of the solutions of boundary value problems suggests a method for their approximate determination. Suppose that a sequence of admissible functions is constructed whose limit minimizes an appropriate functional. Then the function obtained by truncating the sequence after a finite number of terms may provide an approximation to the minimizing function. The approximation is presumably more accurate when more terms in the sequence are retained. Specifically, we select a family of admissible functions

$$u = U(\mathbf{x}; \mathbf{c}) \tag{8}$$

depending on n (unknown) parameters $\mathbf{c} = (c_1, c_2, \ldots, c_n)$. Inserting these functions into the functional and performing the necessary integrations, we obtain

$$I[U(\mathbf{x}; \mathbf{c})] = \Phi(\mathbf{c}), \tag{9}$$

where Φ is a function of the n parameters $\mathbf{c}$. Necessary conditions for Φ to be a minimum are that

$$\frac{\partial \Phi}{\partial c_i} = 0, \qquad i = 1, 2, \ldots, n. \tag{10}$$

The m solutions $\mathbf{c} = \mathbf{c}^j, j = 1, 2, \ldots, m$ of the algebraic equations (10) give the stationary points of Φ. Let $\mathbf{c} = \mathbf{c}^0$ be a stationary point which also furnishes a minimum of Φ. Then we expect that the function $u = U(\mathbf{x}; \mathbf{c}^0)$, which is called a *Ritz approximation* and which minimizes I with respect to all admissible functions of the form (8), is an approximation to an admissible function that minimizes I.[6]

In practice, the family of admissible functions is usually formed by taking a linear combination

$$\bar{u}(\mathbf{x}; \mathbf{c}) = u^0(\mathbf{x}) + \sum_{j=1}^{n} c_j \bar{u}^j(\mathbf{x}), \tag{11}$$

where u^0 is an admissible function and the $\bar{u}^j$ ($j = 1, 2, \ldots, n$) are variations. We shall refer to U in the form (11) as a *trial solution*. For linear problems, the functional I is quadratic in u and its derivatives. Then substitution of (11) into the equations (10) leads to a system of linear algebraic equations for $\mathbf{c}$. Naturally, we should try to choose the functions u^0 and $\bar{u}^j$ so that they approximate the solution as closely as possible. However, there are several practical considerations governing their selection. First of all, they should be chosen so that the integrals necessary to obtain Φ are "easy" to evaluate. Furthermore, the $\bar{u}^j$ must be sufficiently different. If, for example, two of the functions are identical, then the resulting system of linear algebraic equations for $\mathbf{c}$ will have a zero determinant. If two or more of the functions $\bar{u}^j$ differ only slightly, then the determinant may be small and it will be difficult to solve the algebraic equations accurately. If natural boundary conditions are to be satisfied on some portion of the boundary, then, as we have seen in Sec. 1.1, it is not necessary to impose them as part of the admissibility conditions, since the solution of the minimum problem automatically satisfies them. However, if it is easy to select u_0 and $\bar{u}^j$ which satisfy the natural boundary conditions, then it is advantageous to do so in the Ritz method.

To illustrate the application of the Ritz method, consider the boundary value problem consisting of the differential equation (3) and the boundary conditions (2), where the associated functional is (1). For simplicity, we take $u_0 = u_1 = 0$, so that $u^0 \equiv 0$.[7] Then substituting (11) into (1) and

---

[6] Convergence properties and the sense of approximation afforded by the Ritz method have been established in special cases (see [5, 11]).

[7] For the variations we may take, for example,

$$\bar{u}^j = (b - x)(a - x)x^j, \qquad j = 1, 2, \ldots, n,$$

or

$$\bar{u}^j = \sin \frac{j\pi(x - a)}{b - a}, \qquad j = 1, 2, \ldots, n.$$

performing the necessary differentiations, we obtain $I[u] = \Phi(\mathbf{c})$. Applying the stationary conditions (10), we find that the parameters c_j satisfy the system of algebraic equations

$$\sum_{j=1}^{n} A_{ij}c_j + B_i = 0, \qquad i = 1, 2, \ldots, n, \tag{12}$$

where

$$A_{ji} = A_{ij} = \int_a^b (p\bar{u}_i'\bar{u}_j' + q\bar{u}_i\bar{u}_j) \, dx,$$

$$B_i = \int_a^b f\bar{u}_i \, dx, \qquad i, j = 1, 2, \ldots, n.$$

1.2.1. Prove that if the $\bar{u}^j$ are linearly independent functions and the coefficients $p(x)$ and $q(x)$ satisfy the conditions $p(x) > 0, q(x) \geqslant 0$ for all x in $[a, b]$, then the system (12) has a unique solution.

Hint. Show by contradiction that the homogeneous form of the system (12), i.e., with $B_i = 0$ $(i = 1, 2, \ldots, n)$, has only the solution $\mathbf{c} = 0$.

1.2.2. Use the Ritz method to obtain an approximate solution of the boundary value problem

$$u'' + u + x = 0, \qquad u(0) = u(1) = 0$$

for each of the following trial solutions:

$$\text{a) } U = cx(1 - x); \quad \text{b) } U = c_1 x(1 - x) + c_2 x^2(1 - x);$$

$$\text{c) } U = c_1 x(1 - x) + c_2(1 - x^2).$$

Why are these legitimate trial solutions? Compare the approximations so obtained for u and u' with the exact solution.

Ans.

$$\text{a) } c = \frac{5}{18} ; \quad \text{b) } c_1 = \frac{17}{369}, \quad c_2 = \frac{7}{41} ; \quad \text{c) } c_1 = \frac{8}{369}, \quad c_2 = \frac{7}{41}$$

(see [11], p. 269 and [1], p. 220).

1.2.3. Use the Ritz method to obtain an approximate solution of Bessel's equation

$$x^2 u'' + x u' + (x^2 - 1)u = 0$$

in the interval $1 \leqslant x \leqslant 2$, where $u(1) = 1$, $u(2) = 2$. Compare the result with the exact solution.

Hint. First write Bessel's equation in the form (3).

1.2.4. Use the Ritz method to obtain an approximate solution of the boundary value problem

$$(xu')' + u = x, \qquad u(0) = 0, \quad u(1) = 1,$$

of the form $U = x + x(1 - x)(c_1 + c_2 x)$.

Ans.

$$c_1 = \frac{85}{26}, \qquad c_2 = -\frac{35}{13}$$

see [9]).)

1.2.5. Use the Ritz method to obtain an approximate solution of the boundary value problem

$$u'' + (1 + x^2)u + 1 = 0, \qquad u(-1) = u(1) = 0$$

of the form

a) $U = c_1(1 - x^2) + c_2(1 - x^4)$;

b) $U = c_1(1 - x^2) + c_2(1 - x^4) + c_3(1 - x^6)$.

Ans.

$$\text{a) } c_1 = \frac{1050}{1063}, \qquad c_2 = -\frac{231}{4252}$$

$$\text{b) } c_1 \approx 0.966, \quad c_2 \approx -0.00474, \qquad c_3 \approx -0.0297.$$

(see [1], p. 209).

1.2.6. Obtain a Ritz approximation to the solution of the boundary value problem

$$[(2 - x^2)u'']'' + 40y = 2 - x^2, \qquad y''(\pm 1) = y'''(\pm 1) = 0,$$

of the form $U = c_1 + c_2 x^2 + c_3 x^4$.

Ans.

$$c_1 = \frac{143363}{40 \cdot 79301}, \quad c_2 = -\frac{953}{79301}, \quad c_3 = \frac{189}{79301}$$

(see [1], p. 219).

Hint. Use Prob. 1.1.3 to formulate the functional. Note that the boundary conditions are natural boundary conditions. Determine how accurately the boundary conditions are satisfied by the approximate solution.

1.2.7. Use the Ritz method to obtain an approximate solution of the Poisson equation

$$\Delta u = u_{xx} + u_{yy} = -2,$$

subject to the condition $u = 0$ on the boundary of the rectangle $|x| < a$, $|y| < b$, where the trial solution is of the form

a) $U = c(x^2 - a^2)(y^2 - b^2)$;

b) $U = (x^2 - a^2)(y^2 - a^2)[c_1 x + c_2(x^2 + y^2)]$ (for the square $b = a$).

Ans.

$$\text{a) } c = \frac{5}{4(a^2 + b^2)}; \quad \text{b) } a^2 c_1 = \frac{5}{8}\frac{259}{177}, \quad a^4 c_2 = \frac{15}{16}\frac{35}{277}$$

(see [11], p. 281).

Hint. Use Prob. 1.1.4.

1.2.8. Solve Prob. 1.2.7, using the Ritz method with

$$U = \sum_{m=1,3,5,\ldots}^{\infty} \sum_{n=1,3,5,\ldots}^{\infty} c_{mn} \cos\frac{m\pi x}{2a} \cos\frac{n\pi y}{2b}.$$

Show that the constants c_{mn} so obtained coincide with those found by the method of separation of variables.

Ans.

$$c_{mn} = 128\pi^{-4}a^2b^2(-1)^{\frac{1}{2}(m+n)-1}[mn(b^2m^2 + a^2n^2)]^{-1}$$

(see [11]. p. 282).

1.2.9. Apply the Ritz method to construct a solution of $\Delta u = -1$ satisfying the boundary condition $u_n + u = 0$ on the sides of the square $|x| \leqslant 1$, $|y| \leqslant 1$ (see Prob. 1.1.7), where the trial solution is of the form

a) $U = c_1 + c_2(x^2 + y^2)$;   b) $U = c_1 + c_2(x^2 + y^2) + c_3x^2y^2$.

Note that the trial solutions need not satisfy the boundary conditions, since they are natural conditions.

Ans.

a) $c_1 = \dfrac{13}{16}$, $c_2 = -\dfrac{15}{64}$;   b) $c_1 = \dfrac{139}{168}$, $c_2 = -\dfrac{15}{16}$, $c_3 = \dfrac{5}{56}$

(see [1], p. 429).

1.2.10. Solve Prob. 1.2.9 by the Ritz method, selecting trial solutions that satisfy the natural boundary conditions. Make use of the symmetry of the solutions in x and y. Compare with the answer to Prob. 1.2.9.

1.2.11. Find Ritz approximations to the solution of

$$\Delta u + \frac{3}{5 - y} u_y + 1 = 0$$

on the rectangle $|x| \leqslant \frac{1}{2}$, $|y| \leqslant 1$, where $u = 0$ on the edges of the rectangle. As trial solutions, use

a) $U = c(1 - y^2)(1 - 4x^2)(5 - y)^3$;
b) $U = (c_1 + c_2y)(1 - y^2)(1 - 4x^2)(5 - y)^3$.

Ans.

a) $c = \dfrac{7}{7264}$;   b) $10^4c_1 \approx 10.185$, $10^4c_2 \approx 4.84$

(see [1], p. 459).

Hint. Use Prob. 1.1.6.

1.2.12. Obtain a Ritz approximation to the solution of the biharmonic equation $\Delta^2u = 0$, satisfying the following boundary conditions on the edges of the square $|x| \leqslant 1$, $|y| \leqslant 1$:

$$u_{xy} = 0 \quad \text{for} \quad x = \pm 1, \quad y = \pm 1,$$
$$u_{yy} = 1 - y^2 \quad \text{for} \quad x \pm 1,$$
$$u_{xx} = 0 \quad \text{for} \quad y = \pm 1.$$

As trial solutions, use

a) $U = \frac{1}{2}y^2(1 - \frac{1}{6}y^2) + c_1(x^2 - 1)^2(y^2 - 1);$

b) $U = \frac{1}{2}y^2(1 - \frac{1}{6}y^2) + (x^2 - 1)^2(y^2 - 1)^2(c_1 + c_2x^2 + c_3y^2).$

Ans.

    a) $c_1 \approx 0.0425;$ b) $c_1 \approx 0.0404,$ $c_2 = c_3 \approx 0.0117$

(see [17], p. 167).

Hint. Transform the boundary conditions into the form (7), and then use Prob. 1.1.8.

1.2.13. Determine a Ritz approximation to the solution of $\Delta^2 u = f(x, y)$ in the rectangle $0 \leqslant x \leqslant a$, $0 \leqslant y \leqslant b$, satisfying the boundary conditions of Prob. 1.1.10 on the edges of the rectangle. Use a trial solution of the form

$$U = \sum_{m=1}^{\infty} \sum_{n=1}^{\infty} c_{mn} \sin \frac{m\pi x}{a} \sin \frac{n\pi y}{b}.$$

Ans.

$$c_{mn} = \frac{4}{\pi^4 ab}\left[\left(\frac{m}{a}\right)^2 + \left(\frac{n}{b}\right)^2\right]^{-2} \int_0^a \int_0^b f(x, y) \sin \frac{m\pi x}{a} \sin \frac{n\pi y}{b} \, dx \, dy$$

(see [16], p. 345).

1.2.14. Use the Ritz method to obtain an approximate solution of the clamped rectangular plate problem $\Delta^2 u = f$ where f is a constant (see Prob. 1.1.8), subject to the conditions $u = u_n = 0$ on the boundaries of the rectangle $0 \leqslant x \leqslant a$, $0 \leqslant y \leqslant b$. Use a trial solution of the form

$$U = c\left(1 - \cos \frac{2\pi x}{a}\right)\left(1 - \cos \frac{2\pi y}{b}\right).$$

Ans.

$$c = \frac{fa^4}{4\pi^4}\left[3 + 2\left(\frac{a}{b}\right)^2 + 3\left(\frac{a}{b}\right)^4\right]^{-1}$$

(see [18], p. 288).

1.3. KANTOROVICH'S METHOD[8]

Kantorovich's method, which is sometimes called the mixed Ritz method or the method of reduction to ordinary differential equations, is essentially a generalization of the Ritz method. More "freedom" is permitted in the selection of the trial solutions (8) and (11) by allowing the parameters **c** to be functions of one of the independent variables **x**, say x. The functional I then reduces to a functional

$$I[U(\mathbf{x}; \mathbf{c}(x))] = \Psi[\mathbf{c}(x)] \tag{13}$$

of n functions $c_j(x)$, which are determined so as to furnish a minimum of Ψ.

_____

[8] For a general description and analysis of this method, see [11].

Thus the $c_j(x)$ are solutions of a system of n ordinary differential equations which are the Euler equations of Ψ. The solutions of these equations subject to appropriate boundary conditions yield the approximation $U(\mathbf{x}; \mathbf{c}(x))$.

For simplicity, we shall consider Kantorovich's method only for a rectangular region $a_0 \leqslant x \leqslant a_1$, $b_0 \leqslant y \leqslant b_1$ in the xy-plane. However, the method can be applied to regions of more general shape (see [11]). We shall employ trial solutions of the form

$$U(x, y; \mathbf{c}(x)) = u^0(x, y) + \sum_{j=1}^{n} c_j(x)\bar{u}^j(x, y), \tag{14}$$

where u^0 satisfies inhomogeneous boundary conditions and the $\bar{u}^j$ homogeneous boundary conditions on $y = b_0, b_1$. The boundary conditions on $x = a_0, a_1$ yield the values of $c_j(a_0)$ and $c_j(a_1)$, $j = 1, 2, \ldots, n$.

As an example (see [11], p. 304), consider the problem of solving the equation $\Delta u = -1$ for x, y in the square $|x| \leqslant 1$, $|y| \leqslant 1$, subject to the boundary condition $u = 0$ on the edges of the square. As a trial solution, we take $U = (1 - y^2)c(x)$, which satisfies the boundary conditions on $y = \pm 1$. To make the trial solution satisfy the conditions on $x = \pm 1$, we require that $c(-1) = c(1) = 0$. Then the associated functional (see Prob. 1.1.4) reduces to

$$I[u] = \Psi[c(x)] = \frac{8}{3} \int_{-1}^{1} \left(\frac{2}{5} c'^2 + c^2 - c \right) dx.$$

The Euler equation of Ψ is obtained by using (1) and (3), and is given by

$$c'' - \frac{5}{2} c = -\frac{5}{4}.$$

Solving this equation and applying the boundary conditions $c(-1) = c(1) = 0$, we obtain

$$c(x) = \frac{1}{2}\left(1 - \frac{\cosh kx}{\cosh k} \right), \qquad k = \sqrt{\frac{5}{2}},$$

$$U = \frac{1}{2}(1 - y^2)\left(1 - \frac{\cosh kx}{\cosh k} \right).$$

1.3.1. Solve the above boundary value problem by Kantorovich's method, using the trial solution

$$U = (1 - y^2)[c_1(x) + c_2(x)y^2].$$

Compare with the result of the Ritz approximation obtained in Prob. 1.2.7.

Ans.

$$c_1(x) \approx -\frac{1}{2} + 0.516 \frac{\cosh \alpha_- x}{\cosh \alpha_-} - 0.0156 \frac{\cosh \alpha_+ x}{\cosh \alpha_+},$$

$$c_2(x) \approx 0.114\left(-\frac{\cosh \alpha_- x}{\cosh \alpha_-} + \frac{\cosh \alpha_+ x}{\cosh \alpha_+} \right),$$

where $a_{\pm} = (14 \pm \sqrt{133})^{1/2}$ are the roots of the characteristic equation $\xi^4 - 28\xi^2 + 63 = 0$ (see [11], p. 317).

1.3.2. Solve the above boundary value problem by Kantorovich's method, using the trial solution

$$U = \sum_{j=0}^{\infty} c_j(x) \cos (j + \tfrac{1}{2})\pi y.$$

Verify that this yields the infinite series representation of the exact solution.

Ans.

$$c_j(x) = (-1)^{j-1} 2\pi^{-3}(j - \tfrac{1}{2})^{-3} \left[1 - \frac{\cosh (j - \tfrac{1}{2})\pi x}{\cosh (j - \tfrac{1}{2})\pi} \right], \quad j = 1, 2, \ldots$$

(see [11], p. 320).

1.3.3. Use Kantorovich's method to solve the clamped rectangular plate problem, i.e., $\Delta^2 u = 1$ in the rectangle $|x| \leqslant a$, $|y| \leqslant b$, with boundary conditions $u = u_n = 0$ on the edges of the rectangle. Use $U = (y^2 - b^2)^2 c(x)$ as a trial solution.

Ans.

$$24c(x) = A \cosh \alpha\xi \cos \beta\xi + B \sinh \xi \sin \beta\xi + 1,$$

where $\xi = x/b$, $A = d_1/d_0$, $B = d_2/d_0$,

$$d_0 = \beta \sinh \alpha r \cosh \alpha r + \alpha \sin \beta r \cos \beta r,$$

$$-d_1 = \alpha \cosh \alpha r \sin \beta r + \beta \sinh \alpha r \cos \beta r,$$

$$d_2 = \alpha \sinh \alpha r \cos \beta r - \beta \cosh \alpha r \sin \beta r,$$

$r = a/b$, $\alpha \approx 2.075$ and $\beta \approx 1.143$ (see [11], p. 322).

1.3.4. Use Kantorovich's method to obtain an approximate solution of $\Delta^2 u = 0$ on the semi-infinite strip $0 \leqslant x < \infty$, $|y| \leqslant 1$, subject to the following boundary conditions:

$$u_{xx}(x, \pm 1) = u_{xy}(x, \pm 1) = 0, \quad u_{yy}(0, y) = y^2 - \tfrac{1}{3}, \quad u_{xy}(0, y) = 0,$$

$$\lim_{x \to \infty} u_{xy}(x, y) = \lim_{x \to \infty} u_{yy}(x, y) = 0 \text{ uniformly in } y.$$

Use the trial solution

$$U = \frac{(1 - y^2)^2}{12} c(x)$$

(note that U satisfies the boundary conditions on $y = \pm 1$).

Ans.

$$c(x) = e^{-\alpha x}\left(\cos \beta x + \frac{\alpha}{\beta} \sin \beta x \right),$$

where $\gamma = \alpha + \beta i \approx 2.075 + 1.143i$ is a root of $\gamma^4 - 6\gamma^2 + \tfrac{63}{2} = 0$ (see [10]).

2. Related Methods

The application of the Ritz method to the solution of boundary value problems requires a variational principle. However, in some problems there is no such principle, while in others, it is difficult to determine the proper functional or cumbersome to evaluate the integrals needed in the Ritz method. Thus, in this section, we shall discuss three procedures for obtaining approximate solutions which do not require a variational functional, although they lead to approximations related to those obtained by the Ritz method.

For simplicity, consider the following boundary value problem involving a single function $u(\mathbf{x})$:

$$Lu = f \quad \text{for} \quad \mathbf{x} \text{ in } D, \qquad Bu = g \quad \text{for} \quad \mathbf{x} \text{ on } C. \tag{15}$$

Here L is a differential operator defined in a domain D, B is a boundary operator defined only on the boundary C of D, and f and g are prescribed functions. Thus $Bu = g$ is the boundary condition for the single differential equation $Lu = f$.

As in the Ritz method, we seek an approximate solution of (15) of the form

$$u = U(\mathbf{x}; \mathbf{c}),$$

depending on n parameters $\mathbf{c} = (c_1, c_2, \ldots, c_n)$. We shall assume, unless it is otherwise specified, that $\mathbf{c}$ is independent of $\mathbf{x}$. In general, the approximate solution U does not satisfy the differential equation and the boundary condition, and in fact

$$\begin{aligned} LU - f &= e(\mathbf{x}; \mathbf{c}) \quad \text{for} \quad \mathbf{x} \text{ in } D, \\ BU - g &= E(\mathbf{x}; \mathbf{c}) \quad \text{for} \quad \mathbf{x} \text{ on } C, \end{aligned} \tag{16}$$

where e and E, called the *interior error* and the *boundary error*, are algebraic functions of $\mathbf{x}$ and $\mathbf{c}$. If $\mathbf{c}$ is a function of one independent variable, then e will be an ordinary differential operator acting on $\mathbf{c}$, and E will contain initial or boundary conditions for $\mathbf{c}$. If the function U is selected so that $E \equiv 0$ for all $\mathbf{x}$ on C, the procedure used to determine $\mathbf{c}$ is called an *interior method*, while if $e \equiv 0$ for all $\mathbf{x}$ in D, the procedure is called a *boundary method*.

We wish to determine $\mathbf{c}$ so that the errors are, in some sense, as small as possible. Essentially, each of the methods described below amounts to ascribing a definite meaning to the term "small."

2.1. GALERKIN'S METHOD [7]

In Galerkin's method, the n parameters are chosen to make the errors orthogonal to a set of n independent functions $w^1(\mathbf{x})$, $w^2(\mathbf{x})$, $\ldots$, $w^n(\mathbf{x})$, usually taken to be orthogonal. This gives n conditions of the form

$$\int_D e(\mathbf{x}; \mathbf{c})w^j(\mathbf{x}) \, d\mathbf{x} + \int_C E(\mathbf{x}; \mathbf{c})w^j(\mathbf{x}) \, ds = 0, \qquad j = 1, 2, \ldots, n, \quad (17)$$

where ds is an element of area on C. These are n algebraic equations for determining the n parameters $\mathbf{c}$. In fact, the equations are linear if L and B are linear operators and U is chosen in the form

$$U = u^0 + \sum_{j=1}^{n} c_j \bar{u}^j, \qquad (18)$$

as is customary in practice. The *interior Galerkin method*[9] corresponds to choosing u^0 and $\bar{u}^j, j = 1, 2, \ldots, n$ to satisfy the inhomogeneous and homogeneous boundary conditions, respectively. In the applications, it is customary (but not essential) to set $w^j = \bar{u}^j, j = 1, 2, \ldots, n$, and we shall do so in all the problems that follow. If, as $n \to \infty$, the w^j form a complete set of functions, then $e \to 0$ as $n \to \infty$ (being orthogonal to every function of a complete set). Some convergence properties of Galerkin's method are discussed in [12].

Practical selection of the functions $\bar{u}^j$ and w^j is governed by the same considerations as in the Ritz method, i.e., they should make evaluation of the integrals in (17) easy and they should be sufficiently dissimilar (say orthogonal) to lead to a "well-conditioned" system of algebraic equations.

If the boundary value problem (15) can be derived from a variational principle, then, in many cases, it can be shown that Ritz's method coincides with Galerkin's. If the parameters c_j in (18) are permitted to be functions of one variable, we obtain the *Galerkin-Kantorovich method*. The conditions (17) then give ordinary differential equations and boundary conditions for determining $\mathbf{c}$.

2.1.1. Given the differential equation (3) and the boundary conditions (2), with $u_0 = u_1 = 0$, show that the Ritz and Galerkin methods lead to the same system of algebraic equations (12) for determining the coefficients $\mathbf{c}$.

Hint. Use integration by parts.

2.1.2. Given the differential equation

$$(au_x)_x + (bu_y)_y - cu = f$$

(see the answer to Prob. 1.1.6) and the boundary condition $u = 0$ on C, show that the Ritz and Galerkin methods lead to the same system of algebraic equations for the coefficients $\mathbf{c}$.

---

[9] The expression *Galerkin's method* conventionally denotes the interior Galerkin method.

2.1.3. Solve Prob. 1.2.3 by Galerkin's method without transforming Bessel's equation into the form (3). Compare with the exact solution, and also with the Ritz approximation using the same number of parameters.

2.1.4. Use Galerkin's method to obtain an approximate solution of the boundary value problem

$$u'' + xu' + u = 2x, \qquad u(0) = 1, \qquad u(1) = 0,$$

choosing a trial solution of the form

$$U = (1 - x)(1 + c_1 x + c_2 x^2 + c_3 x^3).$$

Ans.

$$c_1 \approx -0.209, \qquad c_2 \approx -0.789, \qquad c_3 \approx 0.209$$

(see [13], p. 115).

2.1.5. Solve Prob. 1.2.4 by Galerkin's method, using the same trial solution. Verify that c_1 and c_2 satisfy the same algebraic equations as in the Ritz method.

2.1.6. Use Galerkin's method to solve the boundary value problem

$$u^{(iv)} + u = 1, \qquad u(0) = u''(0) = u(1) = u''(1) = 0,$$

choosing a trial solution of the form

$$U = c_1 \sin \pi x + c_2 \sin 3\pi x.$$

Ans.

$$c_1 = 4\pi^{-1}(\pi^4 + 1)^{-1}, \quad c_2 = 4[3\pi(81\pi^4 + 1)]^{-1}$$

(see [6], p. 233).

2.1.7. Solve Probs. 1.2.7 and 1.2.12 by Galerkin's method, using the same trial solutions. Verify that the coefficients c_j satisfy the same algebraic equations as in the Ritz method.

2.1.8. Use Galerkin's method to solve Prob. 1.2.9, choosing the following trial solutions which satisfy the (natural) boundary conditions:

a) $U = c[9 - 3(x^2 + y^2) + x^2 y^2]$;

b) $U = c_1[9 - 3(x^2 + y^2) + x^2 y^2]$
$\qquad + c_2[30 - 5(x^2 + y^2) - 3(x^4 + y^4) + x^2 y^2(x^2 + y^2)].$

Ans.

$$\text{a) } c = \frac{5}{54}; \quad \text{b) } 10^3 c_1 \approx 73.3, \quad 10^3 c_2 \approx 5.38$$

(see [1], p. 413).

2.1.9. Use the Galerkin-Kantorovich method to obtain an approximate solution of the heat equation $u_{xx} = u_t$ in the semi-infinite strip $0 \leqslant x \leqslant 1$, $t > 0$. The boundary and initial conditions are

$$u(0, t) - u_x(0, t) - u_x(1, t) = 0, \qquad t > 0,$$

$$u(x, 0) = 1, \qquad 0 \leqslant x \leqslant 1,$$

and u must remain bounded as $t \to \infty$. Use a trial solution of the form

$$U = c_1(t)\left(1 + x - \frac{x^2}{2}\right) + c_2(t)\left(1 + x - \frac{x^3}{3}\right),$$

which satisfies the boundary conditions but not the intial conditions.

Ans.

$$c_1(t) \approx 0.586e^{-0.740t} + 2.45e^{-11.8t}, \qquad c_2(t) \approx 0.144e^{-0.740t} - 2.30e^{-11.86t}$$

(see [6], p. 372).

Hint. In applying (17), set the area integral over the strip and the boundary integral over the initial line separately equal to zero.

2.1.10. Use the Galerkin-Kantorovich method to obtain an approximate solution of the wave equation $u_{xx} = u_{tt}$ in the semi-infinite strip $0 \leqslant x \leqslant 1$, $t \geqslant 0$, where the boundary and initial conditions are

$$u(0, t) = u(1, t) = 0, \qquad t > 0,$$
$$u(x, 0) = x(1 - x), \qquad u_t(x, 0) = 0, \qquad 0 \leqslant x \leqslant 1.$$

Use a trial solution of the form

$$U = x(1 - x)[c_1(t) + c_2(t)x(1 - x)].$$

Ans.

$$c_1 \approx 0.804 \cos \alpha t + 0.197 \cos \beta t,$$

$$c_2 \approx 0.911(\cos \alpha t - \cos \beta t),$$

where $\alpha \approx \pi$, $\beta \approx 10.11$ (see [6], p. 375).

2.2. COLLOCATION

Of all the approximation procedures under consideration, the collocation method is perhaps the simplest to apply. In this method, the n parameters are determined by requiring the errors in (16) to vanish at n points $x_1, x_2, \ldots, x_n$ in $D + C$ called the *collocation points*. Of course, these points must be chosen so that the resulting system of equations has a solution, say $c^0(x_j)$. The ideal collocation points are those for which $c^0(x_j)$ minimizes the maximum error for all x in $D + C$. For example, if we define

$$\mathscr{E}(x_j) \equiv \max_{x \text{ in } D} |e(x; c^0(x_j))| + \max_{x \text{ on } C} |E(x; c^0(x_j))|, \tag{19}$$

then as the collocation points we should take the values $x_j, j = 1, 2, \ldots, n$ for which $\mathscr{E}$ is a minimum. However, no general procedures are presently available for *a priori* selection of points satisfying this criterion; in fact, they are usually determined by intuition or by practical considerations such as computational simplicity. Only interior or boundary points need be considered as collocation points, depending on whether interior or boundary collocation is employed.

A disadvantage of collocation is that the approximate solution may vary considerably with the position of the collocation points. One way to minimize this is to take a sufficient number of points and distribute them over the domain and the boundary.

An obvious generalization of the collocation method is to allow the parameters to be functions of one variable, say x. Then the errors will depend on $c(x)$ and its derivatives, and collocation may yield a system of differential equations and boundary conditions for determining the parameters.

2.2.1. Solving Prob. 1.2.2 by interior collocation, using $U = cx(1 - x)$ as a trial solution and the following collocation points:

a) $x = \frac{1}{4}$; b) $x = \frac{1}{2}$; c) $x = \frac{3}{4}$.

Compare with the Ritz approximation and the exact solution. In each ease, evaluate

$$\mathscr{E} \equiv \max_{0 \leqslant x \leqslant 1} |e|$$

[cf. (19)]. Does the approximation with smallest $\mathscr{E}$ have the smallest deviation from the exact solution?

Ans.

a) $c = \dfrac{4}{29}$; b) $c = \dfrac{2}{7}$; c) $c = \dfrac{\overline{12}}{29}$.

2.2.2. Solve Prob. 1.2.5 by interior collocation, using

$$U = (1 - x^2)(c_1 + c_2 x^2 + c_3 x^4)$$

as a trial solution and the following collocation points:

a) $x = \frac{1}{4}, \frac{3}{4}$ (set $c_3 = 0$); b) $x = \frac{1}{6}, \frac{1}{2}, \frac{5}{6}$.

Ans.

a) $c_1 \approx 0.929$, $c_2 \approx -0.0512$; b) $c_1 \approx 0.932$, $c_2 \approx -0.0341$, $c_3 \approx -0.0302$ (see [1], p. 182).

2.2.3. Solve Prob. 2.1.6 by interior collocation, using the same trial solution and $x = \frac{1}{4}, \frac{1}{2}$ as collocation points. Compare with the approximation obtained by Galerkin's method.

Ans.

$$c_1 = (\sqrt{2} + 1)2^{-1}(\pi^4 + 1)^{-1}, \qquad c_2 = (\sqrt{2} - 1)2^{-1}(81\pi^4 + 1)^{-1}$$

(see [6], p. 233).

2.2.4. Solve Prob. 1.2.7 for the square $b = a$ by interior collocation, using the same trial solutions. For the trial solution a, use $x = y = 0$ as the collocation point, and for the trial solution b, use the points $x = y = 0$ and $x = y = a/2$. Compare with the Ritz approximation and the infinite series solution.

Ans.

$$\text{a) } a^2 c = \frac{1}{2}; \quad \text{b) } a^2 c_1 = \frac{25}{42}, \quad a^4 c_2 = \frac{2}{21}$$

(see [15], p. 437).

2.2.5. Use boundary collocation to solve Prob. 1.2.7 for the square $b = a$. To select trial solutions, it is convenient to introduce polar coordinates

$$r^2 = x^2 + y^2, \qquad \theta = \tan^{-1} \frac{y}{x}.$$

Then the function

$$U = -\frac{r^2}{2} + c_1 + c_2 r^4 \cos 4\theta + c_3 r^8 \cos 8\theta$$

is a solution of the differential equation. Determine the parameters c_1, c_2 and c_3, using the collocation points

$$r = a, \, \theta = 0, \qquad r = \sqrt{5}\frac{a}{2}, \quad \tan \theta = \frac{1}{2}, \qquad r = \sqrt{2}\, a, \quad \theta = \frac{\pi}{4}.$$

Compare with the solutions obtained by interior collocation (Prob. 2.2.4) and by the Ritz method (Prob. 1.2.7). Also compare with the infinite series solution (Prob. 1.2.8).

Ans.

$$c_1 \approx 0.590 a^2, \qquad a^2 c_2 \approx -0.0924, \qquad a^6 c_3 \approx 0.00254$$

(see [2]).

2.2.6. Use boundary collocation to determine an approximate solution of $\Delta u = -2$ where $u = 0$ on the boundary of a regular hexagon with sides of length $2a/\sqrt{3}$ whose vertical sides lie on $x = \pm a$. As a trial solution, use the function

$$U = -\frac{r^2}{2} + c_1 + c_2 r^6 \cos 6\theta + c_3 r^{12} \cos 12\theta,$$

which solves the differential equation. Choose polar coordinates with respect to the center of the hexagon, and use the collocation points

$$r = a, \quad \theta = 0, \qquad r = \frac{2a}{\sqrt{3}}, \quad \theta = \frac{\pi}{6}, \qquad r = \frac{1}{2}\sqrt{\frac{13}{2}}\, a, \quad \tan \theta = \frac{1}{2\sqrt{3}}.$$

Ans.

$$c_1 \approx 0.541 a^2, \quad a^4 c_2 \approx -0.0445, \quad a^{10} c_3 \approx 0.00363$$

(see [2]).

2.2.7. Solve Prob. 1.2.9 by interior collocation, using the trial solutions of Prob. 2.1.8. For the one-parameter approximation, use the collocation points

a) $x = y = 0$;　b) $x = y = \frac{1}{2}$;　c) $x = y = \frac{2}{3}$.

For the two-parameter approximation, use the collocation points

d) $x = \frac{1}{2}$,　$y = \frac{1}{4}$,　　$x = 1$,　$y = \frac{1}{2}$,

e) $x = \frac{1}{2}$,　$y = \frac{1}{4}$,　　$x = \frac{3}{4}$　$y = \frac{1}{2}$.

Compare these approximate solutions with those obtained by the Ritz and Galerkin methods.

Ans.

a) $c = \dfrac{1}{12}$;　b) $c = \dfrac{1}{11}$;　c) $c = \dfrac{9}{92}$;

d) $c_1 = \dfrac{446}{6057} \approx 0.074$,　$c_2 = \dfrac{32}{6057} \approx 0.00528$;

e) $10^3 c_1 \approx 74$,　$10^4 c_2 \approx 5.15$

(see [1], p. 411).

2.2.8. Solve Prob. 1.2.9 by boundary collocation, using the following trial solutions and collocation points:

a) $U = -\frac{1}{4}(x^2 + y^2) + c$,　$x = 1$,　$y = \frac{1}{2}$;

b) $U = -\frac{1}{4}(x^2 + y^2) + c_1 + c_2(x^4 - 6x^2y^2 + y^4)$,
$$x = 1,\quad y = \frac{1}{4},\qquad x = 1,\quad y = \frac{3}{4}.$$

(Both trial functions are solutions of the differential equation.) Compare with the results of Prob. 2.2.7.

Ans.

a) $c \approx 0.813$;　b) $c_1 \approx 0.821$,　$c_2 \approx -0.0144$

(see [1], p. 413).

2.2.9. Use boundary collocation to solve Prob. 1.2.14 for the square $b = a$. Let r and θ be polar coordinates with respect to the center of the square, and use the trial solution

$$U = \frac{fr^4}{64} + c_1 + c_2 r^2 + (c_3 r^4 + c_4 r^6) \cos 4\theta + (c_5 r^8 + c_6 r^{10}) \cos 8\theta$$

and the collocation points

$$r = \frac{a}{2},\quad \theta = 0,\qquad r = \frac{\sqrt{5}\,a}{4},\quad \tan \theta = \frac{1}{2},\qquad r = \frac{a}{\sqrt{2}},\quad \theta = \frac{\pi}{4}.$$

Verify that U is a solution of the differential equation. Compare with the approximate solution obtained by the Ritz method in Prob. 1.2.14.

Ans.

$$c_1 \approx 1.296F, \qquad a^2c_2 \approx -2.256F, \qquad a^4c_3 \approx -0.3603F,$$

$$a^6c_4 \approx 0.3078F, \qquad a^8c_5 \approx 0.01074F, \qquad a^{10}c_6 \approx 0.00207F,$$

where $F = fa^4/64$ (see [2]).

2.3. LEAST SQUARES

In the *method of least squares* we seek an approximate solution in the form $u = U(\mathbf{x}; \mathbf{c})$, as before, but the parameters $\mathbf{c}$ are determined to minimize the "mean square error" of the errors e and E in (16), i.e.,

$$\int_D \omega(\mathbf{x})e^2(\mathbf{x}; \mathbf{c}) \, d\mathbf{x} + \int_C \Omega(\mathbf{x})E^2(\mathbf{x}; \mathbf{c}) \, ds = \text{minimum}, \qquad (20)$$

where the weighting functions $\omega(\mathbf{x}) > 0$ for $\mathbf{x}$ in D and $\Omega(\mathbf{x}) > 0$ for $\mathbf{x}$ on C are at our disposal. Usually it is convenient to take $\omega = \Omega \equiv 1$, and we shall do so in the problems below. Necessary conditions for the mean square error (20) to be a minimum are obtained by differentiating (20) with respect to each c_j:

$$\int_D \omega e \frac{\partial e}{\partial c_j} \, d\mathbf{x} + \int_C \Omega E \frac{\partial E}{\partial c_j} \, ds = 0, \qquad j = 1, 2, \ldots, n. \qquad (21)$$

This gives n algebraic equations for determining the n parameters c_j by the method of least squares.

The method of least squares is usually less convenient than collocation, since the additional integrals in (21) may be difficult to evaluate. On the other hand, the method of least squares is more systematic than collocation, since there is no arbitrariness corresponding to the selection of collocation points.

2.3.1. Solve Prob. 1.2.2 by interior least squares. As the trial solution, use the function

$$U = c_1 x(1 - x) + c_2 x(1 - x^2),$$

which satisfies the boundary conditions. Compare the results with the Ritz and collocation approximations (see Prob. 2.2.1).

Ans.

$$c_1 = \frac{4448}{101 \cdot 2437}, \qquad c_2 = \frac{413}{2437}$$

(see [1], p. 220).

2.3.2. Use interior least squares and the trial solution

$$U = c(x^2 - a^2)(y^2 - a^2)$$

to solve Prob. 1.2.7 for the square $b = a$. Compare the resulting approximation with those obtained by the Ritz and collocation methods (Probs. 1.2.7 and 2.2.4). Also compare with the infinite series representation of the solution obtained by separation of variables.

Ans.

$$a^2 c = \frac{15}{22}$$

(see [15], p. 436).

2.3.3. Solve Prob. 1.2.9 by interior least squares, using the trial solution $U = c[9 - 3(x^2 + y^2) + x^2 y^2]$ which satisfies the boundary conditions. Compare with the approximations obtained by the Ritz, Galerkin and collocation methods (Probs. 1.2.9, 2.1.8 and 2.2.7).

Ans.

$$c = \frac{15}{161}$$

(see [1], p. 414).

2.3.4. Solve the equation $\Delta u = x^2 - 1$ in the rectangle $|x| \leqslant 1$, $|y| \leqslant \frac{1}{2}$ by the boundary least squares method, where $u = 0$ on the edges of the rectangle. Use the trial solution

$$U = \frac{x^2}{2}\left(\frac{x^2}{6} - 1\right) + c_1(x^2 - y^2) + c_2(x^4 - 6x^2 y^2 + y^4),$$

which is a solution of the differential equation.

Ans.

$$c_1 = \frac{16643}{60 \cdot 2443}, \qquad c_2 = \frac{848}{2443}$$

(see [1], p. 417).

3. References[10]

1 Collatz, L., *The Numerical Treatment of Differential Equations*, third edition, Springer-Verlag, Berlin (1960).

2 Conway, H. D., *The approximate analysis of certain boundary value problems*, J. Appl. Mech., **27**, 275 (1960).

3 Courant, R., *Variational methods for the solution of problems of equilibrium and vibrations*, Bull. Amer. Math. Soc., **49**, 1 (1943).

4 Courant, R. and D. Hilbert, *Methods of Mathematical Physics, Volume I*, revised English edition, Interscience Publishers, Inc., New York (1953).

---

[10] These references pertain only to the Supplement, and are completely independent of the main Bibliography.

5 Courant, R. and D. Hilbert, *Methoden der Mathematischen Physik, Volume II*, Springer-Verlag, Berlin (1931).

6 Crandall, S., *Engineering Analysis*, McGraw-Hill Book Co., New York (1956).

7 Galerkin, B. G., *Series solutions of some problems of elastic equilibrium of rods and plates* (in Russian), Vestnik Inzh. i Tekh., **19**, 897 (1915).

8 Gelfand, I. M. and S. V. Fomin, *Calculus of Variations* (translated by R. A. Silverman), Prentice-Hall, Inc., Englewood Cliffs, N.J. (1963).

9 Hildebrand, F. B., *Methods of Applied Mathematics*, Prentice-Hall, Inc., Englewood Cliffs, N.J. (1952).

10 Horvay, G., *The end problem of rectangular strips*, J. Appl. Mech., **20**, 87 (1953).

11 Kantorovich, L. V. and V. I. Krylov, *Approximate Methods of Higher Analysis* (translated by C. D. Benster), Interscience Publishers, Inc., New York (1958).

12 Mikhlin, S. G., *Some sufficient conditions for convergence of Galerkin's method*, Leningrad. Gos. Univ. Uch. Zap., No. 135 (1950).

13 Milne, W. E., *Numerical Solution of Differential Equations*, John Wiley and Sons, Inc., New York (1953).

14 Ritz, W., *Über eine neue Methode zur Lösung gewisser Variationsprobleme der mathematischen Physik*, J. Reine Angew. Math., **135**, 1 (1908).

15 Sokolnikoff, I. S., *Mathematical Theory of Elasticity*, second edition, McGraw-Hill Book Co., New York (1956).

16 Timoshenko, S., *Theory of Plates and Shells*, second edition, McGraw-Hill Book Co., New York (1959).

17 Timoshenko, S. and J. N. Goodier, *Theory of Elasticity*, second edition, McGraw-Hill Book Co., New York (1951).

18 Wang, C. T., *Applied Elasticity*, McGraw-Hill Book Co., New York (1953).

BIBLIOGRAPHY

A1 Akhiezer, N. I. and I. M. Glazman, *Theory of Linear Operators in Hilbert Space* (translated by M. Nestell), Frederick Ungar Publishing Co., New York, *Volume I* (1961), *Volume II* (1963).

B1 Bateman, H., *The solution of the integral equation connecting the velocity of propagation of an earthquake-wake in the interior of the earth with the times which the disturbance takes to travel to the different stations on the earth's surface*, Phil. Mag., **19**, 576 (1910).

B2 Bateman, H., *Partial Differential Equations of Mathematical Physics*, Cambridge University Press, London (1959).

B3 Betz, A., *Konforme Abbildung*, second edition, Springer-Verlag, Berlin (1964).

B4 Bierens de Haan, D., *Nouvelles Tables d'Intégrales Définies*, G. E. Stechert and Co., New York (1939).

B5 Bôcher, M., *Über die Reihenentwicklungen der Potentialtheorie*, B. G. Teubner, Leipzig (1894).

B6 Budak, B. M., A. A. Samarski and A. N. Tikhonov, *A Collection of Problems on Mathematical Physics* (translated and edited by A. R. M. Robson and D. M. Brink), The Macmillan Co., New York (1964).

C1 Campbell, G. A. and R. M. Foster, *Fourier Integrals for Practical Applications*, D. Van Nostrand Co., Inc., Princeton, N.J. (1948).

C2 Carslaw, H. S. and J. C. Jaeger, *Operational Methods in Applied Mathematics*, second edition, Oxford University Press, London (1953).

C3 Carslaw, H. S. and J. C. Jaeger, *Conduction of Heat in Solids*, second edition, Oxford University Press, London (1959).

C4 Churchill, R. V., *Operational Mathematics*, second edition, McGraw-Hill Book Co., New York (1958).

C5   Courant, R. and D. Hilbert, *Methods of Mathematical Physics*, revised English edition, Interscience Publishers, New York, *Volume I* (1953), *Volume II* (1962).

D1   Doetsch, G., *Handbuch der Laplace-Transformation*, Verlag Birkhäuser, Basel, *Volume I* (1950), *Volume II* (1955), *Volume III* (1956).

D2   Dwight, H. B., *Tables of Integrals and Other Mathematical Data*, third edition, The Macmillan Co., New York (1957).

E1   Eisenhart, L. P., *Separable systems of Stäckel*, Annals of Math., **35**, 284 (1934).

E2   Erdélyi, A., W. Magnus, F. Oberhettinger and F. G. Tricomi, *Higher Transcendental Functions* (in three volumes), based, in part, on notes left by Harry Bateman, McGraw-Hill Book Co., New York (1953).

E3   Erdélyi, A., W. Magnus, F. Oberhettinger and F. G. Tricomi, *Tables of Integral Transforms, Volume I*, based, in part, on notes left by Harry Bateman, McGraw-Hill Book Co., New York (1954).

F1   Fock, V. A., *Skin effect in a ring of circular cross section* (in Russian), Zh. Russ. Fiz.-Khim. Ob., **62**, 281 (1930).

F2   Fock, V. A., *Theory of Measurement of the Resistance of Strata by the Method of Electrical Coring* (in Russian), Gos. Tekh.-Teor. Izd., Moscow (1933).

F3   Fock, V. A., *On some integral equations of mathematical physics* (in Russian), Dokl. Akad. Nauk SSSR, **36**, 147 (1962).

F4   Fock, V. A., *On the representation of an arbitrary function by an integral involving Legendre's functions with a complex index*, Dokl. Akad. Nauk SSSR, **39**, 253 (1943).

F5   Fock, V. A., *On some integral equations of mathematical physics* (in Russian), Mat. Sb., **14**, 1 (1944).

F6   Frank, P. and R. von Mises, *Die Differential- und Integralgleichungen der Mechanik und Physik* (in two volumes), second enlarged edition, Dover Publications, Inc., New York (1961).

F7   Franklin, P., *An Introduction to Fourier Methods and the Laplace Transformation*, Dover Publications, Inc., New York (1958).

F8   Friedlander, F. G., *The reflexion of sound pulses by convex parabolic reflectors*, Proc. Camb. Phil. Soc., **37**, 134 (1941).

F9   Fuchs, B. A. and V. I. Levin, *Functions of a Complex Variable and Some of Their Applications* (translated by J. Berry), Addison-Wesley Publishing Co., Inc., Reading, Mass. (1961). (This and the next book form a two-volume set.)

F10 Fuchs, B. A. and B. V. Shabat, *Functions of a Complex Variable and Some of Their Applications* (translated by J. Berry and revised by J. W. Reed), Addison-Wesley Publishing Co., Inc., Reading, Mass. (1964).

G1   Garabedian, P. R., *Partial Differential Equations*, John Wiley and Sons, Inc., New York (1964).

G2   Gray, A. and G. B. Mathews, *A Treatise on Bessel Functions and Their Applications to Physics*, second edition, prepared by A. Gray and T. M. MacRobert, Macmillan and Co., Ltd., London (1952).

G3   Grinberg, G. A., *On the problem of gas emission of metals during heating*, Zh. Eksper. Teor. Fiz., **1**, 245 (1931).

G4   Grinberg, G. A., *A new method for solving certain boundary value problems for the equations of mathematical physics, permitting separation of variables* (in Russian), Izv. Akad. Nauk SSSR, Ser. Fiz., **10**, 141 (1946).

G5   Grinberg, G. A., *Selected Topics in the Mathematical Theory of Electric and Magnetic Phenomena* (in Russian), Izd. Akad. Nauk SSSR, Moscow (1948).

G6   Grinberg, G. A., *On the solution of the equations of mathematical physics with partially or completely separable variables* (in Russian), Collection celebrating the seventieth birthday of Academician A. F. Ioffe, Izd. Akad. Nauk SSSR, Moscow (1950), p. 50.

G7   Gyunter, N. M. and R. O. Kuzmin, *Aufgabensammlung zur Höheren Mathematik, Volume II*, VEB Deutscher Verlag der Wissenschaften, Berlin (1957).

H1   Haentzschel, E., *Studien über die Reduktion der Potentialgleichung auf Gewöhnliche Differentialgleichungen*, G. Reimer, Berlin (1893).

H2   Haentzschel, E., *Rotationscykliden und Lamésche Produkte*, Arch. Math. Phys., **4**, 57 (1902).

H3   Harding, J. W. and I. N. Sneddon, *The elastic stresses produced by the indentation of the plane surface of a semi-infinite elastic solid by a rigid punch*, Proc. Camb. Phil. Soc., **41**, 16 (1945).

H4   Hobson, E. W., *The Theory of Spherical and Ellipsoidal Harmonics*, Cambridge University Press, London (1931).

J1   Jackson, D., *Fourier Series and Orthogonal Polynomials*, Carus Mathematical Monograph No. 6, Mathematical Association of America, State University of New York, Buffalo, N.Y. (1941).

J2   Jeans, J., *The Mathematical Theory of Electricity and Magnetism*, fifth edition, Cambridge University Press, London (1946).

J3   Jeffery, G. B., *Plane stress and plane strain in bipolar coordinates*, Phil. Trans. Roy. Soc. London, Ser. A, **221**, 265 (1921).

J4   Jeffreys, H. and B. S. Jeffreys, *Methods of Mathematical Physics*, third edition, Cambridge University Press, London (1956).

J5   Jolley, L. B. W., *Summation of Series*, second edition, Dover Publications, Inc., New York (1961).

J6   Joos, G., *Theoretical Physics*, third edition, with the collaboration of I. Freeman, Blackie and Son, Ltd., London (1958).

K1   Kantorovich, L. V. and V. I. Krylov, *Approximate Methods of Higher Analysis* (translated by C. D. Benster), Interscience Publishers, Inc., New York (1958).

K2   Kochin, N. E., I. A. Kibel and N. V. Roze, *Theoretical Hydromechanics* (translated and edited by D. Boyanovitch and J. R. M. Radok), John Wiley and Sons, Inc., New York (1964).

K3   Kontorovich, M. I. and N. N. Lebedev, *On a method of solving some problems of diffraction theory and related problems*, Zh. Eksper. Teor. Fiz., **8**, 1192 (1938).

K4   Koshlyakov, N. S., M. M. Smirnov and E. B. Gliner, *Differential Equations of Mathematical Physics* (translation edited by H. J. Eagle), John Wiley and Sons, Inc., New York (1964).

K5   Kupradze, V. D., *Randwertaufgaben der Schwingungstheorie und Integralgleichungen*, VEB Deutscher Verlag der Wissenschaften, Berlin (1956).

L1   Ladyzhenskaya, O. A., *The Mixed Problem for a Hyperbolic Equation* (in Russian), Gos. Izd. Tekh.-Teor. Lit., Moscow (1953).

L2   Lagrange, R., *Les familles de surfaces de révolution qui possèdent des harmoniques*, Acta Math., **71**, 283 (1939).

L3   Lebedev, N. N., *The coefficient of mutual induction between coils wound on a cylindrical core with magnetic permeability μ*, Zh. Tekh. Fiz., **6**, 530 (1936).

L4   Lebedev, N. N., *The functions associated with a ring of oval cross-section*, Tech. Phys. USSR, **4**, 3 (1937).

L5   Lebedev, N. N., *On the application of singular integral equations to the problem of the distribution of electricity on thin non-closed surfaces* (in Russian), Zh. Tekh. Fiz., **18**, 775 (1948).

L6   Lebedev, N. N., *On the expansion of an arbitrary function in an integral with respect to cylinder functions of imaginary order and argument* (in Russian), Prikl. Mat. Mekh., **13**, 465 (1949).

L7   Lebedev, N. N., *Some singular integral equations connected with the integral expansions of mathematical physics* (in Russian), Dokl. Akad. Nauk SSSR **65**, 621 (1949).

L8   Lebedev, N. N. *Some Integral Transformations of Mathematical Physics* (in Russian), Dissertation, Izd. Leningrad. Gos. Univ. (1951).

L9   Lebedev, N. N., *Special Functions and Their Applications* (translated by R. A. Silverman), Prentice-Hall, Inc., Englewood Cliffs, N.J. (1965).

L10   Lebedev, N. N. and M. I. Kontorovich, *On the application of inversion formulas to the solution of some problems of electrodynamics*, Zh. Eksper. Tekh. Fiz., **9**, 729 (1939).

L11   Lense, J., *Reihenentwicklungen in der Mathematischen Physik*, third edition, Walter de Gruyter & Co., Berlin (1953).

L12   Lense, J., *Kugelfunktionen*, second edition, Akademische Verlagsgesellschaft, Geest & Portig K.-G., Leipzig (1954).

L13   Levitan, B. M., *Expansion in Eigenfunctions of Second-Order Differential Equations* (in Russian), Gos. Izd. Tekh.-Teor. Lit., Moscow (1950).

L14   Levitan, B. M., *On expansion in eigenfunctions of the equation $y'' + \{\lambda - q(x)\}y = 0$* (in Russian), Dokl. Akad. Nauk SSSR, **90**, 17 (1953).

L15   Levitan, B. M., *On the asymptotic behavior of the spectral function of a self-adjoint second-order differential equation and on expansion in eigenfunctions* (in Russian), Izv. Akad. Nauk SSSR, Ser. Mat., **17**, 331 (1953).

L16 Lurye, A. I., *Operational Calculus and its Applications to Problems of Mechanics*, second edition (in Russian), Gos. Izd. Tekh.-Teor. Lit., Moscow (1950).

M1 Macdonald, H. M., *The electrical distribution on a conductor bounded by two spherical surfaces cutting at any angle*, Proc. London Math. Soc., **26**, 156 (1895).

M2 MacRobert, T. M., *Spherical Harmonics, An Elementary Treatise on Harmonic Functions with Applications*, second edition, Methuen and Co., Ltd., London (1947).

M3 Magnus, W. and F. Oberhettinger, *Formulas and Theorems for the Functions of Mathematical Physics* (translated by J. Wermer), Chelsea Publishing Co., New York (1954).

M4 Marcuvitz, N., *Waveguide Handbook*, Massachusetts Institute of Technology Radiation Laboratory Series, Vol. 10, McGraw-Hill Book Co., New York (1951).

M5 McLachlan, N. W., *Bessel Functions for Engineers*, second edition, Oxford University Press, London (1955).

M6 Melan, E., *Die Bestimmung des Sicherheitsgrades einfach statisch unbestimmter Fachwerke*, Z. angew. Math. Mech., **12**, 129 (1932).

M7 Mikhlin, S. G., *Integral Equations and Their Applications to Certain Problems in Mechanics, Mathematical Physics and Technology*, second edition (translated by A. H. Armstrong), The Macmillan Co., New York (1964).

M8 Morse, P. M., *Vibration and Sound*, second edition, McGraw-Hill Book Co., New York (1948).

M9 Morse, P. M. and H. Feshbach, *Methods of Theoretical Physics* (in two volumes), McGraw-Hill Book Co., New York (1953).

M10 Muskhelishvili, N. I., *Some Basic Problems of the Mathematical Theory of Elasticity* (translated by J. R. M. Radok), P. Noordhoff Ltd., Groningen (1953).

M11 Muskhelishvili, N. I., *Singular Integral Equations* (translated by J. R. M. Radok), P. Noordhoff N.V., Groningen (1953).

N1 Noble, B., *Methods Based on the Wiener-Hopf Technique for the Solution of Partial Differential Equations*, Pergamon Press, New York (1958).

O1 Ollendorff, F., *Erdströme, Grundlagen der Erdschluss- und Erdungsfragen*, Springer-Verlag, Berlin (1928).

P1 Panovsky, W. K. H. and M. Phillips, *Classical Electricity and Magnetism*, second edition, Addison-Wesley Publishing Co., Inc., Reading, Mass. (1962).

P2 Petrovski, I. G., *Lectures on Partial Differential Equations* (translated by A. Shenitzer), Interscience Publishers, Inc., New York (1954).

R1 Rayleigh, Baron (J. W. Strutt), *The Theory of Sound* (in two volumes), Dover Publications, Inc., New York (1945).

R2 Ryshik, I. M. and I. S. Gradstein, *Tables of Series, Products, and Integrals*, VEB Deutscher Verlag der Wissenschaften, Berlin (1957).

S1   Sagan, H., *Boundary and Eigenvalue Problems in Mathematical Physics*, John Wiley and Sons, Inc., New York (1961).

S2   Sakharov, I. Y., *Bending of a fastened wedge-shaped plate under the action of an arbitrary load* (in Russian), Prikl. Mat. Mekh., **12**, 407 (1948).

S3   Schelkunoff, S. A., *Electromagnetic Waves*, D. Van Nostrand Co., Inc., Princeton, N.J. (1943).

S4   Skalskaya, I. P., *The field of a point current source located on the earth's surface over an inclined layer*, Zh. Tekh. Fiz., **18**, 1242 (1948).

S5   Smirnov, M. M., *Aufgaben zu den Partiellen Differentialgleichungen der Mathematischen Physik*, VEB Deutscher Verlag der Wissenschaften, Berlin (1955).

S6   Smirnov, V. I., *Lehrgang der Höheren Mathematik*, VEB Deutscher Verlag der Wissenschaften, Berlin, *Volume II* (1955), *Volume III, Part 2* (1955), *Volume IV* (1958), *Volume V* (1962).

S7   Smythe, W. R., *Static and Dynamic Electricity*, second edition, McGraw-Hill Book Co., New York (1950).

S8   Sneddon, I. N., *The symmetrical vibrations of a thin elastic plate*, Proc. Camb. Phil. Soc., **41**, 27 (1945).

S9   Sneddon, I. N., *The Fourier transform solution of an elastic wave equation*, Proc. Camb. Phil. Soc., **41**, 239 (1945).

S10   Sneddon, I. N., *Fourier Transforms*, McGraw-Hill Book Co., New York (1951).

S11   Sneddon, I. N., *Elements of Partial Differential Equations*, McGraw-Hill Book Co., New York (1957).

S12   Snow, C., *The Hypergeometric and Legendre Functions with Applications to Integral Equations of Potential Theory*, National Bureau of Standards Applied Mathematics Series, No. 19, U.S. Government Printing Office, Washington, D.C. (1952).

S13   Sobolev, S. L., *Applications of Functional Analysis in Mathematical Physics* (translated by F. E. Browder), American Mathematical Society, Providence, R.I. (1963).

S14   Sommerfeld, A., *Partial Differential Equations in Physics* (translated by E. G. Straus), Academic Press Inc., New York (1949).

S15   Stepanov, V. V., *Sur l'équation de Laplace et certains systèmes triples orthogonaux*, Mat. Sb., **11**, 204 (1942).

S16   Sternberg, W. J. and T. L. Smith, *The Theory of Potential and Spherical Harmonics*, University of Toronto Press, Toronto (1952).

S17   Stratton, J. A., *Electromagnetic Theory*, McGraw-Hill Book Co., New York (1941).

S18   Stratton, J. A., P. M. Morse, L. J. Chu, J. D. C. Little and F. Corbató, *Spheroidal Wave Functions*, John Wiley and Sons, Inc., New York (1956).

S19   Strutt, M. J. O., *Lamésche, Mathieusche und Verwandte Functionen in Physik und Technik*, Springer-Verlag, Berlin (1932).

T1 Tikhonov, A. N. and A. A. Samarski, *Partial Differential Equations of Mathematical Physics* (translated by S. Radding), Holden-Day, Inc., San Francisco, *Volume I* (1964).

T2 Timoshenko, S., *Vibration Problems in Engineering*, second edition, D. van Nostrand Co., Inc., Princeton, N.J. (1937).

T3 Timoshenko, S., *Theory of Plates and Shells*, second edition, McGraw-Hill Book Co., New York (1959).

T4 Timoshenko, S. and J. N. Goodier, *Theory of Elasticity*, second edition, McGraw-Hill Book Co., New York (1951).

T5 Titchmarsh, E. C., *Introduction to the Theory of Fourier Integrals*, second edition, Oxford University Press, London (1950).

T6 Titchmarsh, E. C., *Eigenfunction Expansions Associated with Second-Order Differential Equations*, Oxford University Press, London, *Volume I* (1946), *Volume II* (1958).

T7 Tolstov, G. P., *Fourier Series* (translated by R. A. Silverman), Prentice-Hall, Inc., Englewood Cliffs, N.J. (1962).

T8 Tranter, C. J., *Integral Transforms in Mathematical Physics*, second edition, John Wiley and Sons, Inc., New York (1956).

T9 Tricomi, F. G., *Integral Equations*, Interscience Publishers, Inc., New York (1957).

T10 Tsukkerman, I. I., *Determination of thermal constants by probes* (in Russian), Zh. Tekh. Fiz., **20**, 353 (1950).

U1 Uflyand, Y. S., *Bending of a prismatic rod with a profile bounded by arcs of two intersecting circles* (in Russian), Dokl. Akad. Nauk SSSR, **68**, 17 (1949).

U2 Uflyand, Y. S., *Bipolar Coordinates in the Theory of Elasticity*, Gos. Izd. Tekh.-Teor. Lit., Moscow (1950).

U3 Uflyand, Y. S., *Application of the Mellin transform to the problem of bending of a thin elastic sheet of wedgelike form* (in Russian), Dokl. Akad. Nauk SSSR, **84**, 463 (1952).

V1 Van der Pol, B. and H. Bremmer, *Operational Calculus Based on the Two-Sided Laplace Integral*, Cambridge University Press, London (1955).

W1 Walker, M., *The Schwarz-Christoffel Transformation and its Applications— A Simple Exposition*, Dover Publications, Inc., New York (1964).

W2 Wangerin, A., *Reduction der Potentialgleichung für gewisse Rotationskörper auf eine gewöhnliche Differentialgleichung*, Preisschr. der Jabl. Ges., Leipzig (1875).

W3 Wangerin, A., *Über ein dreifach orthogonales Flächensystem, gebildet aus gewissen Flächen vierter Ordnung*, Borchardt J., **82**, 145 (1876); *Notiz zu dem Aufsatz über ein dreifach orthogonales Flächensystem etc.*, ibid., **82**, 348 (1876).

W4 Watson, G. N., *A Treatise on the Theory of Bessel Functions*, second edition, Cambridge University Press, London (1962).

W5 Webster, A. G., *Partial Differential Equations of Mathematical Physics*, second edition (edited by S. J. Plimpton), Dover Publications, Inc., New York (1955).

W6 Weinel, E., *Das Torsionsproblem für den exzentrischen Kreisring*, Ing.-Arch., **3**, 67 (1932).

W7 Weyl, H., *Über gewöhnliche Differentialgleichungen mit Singularitäten und die zugehörigen Entwicklungen willkürlicher Funktionen*, Math. Ann., **68**, 220 (1910).

W8 Whittaker, E. T. and G. N. Watson, *A Course of Modern Analysis*, fourth edition, Cambridge University Press, London (1963).

W9 Widder, D. V., *The Laplace Transform*, Princeton University Press, Princeton, N.J. (1941).

W10 Wiener, N. and E. Hopf, *Über eine Klasse singulärer Integralgleichungen*, S. B. Preuss. Akad. Wiss., 696 (1931).

NAME INDEX

SUBJECT INDEX

---

[1] Because of the contents of this book (problems and their solutions), the subject index is necessarily eclectic, consisting mainly of first occurrences of key terms.